3D Bioprinting and Nanotechnology in Tissue Engineering and Regenerative Medicine

3D Bioprinting and Nanotechnology in Tissue Engineering and Regenerative Medicine

Lijie Grace Zhang

John P. Fisher

Kam W. Leong

AMSTERDAM • BOSTON • HEIDELBERG • LONDON
NEW YORK • OXFORD • PARIS • SAN DIEGO
SAN FRANCISCO • SINGAPORE • SYDNEY • TOKYO

Academic Press is an imprint of Elsevier

Academic Press is an imprint of Elsevier
32 Jamestown Road, London NW1 7BY, UK
525 B Street, Suite 1800, San Diego, CA 92101-4495, USA
225 Wyman Street, Waltham, MA 02451, USA
The Boulevard, Langford Lane, Kidlington, Oxford OX5 1GB, UK

Notice
No responsibility is assumed by the publisher for any injury and/or damage to persons or property as
a matter of products liability, negligence or otherwise, or from any use or operation of any methods,
products, instructions or ideas contained in the material herein. Because of rapid advances in the
medical sciences, in particular, independent verification of diagnoses and drug dosages should be made

British Library Cataloguing-in-Publication Data
A catalogue record for this book is available from the British Library

Library of Congress Cataloging-in-Publication Data
A catalog record for this book is available from the Library of Congress

ISBN: 978-0-12-800547-7

For information on all Academic Press publications
visit our website at http://store.elsevier.com/

Acquisition Editor: Fiona Geraghty
Editorial Project Manager: Natasha Welford
Production Project Manager: Jason Mitchell
Designer: Christian Bilbow

Typeset by Thomson Digital

Printed and bound in the United States

Working together
to grow libraries in
developing countries

www.elsevier.com • www.bookaid.org

Contents

List of Contributors

Amir Azhari
Department of Mechanical and Mechatronics Engineering, University of Waterloo, Waterloo, ON, Canada

Ahmad Basalah
Department of Mechanical and Mechatronics Engineering, University of Waterloo, Waterloo, ON, Canada

Nathan J. Castro
Department of Mechanical and Aerospace Engineering, The George Washington University, Washington DC, USA

Robert C. Chang
Department of Mechanical Engineering, Stevens Institute of Technology, Hoboken, NJ, USA

Shaochen Chen
Department of NanoEngineering, University of California, San Diego, La Jolla, CA, USA

Boris Chichkov
Nanotechnology Department, Laser Zentrum Hannover e.V., Hollerithallee 8, Hannover, Germany

Douglas B. Chrisey
Department of Physics and Engineering Physics, Tulane University, New Orleans, LA, USA

David T. Corr
Department of Biomedical Engineering, Rensselaer Polytechnic Institute, Troy, NY, USA

J. Lowry Curly
Department of Biomedical Engineering, Tulane University, New Orleans, LA, USA

Darryl D. D'Lima
Shiley Center for Orthopaedic Research and Education at Scripps Clinic, La Jolla, CA, USA

David Dean
Department of Plastic Surgery, The Ohio State University, Columbus, OH, USA

Andrew D. Dias
Department of Biomedical Engineering, Rensselaer Polytechnic Institute, Troy, NY, USA

Amir Dorafshar
Department of Plastic and Reconstructive Surgery, Johns Hopkins University School of Medicine, Baltimore, MD, USA

Erik W. Dorthé
Shiley Center for Orthopaedic Research and Education at Scripps Clinic, La Jolla, CA, USA

John P. Fisher
Fischell Department of Bioengineering, University of Maryland, College Park, Maryland, USA

Warren L. Grayson
Department of Biomedical Engineering, Translational Tissue Engineering Center, Johns Hopkins University School of Medicine, Baltimore, MD, USA

Bagrat Grigoryan
Department of Bioengineering, Rice University, Houston, TX, USA

Shawn P. Grogan
Shiley Center for Orthopaedic Research and Education at Scripps Clinic, La Jolla, CA, USA

Benjamin Holmes
Department of Mechanical and Aerospace Engineering, The George Washington University, Washington DC, USA

John F. Hornick
Finnegan, Henderson, Farabow, Garrett & Dunner, LLP, Washington, DC, USA

Yong Huang
Department of Mechanical & Aerospace Engineering, University of Florida, Gainesville, FL, USA

Ben P. Hung
Department of Biomedical Engineering, Translational Tissue Engineering Center, Johns Hopkins University School of Medicine, Baltimore, MD, USA

Pinar Yilgor Huri
Department of Biomedical Engineering, Translational Tissue Engineering Center, Johns Hopkins University School of Medicine, Baltimore, MD, USA

Rita Kandel
Mount Sinai Hospital, University of Toronto, Toronto, ON, Canada

David M. Kingsley
Department of Biomedical Engineering, Rensselaer Polytechnic Institute, Troy, NY, USA

Lothar Koch
Nanotechnology Department, Laser Zentrum Hannover e.V., Hollerithallee 8, Hannover, Germany

Michael Larsen
Department of Plastic Surgery, The Ohio State University, Columbus, OH, USA

Wei Li
Department of Mechanical Engineering, University of Texas, Austin, TX, USA

Xuanyi Ma
Department of NanoEngineering, University of California, San Diego, La Jolla, CA, USA

Stefanie Michael
Department of Plastic, Hand- and Reconstructive Surgery, Hannover Medical School, Hannover, Germany

Jordan S. Miller
Department of Bioengineering, Rice University, Houston, TX, USA

Michael Miller
Department of Plastic Surgery, The Ohio State University, Columbus, OH, USA

Lee Jia Min
School of Mechanical & Aerospace Engineering, Nanyang Technological University, Singapore

Ruchi Mishra
Department of Plastic Surgery, The Ohio State University, Columbus, OH, USA

Christopher O'Brien
Department of Mechanical and Aerospace Engineering, The George Washington University, Washington DC, USA

JinGyu Ock
Department of Mechanical Engineering, University of Texas, Austin, TX, USA

Theresa B. Phamduy
Department of Biomedical Engineering, Tulane University, New Orleans, LA, USA

Jesse K. Placone
Fischell Department of Bioengineering, University of Maryland, College Park, Maryland, USA

Kai Rajan
Finnegan, Henderson, Farabow, Garrett & Dunner, LLP, Washington, DC, USA

Kerstin Reimers
Department of Plastic, Hand- and Reconstructive Surgery, Hannover Medical School, Hannover, Germany

Tan Yong Sheng, Edgar
School of Mechanical & Aerospace Engineering, Nanyang Technological University, Singapore

Rohan Shirwaiker
Department of Industrial and Systems Engineering, North Carolina State University, Raleigh, NC, USA

S.C. Sklare
Department of Physics and Engineering Physics, Tulane University, New Orleans, LA, USA

Binil Starly
Department of Industrial and Systems Engineering, North Carolina State University, Raleigh, NC, USA

Joshua P. Temple
Department of Biomedical Engineering, Translational Tissue Engineering Center, Johns Hopkins University School of Medicine, Baltimore, MD, USA

Filippos Tourlomousis
Department of Mechanical Engineering, Stevens Institute of Technology, Hoboken, NJ, USA

Ehsan Toyserkani
Department of Mechanical and Mechatronics Engineering, University of Waterloo,
Waterloo, ON, Canada

Mihaela Vlasea
Department of Mechanical and Mechatronics Engineering, University of Waterloo, Waterloo, ON,
Canada

Peter M. Vogt
Department of Plastic, Hand- and Reconstructive Surgery, Hannover Medical School, Hannover,
Germany

Yeong Wai Yee
School of Mechanical & Aerospace Engineering, Nanyang Technological University, Singapore

Lijie Grace Zhang
Department of Mechanical and Aerospace Engineering, The George Washington University,
Washington DC, USA; Department of Medicine, The George Washington University, Washington
DC, USA

Wei Zhu
Department of NanoEngineering, University of California, San Diego, La Jolla, CA, USA;
Department of Mechanical and Aerospace Engineering, The George Washington University,
Washington DC, USA

Zhu Zicheng
School of Mechanical & Aerospace Engineering, Nanyang Technological University, Singapore

Preface

This inaugural edition of *3D Bioprinting and Nanotechnology in Tissue Engineering and Regenerative Medicine* aims to provide an overview of two exciting emergent technologies—3D bioprinting and nanotechnology—for tissue and organ regeneration applications. It includes two main sections: (1) a thorough overview of current advancements in 3D bioprinting and nanomaterials, and (2) 3D bioprinting of complex tissues and the regulatory implications of associated intellectual property. The discerning feature of this text compared to existing titles lies in the breadth of coverage of recent developments of these two technologies separately by leading researchers in the field. In addition, the most beneficial feature to current and future clinicians, students, researchers, and technologists is the depth and emphasis on combinatorial approaches of 3D bioprinting and nanotechnology in addressing complex tissue defects with implications on the fabrication of functional organs.

Historically, scaffold-based approaches have led to a better understanding of the effects of 3D microenvironments and geometric cues on cell fate and resultant tissue/organ formation. Traditional scaffold fabrication methods (i.e., electrospinning, gas foaming, particle-leaching, etc.) have shown to be sufficient for simple, single-tissue regenerative applications, but lack the sophistication and flexibility in design to fabricate more complex and biomimetic 3D structures suitable to support the creation of multicellular tissues. To address biomimetic design, tissue engineers have begun to explore additive manufacturing, particularly 3D printing, as a viable method of fabricating tissue-engineered construct. Innovative applications of existing 3D printing platforms, previously utilized commercially, as well as tissue-engineering-specific systems, have begun to be employed in earnest as tools to address several persistent limitations of traditional tissue construct fabrication methods. 3D bioprinting readily provides greater precision and control over the predesigned internal architecture and composition of a scaffold when compared to the aforementioned traditional techniques. In addition, based on computer-aided designed models reconstructed from noninvasive medical imaging data specific to individual patients' defects, 3D bioprinters can easily fabricate a custom-designed tissue construct in a layer-by-layer manner, which would allow the implant to perfectly integrate with the wound site and expedite tissue regeneration *in vivo*. Through the commoditization of 3D printers and extension of these technologies toward regenerative applications in concert with developing and evolving 3D bioprinting-specific biomaterials, the potential for expedited regulatory oversight can lead to a standardization of biomimetic implantable tissue scaffolding. The current edition will provide an in-depth and up-to-date survey of 3D bioprinting for tissue and organ regeneration.

Complementing the architectural control afforded by 3D printing, nanotechnology offers local control at the site of cell–substrate interactions. Natural tissue, such as extracellular matrix, is full of nanoscale features in the form of nanofibers, nanopores, and nanoridges. These nanostructures play a key role in modulating the repair and regeneration of tissues. Through this book, the readers will learn the most recent effort to exploit the dramatic effects of nanoscale features and novel nanomaterials on cell behavior and subsequent tissue formation. Nanomaterials with feature size below 100 nm in at least one dimension can drastically alter the characteristics of a bulk material even at a low concentration in the form of a composite, ranging from mechanical to electrical properties. In addition to augmenting material properties, many nanomaterials may improve biological properties. The unique properties of these nanobiomaterials offer tissue engineers exciting opportunities to construct a biomimetic micro-environment for cellular studies and regenerative medicine applications.

In essence, tissue engineering thrives at the confluence of engineering and life sciences to solve important medical problems. Tissue engineers must effectively apply cutting-edge and multidisciplinary approaches in an integrated and novel manner. As these research areas evolve and are developed into proven technologies and eventually medical remedies, intellectual property considerations will apply, and have to be addressed. For this reason, a chapter has been included to highlight the role of intellectual property for 3D bioprinting and nanotechnology in tissue engineering. Together, these chapters form a unique and comprehensive book that we believe will be useful to not only understand recent breakthroughs and ongoing challenges in how 3D printing and nanotechnology may impact cell–materials interactions, but also spur the cultivation of new strategies to exploit these two technologies for the translation and commercialization of tissue engineering.

Lijie Grace Zhang
John P. Fisher
Kam W. Leong

NANOTECHNOLOGY: A TOOLKIT FOR CELL BEHAVIOR

Christopher O'Brien[1], Benjamin Holmes[1] and Lijie Grace Zhang[1,2]

[1]*Department of Mechanical and Aerospace Engineering, The George Washington University, Washington DC, USA*

[2]*Department of Medicine, The George Washington University, Washington DC, USA*

1.1 INTRODUCTION

Scientists and researchers have been fascinated with the details of life at small scales ever since Robert Hooke saw the evidence of small structures in cork that he coined cells. This spurred the creation of the compound microscope and the quest of the late 1600s to discover how life operates beneath our very own eyes. That quest has continued even to this day as scientists look for smaller and smaller constituents that contribute to life as we know it; from proteins to functional groups, everything has an important role. The collective scientific gaze looked for finer and finer components to life, and for a short while now has focused on the prevalence of the nano world.

One hundred to one thousand times smaller than Hooke's observed cork cells, researchers have determined that materials and features of less than 100 nm in at least one dimension can have profound effects on the behavior of cells and further tissue and organ regeneration (Zhang and Webster, 2009). When examining nature, using nanotechnology for tissue regeneration becomes obvious. In fact, human cells create and continually interact directly with their natural nanostructured environment, called extracellular matrix (ECM). This momentous discovery spurred many researchers to attempt to more effectively mimic natural biology by creating novel nanobiomaterials and designing nanocomposite scaffolds for improved tissue and organ regenerations (Biggs et al., 2007; Jang et al., 2010; Chopra et al., 2012). Decreasing material size to the nano scale dramatically increases surface roughness and the surface area to volume ratio of materials, and may lead to a higher surface reactivity and many superior physiochemical properties (i.e. mechanical, electrical, optical, catalytic, and magnetic properties) (Zhang and Webster, 2009). The excellent properties of nanobiomaterials make them hold great potential for a wide range of biomedical applications, particularly advanced tissue/organ regeneration.

With the exponential growth in the human population and the similarly rapid increase in lifespan worldwide, there is an enormous market for various tissue and organ transplantations and engraftments. Current treatment options for damaged tissues and organs are nonideal, and often involve severe tissue/ organ shortages, painful surgeries, and long recovery times without offering a complete restoration of the tissue's and organ's function. Simultaneously, in recent years, many health professionals have been advocating for a more active lifestyle with increased exercise and thus an increased risk of injury. These factors, and many others, put a strain on existing treatment methods and hallmark their many weaknesses. For most tissue damage caused by diseases or injuries, many current treatment methods lack the ability to restore the affected area to a level of functionality equivalent to healthy native tissue. They instead provide a stop-gap or temporary solution that either slows the progress of further degeneration or requires sacrifice of other healthy tissue (autografts). Many researchers and doctors hope that by increasing understanding of how cells and tissues interact on the nano scale and creating biomimetic nanostructured tissue constructs to better emulate natural designs, solutions that more effectively treat diseases and injuries can be discovered.

In the following sections, we will focus on the current state of nanotechnology for a series of tissue and organ regeneration. In addition, we will put special emphasis on integrating cutting-edge 3D nano/microfabrication techniques with nanobiomaterials for complex tissue and organ regeneration applications. These nanobiomaterial constituents can be made of nearly any material imaginable, including carbon nanomaterials, self-assembly nanomaterials, natural or synthetic polymers, ceramics, drug-containing spheres, or metal particles. Researchers strive to combine the appropriate nanobiomaterials, cells, and growth factors to create the ideal biomimetic tissue engineered construct that could surmount traditional methods of injury mitigation.

1.2 NANOBIOMATERIALS FOR TISSUE REGENERATION

1.2.1 CARBON NANOBIOMATERIALS

1.2.1.1 Carbon Nanotubes

Carbon and carbon derivatives are some of the most versatile nanomaterials that tissue engineers have in their arsenal (Zhang et al., 2009a; Tran et al., 2009). In addition to constituting 18% of the average human body by mass (Frieden, 1972), carbon is a highly flexible element that can assume many nanometer-sized structures. One of the most well-explored carbon nanomaterials is carbon nanotubes (CNTs) (Figure 1.1). CNTs have several different types, but those used in the tissue engineering field are primarily multiwalled CNTs (MWCNTs) or single-walled nanotubes (SWCNTs). They are one of the strongest materials known (Yu et al., 2000; Terrones, 2004), and can exhibit semiconducting (Jung et al., 2013) and conducting (Lan and Li, 2013) properties, making them interesting media for stimulating tissue regeneration.

One of the most prominent features of CNTs is their ability to significantly influence the electrical conductivity of scaffolds. This trait is of particular interest to groups studying tissues that rely heavily on signaling to perform functions, such as cardiac tissue. In one example, gelatin methacrylate hydrogel scaffolds modified with incorporated CNTs expressed improved cell behavior when seeded with rat cardiomyocytes. The tissue exhibited increased synchronous beating rate and a significantly lower threshold for excitation when compared to control samples without incorporated CNTs (Shin et al., 2013). Furthermore, CNTs also tend to increase cardiomyocyte proliferation and maturation *in vitro* (Martinelli et al., 2013; Shin et al., 2013; Martinelli et al., 2012). Although CNTs are used for several other fields within tissue engineering, they appear to selectively steer mesenchymal stem cells (MSCs) toward a cardiac lineage when introduced into cell culture media and exposed to electrical stimulation (Mooney et al., 2012).

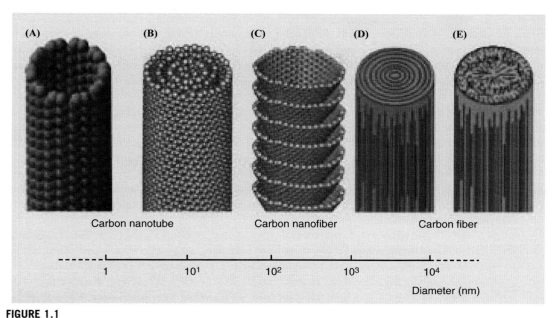

FIGURE 1.1

Comparison of carbon nanotubes and carbon nanofibers showcasing their morphological differences, and relative difference in diameter. Image is from Kim et al. (2013).

Many researchers have also drawn upon the high electrical conductivity exhibited by CNTs to create conductive scaffolds for neural tissue regeneration (Gacem et al., 2013). Improved peripheral nervous system (PNS) (Serrano et al., 2014) and central nervous system (CNS) (Kim et al., 2014) regeneration and stem cell performance (Serrano et al., 2014) have been observed when utilizing CNTs. In particular, a 3D porous scaffold was fabricated from chondroitin sulfate, a biomaterial constituent of native nervous tissue, and MWCNTs via a freeze-drying method, coated with polylysine, and cultured with rat embryonic neural progenitor cells. After 20 days of culture, a viable cell population of more neuron than glial cells was observed, contrasting the 2D, CNT-less controls (Serrano et al., 2014). In another study, MWCNTs were combined with collagen to create a collagen–CNT nanomaterial scaffold that accelerated and directed differentiation of human decidua parieltalis stem cells into a neural lineage (Sridharan et al., 2013). The nanocomposite scaffold elucidated a previously unknown differentiation pathway of parieltalis cells, unique to this scaffold. In addition, CNTs can also be used to reveal important mechanisms of neuronal activity. In this specific study, MWCNTs with a small number of walls, dubbed "few-walled CNTs," were used as a substrate to examine the chloride shift; a hallmark trait of neuronal disorder and injury (Liedtke et al., 2013). Primary CNS neurons were found to have a highly accelerated chloride shift and very high potassium chloride cotransporter 2 expression. The CNTs on the substrate promoted this expression through interaction with voltage-gated sodium channels. In this manner, few-walled CNTs can be employed to reduce chloride concentration and exchange between neurons.

CNTs continue to have great utility across many other tissue types. At first glance, the musculoskeletal system is very different from the cardiac and neuronal systems previously discussed; however, they are alike in that carbon nanotubes can provide important augmentations to biomaterial scaffolds to increase the efficacy of the constructs. In bone tissue engineering, the main concern of researchers attempting to create an ideal tissue-engineered scaffold is modulating the mechanical properties of biomaterials to be similar to that of natural tissue. They seek to not only strengthen scaffolds, but also activate the mechanotransductive pathway that induces osteogenesis in human bone marrow MSCs (Engler et al., 2006). By incorporating MWCNTs researchers were able to not only increase the mechanical strength of poly(caprolactone) (PCL) scaffolds, but also increase adhesion, proliferation, and differentiation of rat MSCs (Pan et al., 2012). In addition, our lab created a new 3D nanocomposite scaffold based on magnetically treated SWCNTs (Figure 1.2A), nanocrystalline hydroxyapatite (nHA), and chitosan hydrogel for improved bone regeneration (Im et al., 2012). Human fetal osteoblasts adhered and proliferated more vigorously on nano scaffolds with magnetically treated SWCNTs over nonmagnetically treated SWCNTs. Notably, the spreading morphology on the magnetically treated SWCNT-augmented scaffolds showed extended filopodia, indicative of strong cell attachment. This effect was further explored by another study in our lab. A nanocomposite coating consisting of magnetically treated SWCNTs and nHA was created, and deposited onto titanium for analysis. Samples coated with SWCNTs exhibited increased MSC and osteoblast adhesion and proliferation when compared to uncoated controls, and samples treated with nonmagnetically treated SWCNTs (Wang et al., 2012). These papers also highlight the possible synergistic effect present when combining multiple nanomaterials within a single orthopedic implant. Moreover, Abarrategi et al. fabricated a scaffold using the freeze-drying method to create fibrous scaffolds consisting of up to 89% MWCNTs. Scaffolds performed well when seeded with myoblastic mouse cell C2C12 (with osteogenic potential) *in vitro* and supported favorable cellular adhesion and proliferation results. The nano scaffolds were then implanted in a mouse subcutaneous muscular pocket defect, and showcased quick degradation and the beginnings of collagen formation at the interface of native tissue and the scaffold (Abarrategi et al., 2008).

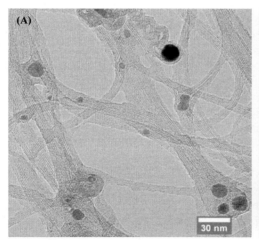 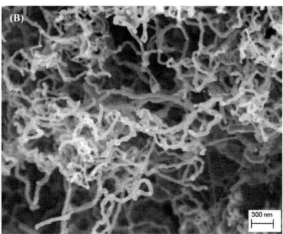

FIGURE 1.2

(A) Scanning electron microscopy (SEM) image of hydrogen-purified MWCNTs. (B) Transmission electron microscopy (TEM) image of magnetically treated SWCNTs.

Carbon nanotubes can also be leveraged to bolster very weak materials and render them usable for tissue engineering applications that otherwise would be less than ideal. Electrospinning (to be described in Section 3.1), a common nano scaffold fabrication technique, can be used to generate tissue-engineered cartilage constructs (Holmes et al., 2012). However, the Young's modulus of the resulting scaffolds is often much lower than autologous tissue. To combat this, Holmes et al. used H_2 purified MWCNTs (Figure 1.2B) to create an electrospun nanocomposite scaffold with a Young's modulus more similar to cartilage, and, with the addition of a polylysine coating, improved chondrogenic differentiation of MSCs significantly when compared to controls (Holmes et al., 2013).

1.2.1.2 Carbon Nanofibers

Similar to CNTs, carbon nanofibers consist of graphene sheets rolled into 3D structures with cone or cylindrical-shaped morphology. A carbon nanofiber can be defined as "sp2-based linear filaments with diameter of 100 nm that are characterized by flexibility and their aspect ratio (above 100)" (Kim et al., 2013) as seen in Figure 1.1. Carbon nanofibers are usually manufactured through vapor deposition with or without a metal catalyst (Endo, 1988), or less commonly, using a mechanical spinning process (Li et al., 2004). Carbon nanofibers have been used throughout tissue engineering to improve mechanical properties and cellular activity, for multiple tissue regeneration applications.

For bone regeneration, Elias et al. reported that osteoblast proliferation and long-term functions (i.e. synthesis of alkaline phosphatase and deposition of extracellular calcium up to 21 days) can be significantly improved on 60 nm diameter carbon nanofibers without a pyrolytic outer layer and 100 nm diameter carbon nanofibers with a pyrolytic outer layer compared to conventional larger carbon fibers (Elias et al., 2002). Since it is well known that surface properties (i.e. surface area, surface roughness, and number of surface defects) of implant materials may have important influences on cell functions (including adhesion, proliferation, differentiation, and mineralization) (Zhang et al., 2008c; Webster et al., 2000), enhanced long term osteoblast functions on carbon nanofibers have been attributed to the special nanometer surface topography of carbon nanofibers which mimics the dimension of inorganic crystalline hydroxyapatite and collagen in natural bone.

Carbon nanofibers are also attractive in neural tissue engineering due to their excellent structural and electrical properties, and from a practical perspective are relatively inexpensive. One approach to utilize carbon nanofibers is to attach or grow them on a conductive polymer to create an array of fibers, and then seed cells onto the construct (Nguyen-Vu et al., 2007). This not only allows for improved electrical conductivity of the material, but the neural cells also demonstrated good electrical contact with the nanofibers. Leveraging these observed advantages, carbon nanofibers were injected into stroke-damaged rat brain defects. The rats that were injected with neural stem cells in tandem with carbon nanofibers formed significantly less glial scar tissue when compared to positive and negative controls, indicating a more successful neural repair (Lee et al., 2006).

1.2.1.3 Graphene

Graphene is the simplest nanomaterial form of carbon, and functions as a base unit for all other carbon nanomaterials. It consists of one monolayer of carbon atoms bonded in sp^2 hybridization orbitals. Sought after as a nanomaterial constituent of composite materials, graphene has a high elastic modulus, high electrical conductivity, and the potential to increase nanotexturization of surfaces. These characteristics apply strongly to many different kinds of tissue engineering, including but not limited to bone, cartilage, and nerve.

Because graphene makes up carbon nanotubes, it is possible to create graphene by "unzipping" carbon nanotubes, a fact that Akhavan et al. took advantage of. They used a silicon dioxide (SiO_2) matrix doped with titanium dioxide (TiO_2) nanoparticles as a photostimulation agent to utilize unzipped MWCNTs to form a graphene nanogrid atop a matrix. The researchers then tested the cellular response to this material using human neural stem cells cultured for 3 weeks. Neural stem cells responded favorably to the growth surface, proliferated faster, and had a strong affinity toward neuronal differentiation lineages when grown on the graphene nanogrids. Additionally, the TiO_2 nanoparticles accelerated differentiation in the neuronal lineage when exposed to flash photo stimulation (Akhavan and Ghaderi, 2013). The experiment showed that graphene can not only improve neuronal differentiation from neural stem cells, but also work in concert with TiO_2 nanoparticles, strongly implying that it is possible to synergistically couple graphene with other nanomaterials for enhanced performance.

The high mechanical strength of graphene leads one to think of it as an ideal material for bone and cartilage tissue engineering, but it inherently lacks sufficient 3D structure for use as a bulk material in tissue engineering. One lab has developed a method to fabricate graphene foams, circumventing this typical problem. First, nickel foam was created and used as a substrate to grow graphene via vapor deposition. The nickel was then dissolved away, leaving behind a 3D structure made entirely of multilayer graphene nanomaterials. Cell viability was observed over 14 days of culture, and osteogenic factors were measured at the conclusion of the 2 week period by fluorescent staining of osteopontin and osteocalcin (Crowder et al., 2013). The graphene foams not only supported MSC attachment, but also spontaneously promoted osteogenesis without the addition of any osteogenic factors in the growth media (Crowder et al., 2013). This is significant as it demonstrates the ability of graphene in particular to upregulate specific differentiation pathways without delivery of additional chemical cues.

A recent study performed in our lab showed that electrospun PCL fibrous scaffolds with incorporated carbon nanotube/graphene could have a powerful effect on MSC fate as well. The scaffolds with carbon nanomaterial exhibited greatly increased MSC growth and glycosaminoglycan synthesis when compared to control, indicating great potential for cartilage repair (Holmes et al., 2014b).

1.2.2 SELF-ASSEMBLING NANOBIOMATERIALS

Since natural tissues are constructed via a bottom-up self-assembly process, scientists are attempting to emulate natural ECM assembly via an emerging class of nanobiomaterials. These nanobiomaterials can self-assemble *in situ* from constituent groups into complex 3D structures on the nano and micro scale, and hold great potential to facilitate the construction of complex, biomimetic tissue environments in a highly reproducible manner (Huebsch and Mooney, 2009). The sheer number of self-assembling nanobiomaterials is quite large and includes collagen, DNA, RNA, peptides, and many more (Zhang, 2003). Several self-assembling nanobiomaterials of particular interest will be discussed next.

1.2.2.1 Self-Assembling Nanotubes

Rosette nanotubes (RNTs, Figures 1.3A-B) are a new class of biologically inspired supramolecular self-assembling nanomaterials. It consists of repeating units of DNA base pairs (Guanine^Cytosine) that assemble into rings (rosettes), that then stack axially to form hollow tubes with controllable 3–4 nm in diameter and lengths up to several microns. They are so versatile that they can be functionalized

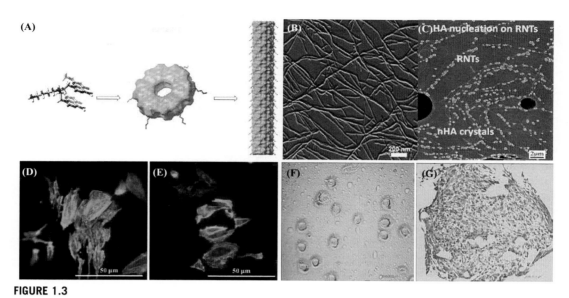

FIGURE 1.3

(A) Schematic illustration of the self-assembly process of RNTs: six twin DNA motifs are self-assembled into rosette-like supermacrocycles and then many of them stack up into stable helical nanotubes with a 3–4 nm diameter and several hundred nanometers long. Atomic force microscopy (AFM) images of (B) RNTs with aminobutane linker (TBL). (C) SEM image of nHA nucleation on RNTs. Fluorescence microscopy images of increased endothelial cell growth on (D) 0.001 mg/ml RNTs coated titanium when compared to (E) uncoated sample. Hematoxylin and eosin staining of (F) controls; and (G) TBL scaffolds for cartilage regeneration at week 1.

with amino acid, peptide, and small molecule side chains. Such side chain groups can be anything, such as cell adhesive lysine, lysine–arginine–serine–arginine (KRSR), and arginine–glycine–aspartic acid–serine–lysine (RGDSK) for enhanced and directed osteoblast, chondrocyte, and MSC function, thus making them intriguing nanomaterials for many types of tissue regeneration (Zhang et al., 2008a, 200 8b, 2009c, 2009d, 2010; Fine et al., 2009; Sun et al., 2012). For instance, it has been demonstrated that RNTs can significantly enhance osteoblast growth and osteogenic differentiation when compared to controls (Sun et al., 2012). It was also observed that RNTs can directly nucleate and align nHA particles along the long axis of the nanotubes (Figure 1.3C) similar to the self-assembled pattern of collagen and nHA in bone, suggesting that our nanotubes can serve as excellent templates for nHA nucleation (Zhang et al., 2009d). In addition, by thoroughly modulating the RNTs peptide side chains, improved endothelial cell growth can be obtained on RNTs when compared to controls (Figures 1.3D–E) (Fine et al., 2009). In our recent work at our lab, we explored MSC adhesion, proliferation, and 4 weeks of chondrogenic differentiation in twin-based RNTs embedded within poly-L-lactic acid (PLLA) scaffolds (Childs et al., 2013). Our results demonstrated that these biomimetic twin-based nanotubes can significantly enhance MSC growth and chondrogenic differentiation, collagen, and protein synthesis (Figures 1.3F–G) when compared to controls without nanotubes. The biomimetic nanostructure and high density of peptides with well-organized architecture contributed the greatly enhanced stem cell functions *in vitro*.

1.2.2.2 Self-Assembling Nanofibers

Besides the aforementioned self-assembling nanotubes, natural peptides have demonstrated the ability to self assemble into a highly ordered, nanofibrous scaffolds in aqueous solution. These peptides possess an ionic self-complementary structure derived from positive and negative side chains on one side of the β-sheet. In addition to this self-complementary system, the amphiphilic character, that is to say containing many hydrophobic and hydrophilic features, of the peptides enables hydrophilic interactions with water molecules. These unique features contribute to the formation of a hygroscopic nanofibrous hydrogel network, with the hydrophobic regions associated into a double sheet, resulting in the formation of the nanofibers (Hauser and Zhang, 2010). For instance, Hartgerink et al. reported that a self-assembly peptide-amphiphile with the cell-adhesive RGD (Arg–Gly–Asp) self-assembled into supramolecular nanofibers and aligned nHA on their long axis for bone application (Hartgerink et al., 2001). Hosseinkhani et al. showed significantly enhanced osteogenic differentiation of stem cells in a 3D peptide-amphiphile scaffold compared to 2D static tissue culture (Hosseinkhani et al., 2006). In addition, Shah et al. designed peptide-amphiphile nanofibers that display a high density of transforming growth factor β1 (TGF-β1) binding sites for improved cartilage regeneration (Shah et al., 2010, Aida et al., 2012).

Simpler peptides can be leveraged as self-assembly nanomaterials for neural applications as well, such as isolucine–lysine–valine–alanine–valine (IKVAV) (Silva et al., 2004), and RADA-16 (Gelain et al., 2006). Because the peptides used in self-assembled scaffolds are derived from biology, the resulting nanofibrous scaffolds are similar to natural ECM and present excellent biomimetic properties. Peptide nanofibers have become the most widely investigated self-assembling nanobiomaterials for tissue engineering.

1.2.3 POLYMERIC AND CERAMIC NANOBIOMATERIALS

1.2.3.1 Polymeric Nanobiomaterials

As the largest biomaterial group, polymers play an important role in complex tissue engineering. Polymeric nanobiomaterials are extremely customizable through a variety of processing methods and chemical modifications, and are common in the clinical environment. This leaves researchers with many practical, clinically ready and FDA-approved options. One very common application of polymers as nanomaterials is in the creation of therapeutic drug-loaded nanoparticles for sustained and targeted delivery to cells and tissue (Parveen et al., 2012). Considering that the process of proving a new biomaterial to the FDA as safe and effective is expensive and time-consuming, many researchers have begun extending the application of current biocompatible polymers already approved by the FDA for other medical devices for use in drug-loaded nanoparticle fabrication. For instance, polylactic acid (PLA), polyglycolic acid (PGA), poly(lactic-co-glycolic acid) (PLGA), and various hydrogels currently are generating huge interest from scientists for their potential for numerous regeneration applications due to their excellent biocompatibility, suitable mechanical properties, biodegradability, and ease of modification for different applications. A myriad of well-designed polymer-based nanoparticles loaded with growth factors or other therapeutics have been created for tissue engineering applications (Dhandayuthapani et al., 2011; Castro et al., 2012a).

As we know, various transforming growth factors (e.g., TGF-β1 and TGF-β3), and bone morphogenic proteins (e.g. BMP-7, BMP-6, or BMP-2) have been shown to improve bone and cartilage regeneration (Noel et al., 2004; Sekiya et al., 2001; Bai et al., 2011; Chim et al., 2012; Kim et al., 2012).

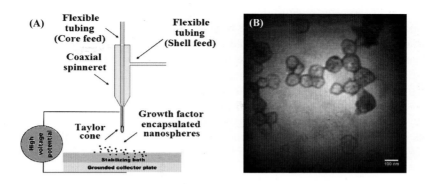

FIGURE 1.4

(A) Graphical representation of a coaxial electrospraying technique for the manufacture of growth factor-encapsulated core-shell nanosphere. (B) TEM image of bone morphogenetic protein-2 encapsulated PDO nanospheres.

However, for *in vivo* applications, these growth factors face ongoing issues related to short-term retention, quick half-life in circulation, and quick loss of biological activity *in vivo* even when administered at higher doses. For example, when delivered directly to tissue defect sites, they rapidly diffuse to adjacent tissues and lose their bioactivity, which limit their potential to promote prolonged tissue formation within a targeted site. In our lab, we have begun exploring the use of coaxial electrospraying (Figure 1.4A) for highly efficient encapsulation of growth factors and therapeutics into a core-shell polymer nanosphere (Figure 1.4B) (Castro et al., 2014; Zhu et al., 2014b). The coaxial electrospraying technique allows easy fabrication of a controllable core-shell nanosphere with intact biologically active growth factor within the core and polymeric outer shell. In addition, it enables the separation of organic and aqueous phases and thus incorporation of biologically active components such as growth factors into the aqueous phase without exposing them to harmful organic solvents. The selected polydioxanone (PDO), known commercially as PDS®, is commonly used as an absorbable suture (Thomas et al., 2009; Venclauskas et al., 2011; Sakamoto et al., 2012). Excellent biocompatibility, mechanical properties as well as a slower degradation rate, render it ideal for controlled drug delivery and tissue engineering applications (Zhu et al., 2011; Wang et al., 2010b; Madurantakam et al., 2009; Smith et al., 2008; Kalfa et al., 2010). Our results showed that the PDO nanospheres exhibited a much slower release of BMP-2 when compared to scaffolds blended with the growth factor, which contributed to improved osteogenic performance of MSCs.

1.2.3.2 Ceramic Nanobiomaterials and Ceramic-Polymer Nanocomposites

Ceramics are nonmetallic inorganic crystalline materials that express excellent cytocompatibility, mechanical properties, and possibly biodegradability in the physiological environment, which makes them attractive for orthopedic and dental applications. Commonly investigated ceramics for bone regeneration can be classified into two categories: bioactive calcium phosphates (such as nHA and tricalcium phosphate (TCP)) or biopassive ceramics (such as alumina and zirconia) (Zhang et al., 2008). As we know, 70% of the human bone matrix is composed of inorganic crystalline nHA, which is typically 20–80 nm long and 2–5 nm thick (Zhang and Webster, 2009; Holmes et al., 2012). Additionally, other components in the bone matrix (such as collagen and noncollagenous proteins (laminin, fibronectin, and vitronectin) are nanometer-scale in dimension. Therefore, novel nanostructured ceramics, which mimic the nanostructure of natural bone, have become quite popular. These nanobiomaterials exhibit increased surface area, and contribute improved surface roughness, surface wettability, and cytocompatibility (Zhang et al., 2008) to tissue-engineered bone constructs. For example, nHA is commonly used in bone tissue engineering, due to its natural abundance in native bone tissue. In one study, nHA and bone marrow aspirate (BMA) were combined to form a paste for use in posterolateral fusion surgeries across 46 patients. When evaluated 12 months later, there was no difference between bone formation rates of iliac crest autografts (the gold standard in this surgery) and the nHA/BMA mixture (Robbins et al., 2014). nHA has been shown to not only improve osteoblast behavior (Webster et al., 2000, Webster, 2001), but also improve bone-marrow-derived MSCs behavior *in vitro* (Castro et al., 2014) and improve new bone formation *in vivo* (Huber et al., 2006, Chang et al., 2001) Another highly studied ceramic material is zirconium dioxide or zirconia. It enhances fracture toughness in other ceramics. Kong et al. studied HA-added zirconia–alumina nanocomposites in load-bearing orthopedic applications (Kong et al., 2005). The HA-added zirconia–alumina nanocomposites contained biphasic calcium phosphates of HA/TCP and had higher flexural strength than conventionally mixed HA-added zirconia–alumina composites. The *in vitro* tests showed that the proliferation and differentiation of osteoblasts on this nanocomposite gradually increased as the amount of added HA increased.

Although nHA and other ceramics are powerful materials when used alone, they can be made more versatile in combination with polymers to create a nanocomposite tissue-engineered scaffold. This approach can effectively allow for the fabrication of biomimetic physical and chemical gradients in bone, cartilage, and osteochondral tissue. Using osteochondral tissue as an example, osteochondral defects, caused by osteoarthritis and trauma, present a common and serious clinical problem. They are notoriously difficult to regenerate due to complex inherent, stratified soft/hard tissue structure, poor cell mobility within the matrix, lack of vascular network, and limited local progenitor cells. Continuous gradients of proteins and collagen fibers are found throughout the cartilage zones and are essential for load transfer and directed cell behavior (Zhang et al., 2009b; Dormer et al., 2012; Dormer et al., 2010). The subchondral bone and calcified zone also contain a gradient of calcified ECM (nHA) (Zhang et al., 2012), which provides osteochondral structural integrity (Zhang et al., 2009b). Considering the unique graded structure of osteochondral tissue, ceramic nanoparticles can be embedded within a polymer or hydrogel scaffold to form a ceramic gradient originating in the rigid subchondral region and terminating with zero ceramic component in the articulating cartilage region of a scaffold (Khanarian et al., 2012). The stiffness of the native microenvironment provides an essential stimulus that helps to shepherd the phenotypic differentiation of pluripotent and multipotent stem cells, such as MSCs, in conjunction with chemical and other stimuli. Therefore, implantable multiphasic scaffolds that leverage spatially controlled stiffness gradients, morphogenetic factors, and configurable geometries are being developed in an effort to direct the differentiation and phenotypic expression of bone-marrow-derived MSCs towards osteogenic and chondrogenic cell types in one construct (Wang et al., 2009; Wang et al., 2010a; Chen et al., 2011). Several researchers also utilize biomimetic spatially controlled gradients (Erisken et al., 2008; Zhang et al., 2005) with graded PCL/nHA composite fiber meshes. The fibrous meshes were subsequently seeded with mouse preosteoblast cells and after a 4-week culture period, a deposited ECM was observed exhibiting gradations of collagen type I and calcium that were similar to the gradients that exist in the osteochondral site.

Similar to osteochondral tissue, the ligament is another highly integrated tissue, which must perform in a complex manner under a high stress environment. In order to best simulate regeneration and repair of this environment, techniques employing two or more disparate biomaterials to create composite scaffolds with considerably improved mechanical and biochemical properties is of great prominence. In ligament tissue engineering, the fibrous nature of the natural tissue has inspired researchers to use electrospun scaffolds with a density gradient, similar to the nHA constituent in osteochondral tissue engineering, generating a stiffness gradient along with a multimaterial approach (Kuo et al., 2010; Samavedi et al., 2011). This multiphasic approach consisted of co-spinning PCL with incorporated nano hydroxyapatite and poly(ester urethane). The result was a graded scaffold that had both a biochemical gradient and a mechanical strength gradient that more accurately mimicked the natural ligament ECM (Samavedi et al., 2011).

1.3 3D NANO/MICROFABRICATION TECHNOLOGY FOR TISSUE REGENERATION

1.3.1 3D NANOFIBROUS AND NANOPOROUS SCAFFOLDS FOR TISSUE REGENERATION

1.3.1.1 Electrospun Nanofibrous Scaffolds for Tissue Regeneration

The most prominent nanofibrous scaffold fabrication method is electrospinning. Similar to the aforementioned electrospraying technique, electrospinning utilizes the same equipment, but with a polymer of higher viscosity, allowing the stream not to break up and form droplets. It is a process by which a charged polymer is dissolved in solvent, and exposed to a large voltage potential of several kilovolts as it is slowly pumped from a needle. The large electrical potential causes the fluid to be drawn out into a fine stream which solidifies into fibers that can be dimensionally controlled by varying the viscosity, voltage potential, flow rate, and working distance between the needle and collector plate during fabrication. Figure 1.5A shows an electrospun highly aligned fibrous scaffold with conductive CNTs fabricated in our lab. This type of nanofibrous scaffold tends to be most effective when used for tissues with similar fibrous morphologies, and have been shown to be effective in many skin, musculoskeletal, and neural tissue regeneration applications (Holmes et al., 2012; Zhu et al., 2014a). Popularity for neural tissue engineering is due to the ability to create highly aligned nanofibrous scaffolds from many biocompatible polymeric materials. These scaffolds have been shown to more effectively promote neural outgrowth to bridge given defect sites (Assmann et al., 2010).

Even though electrospinning normally creates thin constructs, the high surface area to volume ratio, nanometer feature size, and relative ease of fabrication make electrospun nanofibrous scaffolds beneficial for bone and cartilage tissue engineering, which have been thoroughly reviewed in our previous paper (Holmes et al., 2012). For example, Aclam et al. used electrospun PCL nanofibers and collagen type I to create an injectable scaffold that promotes bone regeneration. Briefly, they electrospun PCL fibers, combined them with a cell-laden collagen gel, and allowed the composite to crosslink naturally at physiologic conditions. After 21 days of culture, the injectable scaffolds showed increased total protein, alkaline phosphatase, and calcium concentration when compared to a pure collagen control (Baylan et al., 2013).

1.3.1.2 Other 3D Nanofibrous/Nanoporous Scaffolds for Tissue Regeneration

Besides nanofibrous scaffolds fabricated via electrospinning, other fabrication methods for nanofibrous or nanoporous scaffolds are also commonly utilized in tissue engineering. For instance, for musculoskeletal tissue studies, conventional scaffold fabrication methods such as solvent casting and particle

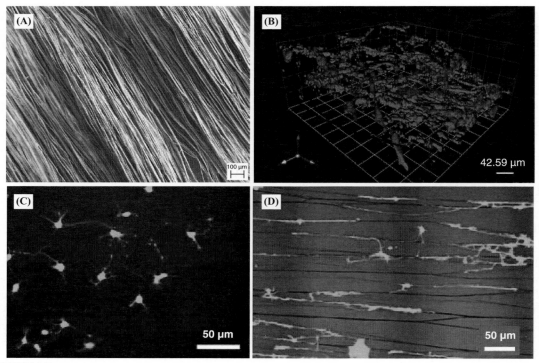

FIGURE 1.5

(A) SEM image of highly aligned electrospun scaffold with 0.5% CNTs. (B) Confocal microscopy image of MSC growth in a cold plasma modified nano bone scaffold. Blue represents cell nuclei stained by DAPI; red represents cytoskeleton stained by rhodamine-phalloidin; gray represents the porous nHA/chitosan scaffold. Image is from Wang et al. (2014). Confocal micrographs of neurons (green) cultured for 5 days on (C) the polystyrene substrate, and (D) the parallel-aligned CNT yarns (black lines) substrate, respectively. Image is from Fan et al. (2012). A color version of this figure can be viewed online.

leaching (Mikos et al., 1994; Jiang et al., 2007), and freeze drying (Whang et al., 1995) have been widely used to fabricate 3D porous tissue scaffolds, which have been shown to influence cell functions and improve cartilage and bone regeneration (Castro et al., 2012b; Zhang et al., 2009b). Recently, 3D porous hydrothermally treated nHA/chitosan nanocomposite scaffolds have been fabricated through a freeze-drying method with cold plasma treatment (Wang et al., 2014). The results revealed that all nHA-embedded, plasma-modified chitosan scaffolds (Figure 1.5B) significantly enhanced MSC growth, migration, and osteogenic differentiation *in vitro*. In addition, for ligament tissue regeneration, it has been shown that fibrous scaffolds employing natural silk are strong and relatively easy to work with, and display biomimetic amino acids on the surface of the material, increasing stem cell performance (Chen et al., 2012). In that study, Chen et al. capitalized on the strength of a macrofibrous knitted silk sponge scaffold coated with self-assembled RADA16 peptide nanofibers in the form of a nanofibrous mesh to increase cell performance (Chen et al., 2012). Scaffolds treated with RADA16 showed increased maximum tensile strength, collagen, and glycosaminoglycan synthesis compared to bare controls. This synergistic effect of combining a microfibrous scaffold with nanomaterial coatings also illustrates the importance of biomimetic nanocomposites in tissue engineering.

Other even more novel methods of scaffold fabrication are constantly being explored, Fan et al. were able to draw out superaligned CNT yarn and apply it to neural tissue engineering. The yarns described are not only biomimetic and able to direct neural growth (Fan et al., 2012), but exist as a transition between traditional fabrication methods and customizable scaffold design. One could imagine the customization of the weave of the yarn (Figures 1.5C and D) to the dimension of a neural defect in a particular subject, enabling researchers and medical professionals to adapt to the individual situation present in each animal model or human patient.

1.3.2 3D PRINTING OF NANOMATERIAL SCAFFOLDS FOR TISSUE REGENERATION

1.3.2.1 3D Printing Techniques for Tissue Regeneration

As an emerging 3D tissue manufacturing technique, 3D printing offers great precision and control of the architecture of a scaffold, and prints complicated structures that closely mirror biological tissues (Derby, 2012). 3D printing has become a driving force in the tissue engineering field with the advent of personalized medicine and the growing interest in complex tissue and organ regeneration (Tasoglu and Demirci, 2013; Lee and Wu, 2012; Cui et al., 2012; Koch et al., 2012; Catros et al.; 2012, Fedorovich et al., 2012b; Shim et al., 2011; Gruene et al., 2011; Catros et al., 2011; Song et al., 2010; Ovsianikov et al., 2010; Detsch et al., 2011; Warnke et al., 2010; Moon et al., 2010; Holmes et al., 2014a). Most 3D printers have micrometer resolutions far above the nano scale, but are still a viable tool for nanomaterial fabrication and manipulation. Unlike traditional manufacturing techniques, 3D printing can deliver materials and cells to precise locations, resulting in constructs that can take advantage of computer-aided designed (CAD) and biomimetic morphology to create shapes that would be difficult or impossible to manufacture traditionally. All 3D printing methods operate upon similar principles. To fabricate a solid object, a CAD model is first input to a program that parses the solid object into a stack of thin axial cross-sections. These cross-sections are then converted into directions that describe the movement of the effector in 3D space. The effector deposits material, solidifies resin, or otherwise performs an action that prints the CAD model itself and is controlled to reproduce each cross-sectional slice delivered to the printer. Finally, all of the individual elements come together and the construct is serially printed from the bottom up.

It is important to note that 3D printing techniques can print materials with or without incorporated cells, and each approach comes with various advantages and disadvantages. The major defining difference between the two is that bioprinting cells must maintain an environment that facilitates cellular survival, and mitigates contamination or infection. 3D printing will be discussed at length later in this book, but briefly, several common methods (Figure 1.6) include inkjet bioprinting (Ferris et al., 2013), bioplotting (Fedorovich et al., 2012a), fused deposition modeling (Kundu et al., 2013), selective laser sintering, and stereolithography (Suri et al., 2011). One of the advantages of 3D printing is the ability to create custom-designed tissue constructs with complex internal architecture and biomimetic external architecture. As illustrated in Figure 1.7A, an MRI image of human osteochondral tissue defects is reconstructed into a unique CAD model. This patient-specific, osteochondral construct can be printed to perfectly integrate with the defect site and expedite tissue regeneration. Figures 1.7C–G show that several custom-designed 3D poly(ethylene glycol) diacrylate (PEG-DA) hydrogel scaffolds with varying pore sizes were fabricated via a stereolithography printer in our lab.

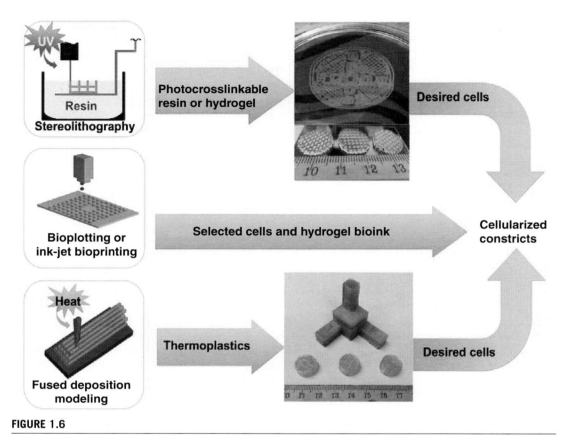

FIGURE 1.6

Several typical 3D printing systems. Image is from O'Brien et al. (2014).

A quickly growing fabrication method for more detailed structures is a technique called two-photon polymerization. Here, two light sources are used: one to excite the material to an intermediate state, and another to initiate crosslinking. Through this method, scaffolds can be created with feature sizes of less than 10 μm, allowing researchers to generate CAD models on a smaller scale, and print constructs in various specific geometries with high resolution. Femtosecond laser two-photon polymerization is one of the highest-resolution 3D printing technologies in use today. Initially, only commercially available polymers (Ovsianikov et al., 2007) could be printed, which may have lacked necessary physical, chemical, and cell favorable characteristics needed for biomimetic tissue engineering. Some materials, such as 4,4′-bis(dimethylamino)benzophenone (SZ2080), can support cell growth and migration (Raimondi et al., 2013), and have effectively demonstrated that cells can react to microgeometries with pores varying from 10 to 30 μm. Scaffolds with graded microporosity outperformed single-dimension microporous scaffolds, and cells tended to migrate to designed niches on the surface of the scaffold, as opposed to flat surfaces. The biocompatible PEG was used in two-photon polymerization printer to produce engineered 3D scaffolds (Torgersen

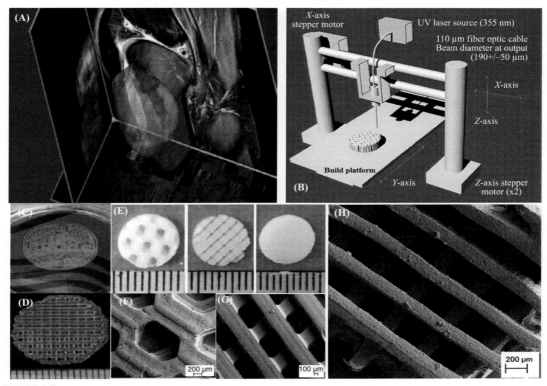

FIGURE 1.7

(A) An MRI image of an osteochondral defect (labeled as red color) in human knee joint. (B) Schematic illustration of a table top stereolithography 3D bioprinter. (C–E) 3D printed PEG-DA hydrogel scaffolds with varied designs. (F–G) SEM images of 3D bioprinted PEG-DA scaffolds with hexagonal and square pores; and (H) with nHA particles for osteochondral regeneration. A color version of this figure can be viewed online.

et al., 2012). This work focused on using two-photon polymerization in proximity to living tissue, and used a near-infrared 100 fs laser instead of a traditional, generally nonbiocompatible UV laser. With a specially developed photoinitiator customized for near-infrared wavelengths, researchers were able to print scaffolds approaching 300 μm^2. Despite Raimondi and Torgeren's success, Torgeren also mentioned that the scaffolds were only 280 μm × 280 μm × 225 μm, and Raimondi mentioned that on average 17 cells were found attached to the scaffolds after 6 days of *in vitro* culture. This serves to highlight the inherent limitations of the two-photon approach; particularly, a small overall scaffold-size that limits the scaffolds dimensions to a few hundred micrometers in each dimension, far below clinical relevance.

1.3.2.2 3D Printing of Nanomaterial Scaffolds for Tissue Regeneration

Nanomaterials for 3D printing are just now being developed. One nanobiomaterial in use is bacterial nanocellulose (BNC). BNC is a naturally occurring nanomaterial synthesized by bacteria that has been used, in this case, with a 3D printing system to produce patient-specific auricular constructs that closely match natural geometries (Nimeskern et al., 2013). This material is nanofibrous and promotes adhesion of endothelial and NIH/3T3 cell lines (Fu et al., 2013), making it an excellent potential biomaterial for 3D fabrication of chondrogenic scaffolds. Other nanomaterials and nanocomposite biomaterials are being developed for use in tissue engineering, and many need only simple modification to be compatible with a number of 3D printing modalities.

As an example of nanobiomaterial utilization, our lab has developed a table-top stereolithography setup using a UV laser as the effector (Figure 1.7B). It has been used to crosslink nHA impregnated photo-crosslinkable PEG-DA hydrogel into 3D osteochondral scaffolds (Figure 1.7H). These scaffolds are unique in that they have a biologically inspired gradient of nHA to guide MSC differentiation to osteogenic and chondrogenic lineages within the same scaffold. This printer has additionally been used to print graphene nanoplatelets suspended in PEG-DA hydrogel into 3D scaffolds for neural regeneration. Both examples showcase the potential of 3D printing to incorporate nanomaterials and traditional biomaterials within the same scaffold to achieve nanoscale features, but some printers attempt to create even higher-resolution constructs without the addition of nanomaterials.

Another nanomaterial used in 3D printing has been the bioceramic TCP. TCP can be processed into a fine powder (Tarafder et al., 2013) and applied in a novel 3D sintering method that uses directed microwaves to selectively heat and sinter the fine powder particles together, layer by layer, into a 3D scaffold. (Wagner et al., 2013; Tarafder et al., 2013). Microwave sintered TCP scaffolds exhibited an increase in compressive strength and more optimal microporosity to macroporosity ratio when compared to scaffolds fabricated with traditional laser energy sources. Furthermore, the microwave sintered scaffolds increased the formation of new bone *in vivo* when compared to constructs fabricated through conventional sintering (Tarafder et al., 2013).

Because it is known that nanobiomaterials are integral to eliciting the optimal response from many cell lines, it may seem beneficial to target nanoscale structures in future 3D printing systems. However, the challenges involved in realizing such precise designs can make fabricating objects with nanoscale precision somewhat impractical. Researchers have employed everything from dynamic masking type procedures to high-resolution digital micromirror array photocuring to highly tuned two-photon femtosecond laser-based printing. Although these methods are extremely high-resolution, the speed at which structures can be fabricated is on the order of hours, and are still limited to a maximum build envelope of several hundred micrometers. These two major factors limit the potential efficacy of these types of systems as clinical tools, in favor of fabrication technology that can rapidly produce defect-sized constructs in a time-efficient manner. Although many challenges are ahead, many are hopeful that with the right combination of hardware, software, and novel nanobiomaterials, advanced nanoscale 3D printing can one day be realized and implemented. More detailed information about integrating 3D bioprinting techniques with nanomaterials for complex tissue and organ regeneration can be referred to (O'Brien et al., 2014).

1.4 CONCLUSION AND FUTURE DIRECTIONS

Current tissue engineering methods can readily create materials, structures, and scaffolds that can support one or a few cell types *in vitro*, but *in vivo* many challenges exist that limit the efficacy of current, single-tissue designs. There will naturally be differences between living tissue and the implanted remedy, so researchers must take this into account and allow for proper integration without major mechanical and biochemical disparities. Furthermore, many injuries that would benefit from a tissue-engineered solution, such as osteochondral, ligament, and nerve damage, occur at the interface of two or more tissue types. This means a fully successful implant must simultaneously support the growth of different cell types and tissues, each with specific mechanical properties, chemical gradients, cell populations, and specific geometric constraints incorporated within the scaffold design. The myriad of complex design constraints limits the effectiveness of many current methods, especially when attempting to regenerate clinically relevant injuries, organs, and other complex tissues and tissue interfaces.

To address the inherent limitations and requirements posed by the increased complexity of interfacial tissue engineering, nanotechnologies (such as nanobiomaterials) are increasingly being utilized. Nanobiomaterials clearly have played an integral role in tissue engineering, and will continue to be an important design consideration for future work. The beauty of incorporating nanomaterials into tissue-engineered constructs is the versatility they contribute almost intrinsically. Many nanomaterials add similarly vast improvements to the constructs they are incorporated within. The cutting-edge of tissue engineering and regenerative medicine research is moving toward customizing therapies to individual patients and individual situations. In order for nanomaterials to be integrated into a patient-specific scaffold, manufacturing techniques need to be employed to allow for further micro- and macroscale customization; exactly what 3D printing excels at.

Now, with the introduction of 3D printing in the tissue engineering field, researchers can begin to experience the benefits of having truly unique solutions to problems not easily solved with traditional fabrication techniques. A robust hobbyist community and open-source movement, RepRap, has been driving down the cost of implementation of many 3D printing technologies, making them available for researchers and clinicians worldwide. This has made 3D printing technology readily available to the research community, and thus 3D bioprinting has been advancing at an exciting pace. The successful implementation of a scaffold with complex requirements, including the utilization of multiple disparate materials and several nanomaterial constituents into a patient-specific geometry with highly defined internal microgeometry, becomes surmountable. By incorporating multiple cell types, biomaterials, and nanomaterials in specific, biomimetic geometries, tissue engineers can expect to develop truly revolutionary medical devices, therapies, and treatments, and potentially usher in a new age of regenerative medicine.

ACKNOWLEDGMENTS

The authors would like to thank the support of NIH Award Number UL1TR000075 from the NIH National Center for Advancing Translational Sciences, NIH Director's New Innovator Award, The George Washington Institute for Biomedical Engineering (GWIBE), and GW Institute for Nanotechnology (GWIN).

REFERENCES

Abarrategi, A., Gutierrez, M.C., Moreno-Vicente, C., Hortiguela, M.J., Ramos, V., Lopez-Lacomba, J.L., Ferrer, M.L., Del Monte, F., 2008. Multiwall Carbon Nanotube Scaffolds for Tissue Engineering Purposes. Biomaterials 29, 94–102.

Aida, T., Meijer, E.W., Stupp, S.I., 2012. Functional Supramolecular Polymers. Science 335, 813–817.

Akhavan, O., Ghaderi, E., 2013. Differentiation of Human Neural Stem Cells into Neural Networks on Graphene Nanogrids. Journal of Materials Chemistry B 1, 6291–6301.

Assmann, U., Szentivanyi, A., Stark, Y., Scheper, T., Berski, S., Drager, G., Schuster, R.H., 2010. Fiber Scaffolds of Polysialic Acid Via Electrospinning for Peripheral Nerve Regeneration. J Mater Sci Mater Med 21, 2115–2124.

Bai, X., Li, G., Zhao, C., Duan, H., Qu, F., 2011. BMP7 Induces the Differentiation of Bone Marrow-derived Mesenchymal Cells into Chondrocytes. Med Biol Eng Comput 49, 687–692.

Baylan, N., Bhat, S., Ditto, M., Lawrence, J.G., Lecka-Czernik, B., Yildirim-Ayan, E., 2013. Polycaprolactone Nanofiber Interspersed Collagen Type-I Scaffold for Bone Regeneration: A Unique Injectable Osteogenic Scaffold. Biomed Mater 8, 045011.

Biggs, M.J., Richards, R.G., Gadegaard, N., Wilkinson, C.D., Dalby, M.J., 2007. Regulation of Implant Surface Cell Adhesion: Characterization and Quantification of S-phase Primary Osteoblast Adhesions on Biomimetic Nanoscale Substrates. J Orthop Res 25, 273–282.

Castro, N., Umanzor-Alvarez, J., Grace ZHANG, L., KeidarF M., 2012a. Nanobiotechnology and Nanostructured Therapeutic Delivery Systems. Recent Patents on Biomedical Engineering 5, 29–40.

Castro, N.J., Hacking, S.A., Zhang, L.G., 2012b. Recent Progress in Interfacial Tissue Engineering Approaches for Osteochondral Defects. Ann Biomed Eng 40, 1628–1640.

Castro, N.J., O'Brien, C., Zhang, L.G., 2014. Biomimetic Biphasic 3D Nanocomposite Scaffold for Osteochondral Regeneration. AICHE Journal 60, 432–442.

Catros, S., Fricain, J.C., Guillotin, B., Pippenger, B., Bareille, R., Remy, M, Lebraud, E., Desbat, B., Amedee, J., Guillemot, F., 2011. Laser-assisted Bioprinting for Creating On-demand Patterns of Human Osteoprogenitor Cells and Nano-hydroxyapatite. Biofabrication 3, 025001.

Catros, S., Guillemot, F., Nandakumar, A., Ziane, S., Moroni, L., Habibovic, P., Blitterswijk, V.A.N., Rousseau, C., Chassande, B., Amedee, O., Fricain, J.C., 2012. Layer-by-layer Tissue Microfabrication Supports Cell Proliferation *in vitro* and *in vivo*. Tissue Eng Part C Methods 18, 62–70.

Chang, C.K., Wu, J.S., Mao, D.L., Ding., C.X., 2001. Mechanical and Histological Evaluations of Hydroxyapatite-coated and Noncoated Ti6Al4V Implants in Tibia Bone. J Biomed Mater Res 56, 17–23.

Chen, J., Chen, H., Li, P., Diao, H., Zhu, S., Dong, L., Wang, R., Guo, T., Zhao, J., Zhang, J., 2011. Simultaneous Regeneration of Articular Cartilage and Subchondral Bone *in vivo* Using MSCs Induced by a Spatially Controlled Gene Delivery System in bilayered integrated scaffolds. Biomaterials 32, 4793–4805.

Chen, K., Sahoo, S., He, P., Ng, K.S., Toh, S.L., Goh, J.C., 2012a. A hybrid silk/RADA-based fibrous scaffold with triple hierarchy for ligament regeneration. Tissue Eng Part A 18, 1399–1409.

Chen, K., Sahoo, S., He, P., Ng, K.S., Toh, S.L., Goh, J.C., 2012b. A hybrid silk/RADA-based fibrous scaffold with triple hierarchy for ligament regeneration. Tissue Eng Part A 18, 1399–1409.

Childs, A., Hemraz, U.D., Castro, N.J., Fenniri, H., Zhang, L.G., 2013. Novel biologically-inspired rosette nanotube PLLA scaffolds for improving human mesenchymal stem cell chondrogenic differentiation. Biomed Mater 8, 065003.

Chim, H, Miller, E., Gliniak, C., Alsberg, E., 2012. Stromal-cell-derived factor (SDF) 1-alpha in combination with BMP-2 and TGF-beta1 induces site-directed cell homing and osteogenic and chondrogenic differentiation for tissue engineering without the requirement for cell seeding. Cell Tissue Res 350, 89–94.

Chopra, K., Mummery, P.M., Derby, B., Gough, J.E., 2012. Gel-cast glass-ceramic tissue scaffolds of controlled architecture produced via stereolithography of moulds. Biofabrication 4, 045002.

Crowder, S.W., Prasai, D., Rath. R, Balikov, D.A., Bae, H., Bolotin, K.I., Sung, H.J., 2013. Three-dimensional graphene foams promote osteogenic differentiation of human mesenchymal stem cells. Nanoscale 5, 4171–4176.

Cui, X., Breitenkamp, K., Finn, M.G., LOTZ, M, D'Lima, F D.D., 2012. Direct human cartilage repair using three-dimensional bioprinting technology. Tissue Eng Part A 18, 1304–1312.

Derby, B., 2012. Printing and prototyping of tissues and scaffolds. Science 338, 921–926.

Detsch, R., Schaefer, S., Deisinger, U., Ziegler, G., Seitz, H., Leukers, B., 2011. *In vitro*: osteoclastic activity studies on surfaces of 3D printed calcium phosphate scaffolds. J Biomater Appl 26, 359–380.

Dhandayuthapani, B., Yoshida, Y., Maekawa, T., Kumar, D.S., 2011. Polymeric Scaffolds in Tissue Engineering Application: A Review. International Journal of Polymer Science, 290602.

Dormer, N.H., Singh, M., Wang, L., Berkland, C.J., Detamore, M.S., 2010. Osteochondral interface tissue engineering using macroscopic gradients of bioactive signals. Ann Biomed Eng 38, 2167–2182.

Dormer, N.H., Singh, M., Zhao. L, Mohan, N., Berkland, C.J., Zdetamore, M.S., 2012. Osteochondral interface regeneration of the rabbit knee with macroscopic gradients of bioactive signals. J Biomed Mater Res A 100, 162–170.

Elias, K.L., Price, R.L., Webster, T.J., 2002. Enhanced functions of osteoblasts on nanometer diameter carbon fibers. Biomaterials 23, 3279–3287.

Endo, M., 1988. Grow carbon fibers in the vapor phase. Chemtech 18, 568–576.

Engler, A.J., Sen, S., Sweeney, H.L., Discher, D.E., 2006. Matrix elasticity directs stem cell lineage specification. Cell 126, 677–689.

Erisken, C., Kalyon, D.M., Wang, H., 2008. Functionally graded electrospun polycaprolactone and beta-tricalcium phosphate nanocomposites for tissue engineering applications. Biomaterials 29, 4065–4073.

Fan, L., Feng, C., Zhao, W., Qian, L., Wang, Y., Li, Y., 2012. Directional neurite outgrowth on superaligned carbon nanotube yarn patterned substrate. Nano letters 12, 3668–3673.

Fedorovich, N.E., Leeuwenburgh, S.C., Van der helm, Y.J., Alblas, J., Dhert, W.J., 2012a. The osteoinductive potential of printable, cell-laden hydrogel-ceramic composites. J Biomed Mater Res A 100, 2412–2420.

Fedorovich, N.E., Schuurman, W., Wijnberg, H.M., Prins, H.J., Van weeren, P.R., Malda, J., Alblas, J., Dhert, W.J., 2012b. Biofabrication of osteochondral tissue equivalents by printing topologically defined, cell-laden hydrogel scaffolds. Tissue Eng Part C Methods 18, 33–44.

Ferris, C.J., Gilmore, K.J., Beirne, S., McCallum, D., Wallace, G.G., In Het Panhuis, M., 2013. Bio-ink for on-demand printing of living cells. Biomaterials Science 1, 224.

Fine, E., Zhang, L., Fenniri, H., Webster, T.J., 2009. Enhanced endothelial cell functions on rosette nanotube-coated titanium vascular stents. Int J Nanomedicine 4, 91–97.

Frieden, E., 1972. The chemical elements of life. Sci Am 227, 52–60.

Fu, L.N., Zhou, P., Zhang, S.M., Yang, G., 2013. Evaluation of bacterial nanocellulose-based uniform wound dressing for large area skin transplantation. Materials Science & Engineering C-Materials for Biological Applications 33, 2995–3000.

Gacem, K., Retrouvey, J.M., Chabi, D., Filoramo, A., Zhao, W., Klein, J.O., Derycke, V., 2013. Neuromorphic function learning with carbon nanotube based synapses. Nanotechnology 24, 384013.

Gelain, F., Bottai, D., Vescovi, A., Zhang, S., 2006. Designer self-assembling peptide nanofiber scaffolds for adult mouse neural stem cell 3-dimensional cultures. PLoS ONE 1, e119.

Gruene, M., Pflaum, M., Hess, C., Diamantouros, S., Schlie, S., Deiwick, A., Koch, L., Wilhelmi, M., Jockenhoevel, S., Haverich, A., Chichkov, B., 2011. Laser printing of three-dimensional multicellular arrays for studies of cell-cell and cell-environment interactions. Tissue Eng Part C Methods 17, 973–982.

Hartgerink, J.D., Beniash, E., Stupp, S.I., 2001. Self-assembly and mineralization of peptide-amphiphile nanofibers. Science 294, 1684–1688.

Hauser, C.A., Zhang, S., 2010. Designer self-assembling peptide nanofiber biological materials. Chemical Society Reviews 39, 2780–2790.

Holmes, B., Castro, N.J., Li, J., Keidar, M., Zhang, L.G., 2013. Enhanced human bone marrow mesenchymal stem cell functions in novel 3D cartilage scaffolds with hydrogen treated multi-walled carbon nanotubes. Nanotechnology 24, 365102.

Holmes, B., Castro, N.J., Zhang, L.G., Zussman, E., 2012. Electrospun fibrous scaffolds for bone and cartilage tissue generation: recent progress and future developments. Tissue Eng Part B Rev 18, 478–486.

Holmes, B., Li, J., Lee, J. D., Zhang, L.G. 2014a. Development of novel three-dimensional printed scaffolds for osteochondral regeneration. *Tissue Engineering Part A*, Epub ahead of print. doi:10.1089/ten.tea.2014.0138.

Holmes, B., Zarate, A., Keidar, M., Zhang, L. G. 2014b. Enhanced Human Bone Marrow Mesenchymal Stem Cell Chondrogenic Differentiation in Electrospun Constructs with Carbone Nanomaterials. *Carbon*, Under review.

Hosseinkhani, H., Hosseinkhani, M., Tian, F., Kobayashi, H., Tabata, Y., 2006. Osteogenic differentiation of mesenchymal stem cells in self-assembled peptide-amphiphile nanofibers. Biomaterials 27, 4079–4086.

Huber, F.X., Belyaev, O., Hillmeier, J., Kock, H.J., Huber, C., Meeder, P.J., Berger, I., 2006. First histological observations on the incorporation of a novel nanocrystalline hydroxyapatite paste OSTIM in human cancellous bone. BMC Musculoskelet Disord 7, 50.

Huebsch, N., Mooney, D.J., 2009. Inspiration and application in the evolution of biomaterials. Nature 462, 426–432.

Im, O., Li, J., Wang, M., Zhang, L.G., Keidar, M., 2012. Biomimetic three-dimensional nanocrystalline hydroxyapatite and magnetically synthesized single-walled carbon nanotube chitosan nanocomposite for bone regeneration. Int J Nanomedicine 7, 2087–2099.

Jang, M.J., Namgung, S., Hong, S., Nam, Y., 2010. Directional neurite growth using carbon nanotube patterned substrates as a biomimetic cue. Nanotechnology 21, 235102.

Jiang, C.C., Chiang, H., Liao, C.J., Lin, Y.J., Kuo, T.F., Shieh, C.S., Huang, Y.Y., Tuan, R.S., 2007. Repair of porcine articular cartilage defect with a biphasic osteochondral composite. J Orthop Res 25, 1277–1290.

Jung, S.H., Song, W., Lee, S.I., Kim, Y., Cha, M.J., Kim, S.H., Jung, D.S., Jung, M.W., An, K.S., Park, C.Y., 2013. Synthesis of graphene and carbon nanotubes hybrid nanostructures and their electrical properties. J Nanosci Nanotechnol 13, 6730–6734.

Kalfa, D., Bel, A., Chen-Tournoux, A., Della Martina, A., Rochereau, P., Coz, C., Bellamy, V., Bensalah, M., Vanneaux, V., Lecourt, S., Mousseaux, E., Bruneval, P., Larghero, J., Menasche, P., 2010. A polydioxanone electrospun valved patch to replace the right ventricular outflow tract in a growing lamb model. Biomaterials 31, 4056–4063.

Khanarian, N.T., Jiang, J., Wan, L.Q., Mow, V.C., Lu, H.H., 2012. A hydrogel-mineral composite scaffold for osteochondral interface tissue engineering. Tissue Eng Part A 18, 533–545.

Kim, M., Erickson, I.E., Choudhury, M., Pleshko, N., Mauck, R.L., 2012. Transient exposure to TGF-beta3 improves the functional chondrogenesis of MSC-laden hyaluronic acid hydrogels. J Mech Behav Biomed Mater 11, 92–101.

Kim, Y., Hayashi, T., Endo, M., Dresselhaus, M., 2013. Carbon Nanofibers. In: Vajtai, R. (Ed.), Springer Handbook of Nanomaterials. Springer Berlin Heidelberg.

Kim, Y.G., Kim, J.W., Pyeon, H.J., Hyun, J.K., Hwang, J.Y., Choi, S.J., Lee, J.Y., Deak, F., Kim, H.W., Lee, Y.I., 2014. Differential stimulation of neurotrophin release by the biocompatible nano-material (carbon nanotube) in primary cultured neurons. J Biomater Appl 28, 790–797.

Koch, L., Deiwick, A., Schlie, S., Michael, S., Gruene, M., Coger, V., Zychlinski, D., Schambach, A., Reimers, K., Vogt, P.M., Chichkov, B., 2012. Skin tissue generation by laser cell printing. Biotechnol Bioeng 109, 1855–1863.

Kong, Y.M., Bae, C.J., Lee, S.H., Kim, H.W., Kim, H.E., 2005. Improvement in biocompatibility of ZrO2-Al2O3 nano-composite by addition of HA. Biomaterials 26, 509–517.

Kundu, J., Shim, J.H., Jang, J., Kim, S.W., Cho, D.W., 2013. An additive manufacturing-based PCL-alginate-chondrocyte bioprinted scaffold for cartilage tissue engineering. J Tissue Eng Regen Med.

Kuo, C.K., Marturano, J.E., Tuan, R.S., 2010. Novel strategies in tendon and ligament tissue engineering: Advanced biomaterials and regeneration motifs. Sports Med Arthrosc Rehabil Ther Technol 2, 20.

Lan, F., Li, G., 2013. Direct observation of hole transfer from semiconducting polymer to carbon nanotubes. Nano Lett 13, 2086–2091.

Lee, J.E., Khang, D., Kim, Y.E., Webster, T.J., 2006. Stem Cell Impregnated Carbon Nanofibers/Nanotubes for Healing Damaged Neural Tissue 915, 17–22.

Lee, M., Wu, B.M., 2012. Recent Advances in 3D Printing of Tissue Engineering Scaffolds. Methods Mol Biol 868, 257–267.

Li, Y.-L., Kinloch, I.A., Windle, A.H., 2004. Direct Spinning of Carbon Nanotube Fibers from Chemical Vapor Deposition Synthesis. Science 304, 276–278.

Liedtke, W., Yeo, M., Zhang, H., Wang, Y., Gignac, M., Miller, S., Berglund, K., Liu, J., 2013. Highly conductive carbon nanotube matrix accelerates developmental chloride extrusion in central nervous system neurons by increased expression of chloride transporter KCC2. Small 9, 1066–1075.

Madurantakam, P.A., Rodriguez, I.A., Cost, C.P., Viswanathan, R., Simpson, D.G., Beckman, M.J., Moon, P.C., Bowlin, G.L., 2009. Multiple factor interactions in biomimetic mineralization of electrospun scaffolds. Biomaterials 30, 5456–5464.

Martinelli, V., Cellot, G., Fabbro, A., Bosi, S., Mestroni, L., Ballerini, L., 2013. Improving cardiac myocytes performance by carbon nanotubes platforms. Front Physiol 4, 239.

Martinelli, V., Cellot, G., Toma, F.M., Long, C.S., Caldwell, J.H., Zentilin, L., Giacca, M., Turco, A., Prato, M., Ballerini, L., Mestroni, L., 2012. Carbon nanotubes promote growth and spontaneous electrical activity in cultured cardiac myocytes. Nano Lett 12, 1831–1838.

Mikos, A.G., Thorsen, A.J., Czerwonka, L.A., Bao, Y., Langer, R., Winslow, D.N., Vacanti, J.P., 1994. Preparation and Characterization of Poly(L-Lactic Acid) Foams. Polymer 35, 1068–1077.

Moon, S., Hasan, S.K., Song, Y.S., Xu, F., Keles, H.O., Manzur, F., Mikkilineni, S., Hong, J.W., Nagatomi, J., Haeggstrom, E., Khademhosseini, A., Demirci, U., 2010. Layer by Layer Three-dimensional tissue epitaxy by cell-laden hydrogel droplets. Tissue Eng Part C Methods 16, 157–166.

Mooney, E., Mackle, J.N., Blond, D.J., O'Cearbhaill, E., Shaw, G., Blau, W.J., Barry, F.P., Barron, V., Murphy, J.M., 2012. The electrical stimulation of carbon nanotubes to provide a cardiomimetic cue to MSCs. Biomaterials 33, 6132–6139.

Nguyen-Vu, T.D.B., Nguyen-Vu, T.D.B., Hua, C., Cassell, A.M., Andrews, R.J., Meyyappan, M., Jun, L., 2007. Vertically Aligned Carbon Nanofiber Architecture as a Multifunctional 3-D Neural Electrical Interface. IEEE Transactions on Biomedical Engineering 54, 1121–1128.

Nimeskern, L., Avila, H.M., Sundberg, J., Gatenholm, P., Muller, R., Stok, K.S., 2013. Mechanical evaluation of bacterial nanocellulose as an implant material for ear cartilage replacement. Journal of the Mechanical Behavior of Biomedical Materials 22, 12–21.

Noel, D., Gazit, D., Bouquet, C., Apparailly, F., Bony, C., Plence, P., Millet, V., Turgeman, G., Perricaudet, M., Sany, J., Jorgensen, C., 2004. Short-term BMP-2 expression is sufficient for in vivo osteochondral differentiation of mesenchymal stem cells. Stem Cells 22, 74–85.

O'Brien, C., Holmes, B., Faucett, S., Zhang, L. G. 2014. 3D Printing of Nanomaterial Scaffolds for Complex Tissue Regeneration. *Tissue engineering Part B*. Epub ahead of print. doi:10.1089/ten.teb.2014.0168

Ovsianikov, A., Gruene, M., Pflaum, M., Koch, L., Maiorana, F., Wilhelmi, M., Haverich, A., Chichkov, B., 2010. Laser printing of cells into 3D scaffolds. Biofabrication 2, 014104.

Ovsianikov, A., Schlie, S., Ngezahayo, A., Haverich, A., Chichkov, B.N., 2007. Two-photon polymerization technique for microfabrication of CAD-designed 3D scaffolds from commercially available photosensitive materials. J Tissue Eng Regen Med 1, 443–449.

Pan, L., Pei, X., He, R., Wan, Q., Wang, J., 2012. Multiwall carbon nanotubes/polycaprolactone composites for bone tissue engineering application. Colloids Surf B Biointerfaces 93, 226–234.

Parveen, S., Misra, R., Sahoo, S.K., 2012. Nanoparticles: a boon to drug delivery, therapeutics, diagnostics and imaging. Nanomedicine (Lond) 8, 147–166.

Raimondi, M.T., Eaton, S.M., Lagana, M., Aprile, V., Nava, M.M., Cerullo, G., Osellame, R., 2013. Three-dimensional structural niches engineered via two-photon laser polymerization promote stem cell homing. Acta Biomaterialia 9, 4579–4584.

Robbins, S., Lauryssen, C., Songer, M. N. 2014. Use of Nanocrystalline Hydroxyapatite With Autologous BMA and Local Bone in the Lumbar Spine: A Retrospective CT Analysis of Posterolateral Fusion Results. *J Spinal Disord Tech.*

Sakamoto, A., Kiyokawa, K., Rikimaru, H., Watanabe, K., Nishi, Y., 2012. An investigation of the fixation materials for cartilage frames in microtia. J Plast Reconstr Aesthet Surg 65, 584–589.

Samavedi, S., Olsen Horton, C., Guelcher, S.A., Goldstein, A.S., Whittington, A.R., 2011. Fabrication of a model continuously graded co-electrospun mesh for regeneration of the ligament-bone interface. Acta Biomater 7, 4131–4138.

Sekiya, I., Colter, D.C., Prockop, D.J., 2001. BMP-6 enhances chondrogenesis in a subpopulation of human marrow stromal cells. Biochem Biophys Res Commun 284, 411–418.

Serrano, M.C., Nardecchia, S., Garcia-Rama, C., Ferrer, M.L., Collazos-Castro, J.E., Del Monte, F., Gutierrez, M.C., 2014. Chondroitin sulphate-based 3D scaffolds containing MWCNTs for nervous tissue repair. Biomaterials 35, 1543–1551.

Shah, R.N., Shah, N.A., Del Rosario Lim, M.M., Hsieh, C., Nuber, G., Stupp, S.I., 2010. Supramolecular design of self-assembling nanofibers for cartilage regeneration. Proceedings of the National Academy of Sciences of the United States of America 107, 3293–3298.

Shim, J.H., Kim, J.Y., Park, M., Park, J., Cho, D.W., 2011. Development of a hybrid scaffold with synthetic biomaterials and hydrogel using solid freeform fabrication technology. Biofabrication 3, 034102.

Shin, S.R., Jung, S.M., Zalabany, M., Kim, K., Zorlutuna, P., Kim, S.B., Nikkhah, M., Khabiry, M., Azize, M., Kong, J., Wan, K.T., Palacios, T., Dokmeci, M.R., Bae, H., Tang, X.S., Khademhosseini, A., 2013. Carbon-nanotube-embedded hydrogel sheets for engineering cardiac constructs and bioactuators. ACS Nano 7, 2369–2380.

Silva, G.A., Czeisler, C., Niece, K.L., Beniash, E., Harrington, D.A., Kessler, J.A., Stupp, S.I., 2004. Selective differentiation of neural progenitor cells by high-epitope density nanofibers. Science 303, 1352–1355.

Smith, M.J., McClure, M.J., Sell, S.A., Barnes, C.P., Walpoth, B.H., Simpson, D.G., Bowlin, G.L., 2008. Suture-reinforced electrospun polydioxanone-elastin small-diameter tubes for use in vascular tissue engineering: a feasibility study. Acta Biomater 4, 58–66.

Song, S.J., Choi, J., Park, Y.D., Lee, J.J., Hong, S.Y., Sun, K., 2010. A three-dimensional bioprinting system for use with a hydrogel-based biomaterial and printing parameter characterization. Artif Organs 34, 1044–1048.

Sridharan, I., Kim, T., Strakova, Z., Wang, R., 2013. Matrix-specified differentiation of human decidua parietalis placental stem cells. Biochem Biophys Res Commun 437, 489–495.

Sun, L., Zhang, L., Hemraz, U.D., Fenniri, H., Webster, T.J., 2012. Bioactive rosette nanotube-hydroxyapatite nanocomposites improve osteoblast functions. Tissue engineering. Part A 18, 1741–1750.

Suri, S., Han, L.H., Zhang, W., Singh, A., Chen, S., Schmidt, C.E., 2011. Solid freeform fabrication of designer scaffolds of hyaluronic acid for nerve tissue engineering. Biomed Microdevices 13, 983–993.

Tarafder, S., Balla, V.K., Davies, N.M., Bandyopadhyay, A., Bose, S., 2013. Microwave-sintered 3D printed tricalcium phosphate scaffolds for bone tissue engineering. J Tissue Eng Regen Med 7, 631–641.

Tasoglu, S., Demirci, U., 2013. Bioprinting for stem cell research. Trends Biotechnol 31, 10–19.

Terrones, M., 2004. Carbon nanotubes: synthesis and properties, electronic devices and other emerging applications. International Materials Reviews 49, 325–377.

Thomas, V., Zhang, X., Vohra, Y.K., 2009. A biomimetic tubular scaffold with spatially designed nanofibers of protein/PDS bio-blends. Biotechnol Bioeng 104, 1025–1033.

Torgersen, J., Ovsianikov, A., Mironov, V., Pucher, N., Qin, X.H., Li, Z.Q., Cicha, K., Machacek, T., Liska, R., Jantsch, V., Stampfl, J., 2012. Photo-sensitive hydrogels for three-dimensional laser microfabrication in the presence of whole organisms. J Biomed Opt, 17.

Tran, P.A., Zhang, L., Webster, T.J., 2009. Carbon nanofibers and carbon nanotubes in regenerative medicine. Adv Drug Deliv Rev 61, 1097–1114.

Venclauskas, L., Grubinskas, I., Mocevicius, P., Kiudelis, M., 2011. Reinforced tension line versus simple suture: a biomechanical study on cadavers. Acta Chir Belg 111, 288–292.

Wagner, D.E., Jones, A.D., Zhou, H., Bhaduri, S.B., 2013. Cytocompatibility evaluation of microwave sintered biphasic calcium phosphate scaffolds synthesized using pH control. Mater Sci Eng C Mater Biol Appl 33, 1710–1719.

Wang, M., Castro, N.J., Li, J., Keidar, M., Zhang, L.G., 2012. Greater osteoblast and mesenchymal stem cell adhesion and proliferation on titanium with hydrothermally treated nanocrystalline hydroxyapatite/magnetically treated carbon nanotubes. J Nanosci Nanotechnol 12, 7692–7702.

Wang, M., Cheng, X., Zhu, W., Holmes, B., Keidar, M., Zhang, L.G., 2014. Design of biomimetic and bioactive cold plasma-modified nanostructured scaffolds for enhanced osteogenic differentiation of bone marrow-derived mesenchymal stem cells. Tissue Eng Part A 20, 1060–1071.

Wang, W., Li, B., Yang, J., Xin, L., Li, Y., Yin, H., Qi, Y., Jiang, Y., Ouyang, H., Gao, C., 2010a. The restoration of full-thickness cartilage defects with BMSCs and TGF-beta 1 loaded PLGA/fibrin gel constructs. Biomaterials 31, 8964–8973.

Wang, X., Wenk, E., Zhang, X., Meinel, L., Vunjak-Novakovic, G., Kaplan, D.L., 2009. Growth factor gradients via microsphere delivery in biopolymer scaffolds for osteochondral tissue engineering. Journal of Controlled Release 134, 81–90.

Wang, X.L., Chen, Y.Y., Wang, Y.Z., 2010b. Synthesis of poly(p-dioxanone) catalyzed by Zn L-lactate under microwave irradiation and its application in ibuprofen delivery. J Biomater Sci Polym Ed 21, 927–936.

Warnke, P.H., Seitz, H., Warnke, F., Becker, S.T., Sivananthan, S., Sherry, E., Liu, Q., Wiltfang, J., Douglas, T., 2010. Ceramic scaffolds produced by computer-assisted 3D printing and sintering: characterization and biocompatibility investigations. J Biomed Mater Res B Appl Biomater 93, 212–217.

Webster, T.J., Ergun, C., Doremus, R.H., Siegel, R.W., Bizios, R., 2000. Specific proteins mediate enhanced osteoblast adhesion on nanophase ceramics. J Biomed Mater Res 51, 475–483.

Whang, K., Thomas, C.H., Healy, K.E., Nuber, G., 1995. A Novel Method to Fabricate Bioabsorbable Scaffolds. Polymer 36, 837–842.

Yu, M.F., Lourie, O., Dyer, M.J., Moloni, K., Kelly, T.F., Ruoff, R.S., 2000. Strength and breaking mechanism of multiwalled carbon nanotubes under tensile load. Science 287, 637–640.

Zhang, C., Hu, Y., Xu, J., 2005. [The effect of bone-related growth factors on the proliferation and differentiation of marrow mesenchymal stem cells *in vitro*]. Zhongguo Xiu Fu Chong Jian Wai Ke Za Zhi 19, 906–909.

Zhang, L., Chen, Y., Rodriguez, J., Fenniri, H., Webster, T.J., 2008a. Biomimetic helical rosette nanotubes and nanocrystalline hydroxyapatite coatings on titanium for improving orthopedic implants. Int J Nanomedicine 3, 323–333.

Zhang, L., Ercan, B., Webster, T.J. 2009a. Carbon Nanotubes and Nanofibers for Tissue Engineering Applications. In: L.I.U., C., (Ed.) *Carbon.* Research Signpost.

Zhang, L., Hemraz, U.D., Fenniri, H., Webster, T.J., 2010. Tuning cell adhesion on titanium with osteogenic rosette nanotubes. J Biomed Mater Res A 95, 550–563.

Zhang, L., Hu. J., Athanasiou, K.A., 2009b. The role of tissue engineering in articular cartilage repair and regeneration. Crit Rev Biomed Eng 37, 1–57.

Zhang, L., Rakotondradany, F., Myles, A.J., Fenniri, H., Webster, T.J., 2009c. Arginine-glycine-aspartic acid modified rosette nanotube-hydrogel composites for bone tissue engineering. Biomaterials 30, 1309–1320.

Zhang, L., Ramsaywack, S., Fenniri, H., Webster, T.J., 2008b. Enhanced osteoblast adhesion on self-assembled nanostructured hydrogel scaffolds. Tissue Eng Part A 14, 1353–1364.

Zhang, L., Rodriguez, J., Raez, J., Myles, A.J., Fenniri, H., Webster, T.J., 2009d. Biologically inspired rosette nanotubes and nanocrystalline hydroxyapatite hydrogel nanocomposites as improved bone substitutes. Nanotechnology 20, 175101.

Zhang, L., Sirivisoot, L., Balasundaram, G., Webster, T. J. 2008c. Nanoengineering for Bone Tissue Engineering. *In* : Khademhosseini, A., Borenstein, J., Toner, M., Takayama, S., (eds.) *Micro and Nanoengineering of the Cell Microenvironment: Technologies and Applications.* Artech House.

Zhang, L.J., Webster, T.J., 2009. Nanotechnology and nanomaterials: Promises for improved tissue regeneration. Nano Today 4, 66–80.

Zhang, S., 2003. Fabrication of novel biomaterials through molecular self-assembly. Nature biotechnology 21, 1171–1178.

Zhang, Y., Wang, F., Tan, H., Chen, G., Guo, L., Yang, L., 2012. Analysis of the mineral composition of the human calcified cartilage zone. Int J Med Sci 9, 353–360.

Zhu, J., Dang, H.C., Wang, W.T., Wang, X.L., Wang, Y.Z., 2011. Cellulose diacetate-g-poly(p-dioxanone) co-polymer: synthesis, properties and microsphere preparation. J Biomater Sci Polym Ed 22, 981–999.

Zhu, W., O'Brien, C., O'Brien, J.R., Zhang, L.G., 2014a. 3D Nano/Microfabrication Techniques and Nanobiomaterials for Neural Tissue Regeneration. Nanomedicine 9, 859–875.

Zhu, W., O'Brien, J. R., Zhang, L. G. 2014b. Highly Aligned Nanocomposite Scaffolds by Electrospinning and Electrospraying for Neural Tissue Engineering. *Nanomedicine: Nanotechnology, Biology and Medicine,* Under review.

3D PRINTING AND NANOMANUFACTURING

2

Wei Zhu[1,*]**, JinGyu Ock**[2,*]**, Xuanyi Ma**[1,*]**, Wei Li**[2] **and Shaochen Chen**[1]

[1]Department of NanoEngineering, University of California, San Diego, La Jolla, CA, USA
[2]Department of Mechanical Engineering, University of Texas, Austin, TX, USA
[*]These authors contributed equally

2.1 INTRODUCTION

Driven by the goal of creating biomimetic microenvironments that manifest natural tissue structures and compositions, developing three-dimensional (3D) constructs that incorporate (a) sophisticated patterning of extracellular matrix (ECM) components, (b) biomolecules, and (c) even cells has been a main research focus for studying biological problems and engineering tissues (Griffith, 2002; Orban et al., 2002; Sharma and Elisseeff, 2004). However, conventional fabrication methods used for manufacturing 3D scaffolds, such as electrospinning (Zong et al., 2005), fiber deposition (Moroni et al., 2006), freeze-drying (O'Brien et al., 2004), gas foaming (Yang et al., 2001; Nazarov et al., 2004), salt leaching (Roy et al., 2003), and porogen melting (Lin et al., 2003), do not allow precise control of the internal structural features and topology. In addition, many of the current 3D scaffolding systems are only capable of achieving either a bulk incorporation of biomolecules within the scaffolding matrix or an exogenous delivery of necessary chemicals, hormones, or growth factors through culture medium (Richardson et al., 2001; Johnstone et al., 1998; Nuttelman et al., 2004). An important step toward achieving the goal of creating precise, spatially patterned 3D microenvironments within a single scaffold for tissue engineering applications is the development of novel scaffold manufacturing techniques by which distributed environmental factors can be incorporated together in a simple yet precise and consistent fashion.

3D printing and nanomanufacturing, in particular laser-assisted direct writing techniques and stereolithography, are capable of efficiently fabricating complex 3D scaffolds with precise microarchitecture to achieve that step (Han et al., 2008; Lu et al., 2006; Mapili et al., 2005). Such fabrication systems provide the possibilities of incorporating cells inside the scaffold walls during fabrication of 3D constructs or seeding cells on the already patterned scaffolding surface. As a promising approach to develop user-defined precise 3D microenvironments with complex biological components, 3D printing has been used in a variety of applications ranging from tissue engineering scaffolding to cancer cell migration study and neural stem cell culture, with the potential of patterning multiple cell types in precise 3D locations.

2.2 3D PRINTING AND NANOMANUFACTURING TECHNIQUES

With the development of computer-aided design (CAD) technology and automation techniques, rapid and automatic additive manufacturing (AM) systems have advanced for the fabrication of complicated 3D structures over the past few decades. In general, the 3D model to be fabricated is first designed using CAD modeling software, such as Solidworks and AutoCAD. Generated 3D CAD data are then processed and sliced into layers of equal thickness, each of which is the cross-section of the 3D model at a certain level. Sliced data are imported into the AM system to fabricate 3D objects layer-by-layer. In this fabrication process, layers are cumulated vertically and fused to form the final physical object (Chua et al., 2010; Tan et al., 2005).

2.2.1 SELECTIVE LASER SINTERING

Selective laser sintering (SLS) (Tan et al., 2005; Duan et al., 2010; Kanczler et al., 2009; Liu et al., 2013; Williams et al., 2005) was developed and patented in the mid-1980s (Deckard, 1989). This technique uses a laser beam (usually CO_2 laser) to sinter slices of powdered materials via repeated process of spreading layers and selectively heating and fusing each powdered layer in order to fabricate three-dimensional structures. Thus, the objects are formed layer-by-layer from sliced CAD data. During the process, the unmelted powders act as the support for the fused object. Figure 2.1 represents the selective laser sintering process.

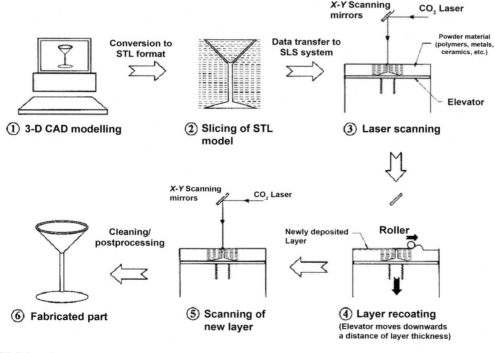

FIGURE 2.1

A schematic layout of the SLS process (Tan et al., 2005).

2.2.2 **LASER-GUIDED DIRECT WRITING**

The laser-guided direct writing (LGDW) (Odde and Renn, 2000; Nahmias et al., 2005; Nahmias and Odde, 2006; Narasimhan et al., 2004; Rosenbalm et al., 2006) technique was first used for micropatterning of embryonic-chick spinal-cord cells (Odde and Renn, 1999). The driving force of the LGDW method arises from the scattering of laser light by microparticles or cells. In contrast to the high-numerical-aperture lens used in optical trapping systems, the LGDW system uses a low-numerical-aperture lens so as to provide an axial propelling force to the particle instead of trapping it in the vicinity of the focal point. Once a particle or cell interacts with the laser, it is drawn to the center of the beam where the intensity is maximal and simultaneously pushed along the axial direction of the laser beam by radiation pressure. The guided object is deposited on a target surface, which is placed vertically at a certain point along the optical axis. By moving the target surface relative to the laser beam, three dimensional patterns of particles can be drawn on the target surface. The basic concept of cell deposition of LGDW method is illustrated in Figure 2.2.

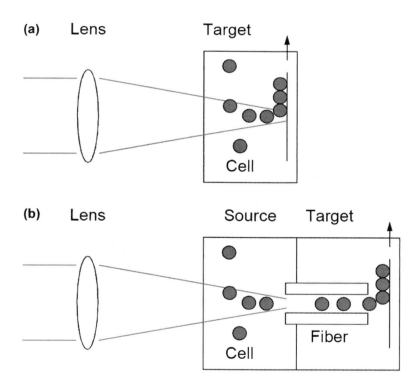

FIGURE 2.2

Laser-guided direct writing system. (a) Laser light is focused weakly into a suspension of particles. The particles are propelled by the light through the fluid and deposited on a target surface. Moving the target relative to the laser beam results in a line of particles being drawn. (b) Light is coupled into a hollow optical fiber and particles are carried through the fiber to the target surface. The process can be observed in real time by optical microscopy (Odde and Renn, 1999).

2.2.3 LASER-INDUCED FORWARD TRANSFER

Laser-induced forward transfer (LIFT) (Koch et al., 2009, 2012; Gruene et al., 2010) mainly employs a high-powered pulse laser and two coplanar glass slides. The experimental setup is described in Figure 2.3. The upper slide, called "donor-slide", is coated with an energy absorption metal layer and a layer of material containing cells. Laser pulses are focused on the metal layer via the glass slide, evaporating the laser absorbing layer locally. During the process, the laser pulse generates a high gas pressure that transfers the underlying cell compound toward the lower slide, referred to as "collector-slide." The biological materials containing cells are usually a culture medium or hydrogel that provides a humid environment, thus preventing cell dehydration. Hydrogel has the additional function of sustaining cell structure.

2.2.4 MATRIX-ASSISTED PULSED LASER EVAPORATION DIRECT WRITING

The setup of matrix-assisted pulsed laser evaporation direct writing (MAPLE DW) (Ringeisen et al., 2004; Patz et al., 2006; Doraiswamy et al., 2007) is similar to the LIFT system shown in Figure 2.3. Instead of using a glass slide for the "donor-slide," MAPLE DW employs an optically transparent quartz support called "ribbon." The ribbon is coated with biological materials, such as Matrigel® or bioceramic. The biomaterials with cells are referred to as "matrix". The substrate, similar to the "collector-slide", may or may not be covered with hydrogel. Laser is focused on the interface of the quartz support and a laser-absorptive layer containing cells. The laser beam causes evaporation of part of the biomaterial layer, which generates gas bubbles locally. The gas bubbles result in the release and propulsion of the cell-seeded matrix to the receiving substrate.

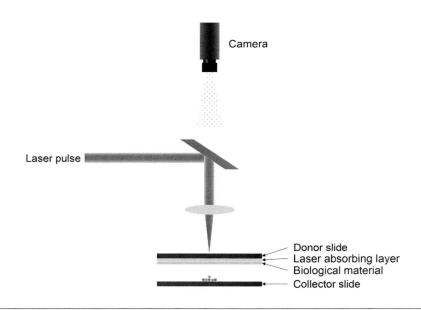

Camera

Laser pulse

Donor slide
Laser absorbing layer
Biological material
Collector slide

FIGURE 2.3

A schematic of LIFT (Gruene et al., 2010; Koch et al., 2012; Ringeisen et al., 2004).

2.2.5 BIOLOGICAL LASER PRINTING

Biological laser printing (BioLP) was developed by Barron et al. (Barron et al., 2004, 2005; Pirlo et al., 2012; Othon et al., 2008). It is similar to the previous two processes, LIFT and MAPLE DW, illustrated in Figure 2.3. BioLP utilizes an optically transparent quartz instead of the glass slide used in LIFT. Metal or metal oxide is coated on the quartz support as the laser absorption layer. The biomaterial layer with cells, such as powder, liquid, or gel, is coated on the laser-absorbing layer. The incident laser energy is focused and absorbed at the interface of the quartz support and laser absorption layer. The heat generated by the laser absorption layer causes vaporization of water in the biomaterial. The biomaterial is then transferred from the ribbon surface to the receiving substrate surface. Thus, this technique reduces potential damage to biological materials.

2.2.6 STEREOLITHOGRAPHY TECHNIQUES

The stereolithography (STL) technique was developed by Hull in 1986 and was described in his patent "Apparatus for Production of Three-Dimensional Objects by Stereolithography" (Hull, 1986). Stereolithography is a technique of producing parts one layer at a time by curing a photoreactive resin with a UV laser or another similar power source. In addition to using a single-point laser, stereolithography can be performed with a digital micromirror-array device (DMD) (Suri et al., 2011; Gauvin et al., 2012; Soman et al., 2012, 2013; Lin et al., 2012). The DMD is an array of up to several millions

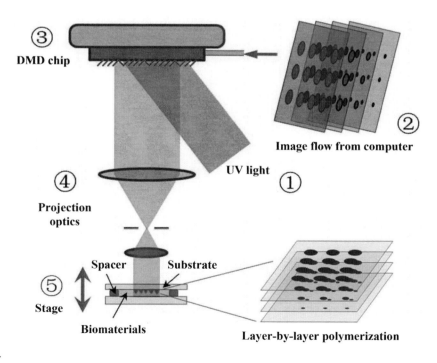

FIGURE 2.4

Schematic setup of the DOPsL system (Zhang et al., 2012).

of microsized mirrors that can be controlled independently to on and off state (Melchels et al., 2010). In this approach, the structure is not created via point-by-point scanning, but in a layer-wise fashion by curing the entire layer simultaneously. After one layer is fabricated, the platform is lowered or raised to cumulate a new layer. The thickness of each layer is controlled by the distance between the surface of the platform and the liquid resin surface. Based on the basic concept just mentioned, Zhang et al. developed a dynamic optical projection stereolithography (DOPsL) system for the rapid fabrication of complex 3D extracellular scaffolds (Zhang et al., 2012). The setup of the DOPsL system is shown in Figure 2.4. Since this DMD-based stereolithography technology simultaneously utilizes a million micromirrors rather than one single focused point, the DOPsL system offers superior processing speed compared to other nanofabrication techniques, thus making it more suitable for manufacturing large structures with complex details with a submicron resolution (Zhang et al., 2012). With many advantages, including rapid fabrication speed, maskless, flexibility, and relatively high resolution, the DOPsL system is an appealing platform for the manufacture of complex 3D designer scaffolds for *in vitro* tissue engineering as well as functional cellular constructs for *in vivo* implantation (Zhang et al., 2012).

A special laser directing technique is two-photon polymerization (2PP) (Zhang and Chen, 2011; Zhang et al., 2013; Ovsianikov et al., 2011a, 2011b; Gebinoga et al., 2013), which has been used to produce nanoscale features due to its high fidelity and resolution. 2PP systems usually utilize a femtosecond laser to induce two-photon absorption (2PA). 2PA is a process by which one molecule is excited to a higher energy electronic state by the simultaneous absorption of two photons (Zhang and Chen, 2011). 2PA was described theoretically in 1931 by Goeppert-Mayer (Göppert-Mayer, 1931) and first demonstrated experimentally in 1961 by Kaiser and Garrett (Kaiser and Garrett, 1961) in a CaF2:Eu2+ crystal, and in 1962 by Abella in caesium vapor (Abella, 1962). Figure 2.5 shows a

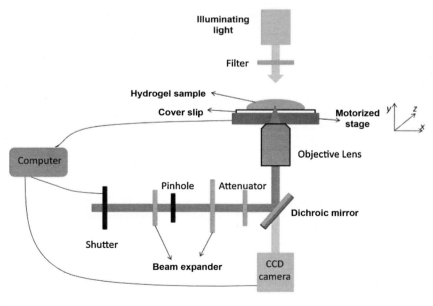

FIGURE 2.5

Schematic setup of a TPP fabrication system (Zhang and Chen, 2011).

schematic of a typical 2PP fabrication system (Zhang and Chen, 2011). The laser used in the system is a Ti:sapphire femtosecond laser (100-fs pulses at a repetition rate of 80 MHz and wavelength of 800 nm). The laser beam is expanded and guided by a group of mirrors into an inverted microscope. It is then focused by an oil-immersion objective lens onto the sample that is mounted on a motorized stage with high resolution (< 20 nm), which moves three-dimensionally to draw a defined 3D nanostructure in the sample. The 3D structure is first designed in AutoCAD and then imported to the software of the motorized stage, which controls the motion of the stage in *xyz* directions. A CCD camera is used to monitor the fabrication process. The laser power can be continuously adjusted by an attenuator. With this femtosecond laser fabrication system, Zhang et al. were able to fabricate defined and complex 3D structures with a resolution of 100 nm (Zhang and Chen, 2011).

2.2.7 CLASSIFICATION OF ADDITIVE BIOMANUFACTURING TECHNIQUES

Based on their working principles, the existing AM systems can mainly be categorized into three groups: (1) powder-fusion-based techniques, such as SLS; (2) particle- or cell-deposition-based techniques, such as LGDW, LIFT, and MAPLE DW; and (3) photo-polymerization-based techniques, such as stereolithography and 2PP (Duan and Wang, 2013).

These additive manufacturing techniques can also be categorized based on their process configurations: top–down and bottom–up. Most forms of laser-assisted additive manufacturing, such as SLS, LG DW, LIFT, MAPLE DW, and BioLP, are classified as bottom–up approaches. The accumulation starts from the bottom of the platform and prepolymer materials are supplied for each layer. This approach has the advantage of fabricating multiple layers with different materials in each layer (Mapili et al., 2005; Nahmias et al., 2005; Odde and Renn, 1999; Koch et al., 2009, 2012; Pirlo et al., 2006; Ovsianikov et al., 2010; Chan et al., 2010). SLA employs both top–bottom and bottom–up approaches, as shown in Figure 2.6. The top–down configuration consists of a container and a movable platform that is located in the container. The platform is immersed just below the surface of a prepolymer solution. The laser beam is focused onto the surface (*x*–*y* plane) of liquid resin to polymerize the resin. Once a layer is photo-polymerized, the platform is lowered by a specific distance to fabricate a new layer. In the bottom–up approach, the container is a movable platform on which a polymerized resin layer is created. Liquid prepolymer is supplied into the container for one layer from the bottom to the top. Table 2.1 lists the type of approach that each technique belongs to.

2.3 BIOMATERIALS USED WITH ADDITIVE BIOMANUFACTURING TECHNIQUES

Over the past two decades, various biocompatible or biodegradable materials, including polymeric materials (Puppi et al., 2010; Nair and Laurencin, 2007), bioceramics (Best et al., 2008), and hydrogels (Fedorovich et al., 2007), have been employed and developed for use with laser-based additive biomanfacturing techniques in bioapplications, such as drug delivery, regenerative medicine, and tissue engineering. In this section, we briefly review typical biomaterials that have been used with laser-based additive biomanufacturing techniques. Table 2.2 provides a summary of the biomanufacturing techniques listed with corresponding biomaterials used.

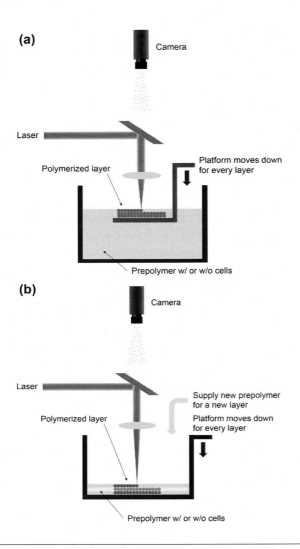

FIGURE 2.6

A schematic representation of SLA: (a) top–down approach, (b) bottom–up approach.

2.3.1 POWDER-TYPE MATERIALS

Selective laser sintering has been used with powdered materials—mostly synthetic polymers, polymer composites, or ceramics—to fabricate 3D solid structures. These materials include PEEK, PVA, PCL, PLLA, and HA (Tan et al., 2005; Duan et al., 2010; Kanczler et al., 2009). Polyetheretherketone (PEEK) is a bio-inert polymer with a melting point of 343 °C and a glass temperature of 143 °C. These properties make the polymer stable to be processed at high temperatures; therefore, PEEK is easily sterilized in autoclave or by radiation without sacrificing its materials. Poly (vinyl alcohol) (PVA) is a semicrystalline polymer and copolymer of vinyl alcohol and vinyl acetate. PVA has a melting point of

Table 2.1   Types of additive manufacturing techniques using laser

Fabrication technique	Top–down	Bottom–up	References
SLS		O	Tan et al., 2005; Duan et al., 2010; Kanczler et al., 2009; Liu et al., 2013; Williams et al., 2005
LGDW		O	Odde and Renn, 2000; Nahmias et al., 2005; Nahmias and Odde, 2006; Narasimhan et al., 2004; Rosenbalm et al., 2006
LIFT		O	Koch et al., 2009; Gruene et al., 2010; Koch et al., 2012
MAPLEDW		O	Ringeisen et al., 2004; Patz et al., 2006; Doraiswamy et al., 2007
BioLP		O	Barron et al., 2004; Barron et al., 2004; Barron et al., 2005; Othon et al., 2008
SLA - laser		O	Mapili et al., 2005; Chan et al., 2012
	O	O	Chan et al., 2010
		O	Lee et al., 2007
SLA - DMD		O	Suri et al., 2011
		O	Gauvin et al., 2012
	O		Soman et al., 2013
		O	Soman et al., 2012
		O	Soman et al., 2012
	.	.	Lin et al., 2012
2PP		O	Zhang and Chen, 2011; Zhang et al., 2013; Ovsianikov et al., 2011a; Ovsianikov et al., 2011b; Gebinoga et al., 2013

Table 2.2   Additive manufacturing technique using laser

Fabrication technique	Unit element	Polymer type	Laser type	Dimension	References
SLS	PEEK, PVA, PCL, PLLA, HA	Powder	CO_2 laser	3D	Tan et al., 2005
	Ca-P/PHBV CHA/ PLLA	Powder (nanocomposite)	CO_2 laser	3D	Duan et al., 2010
	PLA + Carbon Black	Powder	Fiber diode laser	3D	Kanczler et al., 2009
	Titanium	Powder	Nd:YAG laser	3D	Liu et al., 2013
	PCL	Powder	CO_2 laser	3D	Williams et al., 2005

(Continued)

Table 2.2   Additive manufacturing technique using laser *(cont.)*

Fabrication technique	Unit element	Polymer type	Laser type	Dimension	References
LGDW	Laser guided cell	Unit cell	Diode laser	3D	Odde and Renn, 2000; Nahmias et al., 2005; Nahmias and Odde, 2006
	Laser guided cell	Unit cell	Ti: Sapphire laser	2D	Narasimhan et al., 2004; Rosenbalm et al., 2006
LIFT	Alginate + NaCl	Gel-type layer	Nd:YAG laser	3D	Koch et al., 2009; Gruene et al., 2010; Koch et al., 2012
MAPLEDW	Matrigel®	Gel-type layer	ArF excimer laser	3D	Ringeisen et al., 2004; Patz et al., 2006
	Hydroxyapatite and zirconia / glycerol-water solution	Solution layer	ArF excimer laser	3D	Doraiswamy et al., 2007
BioLP	Protein/ cell solution with glycerol	Solution layer	Nd:YAG laser	3D	Barron et al., 2004; Barron et al., 2004
	Cell solution with glycerol	Solution layer	Multigas excimer laser	2D	Barron et al., 2005
	Cell solution with methyl-cellulose	Solution layer	Nd:YAG laser	3D	Othon et al., 2008
SLA - laser	PEGDMA	Liquid	Nd:YAG laser	3D	Mapili et al., 2005
	PEGDA	Liquid	HeCd laser	3D	Chan et al., 2010; Chan et al., 2012
	PPF/ DEF	Liquid	Nd:YVO4 laser	3D	Lee et al., 2007
SLA - DMD	GMHA	Liquid	UV light DMD	3D	Suri et al., 2011
	GelMA	Liquid	UV light DMD	3D	Gauvin et al., 2012; Soman et al., 2013
	PEG	Liquid	UV light DMD	3D	Soman et al., 2012
	PEGDA	Liquid	UV light DMD	3D	Soman et al., 2012
	PEGDA	Liquid	SLA machine	3D	Lin et al., 2012
2PP	PEGDA	Liquid	Ti: Sapphire laser	3D	Zhang and Chen, 2011; Zhang et al., 2013
	GelMOD	Liquid	Ultrafast infrared laser	3D	Ovsianikov et al., 2011a; Ovsianikov et al., 2011b
	Collagen Fibrinogen	Liquid	Femtosecond pulsed laser	3D	Gebinoga et al., 2013

220–240 °C and a glass transition temperature of 58–85 °C (Tan et al., 2005). Polycaprolactone (PCL) is also a semicrystalline polyester with a melting point of 55–60 °C and a glass transition temperature of −60 °C. It could be hydrolytically degraded due to the presence of hydrolytically labile aliphatic ester linkages. The degradation rate is about 2–3 years (Nair and Laurencin, 2007). Poly (L-lactic) acid (PLLA) has a melting point of 172.2–186.8 °C and a glass transition temperature of 60.5 °C (Tan et al., 2005). The mineral component of bone is calcium phosphate. Hydroxyapatite (HA) is one of the synthetic calcium phosphate ceramics. The bioceramic is widely used because it is chemically similar to the inorganic component of hard tissues. HA theoretically consists of 39.68 wt% Ca and 18.45 wt% P. It is more stable compared to other calcium phosphate ceramics within a pH range of 4.2–8.0 (Best et al., 2008).

Nanocomposites of bioceramic and biodegradable polymer are often sintered to facilitate proliferation of and alkaline phosphatase activity expression by human osteoblast-like cells (SaOS-2). Poly (hydroxybutyrate) (PHB) is a naturally occurring polyester by bacteria. The polymer has a melting point of 160–180 °C. The copolymer of PHB and hydroxyvalerate is poly (hydroxybutyrate-co-hydroxyvalerate) (PHBV), which is a semicrystalline polymer with a melting temperature lower than PHB. The glass transition temperature of PHBV is in the range of −5–20°C (Nair and Laurencin, 2007). Calcium phosphate (Ca-P)/PHBV and carbonated hydroxyapatite (CHA)/ PLLA nanocomposites have been employed to fabricate tissue engineering scaffolds (Duan et al., 2010). In addition, titanium powder has been sintered to form bone scaffolds (Liu et al., 2013).

2.3.2 GEL-BASED BIOMATERIALS

Gel-based biomaterials are widely used in additive biomanufacturing processes. Collagen gel or Matrigel® is coated on the substrate to help cells adhere to the substrate; for three-dimensional patterning, Matrigel® is layered on top of the first pattern in the LGDW method (Odde and Renn, 2000; Nahmias et al., 2005; Nahmias and Odde, 2006; Narasimhan et al., 2004; Rosenbalm et al., 2006). Alginate hydrogel is used in the LIFT method to encapsulate cells (Koch et al., 2009, 2012; Gruene et al., 2010). MAPLEDW transfers biomaterials containing cells from transparent quartz support to a receiving substrate. Matrigel® was used as the cell-containing matrix and laser absorptive layer (Ringeisen et al., 2004; Patz et al., 2006). Bioceramic ribbon is also used by solvating hydroxyapatite and zirconia powders in glycerol/water matrices and spin coating this solution (Doraiswamy et al., 2007). BioLP is an improved version of MAPLEDW, since it employs quartz support coated by a metal or metal oxide, which eliminates the direct interaction of laser and biomaterials. A ribbon consists of three layers, optical transparent quartz, metal or metal oxide laser absorptive layer, and cell solution having cells, cell culture medium, and glycerol or methyl-cellulose. Glycerol and methyl-cellulose were used to reduce evaporation of biological materials. Cell solution was transferred onto substrate coated by Matrigel® (Barron et al., 2004, 2005; Pirlo et al., 2012; Othon et al., 2008).

The types of hydrogels utilized for structure fabrication using light-assisted bioprinting include, but are not limited to, poly(ethylene glycol) diacrylate (PEGDA), poly(ethylene glycol) dimethacrylate (PEGDMA), methacrylamide-modified gelatin (GelMOD), gelatin methacrylate (GelMA), and glycidyl methacrylate modified hyaluronic acid (GMHA). Among these, PEGDA, GelMA, and GMHA are the most extensively used (Hribar et al., 2014).

PEGDA hydrogels, with their superior biocompatibility, high water retention ability, and tunable mechanical properties, in particular stiffness retention, serve as excellent candidates for synthetic biomaterials for biomedical applications (Baroli, 2007). The cross-linking extents and thus the material properties (e.g. stiffness, porosity, and osmotic swelling) of polymerized PEGDA hydrogel can be customized by varying the molecular weight of the monomer (typically 700–10000 Da) as well as the fabrication parameters (e.g. exposure time and laser intensity) (Hudalla et al., 2008). PEGDA hydrogels are generally nondegradable and nonbioactive. Nevertheless, such properties can be enhanced

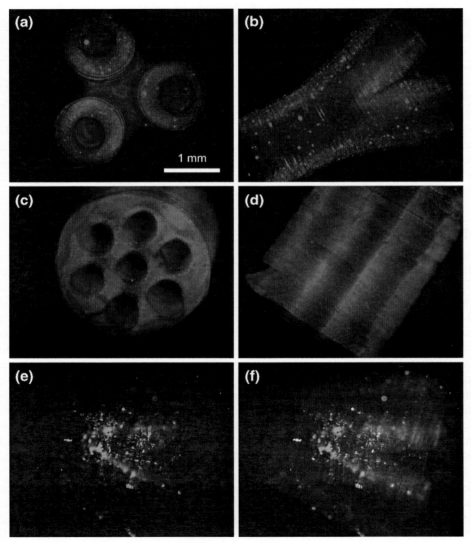

FIGURE 2.7

Fluorescence micrographs of the fabricated scaffolds using DMD-based SLA. (a and b) Top and lateral view of the branched scaffold. (c and d) Top and lateral views, respectively, of the multilumen scaffold. (e and f) Schwann cells seeded inside the scaffold. (Scale bar, 1 mm) (Suri et al., 2011).

by chemical modification, mixing with other degradable materials, or incorporation of adhesive peptide sequence like the tetrapeptide Arg–Gly–Asp–Ser (RGDS) or proteins like laminin (Hribar et al., 2014). GelMA hydrogels, made from xenogenic-modified monomers, belong to the type of naturally derived hydrogels and are bioactive for their peptide sequences, which allow the binding of integrin protein expressed on cell membrane (Hribar et al., 2014). Like most gelatin-based hydrogels, GelMA hydrogels provide high water content and modifiable properties (e.g. swelling and stiffness) based on monomer percentage. GMHA hydrogels, another naturally derived hydrogels from native ECM component, possess superior biocompatibility and are in general nonimmunogenic like natural

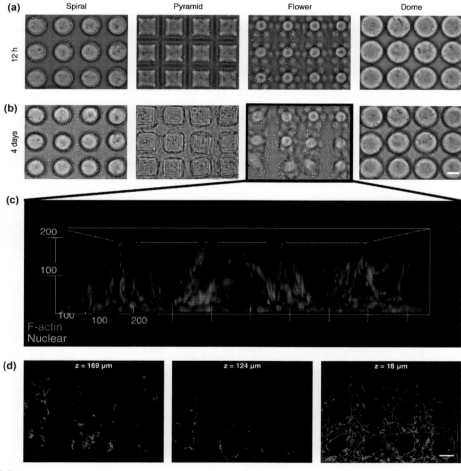

FIGURE 2.8

Complex 3D cell-encapsulated scaffolds fabricated by the DOPsL system. (a and b) Bright field micrographs of GelMA scaffolds with encapsulated NIH/3T3 cells at different time points. (c) 3D reconstruction of confocal fluorescence micrographs of cell–scaffold interaction. (d) Cross-sections of the confocal images in (c). All scale bars are 100 μm (Soman et al., 2013).

hyaluronic acid (Gauvin et al., 2012). Although a certain level of bioactivity has been demonstrated, the surfaces of GMHA scaffold are ready for further modification to incorporate more cell-adhesive proteins like laminin (Hribar et al., 2014).

Laser-based stereolithography (SLA) patterns photocrosslinkable hydrogels or polyesters to create a microenvironment in 3D structure using UV laser. The technology was applied to PEGDMA for 3D scaffold (Mapili et al., 2005), and to PEGDA for cantilever bioactuator (Chan et al., 2012) or cell-encapsulated 3D scaffold (Chan et al., 2010). It was also employed to 3D bone scaffold with a poly(propylene fumarate) (PPF)/diethyl fumarate (DEF) mixture (Lee et al., 2007).

In contrast to the point-by-point processing by laser-based SLA, optical projection stereolithography employs DMD to fabricate 3D hydrogel objects layer-by-layer using UV irradiation. Suri et al. demonstrated the freeform fabrication of nerve regeneration scaffolds with complex microarchitecture using GMHA as shown in Figure 2.7 (Suri et al., 2011). With an improved version of the dynamic optical projection stereolithography (DOPsL) system, Soman et al. succeeded in using GelMA to fabricate

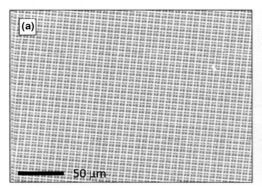
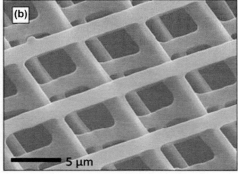

FIGURE 2.9

SEM images of woodpile structures fabricated from PEGDA by TPP: (a) large view, (b) close-up view (Zhang and Chen, 2011).

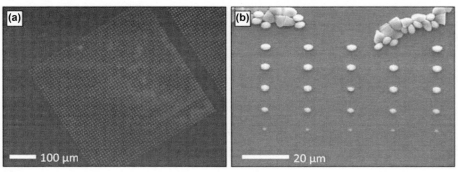

FIGURE 2.10

SEM images of microdot array with various feature sizes fabricated from PEGDA by TPP (Zhang and Chen, 2011).

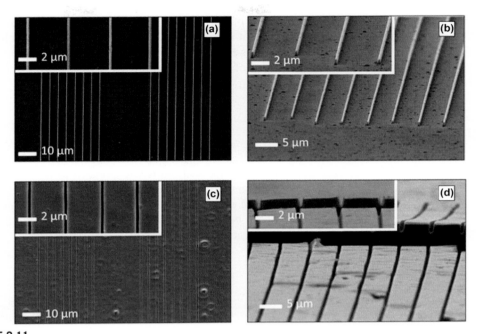

FIGURE 2.11

SEM images of the dipentaerythritol pentaacrylate (DPPA) mold after imprinting (a and b) and the imprinted PEGDA structures (c and d) (Zhang and Chen, 2011).

complex 3D structures encapsulated with cells (Soman et al., 2013) (Figure 2.8). Live cell-captured scaffold was fabricated with PEGDA using commercialized apparatus (Lin et al., 2012).

Femtosecond lasers are usually utilized to achieve 2PP for the fabrication of complex 3D structures with very high resolution (<100 nm) and fidelity. Various types of hydrogels are employed in the 2PP lithography. In the work of Zhang and his colleagues, PEGDA was patterned by a Ti:sapphire femtosecond laser into multiple forms of 3D structures, such as woodpile (Figure 2.9) and microdot array with various feature sizes (Figure 2.10) (Zhang and Chen, 2011). These structures can further be used as a mold for the nano-imprinting of other hydrogels (Figure 2.11) (Zhang and Chen, 2011). Methacrylamide-modified gelatin was utilized to create 3D scaffold (Ovsianikov et al., 2011a, 2011b). A kind of natural hydrogel, collagen, and fibrinogen was dissolved in distilled water and printed layer by layer to construct 3D scaffolds (Gebinoga et al., 2013).

2.3.3 PHOTOCURABLE POLYESTERS

Photocurable resin (PPF/DEF) was polymerized using a stereolithography system (Lee et al., 2007). PPF is an unsaturated linear polyester that has carbon–carbon double bond cross-linking (Peter et al., 1998). It is degradable by simple hydrolysis of the ester bonds and its product is nontoxic (He et al., 2001). Poly (propylene fumarate) is mixed with diethyl fumarate (DEF) and bisacrylphosphrine oxide (BAPO) as solvent and photoinitiator, respectively. The glass transition temperature and viscosity of PPF and

DEF mixture resin is dependent on the weight ratio of PPF and DEF. Glass transition temperature drops dramatically by adding DEF. The viscosity of resin relies on both the weight ratio and the molecular weight of PPF (Lee et al., 2007). Different ranges of laser parameters and resin compositions were employed to analyze the effects on the properties of scaffolds. Considering viscosity and mechanical properties, PPF/DEF mixtures with a 60:40 weight ratio and 1% BAPO might be optimal.

2.4 CELLS AND BIOAPPLICATIONS

Light-assisted biomanufacturing techniques have been used to fabricate tissue constructs with various types of cells and applications. Two approaches exist: the scaffold-based approach where tissue engineering scaffolds are fabricated first and cells and growth factors are loaded in a separate step; and the bioprinting approach, where cells are encapsulated in biopolymers and patterned into 3D structures. Various cells from cancer cells to stem cells have been used for various applications. **Table 2.3** is a summary of the cells used and the applications of each fabrication method.

Table 2.3 Cells and bioapplications				
Fabrication technique	**Cells**	**Bioprinting**	**Application**	**References**
SLS	.	No	3D scaffold	Tan et al., 2005
	Human osteoblast-like cell	No	3D scaffold	Duan et al., 2010
	Fetal femur-derived cell	No	Bone replacement scaffold	Kanczler et al., 2009
	Human osteogenic sarcoma	No	3D scaffold	Liu et al., 2013
	Human gingival fibroblast (HGF)	No	3D scaffold	Williams et al., 2005
LGDW	Embryonic chick Spinal cord cell	Yes	Cell patterned array	Odde and Renn, 2000
	Human Umbilical-Vein endothelial cell (HUVEC) Hepatocytes Multipotent adult progenitor cell (MAPC)	Yes	Tissue architecture	Nahmias et al., 2005
	Endothelial cell Hepatocyte	Yes	Cell patterning and self-assembly	Nahmias and Odde, 2006
	Fibroblast Cardiomyocyte	Yes	Cell patterned array	Narasimhan et al., 2004
	Enbryonic day 7 (E7) chick forebrain neuron	Yes	Cell patterned array	Rosenbalm et al., 2006
LIFT	Fibroblast Keratinocyte Human mesenchymal stem cell (hMSC)	Yes	Skin substitute using cell patterning	Koch et al., 2009

Table 2.3   Cells and bioapplications *(cont.)*

Fabrication technique	Cells	Bioprinting	Application	References
MAPLEDW	Porcine mesenchymal stem cell (MSC)	Yes	Autologous graft	Gruene et al., 2010
	Murine fibroblast Human skin keratinocyte	Yes	Skin substitute using cell patterning	Koch et al., 2012
	Pluripotent embryonal carcinoma cell	Yes	3D cell-seeded scaffold	Ringeisen et al., 2004
	B35 neuronal cell	Yes	3D cell-seeded scaffold	Patz et al., 2006
BioLP	Human osteosarcoma	Yes	3D cell-seeded scaffold	Doraiswamy et al., 2007
	Human osteosarcoma cell	Yes	3D cell-seeded scaffold	Barron et al., 2004; Barron et al., 2004; Barron et al., 2005
	Olfactory ensheathing cell	Yes	3D cell-seeded scaffold	Othon et al., 2008
SLA - laser	Murine bone-marrow stromal cell	No	3D scaffold	Mapili et al., 2005
	Cardiomyocyte	No	Cell-based biohybrid actuator	Chan et al., 2012
	Murine embryonic fibroblast cell	Yes	3D cell-seeded scaffold	Chan et al., 2010
2PP	.	No	3D scaffold	Lee et al., 2007
	.	No	3D scaffold	Zhang and Chen, 2011
	Clonal mouse embryo cell	No	3D scaffold	Zhang et al., 2013
	Human adipose-derived stem cell	No	3D scaffold	Ovsianikov et al., 2011a
	Porcine mesenchymal stem cell	No	3D scaffold	Ovsianikov et al., 2011b
	Fibroblast	Yes	3D scaffold (cell glued)	Gebinoga et al., 2013
SLA - DMD	Schwann cell	No	3D scaffold	Suri et al., 2011
	Human umbilical vein endothelial cell	No	3D scaffold	Gauvin et al., 2012
	Murine embryonic fibroblast Murine mesenchymal progenitor cell	Yes	Cell-laden high-throughput platform	Soman et al., 2013
	Human mesenchymal stem cell	No	Biomedical application	Soman et al., 2012; Soman et al., 2012
	Human adipose-derived stem cell	Yes	3D cell-seeded scaffold	Lin et al., 2012

2.4.1 SCAFFOLD-BASED APPROACH TO TISSUE CONSTRUCTS

Human osteoblast-like (SaOS-2) cells were seeded on nanocomposite scaffolds produced using SLS with Ca-P/PHBV and CHA/PLLA nanocomposite spheres in order to evaluate biomimetic environment (Duan et al., 2010). The Ca-P/PHBV nanocomposite scaffolds were sintered at 15 W laser power, 1257 mm/s scan speed, and part bed temperature of 35°C. For CHA/PLLA scaffolds, the nanocomposites were sintered with a laser power of 15 W using the same scanning speed at a part bed temperature of 45°C. The pore size was designed to be 0.8 mm^3. The porosities of fabricated structures were 62.6 ± 1.3% (Ca-P/ PHBV) and 66.8 ± 2.5% (CHA/PLLA), respectively, although the porosity of the designed model was calculated to be 52.7% using a computer program. Alkaline phosphatase (ALP) activity was measured, since ALP is an enzyme used as a biomarker of the osteogenic phenotype. It can catalyze the hydrolysis of phosphate esters at alkaline pH and influences on the bone matrix mineralization. ALP activity kept increasing until 7 days after seeding, but decreased between 7 days and 14 days. This is because ALP is an early marker for osteogenic differentiation and it decreases with initiation of the mineralization process (Kim et al., 2006). There was no significant difference in cell proliferation and ALP activity between two nanocomposite scaffolds.

Human fetal femur-derived cells were cultured in poly (D,L)-lactic acid (PLA) scaffolds *in vitro* and *in vivo* to evaluate the scaffold fabricated with SLS (Kanczler et al., 2009). PLA powder mixed with carbon black was sintered with a continuous wave fiber diode laser having an energy density of about 100 ± 20 W/cm^2. The scan speed was 3 mm/s. Cell proliferation and histological characteristic were evaluated after 7 days of cell culturing. A PLA scaffold containing human fetal femur-derived cells after 24 h cell culture was subcutaneously implanted in female MF-1 nu/nu immunodeficient mice. After 28 days, the tissue scaffold *in vivo* was evaluated by staining for Acian blue/ Sirius red and for the expression of type I collagen. This study provided a platform for the differentiation of human fetal femur-derived cells to generate new cartilaginous and osteogenic matrices.

A biomedical titanium bone scaffold was fabricated using SLS (Liu et al., 2013). The laser was tuned at laser power of 15 W and a scanning speed of 100 mm/s. The scaffold showed a 142 MPa compressive strength after post-heat treatment at 800°C and proved suitable biocompatibility after cultivation of human osteogenic sarcoma cells (MG63).

A laser-based stereolithography was employed to create predesigned internal architecture and porosities of scaffold with PEGDMA (Mapili et al., 2008). The 3D scaffold was fabricated using Nd:YAG laser having laser energy of ~10 mJ/pulse. The scanning speed on *x–y* direction was about 50 μm/s. Measured pore size and wall thickness of microfabricated scaffolds were ~425 and ~200 μm, respectively. PEG acrylates were modified with the peptide arginine–glycine–aspartic acid (RGD) or the ECM component heparin sulfate, and was later contained within the scaffold to enhance cell adhesion and allow spatial sequestration of heparin-binding growth factor. After the modification, murine bone-marrow stromal cells were seeded and cultured on the scaffolds for 24 h. Cell attachment was evaluated using confocal fluorescence images.

The biohybrid actuator was fabricated using laser-based SLA with PEGDA (Chan et al., 2012). The cantilever-shaped actuator was created by using commercial apparatus. The fabricated cantilever beam measured 2 mm wide, 4 mm long, and 0.45 mm thick. Cardiomyocytes were seeded on the cantilever and bending of the actuator was measured to analyze the traction forces created by cells.

2PP was used to fabricate woodpile scaffold with PEGDA (Zhang and Chen, 2011). The power of the Ti:sapphire femtosecond laser (wavelength 800 nm) was adjusted by rotating a beam attenuator. The woodpile consisted of 1 μm wide lines with spacing of 8 μm. The distance between layers was

3 μm. Woodpile scaffolds could be used for cell transmigration study, where spacing between lines was chosen similar to the typical size of cancer cells. The distance among each line could be changed according to the size of cells. 10T1/2 cells were seeded on the web structures with various shapes for tuning Poisson's ratio, which was hypothesized to change cell response (Zhang et al., 2013).

GelMod was utilized to fabricate a 3D scaffold used for adipose-derived stem cell (ASC) adhesion, proliferation, and differentiation into adipogenic lineage (Ovsianikov et al., 2011a). A cavity-dumped oscillator was employed for femtosecond laser. A layer of scaffolds was produced at a scanning speed of 10 mm/s with a constant average laser power of 3.5 mW. Each layer had a distance of 15 μm. The pore size was a square cross-section of 250 μm by 250 μm. Cell adhesion and proliferation were evaluated using cell staining and fluorescence image analysis. The differentiation into adipogenic lineage was assessed by using Oil Red O staining. Ovsianikov et al. conducted a similar study on fabricated 3D scaffold using 2PP with GelMod (Ovsianikov et al., 2011b). Instead of analysis of cell differentiation, degradation of hydrogels was investigated by measuring mass loss of fabricated scaffolds in the presence of collagenase (Type I, collagenase digestion unit (CDU)). Degradation of scaffolds depended on incubation time. Incubation time of 3–4 h was needed to degrade half of the scaffolds.

Stereolithography using DMD was used to fabricate scaffold for tissue engineering with GMHA. The DMD is composed of an array of micromirrors (1024 by 768) (Suri et al., 2011). For polymerization, the power of UV light was determined to be $\sim$ 8 mW/cm^2 and each layer was exposed under UV light for 30 s. A platform having polymerized layer moved downward 0.5 mm for each layer. The pore sizes were 100–200 μm in one side and diameter with hexagonal and circular geometries, respectively. After the protein grafting process, Schwann cells were seeded and cultured for 24 h. Scaffold degradation was performed in 500 U/ml of hyaluronidase. The longer the UV exposure time the slower the degradation of the scaffold. This is because longer UV exposure could lead to a higher cross-linking density. Cell adhesion was analyzed and the scaffolds represented cell adhesion and retention of cell viability for at least 36 h.

Human umbilical vein endothelial cells (HUVECs) were seeded on 3D scaffolds fabricated with GelMA via DMD-SL (Figure 2.12) (Gauvin et al., 2012). The created scaffold had a dimension of $\sim$2 mm with a micrometer-scale resolution. Unconfined compression test was conducted for scaffolds with different designs, the results of which are biphasic stress–strain curves consisting of low-strain (20–40%) and high-strain (70–90%) compressive moduli (Figure 2.13). At the higher strain stage, the stress–strain relationship showed a linear behavior similar to that of elastic materials. Cell viability, proliferation, and functionality were evaluated by using a confocal microscope after fluorescence staining, which demonstrated that the scaffolds can support cell adhesion and proliferation without damaging the biological function and phenotype of the cells (Figure 2.14).

Elastic modulus and Poisson's ratio are two fundamental mechanical properties that reflect the tissue-engineered scaffolds' ability to handle various loading conditions (Soman et al., 2012). While elastic modulus can be easily tuned by the compositions and fabrication conditions, it is substantially more challenging to tune the Poisson's ratio of the scaffolds (Soman et al., 2012). In the work of Soman and his colleagues, scaffolds with various Poisson's ratio (negative, positive, and zero) were constructed using PEGDA via DMD-SL, as shown in Figures 2.15 and 2.16 (Soman et al., 2012). Human mesenchymal stem cells (hMSCs) were seeded and cultured on the scaffolds to demonstrate the feasibility of these scaffolds for tissue engineering and other biological applications. As long as the scaffolds are deformed in elastic region, Poisson's ratio is solely dependent on the geometry of scaffolds. Thus, the tunable Poisson's ratio property can be imparted to any photocurable materials, which shows great promise for a variety of bioengineering applications (Soman et al., 2012).

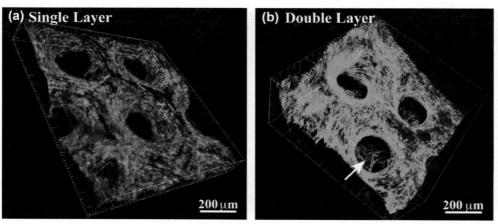

FIGURE 2.12

3D confocal images of the HUVEC-GFP seeded single- (A) and multi- (B) layer scaffolds with precisely defined hexagonal geometries (Gauvin et al., 2012).

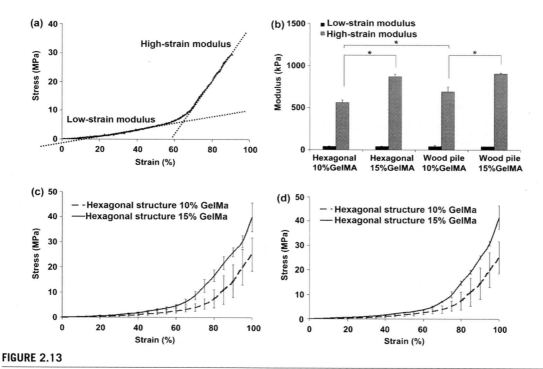

FIGURE 2.13

Mechanical properties of the 3D scaffolds with hexagonal or woodpile structures using different GelMA concentrations (Gauvin et al., 2012).

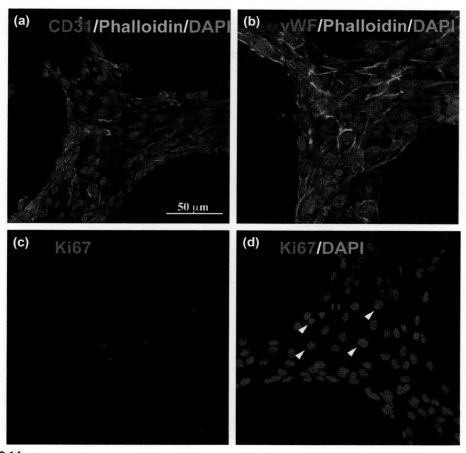

FIGURE 2.14

Immunofluorescence images showing the biological functionality of the HUVEC seeded scaffolds (Gauvin et al., 2012).

2.4.2 BIOPRINTING APPROACH TO TISSUE CONSTRUCTS

Embryonic chick spinal cord cells were patterned on untreated glass surface using LGDW in order to fabricate small arrays of cells for the cell–cell interaction study (Odde and Renn, 2000). A tunable diode laser beam was employed at 450 mW laser power. For long-range transport of cells, the hollow optical fibers were coupled with a laser beam. A laser beam having 800 nm wavelength created gradient force, which propelled cells into the center of the beam at a mean velocity of 11.4 μm/s. The average deposition rate was 2.5 cells/min. In the hollow-fiber experiment, the cells traveled distances of up to 7 mm.

HUVECs were guided on a Matrigel®-coated slide to fabricate two- and three-dimensional cell patterns using LGDW (Nahmias et al., 2005). LGDW employed a weakly focused diode laser beam

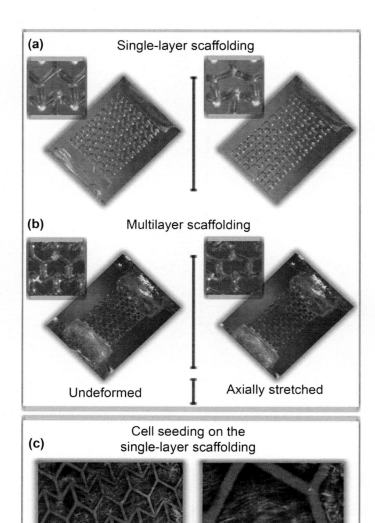

FIGURE 2.15

Optical images showing the deformation of a (A) single- and (B) multilayer ZPR PEG scaffold in response to an axial strain. (C) Fluorescence images of hMSCs seeded on a single-layer ZPR PEGDA scaffold. Green: F-actin. Blue: cell nuclei and scaffold. Scale bars represent (left) 200 μm and (right) 50 μm (Soman et al., 2012). A color version of this figure can be viewed online.

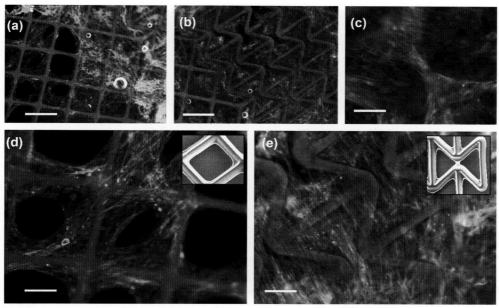

FIGURE 2.16

Fluorescence microscopy images of hMSCs seeded on (A, D) positive Poisson ratio (PPR) region and (B, C, E) negative Poisson ratio (NPR) region. (C) Cells growing in scaffold voids and along scaffold struts in NPR region. (D) Cells growing along scaffold struts (inset: SEM of scaffold struts) in PPR region. (E) Cells seeded on NPR region (inset: SEM of scaffold struts). Scale bars represent (A, B) 250 μm and (C, D, E) 125 μm. Green: actin filaments; blue: nuclei; pink: scaffold struts (Soman et al., 2012). A color version of this figure can be viewed online.

having 200 mW laser power and 830 nm wavelength. The specific cell patterning protocol was developed and described by Nahmias et al. (Nahmias and Odde, 2006). HUVECs were guided in a mixture of culture medium and Percoll solution (40%). Endothelial viability after patterning was 89% ± 7%. The vascular structure of HUVEC (200 μm) was patterned using LGDW and self-assembly of cells. Hepatocytes were then seeded on the fabricated structure to create a sinusoid-like structure, which is one of the structural elements of the liver. The optimal seeding density was 50,000 cells per cm^2 for both endothelial and hepatocytes to create robust formation.

Multiple laser beams were used to pattern fibroblasts on collagen-coated substrate with and without myocyte culture (Narasimhan et al., 2004). Ti:sapphire laser beams had a wavelength of 780 nm. Beam 1 and Beam 2 were perpendicularly aligned on an x–y plane that was vertical to the substrate (z-direction). Beam 2 vertical (x) to z-direction had a higher laser intensity than beam 1 (y) parallel to z-direction. This resulted in a slightly higher velocity for x-direction compared to that for y-direction. The velocity was in the range of 45–150 μm/s.

Embryonic day 7 (E7) chick forebrain neurons were subjected to Ti:sapphire laser having 800 nm wavelength (Rosenbalm et al., 2006). Two experiment parameter settings were compared to control (cell culture without laser exposure): laser intensity of 300 mW at 60 s and one of 100 mW at 10 s. The extension length of neurite was measured to analyze the influence of laser on neurons. There was no significant difference in results. Rosenbalm et al. concluded that LGDW is a safe method for cell patterning.

The LIFT technique was applied to print human skin cells, NIH3T3 fibroblasts, and HaCaT keratinocytes, and human mesenchymal stem cells (hMSC) using Nd:YAG laser (Koch et al., 2009). The donor slide was sputter-coated with a 55–60 nm thick gold and was covered with about 50 μm thick cell-containing layer. Laser having 1064 nm wavelength and 3–6 J/cm^2 fluence transferred a cell-suspended mixture of alginate and blood plasma on the Matrigel®-coated collector slide. Biological material droplets were printed in droplet diameter of 80–140 μm with a speed of 1200 droplets per minute. The cell viabilities were 98% ± 1% for skin cells and 90% ± 10% for human stem cells after laser printing. Apoptosis and genotoxicity were tested to evaluate the damage to cells and DNA, respectively, and the results proved that LIFT has no significant influence on cells and DNA. The result of fluorescence activated cell sorting (FACS) analysis described that the phenotype of hMSC was not affected by LIFT.

Porcine mesenchymal stem cells (MSC) were printed on the hydrogel-covered collector slide using LIFT (Gruene et al., 2010). Gold was sputter-coated on the donor slide in 55–60 μm thickness, and then the cell suspension was covered in ~65 μm thickness on gold layer. Parameters of Nd:YAG laser were fixed with 1046 wavelength and 2.15 J/cm^2 laser fluence. There was no significant difference in viability between laser-printed cells and control (cells without laser exposure). No damage to DNA was observed. ALP activity was measured to evaluate and compare the osteogenic differentiation between laser-printed cells and control. Immunofluorescence staining and Alcian blue staining were conducted to detect the presence of type II collagen and aggrecan, and to quantify sulfated glycosaminoglycan (sGAG). The results supported that MSC differentiated to bone. Koch et al. also three-dimensionally patterned NIH3T3 fibroblasts and HaCaT keratinocytes (Koch et al., 2012). Twenty layers of each cell line were stacked to mimic 3D skin structure.

MAPLEDW was utilized on biomaterials containing pluripotent embryonal carcinoma cells (P19) onto Matrigel®-coated quartz substrate (Ringeisen et al., 2004). Matrigel® was spin-coated on quartz 10–30 μm thickness, and the substrate had a Matrigel® layer on its cell receiving face. An ArF excimer laser was set with 193 nm wavelength and 400 mJ/cm^2 laser fluence. Cell viability was over 95% for 24 h post-transfer. The comet assay was employed to evaluate DNA damage; the results showed no noticeable damage. Cell differentiation was induced via adding retinoic acid or dimethyl sulfoxide (DMSO; 1%). The immunofluorescence staining test proved that P19 cells were differentiated into neuronal and muscle cells.

B35 neuronal cells were transferred onto a Matrigel®-coated quartz substrate using an ArF laser (193 nm wavelength) (Patz et al., 2006). The ribbon consisted of transparent quartz and cell-seeded Matrigel® matrix. The matrix was printed onto the substrate via a laser beam having a laser fluence range of 0.02 to 0.08 J/cm^2. Terminal deoxynucleotidyl transferase biotin-dUTP nick end labeling (TUNEL) immunostaining was used to detect cell apoptosis. The α-tubulin immunofluorescence staining was employed to examine axon morphology. TUNEL staining image showed that 3% of B35 neuronal cells were in apoptosis after 96 h. Analysis of axonal projection showed no significant impairment after the MAPLEDW process. The penetration of cells within the Matrigel® substrate was observed using a confocal microscope and the maximum depth was 75 μm.

Human osteosarcomas (MG-63) were codeposited with bioceramic, hydroxyapatite, and zirconia, using MAPLEDW (Doraiswamy et al., 2007). Hydroxyapatite or zirconia powder solvated in glycerol:water matrices were spin-coated onto a quartz ribbon. ECM was covered onto the bioceramic-coated ribbon then MG-63 cells were seeded on the ribbon. ECM was spin-coated on quartz without any bioceramic in order to compare cell viabilities of ribbon with and without bioceramic. An ArF pulsed excimer laser was set with 193 nm wavelength at 0.22 and 0.18 J/cm^2 laser fluence for hydroxyapatite and zirconia, respectively. Scanning electron microscopy (SEM) and fluorescence images were investigated to analyze morphologies of transferred cells and bioceramics. There was no discernible difference in cell viability between ECM cells deposition and ECM cells bioceramic structure.

BioLP was employed to pattern a multilayer structure with MG-63 cells (Barron et al., 2004). The optical setup was designed with Nd:YAG laser having 266 nm wavelength. The biomaterials consisted of human osteosarcoma cells, culture medium, and 5% of glycerol (v/v). It was transferred onto the Matrigel® substrate with laser fluence of 160 mJ/cm^2 at a deposition rate of 100 spots/s. The patterned cell layer was covered by Matrigel® to print other layers on it. Live/dead fluorescence staining and a confocal microscope were used to evaluate cell viability and morphology of multilayered cell printing. Barron et al. also published a paper concerning BioLP with a similar setup (Barron et al., 2004). The fluorescence image was investigated to evaluate MG-63 cell viability; it showed 95% cell viability. Human osteosarcoma cells were patterned using a similar printing mechanism by Barron et al. (Barron et al., 2005). Immunocytochemical staining was utilized to investigate heat shock protein (HSP) expression. The expression of mouse IgG anti-HSP60 and anti-HSP 70 represented no significant cell impairment.

Olfactory ensheathing cells (OECs) were printed onto the Matrigel® substrate using BioLP in order to fabricate a 3D scaffold (Othon et al., 2008). Titanium or titanium oxide was coated on the transparent quartz ribbon as a laser absorption layer, and then a cell/culture medium solution (0.35% methylcellulose) was applied on the metal or metal oxide layer. Nd:YAG laser was used with a setup of 266 nm wavelength and 4 mJ laser energy. A glass substrate was coated with Matrigel® and translated at the maximum speed of 75 mm/s. Live/dead assay and immunocytochemistry were investigated to evaluate cell viability and cell–cell interaction.

Murine embryonic fibroblast cells (NIH/3T3) were encapsulated in PEGDA hydrogel 3D structures using stereolithography apparatus (Chan et al., 2010). The cells were suspended in the prepolymer solution composed of 20% (w/v) PEGDA, 0.5% (w/v) photoinitiator, and 10% fetal bovine serum (FBS) in Dulbecco's modified Eagle's medium (DMEM). Cell viability and cell spreading were measured and evaluated after cell staining. It was observed that adhesive RGDS peptide sequences have a positive effect on cell viability, proliferation, and spreading.

2PP was used to fabricate a 3D collagen structure containing Fibroblast cells (L929) (Gebinoga et al., 2013). A femtosecond pulsed laser was employed with 780 nm wavelength at a repetition rate of 82 MHz. The laser was focused on a focal area of 280 mm in diameter with 4.5×10^{-10} J pulse energy. Collagenase digestion experiments were conducted to investigate *in vitro* degradation of the collagen structure. Immunofluorescence staining was performed using staining patterns of antibodies. The polymerized line was about 1 μm at a printing speed of 5 μm/s. The collagen structures were not degraded by collagenase I and liberase, but by Terazyme in 24 h. These represented that the laser-patterned collagen structure had a strong cross-link and consisted of native collagen.

Stereolithography was used to print cell-laden hydrogel 3D structures using DMD (Soman et al., 2013). Murine mesenchymal progenitor cells (C3H/10T1/2) and NIH/3T3 cells were encapsulated in GelMA using UV irradiation. UV light had 365 nm wavelength and 11 mW/cm^2 UV intensity. A cell containing GelMA was exposed to UV light for 35 s to pattern each layer. Cell morphology and viability were investigated using immunohistochemical staining. F-actin, α-SM actin, and Hoechst 33258 DNA dye was applied to stain NIH/3T3 cells, C3H/10T1/2 cells, and nuclei, respectively. The resolution in each layer was ~ 17 μm $\times$ 17 μm. There is no significant viability difference between the 3D structure and the control slab. The proliferation was different among locations in the 3D structure due to nutrient concentration gradients.

Human adipose-derived stem cells (hADSCs) were patterned in PEGDA using a projection stereolithography apparatus (Lin et al., 2012). Lin et al. developed an SLA system using a visible light in order to reduce UV light's potential to damage the cellular DNA. The stem cells were suspended in PEGDA solution with photoinitiator (lithium phenyl-2,4,6-trimethylbensoylphosphinate (LAP)) and were captured by a visible light. Minimum pixel size printed was about 68 μm square. The pore size of the fabricated structure was 300 μm $\times$ 300 μm. MTS assay was performed to estimate cell viability, and its result indicated over 90% viability after 7 days.

2.5 DISCUSSION: PROS AND CONS OF EACH TECHNIQUE

Additive manufacturing techniques using laser for bioapplications can be classified into four groups based on the patterning mechanism: laser sintering, laser-writing, laser-transferring, and stereolithography methods. The pros and cons of these techniques are listed in Table 2.4. SLS only prints powder-type materials. It can sinter a variety of materials including biometals, which cannot be achieved by other methods. In addition, the scaffolds fabricated by SLS represent better mechanical strength compared to the ones fabricated by other methods. However, high-power laser is employed to melt and pattern powder-type materials; therefore, SLS can only pattern biomaterials, not cells. The laser-guided direct writing method prints each cell using a gradient force induced by a near-infrared laser. It can achieve a relatively high resolution ($\sim$ 1 μm). In addition, a cell can be guided onto a certain location using multiple lasers. However, scaling up this LGDW technique to fabricate fully 3D structures still remains a challenge. This is because cells are written in a cell by cell fashion at a deposition rate of 2.5 cells/min and cells are seeded on a biological gel layer. Techniques using the laser-transferring mechanism utilize the evaporation of laser absorption layer (typically gold) or cell containing biopolymers. LIFT, MAPLEDW, and BioLP have a higher fabrication rate comparing to LGDW, since they transfer biomaterial drops containing multiple cells. Cell-encapsulated 3D structures can be created using those transferring methods. MAPLEDW employs the simplest ribbon among these techniques. The ribbon is only coated by hydrogels, which is volatilized by UV laser. However, the laser absorption rate of MAPLEDW is lower than LIFT and BioLP. Evaporation of laser absorption layer is used to transfer biological materials onto the collector slide by LIFT. In the case of BioLP, the laser absorption layer is inserted between the transparent quartz and cell suspended biomaterial layer. The volatilization of cell- containing matrix transfers the biomaterial and cells onto the substrate. Stereolithography methods fabricate both 3D scaffolds and 3D cell-laden structures. For 3D scaffolds, the resolution of the stereolithography method, especially 2PP, is on the submicron or nanometer scale. However, in the case of bioprinting, the resolution of stereolithography is limited by the size of cells. Compared to other methods, stereolithography can help create more complex geometries. Stereolithography using DMD has the fastest fabrication rate, since it creates hydrogel patterns in a layer-by-layer fashion (1 layer per 20–35 s). As an expense of high resolution, a more complex experimental setup is required.

Table 2.4   Pros and cons of each technique

Fabrication technique	Advantages	Disadvantages	References
SLS	• High-strength object • Self-support process • Variety of materials	• Limitation to powder form • Rough surface • Hard to remove supporting materials • Impossible to print cells or biological materials	Tan et al., 2005; Duan et al., 2010; Kanczler et al., 2009; Liu et al., 2013
LGDW	• Available cell and biology materials • Single cell resolution • Precise cell printing	• Hard to build full 3D structure • Time consuming • Low cell viability (up to 95%)	Odde and Renn, 2000; Nahmias et al., 2005; Nahmias and Odde, 2006; Narasimhan et al., 2004; Rosenbalm et al., 2006
LIFT	• 3D structure • Available cell and biology materials • High cell viability (~95%)	• Weak structural support	Koch et al., 2009; Gruene et al., 2010; Koch et al., 2012
MAPLEDW	• Available cell and biology materials • High cell viability (near 100%) • Simple donor slide	• Weak structural support • Low reproducibility	Ringeisen et al., 2004; Patz et al., 2006; Doraiswamy et al., 2007
BioLP	• Available cell and biology materials • High efficiency (99% of incident) • High cell viability (~ 95%) • Rapid printing rate	• Weak structural support • Metal or metal oxide layer inserted	Barron et al., 2004; Barron et al., 2004; Barron et al., 2005; Othon et al., 2008
SLA - laser	• Available cell and biology materials • Submicron scale	• Expensive cost • Time consuming • Hard to remove supporting materials • Only available with UV-curable liquid polymer	Mapili et al., 2005; Chan et al., 2010; Chan et al., 2012; Lee et al., 2007
2PP	• Available cell and biology materials • High resolution • Smooth surface	• Expensive cost • Limitation to large scale • Time consuming • Only available with UV-curable liquid polymer • Hard to remove supporting materials	Zhang and Chen, 2011; Zhang et al., 2013; Ovsianikov et al., 2011a; Ovsianikov et al., 2011b; Gebinoga et al., 2013
SLA - DMD	• Available cell and biology materials • Layer-based method • Rapid fabrication	• Only available with UV-curable liquid polymer • Hard to remove supporting materials	Suri et al., 2011; Gauvin et al., 2012; Soman et al., 2013; Soman et al., 2012; Soman et al., 2012; Lin et al., 2012

Laser-based printing is a noncontact printing method. It shares one main advantage with all noncontact printing methods—reduced contamination. During the process, the printing device and the target substrate are always separate thereby reducing the possibility of contamination (Stratakis et al., 2009).

2.6 SUMMARY

Over the past few decades, a variety of 3D printing and nanomanufacturing techniques have emerged with the development of laser technology, CAD techniques, and digital microelectronic devices. Rapid and automatic manufacturing of 3D structures ranging from nanoscale to macroscale has been achieved. The efficiency, flexibility, resolution, and versatility of these 3D printing and nanomanufacturing techniques have generated much excitement in the field of biomedical engineering. AM methods using laser utilize diverse laser sources, materials, and experimental setups. Ultraviolet and near-infrared laser are used to fabricate structures. Donner slide and ribbon are coated by various materials from hydrogel to bioceramics. All structures are designed using CAD layer by layer and printed via laser. These additive manufacturing techniques are applied to a wide range of fields, such as investigation of tumor cell development and progression, regeneration of tissue replacement, and bioactuators, to name a few. Many have reported biocompatibility, resolution, and efficiency of fabricated scaffolds, as well as good cell viability, proliferation, DNA differentiation, and cell–cell interaction. There is no significant change of phenotype, cell damage, and DNA impairment. Compared with other techniques, such as ink-jet and micropen printing, laser-based AMs provide a versatile approach to tissue or organ printing (Mironov et al., 2009). With more and more biomaterials and cell-lines becoming available, tremendous work has been done to apply and make the best use of these 3D printing techniques in both basic biological research and clinical medicine. For instance, engineered functional tissues have shown great promise for *in vitro* drug test as well as *in vivo* transplantation. Medical devices integrated with biomimetic 3D microarchitectures are also revolutionizing traditional healthcare research and industry. While the ultimate goal of 3D printing of functional human organs seems elusive for now, given the limitations of the material, cell-line, and the 3D manufacturing process, the development of 3D bioprinting techniques will continue to expand with potential applications in tissue engineering and regenerative medicine.

REFERENCES

Griffith, L.G., 2002. Emerging design principles in biomaterials and scaffolds for tissue engineering. Annals of the New York Academy of Sciences 961, 83–95.

Orban, J.M., Marra, K.G., Hollinger, J.O., 2002. Composition options for tissue-engineered bone. Tissue engineering 8, 529–539.

Sharma, B., Elisseeff, J.H., 2004. Engineering structurally organized cartilage and bone tissues. Annals of biomedical engineering 32, 148–159.

Zong, X., Bien, H., Chung, C.-Y., Yin, L., Fang, D., Hsiao, B.S., et al., 2005. Electrospun fine-textured scaffolds for heart tissue constructs. Biomaterials 26, 5330–5338.

Moroni, L., De Wijn, J., Van Blitterswijk, C., 2006. 3D fiber-deposited scaffolds for tissue engineering: influence of pores geometry and architecture on dynamic mechanical properties. Biomaterials 27, 974–985.

O'Brien, F.J., Harley, B.A., Yannas, I.V., Gibson, L., 2004. Influence of freezing rate on pore structure in freeze-dried collagen-GAG scaffolds. Biomaterials 25, 1077–1086.

Yang, X., Roach, H., Clarke, N., Howdle, S., Quirk, R., Shakesheff, K., et al., 2001. Human osteoprogenitor growth and differentiation on synthetic biodegradable structures after surface modification. Bone 29, 523–531.

Nazarov, R., Jin, H.-J., Kaplan, D.L., 2004. Porous 3D scaffolds from regenerated silk fibroin. Biomacromolecules 5, 718–726.

Roy, T.D., Simon, J.L., Ricci, J.L., Rekow, E.D., Thompson, V.P., Parsons, J.R., 2003. Performance of degradable composite bone repair products made via three-dimensional fabrication techniques. Journal of Biomedical Materials Research Part A 66, 283–291.

Lin, A.S., Barrows, T.H., Cartmell, S.H., Guldberg, R.E., 2003. Microarchitectural and mechanical characterization of oriented porous polymer scaffolds. Biomaterials 24, 481–489.

Richardson, T.P., Peters, M.C., Ennett, A.B., Mooney, D.J., 2001. Polymeric system for dual growth factor delivery. Nature Biotechnology 19, 1029–1034.

Johnstone, B., Hering, T.M., Caplan, A.I., Goldberg, V.M., Yoo, J.U., 1998. *In Vitro* chondrogenesis of bone marrow-derived mesenchymal progenitor cells. Experimental Cell Research 238, 265–272.

Nuttelman, C.R., Tripodi, M.C., Anseth, K.S., 2004. In vitro osteogenic differentiation of human mesenchymal stem cells photoencapsulated in PEG hydrogels. Journal of Biomedical Materials Research Part A 68, 773–782.

Han, L.-H., Mapili, G., Chen, S., Roy, K., 2008. Projection microfabrication of three-dimensional scaffolds for tissue engineering. Journal of Manufacturing Science and Engineering 130, 021005.

Lu, Y., Mapili, G., Suhali, G., Chen, S., Roy, K., 2006. A digital micromirror device-based system for the microfabrication of complex, spatially patterned tissue engineering scaffolds. Journal of Biomedical Materials Research Part A 77, 396–405.

Mapili, G., Lu, Y., Chen, S., Roy, K., 2005. Laser-layered microfabrication of spatially patterned functionalized tissue-engineering scaffolds. Journal of Biomedical Materials Research Part B: Applied Biomaterials 75, 414–424.

Chua CK, Leong KF, Lim CS. *Rapid prototyping: principles and applications*: World Scientific; 2010.

Tan, K., Chua, C., Leong, K., Cheah, C., Gui, W., Tan, W., et al., 2005. Selective laser sintering of biocompatible polymers for applications in tissue engineering. Bio-medical Materials and Engineering 15, 113–124.

Duan, B., Wang, M., Zhou, W.Y., Cheung, W.L., Li, Z.Y., Lu, W.W., 2010. Three-dimensional nanocomposite scaffolds fabricated via selective laser sintering for bone tissue engineering. Acta Biomaterialia 6, 4495–4505.

Kanczler, J.M., Mirmalek-Sani, S.-H., Hanley, N.A., Ivanov, A.L., Barry, J.J., Upton, C., et al., 2009. Biocompatibility and osteogenic potential of human fetal femur-derived cells on surface selective laser sintered scaffolds. Acta Biomaterialia 5, 2063–2071.

Liu, F.-H., Lee, R.-T., Lin, W.-H., Liao, Y.-S., 2013. Selective laser sintering of bio-metal scaffold. Procedia CIRP 5, 83–87.

Williams, J.M., Adewunmi, A., Schek, R.M., Flanagan, C.L., Krebsbach, P.H., Feinberg, S.E., et al., 2005. Bone tissue engineering using polycaprolactone scaffolds fabricated via selective laser sintering. Biomaterials 26, 4817–4827.

Deckard CR. Method and apparatus for producing parts by selective sintering. In; 1989.

Odde, D.J., Renn, M.J., 2000. Laser-guided direct writing of living cells. Biotechnology and Bioengineering 67, 312–318.

Nahmias, Y., Schwartz, R.E., Verfaillie, C.M., Odde, D.J., 2005. Laser-guided direct writing for three-dimensional tissue engineering. Biotechnology and Bioengineering 92, 129–136.

Nahmias, Y., Odde, D.J., 2006. Micropatterning of living cells by laser-guided direct writing: application to fabrication of hepatic–endothelial sinusoid-like structures. Nature Protocols 1, 2288–2296.

Narasimhan, S.V., Goodwin, R.L., Borg, T.K., Dawson, D.M., Gao, B.Z., 2004. Multiple beam laser cell micropatterning system. In: Proceedings of SPIE, 437–445.

Rosenbalm TN, Owens S, Bakken D, Gao BZ. Cell viability test after laser guidance. In: *Biomedical Optics 2006* International Society for Optics and Photonics; 2006. pp. 608418-608418-608418.

Odde, D.J., Renn, M.J., 1999. Laser-guided direct writing for applications in biotechnology. TRENDS in Biotechnology 17, 385–389.

Koch, L., Kuhn, S., Sorg, H., Gruene, M., Schlie, S., Gaebel, R., et al., 2009. Laser printing of skin cells and human stem cells. Tissue Engineering Part C: Methods 16, 847–854.

Gruene, M., Deiwick, A., Koch, L., Schlie, S., Unger, C., Hofmann, N., et al., 2010. Laser printing of stem cells for biofabrication of scaffold-free autologous grafts. Tissue Engineering Part C: Methods 17, 79–87.

Koch, L., Deiwick, A., Schlie, S., Michael, S., Gruene, M., Coger, V., et al., 2012. Skin tissue generation by laser cell printing. Biotechnology and Bioengineering 109, 1855–1863.

Ringeisen, B.R., Kim, H., Barron, J.A., Krizman, D.B., Chrisey, D.B., Jackman, S., et al., 2004. Laser printing of pluripotent embryonal carcinoma cells. Tissue engineering 10, 483–491.

Patz, T., Doraiswamy, A., Narayan, R., He, W., Zhong, Y., Bellamkonda, R., et al., 2006. Three-dimensional direct writing of B35 neuronal cells. Journal of Biomedical Materials Research Part B: Applied Biomaterials 78, 124–130.

Doraiswamy, A., Narayan, R., Harris, M., Qadri, S., Modi, R., Chrisey, D., 2007. Laser microfabrication of hydroxyapatite-osteoblast-like cell composites. Journal of Biomedical Materials Research Part A 80, 635–643.

Barron, J., Wu, P., Ladouceur, H., Ringeisen, B., 2004. Biological laser printing: a novel technique for creating heterogeneous 3-dimensional cell patterns. Biomedical microdevices 6, 139–147.

Barron, J., Spargo, B., Ringeisen, B., 2004. Biological laser printing of three-dimensional cellular structures. Applied Physics A 79, 1027–1030.

Barron, J.A., Krizman, D.B., Ringeisen, B.R., 2005. Laser printing of single cells: statistical analysis, cell viability, and stress. Annals of biomedical engineering 33, 121–130.

Pirlo, R.K., Wu, P., Liu, J., Ringeisen, B., 2012. PLGA/hydrogel biopapers as a stackable substrate for printing HUVEC networks via BioLP™. Biotechnology and Bioengineering 109, 262–273.

Othon, C.M., Wu, X., Anders, J.J., Ringeisen, B.R., 2008. Single-cell printing to form three-dimensional lines of olfactory ensheathing cells. Biomedical Materials 3, 034101.

Hull CW. Apparatus for production of three-dimensional objects by stereolithography. In: Google Patents; 1986.

Suri, S., Han, L.-H., Zhang, W., Singh, A., Chen, S., Schmidt, C.E., 2011. Solid freeform fabrication of designer scaffolds of hyaluronic acid for nerve tissue engineering. Biomedical Microdevices 13, 983–993.

Gauvin, R., Chen, Y.-C., Lee, J.W., Soman, P., Zorlutuna, P., Nichol, J.W., et al., 2012. Microfabrication of complex porous tissue engineering scaffolds using 3D projection stereolithography. Biomaterials 33, 3824–3834.

Soman, P., Chung, P.H., Zhang, A.P., Chen, S., 2013. Digital Microfabrication of User-Defined 3D Microstructures in Cell-Laden Hydrogels. Biotechnology and Bioengineering.

Soman, P., Fozdar, D.Y., Lee, J.W., Phadke, A., Varghese, S., Chen, S., 2012. A three-dimensional polymer scaffolding material exhibiting a zero Poisson's ratio. Soft Matter 8, 4946–4951.

Soman, P., Lee, J.W., Phadke, A., Varghese, S., Chen, S., 2012. Spatial tuning of negative and positive Poisson's ratio in a multi-layer scaffold. Acta Biomaterialia 8, 2587–2594.

Lin, H., Zhang, D., Alexander, P.G., Yang, G., Tan, J., Cheng, A.W.-M., et al., 2012. Application of visible light-based projection stereolithography for live cell-scaffold fabrication with designed architecture. Biomaterials.

Melchels, F.P., Feijen, J., Grijpma, D.W., 2010. A review on stereolithography and its applications in biomedical engineering. Biomaterials 31, 6121–6130.

Zhang, A.P., Qu, X., Soman, P., Hribar, K.C., Lee, J.W., Chen, S., et al., 2012. Rapid fabrication of complex 3D extracellular microenvironments by dynamic optical projection stereolithography. Advanced Materials 24, 4266–4270.

Zhang, W., Chen, S., 2011. Femtosecond laser nanofabrication of hydrogel biomaterial. MRS Bulletin 36, 1028–1033.

Zhang, W., Soman, P., Meggs, K., Qu, X., Chen, S., 2013. Tuning the poisson's ratio of biomaterials for investigating cellular response. Advanced Functional Materials.

Ovsianikov, A., Deiwick, A., Van Vlierberghe, S., Pflaum, M., Wilhelmi, M., Dubruel, P., et al., 2011a. Laser fabrication of 3D gelatin scaffolds for the generation of bioartificial tissues. Materials 4, 288–299.

Ovsianikov, A., Deiwick, A., Van Vlierberghe, S., Dubruel, P., Moller, L., Drager, G., et al., 2011b. Laser fabrication of three-dimensional CAD scaffolds from photosensitive gelatin for applications in tissue engineering. Biomacromolecules 12, 851–858.

Gebinoga, M., Katzmann, J., Fernekorn, U., Hampl, J., Weise, F., Klett, M., et al., 2013. Multi-photon structuring of native polymers: a case study for structuring natural proteins. Engineering in Life Sciences.

Göppert-Mayer, M., 1931. Über elementarakte mit zwei quantensprüngen. Annalen der Physik 401, 273–294.

Kaiser W, Garrett C. Two-Photon Excitation in CaF_ {2}: Eu^{2 + }. *Physical Review Letters* 1961,**7**:229.

Abella, I., 1962. Optical double-photon absorption in cesium vapor. Physical Review Letters 9, 453.

Duan, B., Wang, M., 2013. Selective laser sintering and its biomedical applications. In: *Laser Technology in Biomimetics*. Edited by Schmidt V. Belegratis MR: Springer Berlin Heidelberg;, 83–109.

Pirlo, R.K., Dean, D., Knapp, D.R., Gao, B.Z., 2006. Cell deposition system based on laser guidance. Biotechnology Journal 1, 1007–1013.

Ovsianikov, A., Gruene, M., Pflaum, M., Koch, L., Maiorana, F., Wilhelmi, M., et al., 2010. Laser printing of cells into 3D scaffolds. Biofabrication 2, 014104.

Chan, V., Zorlutuna, P., Jeong, J.H., Kong, H., Bashir, R., 2010. Three-dimensional photopatterning of hydrogels using stereolithography for long-term cell encapsulation. Lab on a Chip 10, 2062–2070.

Chan, V., Jeong, J.H., Bajaj, P., Collens, M., Saif, T., Kong, H., et al., 2012. Multimaterial biofabrication of hydrogel cantilevers and actuators with stereolithography. Lab on a Chip 12, 88–98.

Lee, K.-W., Wang, S., Fox, B.C., Ritman, E.L., Yaszemski, M.J., Lu, L., 2007. Poly (propylene fumarate) bone tissue engineering scaffold fabrication using stereolithography: effects of resin formulations and laser parameters. Biomacromolecules 8, 1077–1084.

Puppi, D., Chiellini, F., Piras, A., Chiellini, E., 2010. Polymeric materials for bone and cartilage repair. Progress in Polymer Science 35, 403–440.

Nair, L.S., Laurencin, C.T., 2007. Biodegradable polymers as biomaterials. Progress in Polymer Science 32, 762–798.

Best, S., Porter, A., Thian, E., Huang, J., 2008. Bioceramics: past, present, and for the future. Journal of the European Ceramic Society 28, 1319–1327.

Fedorovich, N.E., Alblas, J., de Wijn, J.R., Hennink, W.E., Verbout, A.J., Dhert, W.J., 2007. Hydrogels as extracellular matrices for skeletal tissue engineering: state-of-the-art and novel application in organ printing. Tissue Engineering 13, 1905–1925.

Hribar, K.C., Soman, P., Warner, J., Chung, P., Chen, S., 2014. Light-assisted direct-write of 3D functional biomaterials. Lab on a Chip 14, 268–275.

Baroli, B., 2007. Hydrogels for tissue engineering and delivery of tissue-inducing substances. Journal of Pharmaceutical Sciences 96, 2197–2223.

Hudalla, G.A., Eng, T.S., Murphy, W.L., 2008. An approach to modulate degradation and mesenchymal stem cell behavior in poly (ethylene glycol) networks. Biomacromolecules 9, 842–849.

Peter, S., Miller, M., Yasko, A., Yaszemski, M., Mikos, A., 1998. Polymer concepts in tissue engineering. Journal of Biomedical Materials Research 43, 422–427.

He, S., Timmer, M., Yaszemski, M., Yasko, A., Engel, P., Mikos, A., 2001. Synthesis of biodegradable poly (propylene fumarate) networks with poly (propylene fumarate)–diacrylate macromers as crosslinking agents and characterization of their degradation products. Polymer 42, 1251–1260.

Kim, S.-S., Sun Park, M., Jeon, O., Yong Choi, C., Kim, B.-S., 2006. Poly (lactide- co-glycolide)/hydroxyapatite composite scaffolds for bone tissue engineering. Biomaterials 27, 1399–1409.

Mapili, G., Chen, S., Roy, K., 2008. Projection microfabrication of three-dimensional scaffolds for tissue engineering. Journal of Manufacturing Science and Engineering 130, 021001–021005.

Stratakis, E., Ranella, A., Farsari, M., Fotakis, C., 2009. Laser-based micro/nanoengineering for biological applications. Progress in Quantum Electronics 33, 127–163.

Mironov, V., Trusk, T., Kasyanov, V., Little, S., Swaja, R., Markwald, R., 2009. Biofabrication: a 21st century manufacturing paradigm. Biofabrication 1, 022001.

3D BIOPRINTING TECHNIQUES

3

Binil Starly and Rohan Shirwaiker

Department of Industrial and Systems Engineering, North Carolina State University, Raleigh, NC, USA

3.1 INTRODUCTION

Critical to the success of tissue engineering and regenerative medicine (TERM) approaches, which require the use of a structural matrix, is the design of the scaffold and ensuing tissue construct. Cells involved in the regeneration process are influenced by the macro- and microarchitecture of the constructs. Two primary approaches have been developed to produce scaffolds and tissue constructs: (1) chemically driven processes, such as gas foaming, solvent casting, salt leaching, and freeze casting; and (2) computer-aided layered manufacturing- based approaches. Scaffolds produced through either of these approaches create structural matrices with defined pore architecture in terms of size, shape, and orientation. The three-dimensional (3D) surface area offered by these scaffolds serves as anchoring surfaces for cell adhesion, proliferation, and differentiation to the desired tissue type and function. In these approaches, scaffolds are produced first and the cells of interest are added in a subsequent processing step. The main drawbacks of this approach are the lack of cellular penetration and highly variable cell distribution within the scaffold matrix, primarily seen in scaffold sizes larger than 5 mm in thickness. These limitations arise from two reasons: (1) low cell seeding efficiency at the initial stages to fully inoculate the scaffold itself; and (2) lack of cellular proliferation deep into the scaffold architecture primarily due to rapid drop-off of nutrient concentration within the scaffold core. Several approaches have been developed by the research community over the last decade to mitigate these disadvantages. One popular approach has been the adoption of perfusion-based bioreactor systems to help improve cell seeding efficiency and the transport of nutrients within the scaffold to promote more uniform cellular adhesion and proliferation. An added benefit is that these bioreactors can be customized to provide mechanical and chemical stimulation to accelerate tissue formation. While generally successful, these methods are often limited to small- size defects and thus cannot be translated to thicker tissue constructs.

A more exciting approach is to directly involve cells within the construct design and fabrication process. This offers the advantage of combining multiple processing steps together through which a cellular construct is directly achieved. This approach helps to overcome the seeding efficiency and cell distribution problem. It opens up newer opportunities to customize and regulate the cellular microenvironment by the controlled placement of cells and other biological molecules in defined spatial orientation. In recent years, this approach has rapidly taken off, with several fabrication processes being developed to build *in situ* cellular constructs. These processes include photopolymerization-based processes, laser-based patterning, contact stamping, and cellular microencapsulation to form multicellular tissue spheroids and casting-based biocompatible processes to help achieve complex micro- and macroarchitecture (Valerie Liu Tsang, 2004; Derby, 2012). Perhaps one of the more exciting approaches is "3D bioprinting"-based methods due to the advantages they offer over competing methods to fabricate tissue and organ constructs. Among the advantages of this approach are

(1) Capability to build both 2D and 3D structures
(2) Ability to directly incorporate two or more different cell types in a defined spatial architecture in multi-scale patterns
(3) Flexibility to achieve hybrid processes simply by switching out "tool" options as seen in conventional manufacturing machines, such as CNC-based systems. This suggests that multiple nozzle types or print heads can be incorporated to achieve heterogeneous structures.
(4) Digitally enabled processes allow faster clinical translation and eventual approval by regulatory agencies. The control systems accompanying these processes offer the ability to control process parameters to help achieve desired cellular construct characteristics.

3.2 DEFINITION AND PRINCIPLES OF 3D BIOPRINTING

3D bioprinting is the process of automated deposition of biological molecules on a substrate to form a 3D heterogeneous functional structure with data derived from a digital model. The "print" material used in bioprinting techniques, also known as bioinks, often include a judicious combination of living biological cells, polymers, chemical factors, and biomolecules to form a physical and functional 3D living structure. The substrate is typically planar solid surfaces such as those of Petri dishes, glass slides, or wells of culture plates, although the concept can be extended to nonplanar, nonsolid, and flexible substrates. Living biological cells can be mammalian, insect, and plant-based as well as viruses and bacteria. 3D bioprinting has its roots in the conventional "ink-jet" process developed in the early 1950s that reproduces text and images from a computer file through droplets of ink deposited on a substrate such as paper. Much of the 3D bioprinting techniques have also grown from conventional additive manufacturing (AM) or layered manufacturing (LM) approaches. The complexity of the 3D bioprinting techniques, when compared to AM-based methods of scaffold fabrication, is attributed to the direct involvement of biological living materials during the fabrication process.

Hod Lipson in his book "Fabricated: The New World of 3D Printing" highlights 10 principles of 3D printing as guiding beacons to disrupt the current notion of manufacturing by reducing key barriers of time, cost, and skill level (Lipson, 2013). We have highlighted 10 principles of 3D bioprinting that will help shape the future of printing living tissue and organs for applications in regenerative medicine and tissue engineering.

Principle 1: Physical replication of the living construct from a digital blueprint file.

Bioprinting machines receive initial manufacturing process data input from a digital model of the construct. This digital model must serve as a repository of information that informs upstream and drives downstream manufacturing process activities.

Principle 2: Product customization with high degree of feature variety

Decision-makers and end-users of the printed construct must have the freedom to specify feature sets and functionalities of the construct with minimal complexity in manufacturing process activities.

Principle 3: Structurally heterogeneous product spanning more than two dimensional scales

Bioprinting processes must enable the realization of constructs with structural properties that vary in more than two dimensions. This is necessary to mimic the complexity of nature's own tissue and organ architecture. Processes must take advantage of repeating functional units often seen in complex organs, such as the liver and kidney.

Principle 4: Precise spatial patterning of "bio-ink" materials

The drop-on-demand and the continuous print capability combined with robotic automation in bioprinting processes enable the precise spatial patterning of biological entities in two and three dimensions.

Principle 5: Minimal handling and manipulation of living cells

All bioprinting processes will need to ensure minimal mechanical and environmental stress on non-living biological molecules, chemical factors, and living biological cells. This is to ensure protein and growth factor stability and maximum cellular viability during fabrication.

Principle 6: Conducive microenvironment for functional cells before and after printing

Processes must provide a suitable microenvironment for cells to sustain themselves both prior to and post-printing. Both viability and functionality of cells must be minimally altered during the printing process.

Principle 7: Construct must be functionally stable for downstream operations.

Any printed construct that involves biomaterials and cells must be mechanically, chemically, and/or biologically stable for use in downstream applications. Structurally weak constructs cannot be handled by either humans or automated equipment. Chemically unstable structures will lead to disintegration of properties, while biologically unstable constructs simply cannot be used reliably for downstream processes.

Principle 8: Plug-n-Play bioprinting machines with minimal operator assistance

The machines should be relatively easy to operate with minimal input from operators and should be capable of being installed at any qualified manufacturing facility or hospital. This capability is essential to economically produce the final construct at a price acceptable to the end user. If the machines require very high skilled operation with labor-intensive monitoring, manufacturing costs will be untenable.

Principle 9: Cellular construct, engineered tissue and organs on demand

The digitally enabled technology provides the capability for decision-makers and end-users to custom specify features of the product and request them on-demand, reducing the need for large inventory stock-up of biomaterials and final product storage solutions.

Principle 10: Repeatable and assured functional quality of the bioprinted structure

Bioprinting processes must be accurate to initial design specifications, precise in terms of structural/biological variability, and repeatable in terms of its operation. This is an essential requirement for any process to be qualified by regulatory agencies for any targeted biomedical application.

Several bioprinting processes have achieved varying degrees of adherence to each of the 10 listed principles. The processes themselves are continually being improved by several research groups through expanded biomaterial selection, improved accuracy of spatial arrangement, degree of automation, and complexity of structures generated. Processes successfully meeting all 10 principles will yield commercially viable fabrication processes for the scale-up manufacturing of engineered tissue and organoid systems. If developed well, processes will also find limited challenges to being integrated into any upstream operation that would define the entire manufacturing process cycle for TERM. In the following sections, we will go through some of the most widely investigated 3D bioprinting processes, their fundamental working principles, their application toward the fabrication of cellular constructs, and differences between each process.

3.3 3D BIOPRINTING TECHNOLOGIES

Scaffold-based regenerative medicine therapies involve the fabrication of scaffolds followed by seeding them with cells, preconditioning them inside an incubator, and then conditioning the construct inside a bioreactor to achieve adequate cell proliferation and function. Traditionally, both scaffold fabrication and cell seeding are two mutually separate and distinct steps in the process cycle. 3D Bioprinting techniques essentially combine both the scaffold fabrication and cell placement steps in an *in situ* layered manufacturing process step. The layer by layer bioprinting process enables the direct realization of scaffolding biomaterials, chemical molecules, and living cells in a desired spatial pattern to form the heterogeneous 3D construct. This feature was previously impossible to achieve with any of the conventionally produced tissue scaffolds or even with biopatterning approaches such as UV photopolymerization. The fundamental ability to "bioprint" essentially means that it is possible to accurately control the amount of "bioink" ejected out of the printhead or delivered on to a substrate. Figure 3.1 gives the generic architecture for a bioprinting machine with information and material flow paths highlighted among the entities contained with any system.

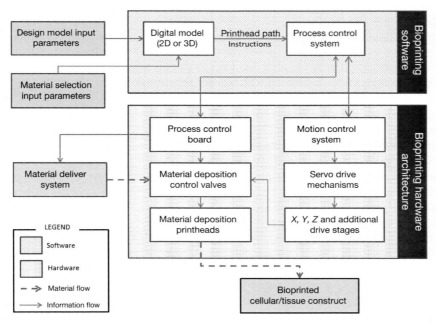

FIGURE 3.1

General 3D bioprinting hardware and software information and material flow.

3D Bioprinting processes begin with a digital model definition of the cellular and tissue construct architecture to be fabricated. In the case of 2D patterns, the digital files can be directly coded into the process control interface to drive the motion of printheads to help physically reproduce the architecture. This digital modeling of a complex 3D construct is typically performed in a computer-aided design (CAD) software environment. Both the external and internal architecture for the 3D construct can be designed with initial data obtained from CT/MRI images of the patient in need of a tissue replacement strategy. Image-based 3D reconstruction procedures can be carried out to help define the 3D digital model of the tissue replacement construct. Tools available in the software environment can help identify different material regions which specify the placement of the biomaterial matrix, biological molecules, and living cells. Process algorithms are written to convert the digital model to printhead path instructions necessary to drive the hardware systems. The exact format for machine instructions in such computer-aided manufacturing (CAM) depends on the printing technique and hardware configuration utilized. Once the signal for printing is activated, the process control system drives the bioprinting system hardware components for the physical realization of the printed construct.

Complex engineered tissue and cellular construct-based products will be made from spatially patterned cellular layers which ultimately will become large aggregates for a specialized tissue function. The entire operation must take place in sterile conditions to limit contamination of both source raw material and the final construct. If cells are involved in the fabrication process, the total time needed to produce a construct can be critical. The amount of time available is dependent on the cell type used. Unless printing conditions are well suited for cell maintenance, time to fabricate constructs should not exceed an hour. Longer times will result in reduced cellular viability and abnormally higher cellular stress, which will lead to degraded function.

We describe the main bioprinting techniques used by the research community to print biomaterials including cells and biomolecules to form 3D constructs. These 3D constructs can be used as tissue models for drug screening, as disease models to study cancer, and as constructs meant for animal or human implantation. Due to the size scale of cells, which are generally in the 5–20 μm size range, all bioprinting processes work at dimensional scale levels larger than the cell type utilized.

3.3.1 INK-JET-BASED BIOPRINTING

Ink-jet bioprinting is a noncontact printing process involving the precise deposition of picoliter to nanoliter droplets of "bioink" (a low-viscosity suspension of living cells, biomolecules, growth factors, etc.) onto a "biopaper" (a hydrogel substrate, culture dish, etc.) in a digitally controlled pattern. It is a direct adaptation of the conventional ink-jet printing process, and a majority of current ink-jet bioprinting activities continue to be conducted using partially modified commercially available desktop ink-jet printers. There are two fundamental approaches to ink-jet printing: continuous (CIJ) and drop-on-demand (DOD). In the CIJ approach, an uninterrupted stream of droplets is produced by forcing the ink through a microscopic nozzle orifice under pressure and deflecting it onto the substrate using an electrostatic field. Where droplet deposition is not required in the digital pattern, the droplets are steered into a gutter and collected for reuse. In the DOD approach, the ink droplets are ejected through the nozzle orifice by creating a pressure pulse inside a microfluidic chamber only when required. The DOD approach is of primary interest in bioprinting due to the pulsed nature of printing. The CIJ approach is not well suited to bioprinting due to the need for conductive ink formulations and the risk of contamination due to ink recirculation, among other reasons.

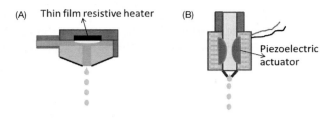

FIGURE 3.2

Inkjet printing mechanisms. (A) Thin film resistive heater generating a vapor bubble that ejects the bio-ink material; (B) A piezoelectric driven actuator that squeezes out a defined quantity of the bio-ink upon pulsed activation. Both mechanisms can either work in continuous or drop-on-demand jetting mode.

The DOD approach can be further categorized into thermal (heat) or piezoelectric (mechanical compression) based on the droplet actuation mechanism. A schematic of DOD ink-jet printing based on both mechanisms is presented in Figure 3.2. In thermal DOD, an electric current pulse applied to the heating element (thin film resistor) rapidly vaporizes a small pocket of ink in the microfluidic chamber. The resulting vapor bubble creates the pressure pulse that propels the ink droplet through the nozzle orifice and onto the substrate. In piezoelectric DOD, a microfluidic chamber above the nozzle contains a piezoelectric transducer for droplet actuation instead of a heating element. A voltage pulse applied to the transducer causes it to expand, creating the transient pressure that results in droplet ejection. For both forms of DOD, the rheological and surface tension properties of the ink govern their ability to be printed. The ink viscosity requirements vary from system to system, but a typical threshold is around 30 mPa/s (Reis et al., 2005; Seerden et al., 2001; Derby, 2008). In addition to the ink characteristics, the orifice size, the distance between nozzle and substrate, the frequency of the current pulse and resulting temperature gradient (thermal DOD), and the frequency of the voltage pulse and piezo-deformation characteristics of the transducer (piezoelectric DOD) have an effect on the ejected droplet size and spatial resolution in ink-jet printing.

In addition to creating structures of nonliving biomolecules, such as DNA (Okamoto et al., 2000) and proteins (Delaney et al., 2009), DOD ink-jet bioprinters have been successfully used to print and pattern live mammalian cells, opening up new and exciting avenues in the field of tissue engineering and regenerative medicine. Ink-jet printing also offers the capability to print multiple cell types, biomaterials, or their combinations from different printheads in a single fabrication operation, allowing for complex multicellular patterns and constructs. The concept of tissue and organ 3D printing, which is being widely explored today, has evolved over years starting with ink-jet bioprinting. Both thermal and piezoelectric DOD printers have been explored for cell bioprinting, but the use of thermal ones has been more prevalent (Cui et al., 2012). In thermal ink-jet printing, while the localized temperature around the heating element can reach between 200–300°C, it lasts for only a few microseconds, and the ejected cells are subjected to a temperature rise of only a couple of degrees above ambient for 2 μs (Cui et al., 2012; Roth et al., 2004; Cui et al., 2010).

Xu et al. (2005) have highlighted the initial challenges in adapting the piezoelectric mechanism for ink-jet cell printing, which revolve around their higher ink viscosity. Primarily, the frequencies and power employed by commercial piezoelectric printers lie within the same range of vibrating frequencies (15–25 kHz) and power (10–375 W) that are known to disrupt cell membrane and cause cell lysis during sonication (Cui et al., 2012; Xu et al., 2005; Simons et al., 1989; Hopkins, 1991). Adapting piezoelectric printers for less viscous ink to lower the frequency and power would be challenging, since the resulting ink leakage and mist formation during printing would affect the spatial and feature resolution (Cui et al., 2012; Xu et al., 2005). However, it should be noted that in recent years, other research groups have successfully demonstrated more than 90% viability for piezoelectrically deposited mammalian cells including human osteoblasts, fibroblasts, and bovine chondrocyte cells (Saunders et al., 2005; Saunders et al., 2008; Saunders et al., 2004). Nonetheless, the original investigations into the feasibility of ink-jet bioprinting of live cells were performed on a piezoelectric system. (Wilson and Boland, 2003) used a bioprinter derived from commercially available HP 660C piezoelectric ink-jet printer and custom designed piezoelectric printheads to print viable line patterns (2D) of bovine aortal endothelial cells (BAEC) and smooth muscle cells. In a subsequent study, they printed BAEC aggregates layer-by-layer (3D) in thermosensitive gels with the same printer, and demonstrated that the closely placed cell aggregates could fuse together, which is critical for tissue formation (Boland et al., 2003). Later, they also became the first to use a commercial thermal ink-jet printer (HP 550C and a modified HP 51626a ink cartridge) to create viable patterns of mammalian cells (Chinese Hamster Ovary (CHO) and embryonic motoneuron cells) onto gel substrates (Cui et al., 2010). The viability of printed mammalian cells was found to be in the range of 85–95% for different cell concentrations. The difference in apoptosis ratio and heat shock protein expression level between printed and non-printed cells was reported to be not statistically significant. Other studies have further investigated the fundamental effects of ink-jet process parameters on viability of different types of cells, and also developed multimaterial composite strategies that combine cells with other biomaterials including proteins, growth factors, and scaffolding polymers (Pepper et al., 2012a; Pepper et al., 2012b; Cui and Boland, 2009; Xu et al., 2006; Chahal et al., 2012). Binder et al. (2011) have provided a primer on DOD ink-jet bioprinting for the research community. Cui et al. (2012) have discussed thermal ink-jet printing from the tissue engineering and regenerative medicine perspective. Derby (2008) has provided a detailed review of ink-jet bioprinting of proteins and hybrid cell-based biomaterials.

3.3.2 PRESSURE-ASSISTED BIOPRINTING

Pressure-assisted bioprinting (PAB) refers to a set of extrusion-based layered manufacturing processes capable of creating digitally controlled 3D patterns and constructs. Biomaterials including polymers and ceramics, proteins and biomolecules, living cells, and growth factors as well as their hybrid structures can be printed using PAB. For printing cells, the bioink is essentially a cell-laden hydrogel of the appropriate viscosity capable of being extruded under pressure through a microscale nozzle orifice or a microneedle at temperatures around 37°C to maintain cell viability. The mechanical integrity of the extruded structures can be controlled through thermal or chemical cross-linking, or multimaterial channel approaches postdeposition. During the process, the biomaterial is contained in a temperature controlled cartridge inside a three axis robotic printhead with a nozzle or microneedle. Deposition takes place by pneumatic pressure, plunger or screw-based extrusion of the material as a continuous filament through the nozzle or microneedle orifice onto a substrate. The substrate can be solid (e.g. culture dish), liquid (e.g. growth media) or a gel based substrate material. The substrate as well as the deposition setup can be contained within a sterile and climate-controlled environment further enabling the use of temperature-sensitive cells and biomaterials. The printhead trajectory is guided by layered data obtained from the digital model of the construct to be laid out. A schematic of the PAB setup is presented in Figure 3.3. The rheological properties of the biomaterial, extrusion temperature, nozzle type used, and applied pressure are the critical parameters that affect the physical and biological characteristics of the printed construct.

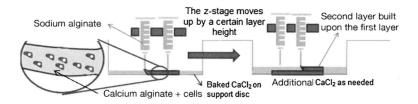

FIGURE 3.3

Pressure-assisted bioprinting using sodium alginate and calcium chloride as the cross-linking agent to fabricate calcium alginate-based hydrogels. The hydrogels can encapsulate or immobilize any desired cell type within the gel. The precursor solution of sodium alginate and water can be modified to include collagen or other peptides to enhance cellular attachment to promote growth within the gel.

Multiple commercially available multinozzle systems that can be used to deposit biopolymers, cells, and growth factors are currently being marketed by Envisiontec GmbH, Germany (3D Bioplotter) and nScyrpt, Florida, USA. Other low-cost 3D printing systems, such as those available from Fab@Home, are being modified to fabricate cellular constructs. Traditionally, pressure-assisted deposition processes were being utilized for the fabrication of tissue-engineered scaffolds, since the controlled pore architecture due to the CAD/CAM-controlled layer-by-layer approach was not achievable by traditional scaffold fabrication processes, such as salt leaching and solvent casting. Several studies have focused on the design and optimization of scaffolds using polymers, such as PCL, PLGA, PLLA, PEG, and their blends, and their composites with ceramics, such as hydroxyapatite (HA) and tricalcium phosphate (TCP), have been reported (Vozzi et al., 2002; Wang et al., 2004; Xiong et al., 2001; Landers and Mulhaupt, 2000; Park et al., 2011).

In recent years, with the emergence of the concept of organ printing, the focus has shifted toward direct printing of cell-encapsulated hydrogels. (Yan et al., 2005a; Yan et al., 2005b; Wang et al., 2006; Cheng et al., 2008), and (Xu et al., 2007) used pressure-assisted multisyringe deposition systems to fabricate liver constructs by encapsulating rat hepatocytes in hydrogels including gelatin in conjunction with chitosan, alginate, and fibrinogen. The initial structural support was achieved by thermal cross-linking of gelatin extruded from a low-temperature syringe onto a warmer stage, and the constructs were further strengthened by chemical cross-linking. Favorable cell viability and function results were obtained based on liver tissue markers (urea and albumin production). Although there was difficulty in stabilizing the 3D structure due to enzymatic degradation of the gelatin/chitosan constructs, the process allowed simultaneous deposition of living cells within a biomaterial. Furthermore, (Xu et al., 2009) and (Li et al., 2009) fabricated biomimetic 3D constructs by simultaneous deposition of adipose-derived stem cells and hepatocytes encapsulated in gelatin-based hydrogels. Similarly, Fedorovich et al. have demonstrated the feasibility of multicellular bioprinted constructs incorporating goat multipotent stromal cells (MPSCs), endothelial progenitor cells in Matrigel®, alginate-based materials for bone grafts (Fedorovich et al., 2001), human mesenchymal stem cells (MSCs), and articular chondrocytes for osteochondral grafts (Fedorovich et al., 2011).

Most recent approaches with pressure-assisted deposition have focused on creating multimaterial multifunctional constructs. For example, (Schuurman et al., 2011) have demonstrated fabrication of hybrid constructs of polycaprolactone (PCL) and chondrocytes-laden alginate. The alginate was cross-linked with calcium chloride postdeposition, but showed relatively good cell viability. Using the same strategy, (Shim et al., 2011) successfully fabricated hybrid constructs containing a PCL/PLGA blend as the structural polymer alongside collagen containing preosteoblast cells. Using the same polymer and cell-laden hydrogel composite strategy, (Lee et al., 2014a) fabricated a viable auricle using PCL for the structural framework, water-soluble poly-ethylene-glycol (PEG) as a sacrificial material, and chondrocytes/ adipocytes differentiated from adipose-derived stromal cells encapsulated in alginate hydrogel as the biological component. Quantitative analyses showed that the auricular cartilage and earlobe fat can be regenerated while maintaining their inherent functions in different regions of the same structure at the same time by printing chondrocytes and adipocytes separately. (Mannoor et al., 2013) fabricated a bionic ear in the anatomic geometry of a human auricle using silicone as the structural component, bovine chondrocytes-laden alginate as the biological component, and silver nanoparticles infused silicone for electronics. They were able to demonstrate structural integrity and shape retention, >90% viability of the printed chondrocytes, and enhanced auditory sensing for radio frequency reception, all within a single simultaneously printed 3D construct.

3.3.3 LASER-ASSISTED BIOPRINTING

Laser-assisted bioprinting (LAB) is a set of noncontact direct writing processes that utilize a pulsed laser beam to deposit biological materials using cells onto a substrate. Three components are central to most LAB systems—a pulsed laser source, a bioink-coated "ribbon," and a receiving substrate. Nanosecond lasers with UV or near UV wavelengths are used as the energy source. The "ribbon" is a glass or quartz target plate that is transparent to the laser radiation wavelength and has one side coated with a heat-sensitive bioink consisting of cells either adhered to a biological polymer or uniformly encapsulated within a thin layer of hydrogel. Depending on the optical characteristics of the bioink and the laser wavelength, the system may also contain a laser-absorbing interlayer between the target plate and the bioink to allow viable cell transfer. The receiving substrate positioned below the bioink-coated side of the ribbon is coated with a biopolymer or cell culture medium to maintain cellular adhesion and sustained growth after cell transfer from the ribbon. The pulse of laser causes rapid volatilization at the ribbon's plate-bioink interface and propels a high-speed jet of the cell-laden bioink onto the receiving substrate.

A schematic of the LAB approach is presented in Figure 3.4. Commonly used LAB processes based on this fundamental working principle include absorbing film-assisted laser-induced forward transfer (AFA-LIFT) or biological laser processing (BioLP) (Hopp et al., 2004; Barron et al., 2005)

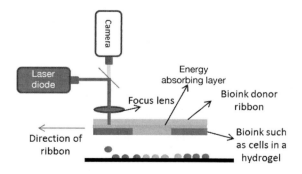

FIGURE 3.4

Laser-assisted bioprinting

and matrix-assisted pulsed laser evaporation direct writing (MAPLEDW) (Barron et al., 2004; Patz et al., 2006). The AFA-LIFT and BioLP are bioprinting versions of the laser-induced forward transfer (LIFT) technique which was originally developed for the direct writing of metal features using a high-energy pulsed laser to deposit a metal film on an optically transparent substrate and has also been employed for direct writing of biomolecules (Duocastella et al., 2007). In AFA-LIFT and BioLP, a high-powered pulsed laser is used to vaporize and deposit the bioink onto the substrate. To protect cells from direct exposure to the high-energy laser beam, a sacrificial metal or metal oxide (e.g. Au, Ag, Ti, and TiO_2) thin film (~ 100 nm) is included at the interface between the target plate and the bioink. The rapid thermal expansion of this interfacial layer due to the high-energy laser pulse propels a small volume of bioink onto the substrate, but with minimal heating of the bioink to prevent cell damage. The BioLP process also utilizes computer-controlled motorized stages and a CCD camera to visualize and focus the laser. Unlike AFA-LIFT and BioLP, the MAPLE-DW process uses a low-power pulsed laser and an interfacial layer of a sacrificial hydrogel such as Matrigel® instead of a metal thin film to accomplish cell transfer. This interfacial layer that acts as an attachment layer for the cells also absorbs the laser energy to prevent it from affecting them. Similar to BioLP, computer-controlled manipulation of the stages coupled with a CCD camera allows for selective cell patterning. Laser-guided direct writing (LGDW) is another commonly used LAB process, but unlike other LAB processes, it does not use a pulsed laser or a print ribbon (Nahmias et al., 2005). Instead, the optical energy of a weakly focused continuous laser is directly used to target and manipulate individual cells from liquid cell suspension onto the substrate. Several parameters related to the laser source, bioink, substrate, interfacial ribbon layer, among others, affect the resolution and performance of LAB processes, and have been described by (Guillemot et al., 2010).

MAPLEDW was one of the first processes used to successfully demonstrate the feasibility of laser-based printing of mammalian cells. Bu et al. (2001) first demonstrated the patterning of Chinese hamster ovary (CHO) cells using the process. Later, Ringeisen et al. (2004) and Barron et al. (2005) printed mouse carcinoma cells (P19), human osteosarcoma (MG-63), and rat cardiac cells with the same process with viability greater than 95%. Recently, (Schiele et al., 2010) have used the MAPLEDW to create viable patterns of human dermal fibroblasts, mouse C2C12 myoblasts, bovine pulmonary artery endothelial (BPAEC), breast cancer (MCF-7), and rat neural stem cells. The AFA-LIFT and BioLP processes have also been successfully used to print various cell types including human osteosarcoma (MG-63), rat Schwann and astroglial and pig lens epithelial cells, BAEC, human umbilical vein endothelial cells (HUVECs), and human umbilical vascular smooth muscle cells (HUVSMC). Invalid source specified. *Schiele* et al. (2010b) have recently provided a thorough topical review of LAB processes and their applications.

3.3.4 SOLENOID VALVE-BASED PRINTING

Microdispensing using solenoid valves has seen applications in depositing solder and adhesives onto electronic boards, deposition of optical and electrical polymers, and deposition of biomolecules such as DNA, proteins, and diagnostic reagents. The system has shown itself capable of printing live biological cells for dermal repair, printing mesenchymal stem cells onto tissue well plates, and printing of constructs within a controlled environment. A complete system consists of a fluid reservoir, a solenoid-based dispensing device with droplet volumes ranging from 5 pl to 1 nl droplet quantities, heating elements to control the nozzle head temperature, connections to the pneumatic controller, and

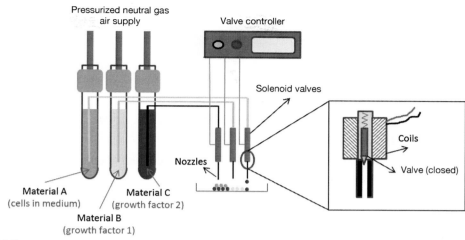

FIGURE 3.5

Solenoid valve-based bioprinting capable of depositing 20 pl or higher droplets of living cells and biological molecules.

an inert gas source. Multiple print head assemblies can be fitted together to improve the throughput of the system. Figure 3.5 shows the schematic of a solenoid-based jetting nozzle. The droplet volume of the printed materials can be controlled by the applied air pressure and the frequency of the solenoid valve open time. Electrical pulse signals sent from the computer can engage or disengage the solenoid leading to droplet ejection from the nozzle. Different nozzle diameters can be attached to the print head to deliver precise quantities of the fluid. The solenoid valve system does not involve heat and is capable of accepting viscous polymers such as collagen and 1–2% sodium alginate. Multiple nozzles can be fitted to the robotic stage to print multiple materials to form a complex heterogeneous construct.

Yoo and coworkers (Lee et al., 2010) reported using this technique for the on-demand fabrication of cellular constructs containing a neural cell line, a fibrin matrix containing a vascular endothelial growth factor (VEGF), and a collagen hydrogel. Since fibrin gel cannot be preloaded into the cartridge, its constituents—fibrinogen, thrombin, and heparin—were separated into two different material cartridges. A third cartridge contained the neural cells to be printed and a fourth cartridge contained a sodium bicarbonate to help in the cross-linking of the collagen gel. With an initial cell density of 1×10^6 cells/ml, each printed droplet of volume 11 ± 0.6nl contained about 56 ± 9 cells. Reported viability of cells within the 500 μm thick collagen construct was greater than 93% soon after printing. This work demonstrates the feasibility of precisely placing desired concentrations of VEGF within spatial locations inside the construct to affect cellular behavior, namely proliferation and differentiation, by controlling the time release behavior of these growth factors. In another study using a similar technique, Karande and his team demonstrated the solenoid-based printing technique to engineer human skin. Fibroblasts and keratinocytes representing the epidermis and dermis respectively, along with collagen were bioprinted to showcase the capability of fabricating a complex living system. The printed skin tissue provides applications in topical drug formulation discovery and screening along with designing autologous grafts for wound healing (Lee et al., 2014b).

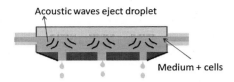

FIGURE 3.6

Acoustic wave-based bioprinting.

3.3.5 ACOUSTIC-JET PRINTING

For applications that require precise placement of single cells, acoustic-based bioprinting is a viable method to deposit picoliter quantities of the medium or hydrogel encapsulating a single cell in a droplet. The ejection of the fluid medium is by focused surface acoustic waves generated by a piezoelectric actuator (such as quartz, lithium tantalite, and lithium niobate) on interdigitated and periodically spaced gold rings. Upon activation by a sinusoidal electrical signal at the same resonant frequency of the device, surface acoustic waves are generated (Figure 3.6). With specially designed printheads, these waves pass through the fluidic environment, in this case, a biomaterial with cells, and are focused to a single point at the fluid-air interface. At the focal point, the waves interfere constructively at the point and the forces exerted by the acoustic radiation will be larger than the surface tension of the fluid leading to the ejection of a droplet from the printhead. The higher the applied frequency, the smaller the droplets generated from a single nozzle. The droplet diameter also depends on the viscosity of the solution contained within the device. For example, a sucrose-dextrose solution can generate droplets ranging from 200 to 3 μm with a respective frequency range of 10 to 100 MHz at a prescribed operational wavelength. The droplet diameter can be customized to the biomolecule being printed. The drop-on-demand feature of the actuation makes for fast deposition speeds of up to 100,000 droplets per second, making it one of the fastest printing methods available. Multiple nozzles and associated acoustic wave generators can be placed periodically to improve throughput rates of fluid ejection. Hence, this method can enable high-throughput studies by patterning a range of biomolecules including RNA, DNA, extracellular matrix proteins, drugs, growth factors, and living cells in microwells (Moon et al., 2010; Gurkan et al., 2014).

3.4 CHALLENGES AND FUTURE DEVELOPMENT OF 3D BIOPRINTING

The eventual goal of all biomedical technologies is to make them clinically available so that patients can benefit from them. All bioprinting processes discussed in this chapter are at different stages of development along that path, but none of them have been translated to FDA-approved clinical applications yet (Table 3.1). By allowing the integration of living and nonliving biomaterials in hybrid structures, bioprinting overcomes several drawbacks associated with previous approaches to tissue fabrication. But despite the significant effort in developing and improving process principles and demonstrating their feasibility for viable cell printing, several supplementary and complementary challenges will need to be addressed to accelerate the successful clinical translation of bioprinting technology, first *in vitro*, and then *in vivo*. As several researchers and experts have noted, thorough characterization of the underlying process mechanisms, their inputs and their outputs will help in the development of machines and systems with low output variability, eventually leading to lower processing costs and improved commercial and clinical potential.

Table 3.1   Comparison between bioprinting processes

Process	Common subprocesses/ mechanisms	Mammalian cell/ tissue types	Typical hydrogels/ bioinks/ biopaper	Suitability for applications			Salient features		
				Therapeutic	Drug-testing and disease modeling	Cell interaction studies	Spatial resolution	Single cell control	Throughput
Ink-jet Bioprinting	• Thermal • Piezoelectric	• Ovary cells (Delaney et al., 2009; Cui et al., 2012) • Motoneurons (Cui et al., 2012) • Endothelial cells (Saunders et al., 2005) • Osteoblasts (Xu et al., 2005; Hopkins, 1991) • Fibroblasts (Simons et al., 1989; Hopkins, 1991) • Chondrocytes (Xu et al., 2005; Hopkins, 1991) • Smooth muscle cells	• DPBS (Delaney et al., 2009, Cui et al., 2012) • Collagen (Delaney et al., 2009; Cui et al., 2012; Saunders et al., 2005) • Soy agar (Cui et al., 2012) • Culture Media (Xu et al., 2005; Simons et al., 1989; Saunders et al., 2005) • Matrigel (Saunders et al., 2005) • Alginate	L	H	M	M	L	H

(Continued)

Table 3.1 Comparison between bioprinting processes (cont.)

Process	Common subprocesses/ mechanisms	Mammalian cell/ tissue types	Typical hydrogels/ bioinks/ biopaper	Suitability for applications			Salient features		
				Therapeutic	Drug-testing and disease modeling	Cell interaction studies	Spatial resolution	Single cell control	Throughput
Pressure-assisted Bioprinting	• Air-pressure • Screw-extrusion	• Hepatocytes (Xiong et al., 2001; Landers and Mulhaupt, 2000; Park et al., 2011; Yan et al., 2005a; Yan et al., 2005b; Cheng et al., 2008; Fedorovich et al., 2001) • Adipose-derived stromal cells (Wang et al., 2006; Cheng et al., 2008) • Multipotent stromal cells (Xu et al., 2007; Xu et al., 2009) • Endothelial progenitor cells (Xu et al., 2007) • Chondrocytes (Xu et al., 2009; Li et al., 2009; Schuurman et al., 2011) • Pre-osteoblasts (Fedorovich et al., 2001) • Adipocytes (Fedorovich et al., 2001) • Ovary cells • Smooth muscle cells • Fibroblasts	• Gelatin (Xiong et al., 2001; Landers and Mulhaupt, 2000; Park et al., 2011; Yan et al., 2005a; Yan et al., 2005b; Wang et al., 2006; Cheng et al., 2008; Fedorovich et al., 2001) • Alginate (Xiong et al., 2001; Landers and Mulhaupt, 2000; Park et al., 2011; Xu et al., 2007; Xu et al., 2009; Li et al., 2009; Fedorovich et al., 2011; Schuurman et al., 2011) • Chitosan (Landers and Mulhaupt, 2000; Yan et al., 2005a; Cheng et al., 2008) • Fibrinogen (Yan et al., 2005b; Wang et al., 2006; Cheng et al., 2008) • Matrigel Xu et al., 2007 • Collagen (Fedorovich et al., 2001) • Hyaluronic acid (Fedorovich et al., 2001) • Agarose	H	H	L	L	L	H

			L	H	H	M	M	M
Laser-assisted Bioprinting	• MAPLE-DW • AFA-LIFT/BioLP • LG-DW	• Epithelial cells (Lee et al., 2014a) • Osteosarcoma (Mannoor et al., 2013; Hopp et al., 2004; Barron et al., 2005) • Cardiac cells (Barron et al., 2005) • Neurons (Barron et al., 2004) • Spinal cord cells (Ringeisen et al., 2004) • Endothelial cells (Schiele et al., 2010a; Moon et al., 2010) • Hepatocytes (Schiele et al., 2010a) • Ovary cells (Lee et al., 2010) • Carcinoma (Lee et al., 2014b) • Fibroblasts (Moon et al., 2010; Gurkan et al., 2014) • Myoblasts (Moon et al., 2010) • Neural stem cells (Moon et al., 2010) • Smooth muscle cells	• Gelatin (Lee et al., 2014a; Schiele et al., 2010a; Gurkan et al., 2014) • Matrigel (Mannoor et al., 2013; Hopp et al., 2004; Barron et al., 2005; Barron et al., 2004; Schiele et al., 2010a; Lee et al., 2010; Lee et al., 2014b; Moon et al., 2010)				

(Continued)

Table 3.1 Comparison between bioprinting processes *(cont.)*

Process	Common subprocesses/ mechanisms	Mammalian cell/ tissue types	Typical hydrogels/ bioinks/ biopaper	Suitability for applications			Salient features		
				Therapeutic	Drug-testing and disease modeling	Cell interaction studies	Spatial resolution	Single cell control	Throughput
Solenoid-based Printing	N/A	• Neural cells (Lee et al., 2010) • Fibroblasts (Lee et al., 2014b) • Keratinocytes (Lee et al., 2014b)	• Fibrin (Lee et al., 2010) • Collagen (Lee et al., 2010; Lee et al., 2014b)	H	H	L	M	M	M
Acoustic-jet Printing	N/A	• Embryonic stem cells (Moon et al., 2010) • Fibroblasts (Moon et al., 2010) • Cardiomyocytes (Moon et al., 2010) • Raji cells (Moon et al., 2010) • Breast cancer cells (Gurkan et al., 2014) • Embryonic kidney cells [B]	• Agarose (Moon et al., 2010) • PBS (Moon et al., 2010) • Dextran (Gurkan et al., 2014)	L	M	H	M	H	H

L: Low; M: Medium; H: High.

We have highlighted here several issues that must be addressed by the 3D bioprinting research community for the wide-scale adoption of this base technology for several therapeutic and nontherapeutic applications.

Lack of digital models: Data drive the bioprinting machines. Currently there is a lack of software architecture with necessary design tools to engineer tissue and organ systems. New data structures must be developed to capture the heterogeneous information necessary to define such living products. Virtually defining the placement of cells, biomaterials, and biological molecules will lead to designs that are robust. The digital definition can also translate to drive all downstream manufacturing operations. Conventional design software revolutionized methods through which automobiles, airplanes, and consumer and medical products are designed and manufactured. Similar software design platforms must be available if the benefits of bioprinting are to be widely adopted by the research and industrial communities. On a similar note, designers should also be presented with tools to simulate engineered tissue/organoids to pursue what-if analysis scenarios. Tissue systems constantly remodel and change over time. Tools to help predict end function, stability, and efficacy will be necessary for the widespread adoption of bioprinting technologies.

Biomaterials are limited, proprietary, and expensive: Another key limitation is the limited class of biomaterials in which the cells are encapsulated. While cell-free scaffolds are printed in a variety of biomaterials, there has not been much development into designing newer biomaterials able to accommodate the encapsulated cells and the printing process in general. While it is generally understood the cells will secrete their own extracellular matrix over time, the initial biomaterial plays an important role in producing the right microenvironment for cells to be accommodated in their new engineered environment. Polymers such as calcium alginate, poly-lactic-glycolic acid (PLGA), and poly-ethylene glycol-diacrylate (PEGDA) are some of the most common polymers used in the 3D bioprinting process. Most polymers used for bioprinting are synthesized in laboratories or are available in commercial quantities that can be cost-prohibitive. There is a need to expand the library of biomaterials to be used for the several bioprinting machines available. Similar to established metals and polymers with their defined properties, new classes of printable biomaterials must be developed to produce engineered tissue and organs.

Lack of adequate cell loading and uniform cell distribution: Engineered constructs of the heart or liver will require millions of cells packed in given volume of a printed construct if they are to function in a physiological manner. Current bioprinting techniques are limited to less than 10 M cells/ml. Improving cell density to reach greater than 50 to 100 M cells/ml will be necessary for proper tissue and organ function. Long fabrication times can lead to settling down of the cells in the printhead chamber. This will lead to nonuniform distribution of cells resulting in inconsistent results. For bioprinting technologies that specify one or two cells per droplet, improving the reliability of cells contained within a droplet is necessary for improving the robustness of the process.

Material development and standardization: Bio-inks are an integral part of the bioprinting technology. It is not the bioprinting process parameters alone, but the material–process interactions that govern the viability and success of the resultant constructs. Hence, developing appropriate bioinks and comprehensively characterizing their rheological, mechanical, and biological characteristics is critical to the success of bioprinting. It is accepted that this development and characterization will have to be cell/tissue- and process-specific. Standardization of the living as well as nonliving bioink components and their sources is also of equal importance, not only from the final production perspective, but also during the process development stage, since any variability in the bioink characteristics ultimately affects the variability of the bioprinting process output. For example, lot-to-lot variability in the growth media can affect the viability of living and biomolecular components of the bioink post printing. Similarly, the viability of fabricated structures can be affected by the source of cells and their storage method prior to printing, even when the bioprinting process parameters are held constant. During the development and characterization of a bioprinting process for allogenic applications, it is critical that the appropriateness of the selected cell source be determined from an application perspective, and its characteristics be thoroughly analyzed and documented. Concepts such as crowd sourcing can be used during the research and development phase to eliminate variability caused by the bioink material.

Measurement and in-process monitoring technologies: A key technology gap to advance bioprinting technologies for engineered tissue is the ability to monitor in real time the process of printing cells and biological moieties. While optical methods are being utilized to monitor droplets and printed constructs, currently there are limited options available to nondestructively evaluate the "living" components within the printed constructs. For example, in the scale-up production of printed cellular constructs, how do we monitor both the placement and functionality of cells within the biomaterial? The problem is even more challenging when encapsulated in 3D hydrogel. Real-time label-free monitoring methods must be developed to advance the scalability and integration of the bioprinting process to economically viable production. The data generated can also help to identify critical process parameters and help implement statistical process control for the bioprinting processes. Ultimately these technologies will be needed to help scale-up or scale-out the process to help meet the demand for customized tissue- and organs-on-demand.

3.5 CONCLUSION

Undoubtedly, the ability to accurately place cells and cellular constructs to form engineered tissue and organoid systems by the definition of a digital model is powerful. The technologies presented offer viable and high-throughput approaches to printing cells and hydrogels in a biocompatible and cell-friendly environment. Spatial control, precise placement of multiple cell types, and high-throughput speed when compared to manual methods are clearly the biggest advantages of the 3D bioprinting technology. When compared to microfabrication techniques such as microstamping and micromolding, the biggest advantage of 3D bioprinting is the ability to define complex interior architecture in true three-dimensions due to the layer by layer additive manufacturing approach taken by most bioprinting technologies. This bottom-up technology bridges microscale and mesoscale definition of engineered tissue, thereby offering the possibility of addressing the challenge of building thick tissues and organ systems.

An immediate application of 3D bioprinting are the use of these systems to create disease models to study pathophysiology, as complex *in vitro* tissue models to screen for new therapeutic drugs and as systems to help achieve engineered meat and leather products. Maturation of these applications of bioprinting will see some of the groundwork being laid to address the challenges highlighted in the previous section. Lab-grown organs and functional tissue intended for human implantation will likely see another decade of fundamental research activity before they have the possibility to become real application scenarios. The future success and commercial viability of the process will depend on efforts to address the challenges to adoption of bioprinting. The systems have the potential to be widely adopted and integrated within conventional cell culture laboratories. They provide a new avenue for researchers to ask research questions in the third dimension, thereby replicating the dynamics of *in vivo* cellular physiology and microenvironment. Bioprinting also has clear applications in regenerative medicine with the ability of these machines to be used in clinical room settings for achieving personalized tissue-on-demand and replacement organs. There are challenges to achieving this vision, but as advancements are made in hardware and software technology coupled with an increased understanding of cellular behavior in the engineered environment, bioprinting will be a viable new biomanufacturing technology.

REFERENCES

Barron, J.A., Ringeisen, B.R., Kim, H., Spargo, B.J., Chrisey, D.B., 2004. Application of laser printing to mammalian cells. Thin Solid Films 1, 383–387.

Barron, J., Wu, P., Adouceur, H., Ringeisen, B., 2005. Biological laser printing: a novel technique for creating heterogeneous 3-dimensional cell patterns. Ann Biomed Eng 6 (2), 121–130.

Binder, K.W., Allen, A.J., Yoo, J.J., Atala, A., 2011. Drop-on-demand inkjet bioprinting: a primer. Gene Therapy and Regulation 6 (1), 33–49.

Boland, T., Mironov, V., Gutowska, A., Roth, E.A., Markwald, R.R., 2003. Cell and organ printing 2: fusion of cell aggregates in three-dimensional gels. Anatomical record Part A, Discoveries in Molecular, Cellular, and Evolutionary Biology 272, 497–502.

Bu, R., Wu P, C.J., Brooks, M., Bubb, D., Wu, H., Piqué, A., B, S., R, M., Chrisey, D.B., 2001. The deposition, structure, pattern deposition, and activity of biomaterial thin-films by matrix-assisted pulsed-laser evaporation (MAPLE) and MAPLE direct write. Thin Solid Films, 607–614.

Chahal, D., Ahmadi, A., Cheung, K.C., 2012. Improving piezoelectric cell printing accuracy and reliability through neutral buoyancy of suspensions. Biotechnology and Bioengineering 109, 2932–2940.

Cheng, J., Lin, F., Liu, H., Yan, Y., Wang, X., Zhang, R., Xiong, Z., 2008. Rheological properties of cell-hydrogel composites extruding through small-diameter tips". Journal of Manufacturing Science and Engineering 130, 021014.

Cui, X., Boland, T., 2009. Human microvasculature fabrication using thermal ink-jet printing technology. Biomaterials 30, 6221–6227.

Cui, X., Dean, D., Ruggeri, Z.M., Boland, T., 2010. Cell damage evaluation of thermal ink-jet printed Chinese hamster ovary cells. Biotechnology and Bioengineering 106, 963–969.

Cui, X., Boland, T., D'Lima, D. D., Lotz, M. K., 2012. Thermal inkjet printing in tissue engineering and regenerative medicine, Recent Patents on Drug Delivery & Formulation, 6(2), 149–155.

Delaney, Jr., J.T., Smith, P.J., Schubert, U.S., 2009. Inkjet printing of proteins. Soft Matter 5, 4866–4877.

Derby, B., 2008. Bioprinting: ink-jet printing proteins and hybrid cell-containing materials and structures. Journal of Materials Chemistry 18, 5717–5721.

Derby, B., 2012. Printing and prototyping of tissues and scaffolds. Science 338, 921.

Duocastella, M., Colina, M., Fernandez-Pradas, J.M., Serra, P., J.L, M., 2007. Study of the laser-induced forward transfer of liquids for laser bioprinting. Applied Surface Science 253, 7855–7859.

Fedorovich, N.E., Wijnberg, H.M., Dhert, W.J., Alblas, J., 2011. Distinct tissue formation by heterogeneous printing of osteo- and endothelial progenitor cells. Tissue Engineering Part A 17, 2113–2121.

Fedorovich, N.E., Schuurman, W., Wijnberg, H.M., Prins, H.J., Van Weeren, P.R., Malda, J., 2011. J. Alblas and W. J. Dhert, Biofabrication of osteochondral tissue equivalents by printing topologically defined, cell-laden hydrogel scaffolds. Tissue Engineering Part C: Methods 18 (1), 33–44.

Guillemot, F., Souquet, A., S, C., B, G., 2010. Laser-assisted cell printing: principle, physical parameters versus cell fate and perspectives in tissue engineering. Nanomedicine 5 (3), 507–515.

Gurkan, U.A., Assal, R.E., Yildiz, S.E., Sung, Y., Alexander, W.P.K., Trachtenberg, J., Demirci, U., 2014. Engineering anisotropic biomimetic fibrocartilage microenvironment by bioprinting mesenchymal stem cells in nanoliter gel droplets. Mol. Pharmaceutics.

Hopkins, T.R., 1991. In: Seetharam, R., Sharma, S.K. (Eds.), Purification and analysis of recombinant proteins. Marcel Dekker, New York, p. 69.

Hopp, B., Smausz, T., Antal, Z., Kresz, N., Bor, Z., D.B, C., 2004. Absorbing film assisted laser induced forward transfer of fungi (Trichoderma conidia). Journal of Applied Physics 96, 3478–3481.

Landers, R., Mulhaupt, R., 2000. Desktop manufacturing of complex objects, prototypes, and biomedical scaffolds by means of computer-assisted design combined with computer-guided 3D plotting of polymers and reactive oligomers. Macromolecular Materials and Engineering 282, 17–21.

Lee, Y.-B., Polio, S., Lee, W., Dai, G., Menon, L., Carrol, R.S., Yoo, S.-S., 2010. Bioprinting of collagen and VEGF-releasing fibrin gel scaffolds for neural stem. Experimental Neurology 223, 645–652.

Lee, J.-S., Hong, J.M., Jung, J.W., Shim, J.-H., Oh, J.-H., Cho, D.-W., 2014a. 3D printing of composite tissue with complex shape applied to ear regeneration. Biofabrication 6, 024103.

Lee, V., Singh, G., Trasatti, J., Bjornsson, C., Xu, X., Tran, T.N., Yoo, S.-S., Dai, G., Karande, P., 2014b. Design and fabrication of human skin by 3D bioprinting. Tissue Engineering Part C: Methods 20 (6), 473–484.

Li, S., Xiong, Z., Wang, X., Yan, Y., Liu, H., Zhang, R., 2009. Direct fabrication of a hybrid cell/hydrogel construct by a double nozzle assembling technology. Journal of Bioactive and Compatible Polymers 24, 249–265.

Lipson, H., 2013. Fabricated: The New World of 3D Printing. Wiley.

Mannoor, M.S., Jiang, Z., James, T., Kong, Y.L., Malatesta, K.A., Soboyejo, W.O., Verma, N., Gracias, D.H., McAlpine, M.C., 2013. 3D printed bionic ears. Nano Letters 13, 2634–2639.

Moon, S., Hasan, S.K., Song, Y.S., Khademhosseini, A., Demirci, U., 2010. Layer by layer three-dimensional tissue. Tissue Engineering: Part C 16 (1), 157–165.

Nahmias, Y., Schwartz, R.E., Verfaillie, C.M., D.J, O., 2005. Laser-guided direct writing for three-dimensional tissue engineering. Biotechnology and Bioengineering 92, 129–136.

Okamoto, T., Suzuki, T., Yamamoto, N., 2000. Microarray fabrication with covalent attachment of DNA. Nature Biotechnology 18, 438–441.

Park, S.A., Lee, S.H., Kim, W.D., 2011. Fabrication of porous polycaprolactone/hydroxyapatite (PCL/HA) blend scaffolds using a 3D plotting system for bone tissue engineering. Bioprocess and Biosystems Engineering 34 (4), 505–513.

Patz, T.M., Doraiswamy, A., Narayan, R.J., He, W., Zhong, Y., Bellamkonda, R., Modi, R., Chrisey, D.B., 2006. Three-dimensional direct writing of B35 neuronal cells. Journal of Biomedical Materials Research Part B: Applied Biomaterials 78 (1), 124–130.

Pepper, M.E., Seshadri, V., Burg, T.C., Burg, K.J., Groff, R.E., 2012a. Characterizing the effects of cell settling on bioprinter output. Biofabrication 4, 011001.

Pepper, M.E., Groff, R.E., Cass, C.A., Mattimore, J.P., Burg, T., Burg, K.J., 2012b. A quantitative metric for pattern fidelity of bioprinted cocultures. Artificial Organs 36, E151–162.

Reis, N., Ainsley, C., Derby, B., 2005. Ink-jet delivery of particle suspensions by piezoelectric droplet ejectors. Journal of Applied Physics 97 (9), 094903.

Ringeisen, K.H., Barron, B.R.J., Krizman, D., Chrisey, D., Jackman S, A.R.S.B., 2004. Laser printing of pluripotent embryonal carcinoma cells. Tissue Eng 10 (3–4), 483–491.

Roth, E.A., Xu, T., Das, M., Gregory, C., Hickman, J.J., Boland, T., 2004. Inkjet printing for high-throughput cell patterning. Biomaterials 25, 3707–3715.

Saunders, R.E., Bosworth, L., Gough, J.E., Derby, B., Reis, N., 2004. Selective cell delivery for 3D tissue culture and engineering. European Cells and Materials 7 (S1), 84–85.

Saunders, R., Gough, J., Derby, B., 2005. Inkjet printing of mammalian primary cells for tissue engineering applications, in: Nanoscale materials science in biology and medicine, Boston.

Saunders, R.E., Gough, J.E., Derby, B., 2008. Delivery of human fibroblast cells by piezoelectric drop-on-demand ink-jet printing. Biomaterials 29 (2), 193–203.

Schiele, N.R., Chrisey, D.B., D.T, T., 2010a. Gelatin-based laser direct-write technique for the precise spatial patterning of cells. Tissue Engineering Part C: Methods 17 (3), 289–298.

Schiele, N.R., Corr, D.T., Huang, Y., Raof, N.A., Y, X., Chrisey, D.B., 2010b. Laser-based direct-write techniques for cell printing. Biofabrication 2 (3).

Schuurman, W., Khristov, V., Pot, M.W., Van Weeren, P.R., Dhert, W.J., Malda, J., 2011. Bioprinting of hybrid tissue constructs with tailorable mechanical properties. Biofabrication 3, 021001.

Seerden, K.A.M., Reis, N., Evans, J.R.G., Grant, P.S., Halloran, J.W., Derby, B., 2001. Ink-jet printing of wax-based alumina suspensions. Journal of the American Ceramic Society 84 (11), 2514–2520.

Shim, J.-H., Kim, J.Y., Park, M., 2011. J. Park and D.-W. Cho, Development of a hybrid scaffold with synthetic biomaterials and hydrogel using solid freeform fabrication technology. Biofabrication 3, 034102.

Simons, K.R., Morton, R.J., Mosier, D.A., Fulton, R.W., Confer, A.W., 1989. Comparison of the Pasteurella haemolytica A1 envelope proteins obtained by two cell disruption methods. Journal of Clinical Microbiology 27 (4), 664–667.

Valerie Liu Tsang, S.N.B., 2004. Three-dimensional tissue fabrication. Advanced Drug Delivery Reviews 56, 1635–1647.

Vozzi, G., Previti, A., De Rossi, D., Ahluwalia, A., 2002. Microsyringe-based deposition of two-dimensional and three-dimensional polymer scaffolds with a well-defined geometry for application to tissue engineering. Tissue Engineering 8 (6), 1089–1098.

Wang, F., Shor, L., Darling, A., Khalil, S., Sun, W., Güçeri, S., Lau, A., 2004. Precision extruding deposition and characterization of cellular poly-ε-caprolactone tissue scaffolds. Rapid Prototyping Journal 10 (1), 42–49.

Wang, X., Yan, Y., Pan, Y., Xiong, Z., Liu, H., Cheng, J., Liu, F., Wu, R., Zhang, R., Lu, Q., 2006. Generation of three-dimensional hepatocyte/gelatin structures with rapid prototyping system. Tissue Engineering Part A 12, 83–90.

Wilson, Jr., W.C., Boland, T., 2003. Cell and organ printing 1: protein and cell printers. The Anatomical Record Part A: Discoveries in Molecular, Cellular, and Evolutionary Biology 272A (2), 491–496.

Xiong, Z., Yan, Y., Zhang, R., Sun, L., 2001. Fabrication of porous poly(l-lactic acid) scaffolds for bone tissue engineering via precise extrusion. Scripta Materialia 45 (7), 773–779.

Xu, T., Jin, J., Gregory, C., Hickman, J.J., Boland, T., 2005. Inkjet printing of viable mammalian cells. Biomaterials 26 (1), 93–99.

Xu, T., Gregory, C.A., Molnar, P., Cui, X., Jalota, S., Bhaduri, S.B., Boland, T., 2006. Viability and electrophysiology of neural cell structures generated by the ink-jet printing method. Biomaterials 27, 3580–3588.

Xu, W., Wang, X., Yan, Y., Zheng, W., Xiong, Z., Lin, F., Wu, R., Zhang, R., 2007. Rapid prototyping three-dimensional cell/gelatin/fibrinogen constructs for medical regeneration. Journal of Bioactive and Compatible Polymers 22, 363–377.

Xu, M., Yan, Y., Liu, H., Yao, R., Wang, X., 2009. Controlled adipose-derived stromal cells differentiation into adipose and endothelial cells in a 3D structure established by cell-assembly technique. Journal of Bioactive and Compatible Polymers 24, 31–47.

Yan, Y., Wang, X., Xiong, Z., Liu, H., Liu, F., Lin, F., Wu, R., Zhang, R., Lu, Q., 2005a. Direct construction of a three-dimensional structure with cells and hydrogel. Journal of Bioactive and Compatible Polymers 20, 259–269.

Yan, Y., Wang, X., Pan, Y., Liu, H., Cheng, J., Xiong, Z., Ln, F., Wu, R., Zhang, R., Lu, Q., 2005b. Fabrication of viable tissue-engineered constructs with 3D cell-assembly technique. Biomaterials 26 (29), 5864–5871.

THE POWER OF CAD/CAM LASER BIOPRINTING AT THE SINGLE-CELL LEVEL: EVOLUTION OF PRINTING

S.C. Sklare[1], Theresa B. Phamduy[2], J. Lowry Curly[2], Yong Huang[3] and Douglas B. Chrisey[1]

[1]Department of Physics and Engineering Physics, Tulane University, New Orleans, LA, USA
[2]Department of Biomedical Engineering, Tulane University, New Orleans, LA, USA
[3]Department of Mechanical & Aerospace Engineering, University of Florida, Gainesville, FL, USA

4.1 INTRODUCTION

Single-cell deposition onto homogeneous two-dimensional (2D) and into three-dimensional (3D) environments with high accuracy and reproducibility is currently entering a new stage of maturity by way of systems engineering, based on a critical mass of technological developments. Coupling computer-aided design (CAD) and manufacturing (CAM) principles with the benefits of laser systems enables researchers to create reproducible or iteratively varying biological constructs with single-cell precision. This chapter illustrates the importance of single-cell deposition and introduces laser-assisted bioprinting as a viable method of printing individual cells. Specifically, the following topics will be discussed: advantages of laser-assisted bioprinting systems; examples of laser systems adapted for cell printing; the technology and basic physics behind these systems; mechanistic modeling of laser-assisted cell transfer; and several case studies illustrating the importance of printing on an individual cell basis. In particular, we showcase matrix-assisted pulsed-laser evaporation direct write (MAPLE-DW) as the premier laser-assisted, nozzle-free, and contactless printing method. We focus on recent enhancements in MAPLE-DW: potential scalability; single-cell deposition; ease-of-use; and environmental controls (e.g. temperature and humidity control).

4.1.1 DIRECT-CONTACT VS. DIRECT WRITE FOR SINGLE-CELL PRINTING

Laser-assisted bioprinting possesses inherent advantages for single-cell applications, but it is not the only method to demonstrate single-cell resolution deposition. Four of the principal categories of bioprinting methods, (1) micropatterning, (2) ink-jet printing, (3) nanocontact bioprinting, and (4) laser deposition, can be further classified as direct-contact (DC) or direct-write (DW) mechanisms. DC methods, such as microcontact patterning, load cells suspended in media onto a "stamp" surface and then bring that "inked" surface in contact with the desired substrate surface. This prints an entire pattern at once, as opposed to being built over time. Stamps are *a priori* patterns that require new molds for new patterns. After the solution is applied to the stamp, there is no way to select certain cells or groups of cells for printing. Thus, the nature of cell distribution in suspension determines the probabilistic cell number per print area. DW methods, such as ink-jet printing and laser-assisted transfer, use "bottom-up techniques" that generate whole constructs on a subunit-by-subunit basis. DW methods can be combined with CAD/CAM techniques to facilitate the controlled transfer of biological "voxels" (volume pixels) containing desired cells. By designing blueprints and depositing biological voxels accordingly, researchers can generate 2D arrays or layered 2D patterns to create 3D constructs. However, laser-assisted cell transfer is distinct from other DW methods for single-cell printing applications.

Unambiguous scientific conclusions rely on reproducibility of experiments to test theories and prove statistical significance. Single-cell printing reduces sample-to-sample variability and permits clear quantification in biological studies by repeating addressable units (voxels) to create cell or tissue constructs. Inkjet methods that rely on nozzle-ejected droplets and blind laser-assisted methods that depend on laser-material interactions for droplet formation are capable of printing one cell at a time on average. This average and reproducibility depend on the probabilistic localization of only a single cell within the ejection volume (Liberski et al., 2011; Barron et al., 2005). Caution should be taken, however, as voxel-to-voxel reproducibility necessarily varies due to inherent cell-to-cell differences. In addition, the ability to select cells during DW procedures produces less variance compared to blind DW methods simply by enabling the end-user to visibly target cells. Moreover, the impact of scientific conclusions increases inversely with the number of cells written with single-cell deposition being paramount.

Rather than rely on distribution statistics, MAPLE-DW and other select DW techniques provide *in situ*, real-time optical monitoring and recording to enable users to select cells for printing and confirm deposition onto the substrate. This distinguishing feature provides feedback to determine precision in cell placement, relative to target location. MAPLE-DW spatial resolution is ±5 μm (Schiele, 2010). By using a camera-equipped, laser-assisted DW system, one is able to reduce voxel-to-voxel variation, monitor deposition in real time, and achieve tight control in spatial printing. This platform for cell printing enables researchers to explore numerous biological systems, including stem niches, cancer invasion, and neuron manipulation/functionalization using functional testing platforms, such as isolated-node single-cell arrays, network-level single- cell arrays, integrated single cells, and 3D cellular invasion models.

4.2 BASICS OF LASER-ASSISTED PRINTING: OVERVIEW OF SYSTEMS AND CRITICAL ANCILLARY MATERIALS

4.2.1 LASER-ASSISTED CELL TRANSFER SYSTEM COMPONENTS

Variations and generational iterations of laser-based cell transfer bioprinting systems are based on similar principles to those of laser-induced forward transfer (LIFT) for inorganic electronic materials (Piqué et al., 1999). The four most common laser bioprinting systems are LIFT, absorbing film-assisted LIFT (AFA-LIFT), biological laser processing (BioLP), and MAPLE-DW. They all utilize optically-transparent "ribbon" disks, biopolymer-coated receiving substrates, beam delivery optics, and imaging and pulsed laser system, but contain variations in ribbon processing and laser wavelength for optimal laser-material interaction (Figure 4.1). Auxiliary components include *in situ* imaging devices, laser beam energy meters, and up to three computer-controlled stages. A complete survey of laser-assisted printing systems is provided by Phamduy et al. (2010). Two specific system schemes, AFA-LIFT and MAPLE-DW, are highlighted here.

The laser-transparent disks, termed "print ribbons," are analogous to ink ribbons historically utilized in typewriters. By analogy, ribbons are loaded on one side with ink, as the optically transparent print ribbon disks are coated on one side with cells. Cells either adhere to a biopolymer coating on the disk ribbon or are suspended in a sacrificial matrix layer, such as cell culture media or low viscosity biopolymer (e.g. hydrogel). In AFA-LIFT, there may be an additional metallic layer between the ribbon surface and cell-embedded medium. This sacrificial layer acts as an absorption medium intended to reduce the effects of laser radiation on cells and facilitate the conversion of light energy to mechanical energy. In systems with UV or near-UV lasers this is a laser-absorbing biopolymer layer. However, in high-power (longer wavelengths with longer pulse widths) laser systems, this layer is often metallic, gold, or titanium. Potential cytotoxic effects exist from volatilization of the sacrificial metal layer, which leads to potential nanoparticulate inclusion during droplet ejection (Smausz et al., 2006; Lewinski et al., 2008). In addition, the metal layer blocks visualization of cells on the ribbon.

MAPLE-DW positions the print ribbon on planar motorized stages in conjunction with a coordinated imaging system, which allows researchers to traverse the ribbon and select individual cells for printing in real time. For single-cell transfer applications, print ribbons must have enough intercellular spatial separation so that only one cell is situated within the transfer area. The ability to visually target individual cells by way of coordinated imaging and a low cell density on print ribbon removes the element of volumetric probability, reducing droplet-to-droplet variation.

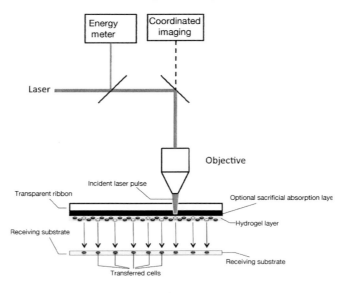

FIGURE 4.1

General schematic for laser-assisted bioprinting.

In MAPLE-DW, the receiving substrate is mounted on a triaxial motorized stage positioned below the ribbon, separated by a small gap approximately 700–2000 μm wide (Patz et al., 2006). There is a critical medium coating on the receiving face of the substrate (whichever side faces the ribbon). This coating is usually a low-viscosity biopolymer (e.g. hydrogel) or gelatin mixture containing cell culture medium. The substrate coatings are carefully chosen to cushion cell impact, maintain a moist environment for cells during printing at ambient conditions, and promote cell adhesion when appropriate. Additional requirements of substrate coatings are discussed later in this section.

Single laser beam pulses are focused at the print ribbon-cell suspension medium (or intermediate absorption layer) interface to achieve noncontact transfer. Each laser pulse causes localized, rapid evaporation of the support medium and the formation of a vapor bubble. The expanding bubble forces a volume of material to be ejected from the cellular suspension layer. The mechanism of bubble formation and material ejection are detailed later in this chapter and a schematic representation is given in Figure 4.2. In general, systems that use metallic absorption layers (e.g. LIFT, AFA-LIFT, and BioLP) also utilize high-power lasers (typically KrF). Low-power, high-energy excimer UV lasers (typically ArF) are used in MAPLE-DW. In general, higher laser pulse energies lead to increased impact force during cell deposition onto the receiving substrate (Patz et al., 2006). When coupled with a viscous receiving medium, this method can be used to create 3D constructs with adherent cells initially deposited at variable depths. As cells may be damaged due to impact and DNA irradiation from laser radiation, the variation in laser energy must be minimized.

4.2.2 AFA-LIFT

AFA-LIFT is a modified version of LIFT, which incorporates a thicker (50–100 nm) metal film to protect cells from laser-related damage rather than a dynamic release layer as used in traditional LIFT. This metallic layer is deposited by vacuum evaporation and interacts with the laser at the print ribbon interface. As KrF excimer lasers ($\lambda = 248$ nm) used for AFA-LIFT have a penetration depth of 10–20 nm, the thick metal layer protects cells from UV irradiation effects. Depending on the laser wavelength, the pulse width at half max pulsing frequency can range from 500 fs–30 ns. Femtosecond lasers impose more unfavorable mechanical force on the cells. In AFA-LIFT, the cell suspension medium, 140–160 μm thick, is spread onto the metal layer. Distance between the ribbon and receiving substrate is typically between 600–1000 μm. Spot size and fluence vary, but the apparatus described by Hopp et al. (2010), uses 250–300 μm diameter spot size and 200–350 mJ/cm² fluence. Thermal expansion of the metallic layer expels a volume of cell suspension to the receiving substrate rather than force generation via vaporization of the cell suspension medium. Some AFA-LIFT systems use charge

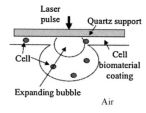

FIGURE 4.2

Droplet formation and expanding bubble schematic (Wang et al., 2009).

couple device cameras to observe and record the transfer process, and motorized stages to enhance laser focusing and cell placement (Hopp et al., 2010).

4.2.3 MAPLE-DW

MAPLE-DW facilitates forward transfer of biological materials in a controlled environment by utilizing two 3D motorized computer-controlled stages and automated features. Additionally, MAPLE-DW is an all-optical imaging and printing system, with environmental controls in the current version. These are among the features that distinguish it from AFA-LIFT and other laser-assisted transfer systems. The current system is the culmination of improvements from previous generations; the first-generation MAPLE-DW system had no motorized stages, the second generation added motorized stages and incorporated CAD/CAM features, and the contemporary third-generation system has subsequently added environmental control. Previously, during printing, it has been difficult to keep small volumes hydrated and trypsinized cells alive for attachment. Environmental control, which includes control and monitoring of temperature and relative humidity, mollifies this without disrupting the print pattern or changing other print parameters.

Screen captures from the graphical user interface (GUI) of the third-generation MAPLE-DW system is shown in Figure 4.3. This GUI enables researchers to control ribbon and substrate stage positioning, laser fluence, pulse trigger, local temperature, and relative humidity. Live video feed of the print ribbon allows researchers to inspect and select cell groups for printing. The live video feed is shown on the right side of the first screen in Figure 4.3. Geometric arrays, 2D patterns and 3D constructs can be printed in a semi-autonomous way thanks to incorporated CAD and CAM tools. Cell selection, the rate-limiting step, still needs to be done manually, significantly decreasing the process speed. In order to automate single-cell printing, "machine vision" (MV) capability was incorporated into the latest generation of MAPLE-DW. The MV image analysis algorithms identify and locate individual cells for transfer into user-defined patterns, thereby increasing the fabrication rate and achieving fully automated printing.

MAPLE-DW print ribbons are coated with a thick ($\sim$30 μm) sacrificial biopolymer layer, frequently Matrigel® or gelatin, to absorb laser energy instead of the metal layer used in AFA-LIFT. Trypsinized cells are partially encapsulated into the hydrogel layer, leaving a sacrificial zone between

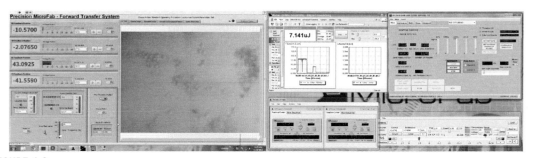

FIGURE 4.3

Dual screen GUI for MAPLE-DW. **GUI Screen 1:** (Left side top to bottom) Ribbon and substrate X and Y stage position feedback and motion controls; temperature and humidity control; laser fire type controls. (Right side top to bottom) Tabs to select display below (live feed, original image, processed image, humidity and temperature graph, system parameters); system shutdown; display window (live feed pictured here). **GUI Screen 2:** Counter-clockwise from top left: energy meter; ribbon and substrate Z stage position feedback, motion controls, and focus control; automated iris control; laser control software.

the ribbon-biopolymer interface and cell layer. MAPLE-DW uses low-powered lasers in UV or near-UV range, typical pulse duration is 8 ns. Single pulses focused at the quartz ribbon-absorption layer interface cause volatile bubble formation that enables transfer. Bubble formation and the beginning of material transfer are schematically represented in Figure 4.2. Shallow penetration depth associated with low-power UV lasers prevents direct effects and interaction between the cells and laser, preventing adverse effects on the cells (Riggs et al., 2011).

4.2.4 ANCILLARY MATERIALS

Materials chosen for the print ribbon and receiving substrate can maintain cell viability before printing and promote cell proliferation, movement, and differentiation after printing. Poor choices for these mediums can lead to cell damage and death. In general, hydrogel biopolymers are used for print ribbons and substrates, but the amounts and biopolymer materials used should be carefully considered.

Cells are embedded in biopolymer hydrogel precursors on the print ribbon. The print ribbon requires a biopolymer that allows cell adherence and admits cell ejection with a single laser pulse. For nonfilm assisted transfer methods, such as MAPLE-DW, this layer acts as a volatile sacrificial medium. Shear thinning and controllable viscosity are in new biopolymers for print applications, because these qualities permit adjustments to compensate for altered printing parameters and while maintaining single pulse ejection. Less viscous biopolymers (e.g. hydrogels) reduce the propagation and magnitude of the stress wave generated, and thus subsequent cell damage, during printing (Wang et al., 2009).

Gelatin has been an effective biopolymer, used for both film-assisted and non-film-assisted methods (Hopp et al., 2005; Ringeisen, Kim, et al., 2004). Gelatin is amenable to thermal manipulation and can be partially cross-linked using heat, which reduces pressure on the embedded cells during printing. Thickness of this coating affects cell viability and varies based on the system and cells of interest. Ringeisen, Kim, et al. (2004) found that cell viability increased from 50% to more than 95% when their Matrigel® biopolymer coating increased from 20 μm to 40 μm for pluripotent embryonal carcinoma cells transferred using the MAPLE-DW platform.

Biopolymers with various growth factors and/or extracellular matrix (ECM) keep cells moist and promote cell adherence. However, growth factors and ECM can introduce additional variability and complications. ECM causes cells to bind firmly to the coated print ribbon, and thus requires more power to achieve cell ejection. This in turn causes cells to be exposed to more irradiation and suffer greater eventual impact damage. Matrigel®, one such biopolymer, is often a good choice of biopolymer and has been extensively utilized by laser DW researchers, even though it has been shown to have unintended effects on cells, such as unintended stem cell differentiation (Riggs et al., 2011; Vukicevic et al., 1992).

The receiving substrate serves to cushion impact-induced stress, maintain moisture, and provide the appropriate growth environment for post-transfer cells. Gelatin and Matrigel® are both used as receiving substrate biopolymers. Impact-induced stress and modeling is discussed in the mechanistics section of this chapter.

4.3 MAPLE-DW MECHANICS

The cell transfer process in MAPLE-DW occurs in three sequential events: cellular droplet formation, cellular droplet travel, and cellular droplet landing. Analytical modeling of the two main events—cellular droplet formation and landing—and their effect on postdeposition cell viability are discussed in detail next.

4.3.1 MODELING CELLULAR DROPLET FORMATION

External pressure and internal stresses are exerted on cells during rapid evaporation (normal boiling and phase explosion) and thermoelastic expansion of viscous droplets. The transformation of a super-heated liquid to an equilibrium state of mixed phases is called a "phase explosion," which eventually leads to a pressure pulse. The pressure pulses are systematically exploited by MAPLE-DW to generate printable droplets on demand. Although necessary for the formation of cellular droplets, the pressure pulse and thermoelastic expansion can injure printed cells, reducing viability after deposition. As such, both bubble formation and thermoelastic stress should be investigated via computational modeling to understand and minimize possible sources of damage to printed cells.

4.3.1.1 Modeling Bubble-Formation-Induced Process Information

Figures 4.3 and 4.4 are schematic representations of the laser-induced bubble formation and expansion process in a typical MAPLE-DW setup. While the MAPLE-DW scheme is shown here, the proposed modeling approach is applicable to other laser-based printing methods that utilize a sacrificial energy-absorbing layer. The modeling assumes that the energy conversion thickness ($<$ 100 nm) at the ribbon-hydrogel interface is negligible.

During the bubble expansion process, a high-pressure pulse is generated, which ejects a droplet volume containing the cells. The bubble expansion process can be modeled using a computational domain as shown in Figure 4.4. The materials involved consist of (1) vaporized gas bubble, (2) air, (3) hydrogel (used here as a coating material), and (4) the cell. Typically, the cell is modeled as a solid type material and applied a Lagrangian mesh for simplicity. The bubble, coating material, and air are modeled using the Eulerian mesh to avoid any extreme element distortion of these materials during ejection. The cell/hydrogel interaction is modeled using the appropriate Euler/Lagrange coupling to capture the effect of viscosity at the cell boundary layer. In addition, the interaction among the hydrogel, bubble gas, and air is modeled by defining the borders in a multimaterial grouping.

The pressure pulse accelerates resting cells from the ribbon until the droplet is ejected at a critical ejection velocity. Ejection velocity largely determines the initial velocity at which the cell droplet encounters the receiving substrate, and should be controlled to minimize cell injury during landing. Figure 4.5 shows the cell-center velocity evolution during the ejection process. It can be seen that the cell velocity oscillates initially and then gradually smoothes out to a constant ejection velocity, in this case 107 m/s. This velocity oscillation is attributed to the elasticity of the cell, implying a negative acceleration. Due to the compressibility of hydrogel, there is a delay in the velocity response to the bubble expansion as seen from Figure 4.5. After approximately 2 µs, the cell droplet has a very weak connection with the coating material and starts to separate from the coating material with a constant velocity.

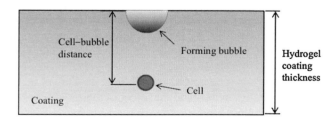

FIGURE 4.4

Modeling domain for the bubble expansion-induced cell deformation (Wang et al., 2009).

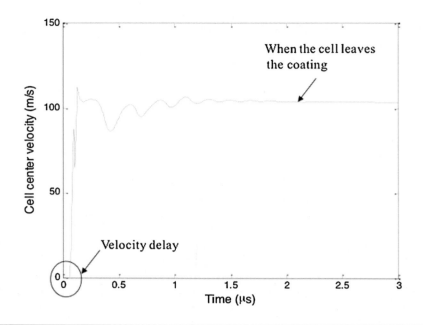

FIGURE 4.5

Cell center velocity during printing process (Wang et al., 2009).

It has been found that the cell can initially accelerate as high as 10^9 m/s² at the beginning period of bubble expansion and then quickly approach zero in an oscillatory manner. Fortunately, this period of high acceleration is very brief, only lasting about 0.1 µs. The pressure that the cell experiences can also be very high at the beginning period of bubble expansion, but quickly decreases to zero in an oscillatory manner, as seen from the cell acceleration evolution. The top surface region of the cell usually experiences the highest pressure level, followed by the bottom surface and then middle regions (Wang et al., 2009).

4.3.1.2 *Modeling Laser–Matter Interaction Induced Thermoelastic Stress*

In general, and during cell deposition, localized heating and thermal expansion of a material cause thermoelastic stress. Two confinement conditions are necessary for the prominent generation of the thermoelastic stress: (1) the laser pulse duration should be much shorter than the characteristic thermal relaxation/diffusion time, and (2) the laser pulse duration should also be shorter than the characteristic acoustic relaxation time to achieve a high-amplitude thermoelastic stress wave. If the laser beam size is taken as finite (laser spot diameter is comparable to the optical penetration depth), the wave generation becomes 3D, which can be solved analytically using Green's function. Unfortunately, however, this approach usually assumes the wave propagation is within a homogenous infinite medium. The image source method has also been explored to model this wave propagation challenge when one of the boundaries is rigid (Paltauf et al., 1998). However, the coating layer during MAPLE-DW is usually very thin. Consequently, this layer cannot be treated as an infinite medium as in a 2D case and the wave is reflected at the free surface. To better understand the effect of thermoelastic stress on the cell injury, the thermoelastic stress wave propagation is modeled here by considering the unique boundary conditions which are different from other previous studies, such as Paltauf et al. (1998).

As shown in Figure 4.6, the computational domain used to simulate the thermoelastic stress generation is treated as 2D. This domain is due to the axisymmetric characteristic of the laser bioprinting process under a typical round laser spot. For the stress wave governing equation, the second order central difference scheme is used to approximate spatial derivatives, and the backward difference scheme is used for the time derivative computation. The Crank-Nicolson method, which has second-order time accuracy, is used to solve the stress wave governing equation. For a finite thickness coating, the pressure wave reflection at the coating-air and the coating-glass interfaces has to be considered. Pressure reflection occurs at the coating-air and the coating-glass interfaces due to their acoustic impedances. The interface reflectivity is equal to -1 for a free surface (interface) and 1 for a rigid surface (interface). Thus the reflected stress at the rigid transparent support does not change the sign of stress due to the very high acoustic impedance in the rigid support, while it does change its sign due to the reflection at the free surface. The stress wave may be canceled by the reflected stress wave in the near vicinity of the coating-air free surface since the reflected stress wave has an opposite sign.

Figure 4.7 shows the pressure profile at a fixed location, 50 μm below the laser spot center, for the first 140 ns (Wang et al., 2011). It is found that a bipolar pressure pulse is developed. A bipolar pulse such as this was also observed in the study of the acoustic wave field generated in front of a submerged fiber tip by Paltauf et al. (1998). At about 33 ns after laser radiation, a positive compressive pressure arrives at this fixed location, immediately followed by a negative tensile pressure. The first pressure peak (13.9 MPa magnitude) originates from the compressive pressure of a plane wave, and the subsequent tensile pressure (-14.4 MPa magnitude) emits from the edge of the laser spot. Both compressive and tensile components physically coexist for the sake of the momentum conservation (Vogel and Venugopalan, 2003). They are experienced 4.6 ns apart on the order of 10 MPa at this fixed location, 50 μm below the center of the laser spot. At approximately 66 ns, the compressive pressure wave reaches the free surface and is reflected back into the coating medium as a tensile stress wave. Then, at approximately 100 ns, the first reflected wave reaches the fixed location with a peak magnitude of -6.4 MPa, and another compressive wave is observed with an even higher peak magnitude of 10.3 MPa; that is, a negative tensile pressure is followed by a larger positive compressive pressure. The second pressure pair is formed due to pressure reflection at the coating-air free surface, which changes the sign of pressure upon reflection. Since the wave energy is transmitted into the surrounding coating during propagation, both components of the second pressure pair decrease in magnitudes, as seen in Figure 4.7.

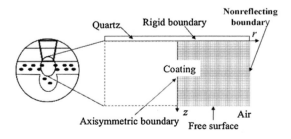

FIGURE 4.6

Schematic of the computational domain (Wang et al., 2009).

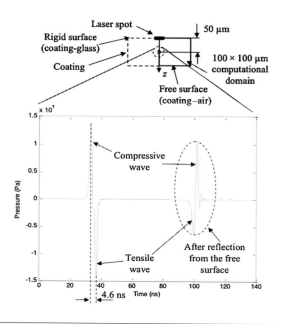

FIGURE 4.7

A representative pressure profile below the laser spot center (z = 50 μm) (Wang et al., 2011).

4.3.2 MODELING OF DROPLET LANDING PROCESS

During landing, cell droplets undergo significant deceleration and impact(s), surviving a much higher external force than they are capable of surviving under steady state conditions. This landing process and its induced impact can be modeled using the mass, momentum, and energy conservation equations (Wang et al., 2007, 2008). These equations hold true for cells, hydrogel in the droplet, and the substrate coating. In addition to boundary and initial conditions, proper material models, which include the equation of state, constitutive model, and failure criteria, are also indispensable in solving these equations. The equation of state is used to define the functional relationship between pressure, density, and internal energy. The constitutive model defines the stress dependence of related strain, strain rate, and temperature. Generally, a material model also includes a failure criterion to determine whether and when the material fails and loses its ability to support certain stress/strain.

A representative result of simulated landing is presented in Figures 4.8 and 4.9, when a cell droplet with a velocity of 50 m/s hits a rigid substrate coated with a 30 μm thick layer of hydrogel. A cell droplet with a cell in the center is modeled using a mesh-free smooth particle hydrodynamic (SPH) method. It can be seen that there are two different impacts during the process under the specified conditions. The first impact is between the cell droplet and the hydrogel coating, and the second impact is between the cell and the rigid substrate after the cell passes through the coating after the first impact. As the landing progresses, the hydrogel-enclosed cell droplet gradually merges into the substrate coating. After the cell is immersed in the coating (Figure 4.8), the outside hydrogel enclosure and the coating bear relatively less stress even though the cell experiences more stress.

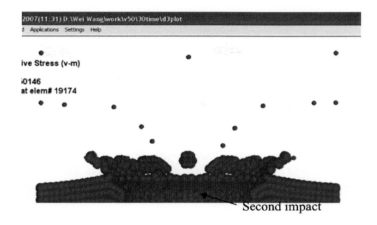

FIGURE 4.8

Landing process at 2.4865 μs.

To study the von Mises stress and shear strain information during the landing process, three particles, the top particle 19139, the inner particle 19144 (one of the four center particles), and the bottom particle 19150, are selected as the representative positions to better understand the overall cell response during the landing process. The simulation is performed using a coating thickness of 30 μm and initial velocity (V_0) of 50 m/s. The particles' von Mises stress responses are shown in Figure 4.9. The stress profiles

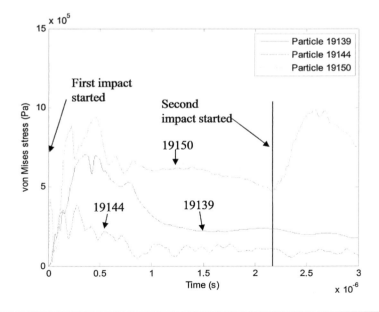

FIGURE 4.9

Particle von Mises stress information (coating thickness = 30 μm and V0 = 50 m/s) (Wang et al., 2008).

show that two different impacts occur during the process under the specified conditions. First impact happens at the computation start time, and second impact happens at approximately 2.2 μs. During the process, the peripheral particles 19139 (top) and 19150 (bottom) are subject to a higher stress level than the inner particle 19144, which indicates that the cell membrane has a higher impact-induced mechanical stress during laser bioprinting. Also, the bottom particle 19150 undergoes a higher stress than the top particle 19139. Figure 4.9 also shows that the second impact has a negligible effect on the particles 19139 (top) and 19144 (inner). However, the bottom particle 19150 experiences higher stresses during the second impact compared to the first impact (1.33 MPa vs. 0.96 MPa), which means that it is important to study the stress information of the bottom particles during both impacts. In this simulation, the bottom particle 19150 experiences the first impact-induced stress peak at 0.2 μs and the second peak at approximately 2.6 μs. The probability of experiencing a second impact is highest for the bottom peripheral particles, lowest for the inner particles with the top peripheral particles falling somewhere in between.

Through modeling studies (Wang et al., 2007, 2008), it has been found that cell peripheral regions, especially the bottom peripheral region, usually experience a higher stress level than the inner regions. This indicates that the cell membrane can be adversely affected by the impact-induced mechanical injury during laser bioprinting. Additionally, the cell mechanical loading profile and the post-transfer cell viability depend on the cell droplet initial velocity and the substrate coating thickness. Generally, a larger initial velocity poses a higher probability of cell injury, and substrate coating can significantly relieve the cell mechanical injury severity. Furthermore, two important impact processes occur during the cell droplet landing process: first impact between the cell droplet and the substrate coating, and second impact between the cell and the substrate. It is assumed that impact-induced cell injury depends on the magnitudes of stress, acceleration, and/or shear strain, and also the cell loading history. In fact, over the entire impact duration, the collective cell momentum change, rather than peak stress, acceleration and/or strain, appears to be critical in determining the cell viability during laser bioprinting and deposition.

4.4 CELL VIABILITY

Post-deposition cell viability has been demonstrated for laser-assisted transfer techniques in terms of cell survival, adhesion, mobility, proliferation, and differentiation. Damage to the cells can be due to mechanical, thermal, or irradiative sources. Modeling of mechanical forces is detailed earlier. Due to the properties of the sacrificial hydrogel layer, thermal and irradiative injury is considered negligible. For example, in MAPLE-DW, for a print ribbon coating of 100 μm, approximately 5 μm depth would be affected by UV laser irradiation. In addition, thermal cell injury does not occur due to disparate time scales. During a typical 8–12 ns laser pulse, cell ejection from the print ribbon will occur within a few μs (as previously shown) and heat conduction does not permeate through the first biopolymer layer to cause cell damage. As explained earlier in the discussion of ancillary materials, the choice of medium for both ribbon and substrate coating are crucial to cell viability (Schiele et al., 2010; Lin et al., 2010; Riggs et al., 2011).

Similarly, in a study on the transfer of "vulnerable cell types" via AFA-LIFT, Hopp et al. (2005) demonstrated successful transfer of rat Schwann and astroglial cells as well as pig lens epithelial cells. Two weeks after transfer, the cells survived, proliferated, differentiated, and regained their original phenotypes. The three cell types were each isolated and then cultured in HEPES-buffered Dulbecco's modified Eagle's medium (HDMEM). Cells were then harvested from culture and prepared for

AFA-LIFT printing. Cells were printed onto gelatin using an LLG two-pass amplification KrF excimer laser. Each target was irradiated with a single 30 ns pulse from the laser. Cell viability was inspected using trypan blue dye, before and after printing, showed that 98–99% of each cell type was alive before printing and 80–85% of each type of the transferred cells remained alive after printing. In addition, the cells retained the ability to proliferate and differentiate 2 weeks after initial printing.

Greater survival rates have been demonstrated with MAPLE-DW. Ringeisen et al. (2004) demonstrated 95% cell survival rate when printing with pluripotent embryonic carcinoma cells. This MAPLE-DW array used an ArF excimer laser from Lambda Physik, model LPX 305, wavelength of 193 nm, and 30 ns full width at half maximum. The target spot on the print ribbon was 100×125 μm^2 and incident laser fluence was in a controlled range from 100 to 500 mJ/cm^2. P19 (pluripotent embryonal carcinoma) cells were cultured and differentiated to neural and muscle cell phenotypes. Cells were prepared on the gelatin-coated ribbon. Printing was done in a humidity-controlled environment, as evaporation of droplet volume reduces cell survivability.

Cell viability was tested 6 h following transfer using live/dead visibility kit. Cells printed onto a 40 μm layer of hydrogel resulted in greater than 95% post-transfer viability. Furthermore, the P19 cells retained their ability to differentiate in induction media.

This study also investigated DNA damage due to ultraviolet light exposure. Comet assays were performed to detect single- and double-strand breaks. MAPLE-DW was used to perform noncontact cell transfer from the quartz ribbon directly into α-MEM. After comparison with control groups, there was no statistically significant damage to the cell DNA as a result of the ultraviolet laser interaction with the cell. This mitigation is due to the energy absorption by the sacrificial biopolymer on the quartz print ribbon.

With laser-assisted cell transfer methods, researchers benefit from soft cell deposition into programmable pattern positions to study cell behavior.

4.5 CASE STUDIES AND APPLICATIONS ILLUSTRATING THE IMPORTANCE OF SINGLE-CELL DEPOSITION

High-throughput, parallel analysis of single cell behavior within the context of biological screening tools or structured cell–cell interaction interfaces necessitates the ability to isolate and position individual cells into engineered arrays. The rationale for intentional cell isolation is to identify significant singularities that may be lost when averaging response across an entire population of heterogeneous cells (Birtwistle et al., 2012). Resolving single-cell characteristics allows researchers to identify cells with the greatest differentiation potential—for example, pluripotent stem cells or cancer stem cells (Shackleton et al., 2006). Similarly, as tissues are composed of various cell types, the dynamics of cells from an explant or whole organism can drown out the finer nuances of the cell–cell interaction of interest. Reintegrating individual cells into a spatially controlled, construct-containing discrete voxel of single cells simplifies the cellular cross-talk. In this section, we explore case studies of single-cell patterns, organized by general application.

4.5.1 ISOLATED-NODE, SINGLE-CELL ARRAYS

The simplest single-cell arrays are ones wherein each cell is a standalone node for independent analysis. Neighboring cells, isolated by physical or spatial means, cause negligible effects on an individual cell's response to presented external cues. These types of arrays are particularly useful for studying cell plasticity in response to external stimuli or as screening tools to identify unique cells in a heterogeneous population. In the following cases, cell isolation and immobilization may be done in parallel. The ability to perform subsequent analysis in a high-throughput manner is critical. By arranging cells into organized arrays, high-throughput analysis can be done in an efficient manner.

In cases where arrays are populated by clonally similar cells, studies focus on individual cell response to combinatorial external stimuli (e.g. extracellular matrix, growth factors, and therapeutics). For example, correlation of cell-matrix interactions to biomechanical transduction events caused by *in situ* stresses localized at the cellular boundaries benefits from single-cell resolving technologies. These fundamental studies utilize chemical micropatterning of surfaces to dually generate attachment islands of various geometries and present combinatorial extracellular matrix cues for assays that profile cell adhesion, morphology, and cytoskeletal stresses (Kuschel et al., 2006; Gallant et al., 2005; Théry 2010). In general, these differential-surface assays contain islands 2–20 μm in diameter to bind single cells. Recent work by Ma et al. (2013) constrained cells in 3D volumes by laser-guided micropatterning to study contractile functions of differentiated mesenchymal stem cells. Subsequent analysis, ranging from fluorescence image-based analysis of mechanical stress fibers to AFM-tip force probing, identified distinctive extracellular matrix components that may serve therapeutic targets in disease states.

Although static adhesion islands are suitable for cell-matrix adhesion assays, additional complexity is required for studies that investigate dynamic cell interaction with soluble cues in an external microenvironment, such as growth factors or therapeutic drugs. Microfluidic device-based assays provide the means to organize individual cells and directly interrogate them with a combinatorial library of soluble factors. High-throughput, single-cell resolving is achieved by physically catching cells in sized chambers against the flow stream. Because microfluidic platforms utilize flow rates in the laminar region (Reynolds number < 1000), gradual mixing allows for continuous presentation of defined chemical gradients to cell arrays (Carlo et al., 2006; ChunHong Zheng et al., 2012). To date, single-cell arrays have been used to monitor cellular-level response to growth, cytokines signaling, migratory, and metabolite factors (Grecco et al., 2010; Köstler et al., 2013; Cheng et al., 2010). As a natural extension and spin-off, such devices have also been used for drug screening and resistance studies (Kuss et al., 2013).

In cases where heterogeneous cell populations must be sorted, single-cell arrays provide ideal platforms for identification and subsequent processing. For example, adult stem and cancer stem cell populations consist of heterogeneous constituent cells. Some cells differentiate into new tissues while others are more likely to differentiate into metastatic lesions. Fractionation of such heterogeneous populations down to the single-cell level elucidates the internal cellular mechanism that gives cells their unique phenotype. Nontrivial isolation and identification of pluripotent stem cells requires high-throughput genetic analysis to process clonally unique cells. Generally, microfluidic devices are used to capture single stem or cancer cells and allow for *in situ* lysing for genetic analysis (Wilson et al., 2014; Liang et al., 2013; Wood et al., 2010; Czyż et al., 2014; Liu et al., 2013).

4.5.2 NETWORK-LEVEL, SINGLE-CELL ARRAYS

Homotypical and heterotypical cell–cell communication dictates events that regulate tissue function, and disruption of this cellular signaling mechanisms leads to disease states (Pirlo et al., 2011). However, interrogation of desired cell–cell interactions may be hampered by background noise produced by extraneous cellular regulating events. Omission of these irrelevant processes and deposition of desired cells into a pattern drastically simplifies the nature of studied cross-talk into well-defined, intentional reciprocal interactions. Unlike single-cell arrays in which each node is independent contact with the external environment (previous section), in network-level single-cell arrays, cells interface with each other and produce interdependent responses that cumulate in network-level behavior. Thus the two most important parameters are: (1) identity of the neighboring cells, as the different cells elicit different interactions, and (2) distance between individual cells, as distance defines the mode of cellular communication (direct contact vs. paracrine signaling). Two research areas that utilize single-cell arrays are cancer invasion and neural networks.

Tumor invasion is a critical step in cancer progression to lethal metastatic disease, and understanding the cancer-stroma interface will provide clues to new therapeutic targets. Stroma is connective and supportive tissue. Fabricating artificial tissue interfaces at the single-cell level allows for the study of high-resolution cell–cell interactions at the micrometer length scale. Microfluidic and micropatterned devices have been fabricated to study both paracrine signaling (Hong et al., 2012) and direct contact events (Frimat et al., 2011; Zhang et al., 2014; Nikkhah et al., 2011), but are limited to isolated pair-wise interactions. Direct-write tools, such as laser-based bioprinters, are advantageous in depositing cells over a homogeneous area to study network-level behavior. Printing cells onto homogeneous substrates ensures that the development of the construct is influenced solely by guidance cues from neighboring cells, rather than limited by physical chambers or adhesion islands. In addition, there has been a push toward 3D tissue models to better study the *in vivo* cancer-stroma cross-talk. Several researchers have demonstrated the patterning of single hydrogel microbeads containing cancer cells into small-scale patterns (Dolatshahi-Pirouz et al., 2014; Phamduy et al., 2012). Larger patterns have also been generated by coprinting 3D hydrogel biomaterials and cells (Gruene et al., 2011; Kingsley et al., 2013).

In monoculture neural networks, guiding spontaneous synaptic connectivity requires defined spacing and pathways between individual cells. Several schemes have been developed toward this end. Dinh et al. (2013) generated high-density interconnected circuits of compartmentalized neurons in separate microfluidic chambers, with a biomaterials-guided outgrowth path between pairs of cells. Sanjana and Fuller (2004) and Macis et al. (2007) printed adhesion islands onto nonadherent substrates to allow for both cell attachment and directed outgrowth. In such examples, cell bodies are stationary and outgrowth occurs due to physical guides. Difato et al. (2011) utilized a laser-based optical tweezers technique to generate neural network patterns on homogeneous surfaces, which allows for cell migration in addition to synaptic development due to soluble guidance cues. Extracellular recording of action potential stimulation reveals that increasing internodal distance accelerates signal propagation (Wu et al., 2012), but decreases signal propagation efficiency (James et al., 2004). Single-neuron patterns with defined synaptic networks open up new avenues for fundamental biology studies and neural sensor/actuator applications.

4.5.3 NEXT-GENERATION SINGLE-CELL ARRAYS: INTEGRATED, COMPUTATION-DRIVEN ANALYSIS

As single-cell arrays are generally compared against a combinatorial library of soluble cues or first-neighbor cells, computational tools could be used to predict or validate empirical data in next-generation platforms. Standalone technology for high-throughput single-cell analysis currently exists. However, integration of existing technologies with single-cell arrays, especially in the context of network arrays, would produce a synergistic platform that is greater than the sum of its parts. Such improved platforms would provide *in situ*, multiplexed analysis of cell behavior, including morphology, biochemical state (Burguera et al., 2010), biomechanical stresses, genetic (Vanneste et al., 2012), metabolic, and migration (Chunhong Zheng et al., 2012). Information gleaned from arrays can then be compared to *in vivo* observations by computation analysis simply by compiling a database of single-cell array responses. This is the type of unambiguous testing that drives research scientists to push the boundaries of precision bioprinting technologies.

4.5.4 EXAMPLE SINGLE-CELL ARRAY VIA MAPLE-DW

Figure 4.10 is a 4 × 4 array pattern (800 µm center-to-center spacing) generated by MAPLE-DW and populated by single MDA-MB-231 human breast cancer cells (Phamduy, in press). The phase contrast image was taken immediately after the printing process, so that the transferred gelatin droplets appear raised on the gelatin-coated substrate. In addition, each droplet contains single cells in the trypsinized state that will eventually attach to the underlying substrate surface.

4.5.5 LASER DIRECT WRITE FOR NEURONS

The nervous system represents the most complicated organ in the body, controlling all higher- order faculties. As Tim Berners-Lee stated in 1999, "All that we know, all that we are, comes from the way our neurons are connected." Its fascinating complexity makes it one of the most difficult systems to study. Many advances have been made in studying population level events in the brain (e.g. fMRI,

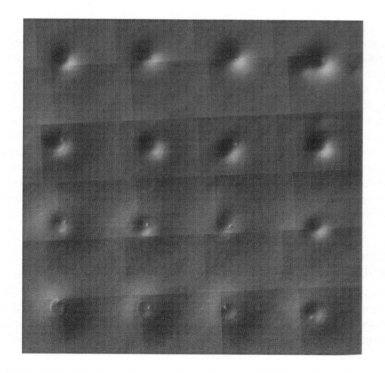

FIGURE 4.10

Single-cell array via LDW.

MEG, and PET), but mechanistic approaches to ascertaining fundamental neurobiology at the cellular level are limited, preventing a complete understanding of the governing order.

As previously mentioned, the ability to study neuronal cell behavior in a separated manner has value to understanding both environmental and cell–cell interactions that regulate mammalian systems. Interest in such biological mechanisms spans the breadth of subjects from developmental neuroscience to functioning electrophysiology, with the added benefit of understanding and treating pathological disease states. Not until researchers can tease apart each component will we be able to truly appreciate how all of the nervous system's complex pieces fit together. Currently, to examine neurons *in vitro*, the most commonly available methods involve organotypic slices or dissociated cell cultures. Studying individual neuronal characteristics within the context of a defined slice of living brain circuitry has led to some incredible insights, but the inability to parse out individual influences undermines the capacity to examine distinct mechanistic components (Gahwiler et al., 1997; Shankar et al., 2007; Yamaguchi et al., 2003). Dissociated cell cultures, on the other hand, offer more control over cell types and the influences of extracellular factors (Potter and DeMarse, 2001; Shahaf and Marom, 2001), while inherently negating the complex spatial organization present *in vivo*. While some control of cell-level interactions can be modulated through seeding density, the random nature of dissociated cultures limits a researcher's ability to define specific cell–cell interactions in a contextual manner.

Some of the techniques to engineer both isolated node and network-level single-cell arrays were briefly described earlier. These technologies involve either "trapping" cells through patterned adhesion molecules or microfluidics, or "printing" cells through extrusion or ink-jet deposition. Again, limitations currently exist on spatial accuracy and cell specificity, with study of nuanced interactions hinging on the ability to control the manner in which distinct cell types interact. To that end, organized patterns of both 2D and 3D printed neuronal cells using laser direct write has been demonstrated, with cells maintaining the ability to sprout axons (Patz et al., 2006). Laser direct write addresses advantages for both shortcomings in a manner that has so far been underutilized for neural applications, despite the apparent need.

4.5.1.1 Neural Development

Early development of the nervous system is a highly organized and multifaceted process in which environmental and extracellular inputs, including biochemical and mechanical cues, govern the emergence of increasingly differentiated cell types. The first experiment to show that cells influence the fate of other cells was performed in 1924 using a crude method to separate and grow developing neuronal cells (Spemann and Mangold, 2001). Since then, the use of bioprinting for precise control of cell-environment interaction has facilitated studies of neural stem cell differentiation into neurons, astrocytes, and oligodendrocytes. Ikhanizadeh et al. demonstrated that guidance of neural stem cell fate was possible through ink-jet printing (Ilkhanizadeh et al., 2007). Additionally, 3D printed microarrays of embryonic stem cells were able to characterize factors influencing neural commitment in a high-throughput capacity (Fernandes et al., 2010). Lastly, laser micromachining of growth substrates was shown to grow, proliferate, and differentiate neuroblasts in reproducible and controllable patterns (Doraiswamy et al., 2006). The combination of environmental manipulation with single-cell printing would allow combinatorial studies of cell–cell and cell-environment interactions on differentiation pathways.

Similarly, targeted guidance of neuronal processes to their proper functional targets is an amazingly precise and intricate process governed by attractive and repulsive signals. The sprouting axon of each developing neuron samples its surroundings and makes directional decisions through its growth cone. By engineering growth substrates, an increased understanding of how axons interact with the extracellular environment has begun to develop, including how individual molecular guidance cues affect guidance (McCormick and Leipzig, 2012; Yu et al., 2008). Where single-cell printing can add value is recapitulating the numerous external cellular "guideposts" each axon encounters along its journey, to study the multitude of cellular interactions that coalesce during development.

4.5.1.2 Engineered Circuits

The ability to fabricate highly specific patterns of neuronal cells has the potential to lend insight into many of the functional units of the brain and beyond. Xu et al. were the first to demonstrate that printed neuronal cells retained healthy electrophysiological characteristics (Xu et al., 2006). The subsequent ability to engineer neuronal circuits at the single-cell level was demonstrated by Edwards et al. with hippocampal neurons (Edwards et al., 2013). In a similar vein, techniques have been demonstrated to control individual synaptic connectivity of larger homogenous neuronal populations, enabling specific control over network and circuit properties (Staii et al., 2009; Vogt et al., 2005).

These methods do not allow for incorporation of multiple cell types, severely limiting their ability to recapitulate critical heterogeneous cell synapses. The advantages of laser bioprinting to control the precise location and association of multiple cell types, in combination with hydrogel and biomolecular-based models for controlled axonal outgrowth, would allow the creation of mono and polysynaptic circuits between heterogeneous neuronal populations (Curley and Moore, 2011; Horn-Ranney et al., 2013). Laser direct write represents the only technology currently available to accomplish this complicated feat, and the ability to recreate neural tracts of virtually any system at the single- and multicell level would allow for unprecedented experiments into the function of higher-order activities such as memory, learning, cognition, locomotion, and pain.

Highly specific brain regions are known to integrate and communicate through transient synapses, facilitating the signaling that encodes and dictates all activity. Further complicating matters, these neuronal connections are constantly being strengthened, weakened, or lost based on external outputs. For example, within the hippocampus, multiple cell types are known to control memory and spatial navigation (Dickerson and Eichenbaum, 2010; Turrigiano, 2012). Similarly, thalamocortical circuits govern cognition, behavior, and consciousness, with clinical implications in autism, schizophrenia, and attention deficit disorder (Buonanno, 2010; Sudhof, 2008). Sensorimotor neuron integration is another wide ranging pathway that influences motor control and proprioception as well as all of the five classic senses (Goulding et al., 2002). The ability to understand and engineer highly specific sensorimotor circuits also has implications in advancing prosthetic design to interface with native tissue and restore near-natural function (Guo et al. 2012; Raspopovic et al., 2014). Another system with implication in limb prostheses is the neuromuscular junction. Das et al. have defined and explored *in vitro* models for synapse formation, though to date only in heterogeneous dissociated cultures (Das et al., 2010, 2007). In both cases, precisely how synaptic plasticity influences information processing is not yet understood. All of these systems are highly complex and divergent across applications, but laser direct write could be used to manipulate long-term studies of cellular, genetic, and molecular influences on individual synapse formation and maintenance. Due to the stochastic nature of these systems, the high-throughput potential of laser direct write will be critical to increasing our understanding of normal function and will inform treatment of pathological states.

4.5.1.3 Non-neuronal Interactions

The consequences of studying engineered interactions between neuronal and non-neuronal cells also have application outside of synaptic function. Though the number is still contested, there are at least as many non-neuronal as neuronal cells in the brain (Azevedo et al., 2009). Clearly supportive glial cells are critical to nervous system function, but the intricacies of their interactions are not well characterized. Schwann cell and oligodendrocyte myelination of axons is required for physiologically relevant conduction of nerve impulses, and astrocyte–neuron communication influences both neuronal differentiation and synapse formation (Sherman and Brophy, 2005; Sofroniew and Vinters, 2010). Mirroring the sentiments of previous applications, glia cells are both necessary for natural neuronal function and responsible for many pathogenic states. Implications in Alzheimer's, Parkinson's, multiple sclerosis, neuropathy, and chronic pain have been tied to aberrant glial function (Antony, 2014; Block et al., 2007; Luongo et al., 2014; McMahon and Malcangio, 2009). In order to study such interactions, Park et al. constructed a microfluidic device to isolate axonal processes and enable coculture with both oligodendrocyte precursor and astrocyte cells (Park et al., 2012). They then saw evidence of healthy oligodendrocyte differentiation and myelination, alongside induced astrocytic and biomolecular damage. Goudriaan et al. also demonstrated the reciprocal nature of genetic disturbances in astrocyte–neuron interactions through a novel cellular isolation method (Goudriaan et al., 2014). Both tools represent powerful examples of potentially high-throughput studies in neuron–glia interaction. However, the speed and spatial resolution of laser direct write and CAD/CAM technology would allow for the rapid manipulation of multiple-cell juxtaposition to define mechanisms for axon–glia communication and interaction. The combination of throughput, flexibility, and accuracy would enable experiments to define and explore contact- and signaling-related influences toward understanding both healthy and pathological states.

4.5.1.4 Outlook

Through the examples given, it is clear that advances in technology have fundamentally increased our understanding of the nervous system. By incrementally deconstructing the intricate milieu of factors which govern the organization of the brain, both from a developmental and functional standpoint, researchers are beginning to recognize and understand the consequences of particular biological mechanisms. Engineered models utilizing laser direct write techniques would enable the manipulation of molecular, cellular, and synaptic contributions from a single-cell level, representing the next advancement required to decipher implication on emergent behavior and inform potential treatment strategies.

4.6 CONCLUSION

Laser bioprinting has become a proven method for biomaterial transfer and it is the only bioprinting approach having single-cell resolution. Because of the unique combination of optical imaging and CAD/CAM control, researchers have created a platform capable of fabricating complex constructs from the bottom up with the potential for high-throughput biological construct fabrication, for example, reproducible tissue constructs to the single-cell-level resolution to test different drugs with iterative composition and/or spacing manipulations allowing unambiguous conclusions about the influence of the cellular microenvironment. The MAPLE-DW platform, especially, has demonstrated greater than 5 μm spatial precision with post-transfer cell viability consistently greater than 90%. Researchers have achieved viable transfers with all types of cells enabling an even greater number of applications. Moreover, the short time required to make constructs coupled with conventional cell culture protocol and ease of printing make laser bioprinting with MAPLE-DW an efficient, feasible, and reliable way to isolate and study cellular interactions and behaviors. Research models created by laser direct-write for cancer, stem cell, and neural studies show promise for the future, and are a disruptive technology in tissue engineering.

Some of the major challenges facing bioprinting and tissue engineering, which must be addressed in the next generation of bioprinting systems, include optimizing the protocol, scale up, and imaging. Laser-assisted single-cell printing enables more reproducible studies, but also hinders large-scale tissue engineering until comprehensive pattern recognition and automation are realized. Next-generation laser-assisted bioprinters will have single-cell print resolution and demonstrate parallelization, to show that scaling up these systems is possible and eventually can be used for large-scale tissue engineering as well as clonal studies.

Improved imaging techniques will not only help researchers monitor and analyze deposition *in situ*, but also will be incorporated into automated data mining systems that will track deposition successes and failures, and associate each transfer attempt with printing metadata. Bioprinting metadata will include print conditions (e.g. temperature and humidity) and parameters (e.g. laser fluence and pulse duration) and be stored in a database with information on cell phenotypes, transfer precision, and morphological changes. Using data analysis, mining, and visualization techniques, users will be able to learn from each cell transfer, accelerate system improvements, and incorporate embedded system and construct statistics in subsequent studies with the cell constructs.

Next-generation laser bioprinting systems need to build on current success and ***continue*** to look to other disciplines for inspiration. This approach will help push single-cell laser-assisted bioprinting to a point of enhanced functionality, reproducibility, and parallelization. Single-cell printing is crucial but needs to be done with a systems engineering approach that has an eye turned toward future applications and problems, such as vascularization in tissue engineering. The MAPLE-DW platform advances this effort beyond other currently available technologies and demonstrates how generations of laser bioprinting systems evolve and improve to enact cutting-edge research technology.

REFERENCES

Antony, J.M., 2014. Neuropathogenesis: rogue glia cause mayhem in the brain. *Translational Neuroscience*, 5(1), pp.91-98. Available at: <Go to ISI > ://000333709000011.

Azevedo, F.A.C. et al., 2009. Equal numbers of neuronal and nonneuronal cells make the human brain an isometrically scaled-up primate brain. *Journal of Comparative Neurology*, 513(5), pp.532–541. Available at: <Go to ISI > ://000263981100007.

Barron, J. a., Krizman, D.B. & Ringeisen, B.R., 2005. Laser printing of single cells: statistical analysis, cell viability, and stress. *Annals of Biomedical Engineering*, 33(2), pp.121–130. Available at: http://link.springer.com/10.1007/s10439-005-8971-x. [Accessed February 26, 2014].

Birtwistle, M.R. et al., 2012. Emergence of bimodal cell population responses from the interplay between analog single-cell signaling and protein expression noise. *BMC Systems Biology*, 6(1), p.109. Available at: http://www.biomedcentral.com/1752-0509/6/109 [Accessed May 23, 2014].

Block, M.L., Zecca, L. & Hong, J.S., 2007. Microglia-mediated neurotoxicity: uncovering the molecular mechanisms. *Nature Reviews Neuroscience*, 8(1), pp.57–69. Available at: <Go to ISI > ://000242994200016.

Buonanno, A., 2010. The neuregulin signaling pathway and schizophrenia: from genes to synapses and neural circuits. *Brain Research Bulletin*, 83(3-4), pp.122–131. Available at: <Go to ISI > ://000283039300006.

Burguera, E.F., Bitar, M. & Bruinink, A., 2010. Novel *in vitro* co-culture methodology to investigate heterotypic cell-cell interactions. *European Cells & Materials*, 19, pp.166-79. Available at: http://www.ncbi.nlm.nih.gov/pubmed/20419629.

Carlo, D. Di, Wu, L.Y. & Lee, L.P., 2006. Dynamic single-cell culture array. *Lab on a Chip*, 6(11), pp.1445–1449. Available at: http://dx.doi.org/10.1039/B605937F.

Cheng, W. et al., 2010. Microfluidic cell arrays for metabolic monitoring of stimulated cardiomyocytes. *Electrophoresis*, 31(8), pp.1405–13. Available at: http://www.ncbi.nlm.nih.gov/pubmed/20333720 [Accessed June 13, 2014].

Curley, J.L. & Moore, M.J., 2011. Facile micropatterning of dual hydrogel systems for 3D models of neurite outgrowth. *Journal of Biomedical Materials Research. Part A*, 99(4), pp.532–43. Available at: http://www.ncbi. nlm.nih.gov/entrez/query.fcgi?cmd(Retrieve&db(PubMed&dopt(Citation&list_uids(21936043   [Accessed June 15, 2014].

Czyż, Z.T. et al., 2014. Reliable single cell array CGH for clinical samples. *PloS one*, 9(1), p.e85907. Available at: http://www.pubmedcentral.nih.gov/articlerender.fcgi?artid(3897541&tool(pmcentrez&rendertype(abstract [Accessed June 6, 2014].

Das, M. et al., 2010. A defined long-term *in vitro* tissue engineered model of neuromuscular junctions. *Biomaterials*, 31(18), pp.4880–4888. Available at: <Go to ISI > ://000277783100012.

Das, M. et al., 2007. Embryonic motoneuron-skeletal muscle coculture in a defined system. *Neuroscience*, 146(2), pp.481–488. Available at: <Go to ISI > ://000246535000001.

Dickerson, B.C. & Eichenbaum, H., 2010. The episodic memory system: neurocircuitry and disorders. *Neuropsychopharmacology*, 35(1), pp.86–104. Available at: <Go to ISI > ://000272784600006.

Difato, F. et al., 2011. Combined optical tweezers and laser dissector for controlled ablation of functional connections in neural networks. *Journal of Biomedical Optics*, 16(5), p.051306. Available at: http://www.ncbi. nlm.nih.gov/pubmed/21639566 [Accessed June 13, 2014].

Dinh, N.-D. et al., 2013. Microfluidic construction of minimalistic neuronal cocultures. *Lab on a Chip*, 13(7), pp.1402–12. Available at: http://www.ncbi.nlm.nih.gov/pubmed/23403713 [Accessed June 5, 2014].

Dolatshahi-Pirouz, A. et al., 2014. A combinatorial cell-laden gel microarray for inducing osteogenic differentiation of human mesenchymal stem cells. *Scientific Reports*, 4, p.3896. Available at: http://www.pubmedcentral.nih. gov/articlerender.fcgi?artid(3905276&tool(pmcentrez&rendertype(abstract [Accessed June 13, 2014].

Doraiswamy, A. et al., 2006. Two-dimensional differential adherence of neuroblasts in laser micromachined CAD/CAM agarose channels. *Applied Surface Science*, 252(13), pp.4748-4753. Available at: http://www. sciencedirect.com/science/article/pii/S0169433205013425.

Edwards, D. et al., 2013. Two cell circuits of oriented adult hippocampal neurons on self-assembled monolayers for use in the study of neuronal communication in a defined system. *Acs Chemical Neuroscience*, 4(8), pp.1174–1182. Available at: <Go to ISI > ://000323535900006.

Fernandes, T.G. et al., 2010. Three-dimensional cell culture microarray for high-throughput studies of stem cell fate. *Biotechnology and Bioengineering*, 106(1), pp.106-118. Available at: <Go to ISI > ://000276844500011.

Frimat, J.-P. et al., 2011. A microfluidic array with cellular valving for single cell co-culture. *Lab on a Chip*, 11(2), pp.231–7. Available at: http://www.ncbi.nlm.nih.gov/pubmed/20978708 [Accessed June 2, 2014].

Gahwiler, B.H. et al., 1997. Organotypic slice cultures: a technique has come of age. *Trends in Neurosciences*, 20(10), pp.471–477. Available at: <Go to ISI > ://A1997XZ02800012.

Gallant, N., Michael, K. & García, A., 2005. Cell adhesion strengthening: contributions of adhesive area, integrin binding, and focal adhesion assembly. *Molecular Biology of the Cell*, 16(September), pp.4329–4340. Available at: http://www.molbiolcell.org/content/16/9/4329.short [Accessed June 13, 2014].

Goudriaan, A. et al., 2014. Novel cell separation method for molecular analysis of neuron-astrocyte co-cultures. *Frontiers in Cellular Neuroscience*, 8. Available at: <Go to ISI > ://000331054000001.

Goulding, M. et al., 2002. The formation of sensorimotor circuits. *Current Opinion in Neurobiology*, 12(5), pp.508–515. Available at: <Go to ISI > ://000178693000006.

Grecco, H.E. et al., 2010. *In situ* analysis of tyrosine phosphorylation networks by FLIM on cell arrays. *Nature Methods*, 7(6), pp.467–72. Available at: http://www.ncbi.nlm.nih.gov/pubmed/20453867 [Accessed June 6, 2014].

Gruene, M. et al., 2011. Laser printing of three-dimensional multicellular arrays for studies of cell-cell and cell-environment interactions. *Tissue Engineering. Part C, Methods*, 17(10), pp.973–82. Available at: http://online.liebertpub.com/doi/abs/10.1089/ten.TEC. 2011.0185.[Accessed June 9, 2014].

Guo, X.F. et al., 2012. Tissue engineering the monosynaptic circuit of the stretch reflex arc with coculture of embryonic motoneurons and proprioceptive sensory neurons. *Biomaterials*, 33(23), pp.5723–5731. Available at: <Go to ISI > ://000305597100007.

Hong, S., Pan, Q. & Lee, L.P., 2012. Single-cell level co-culture platform for intercellular communication. *Integrative Biology*, 4(4), pp.374–380. Available at: http://dx.doi.org/10.1039/C2IB00166G.

Hopp, B. et al., 2005. Survival and proliferative ability of various living cell types after laser-induced forward transfer. *Tissue Engineering*, 11(11-12), pp.1817–23. Available at: http://www.ncbi.nlm.nih.gov/pubmed/16411827.

Hopp, B., Smausz, T. & Nógrádi, A., 2010. Absorbing-film assisted laser induced forward transfer of sensitive biological subjects. In B. R. Ringeisen, B.J. Spargo, & P. K. Wu, eds. *Cell and Organ Printing SE - 7*. Springer Netherlands, pp. 115–134. Available at: http://dx.doi.org/10.1007/978-90-481-9145-1_7.

Horn-Ranney, E.L. et al., 2013. Structural and molecular micropatterning of dual hydrogel constructs for neural growth models using photochemical strategies. *Biomedical Microdevices*, 15(1), pp.49–61. Available at: http://www.ncbi.nlm.nih.gov/entrez/query.fcgi?cmd(Retrieve&db(PubMed&dopt(Citation&list_uids(22903647 [Accessed June 15, 2014].

Ilkhanizadeh, S., Teixeira, A.I. & Hermanson, O., 2007. Inkjet printing of macromolecules on hydrogels to steer neural stem cell differentiation. *Biomaterials*, 28(27), pp.3936–3943. Available at: http://www.ncbi.nlm.nih.gov/entrez/query.fcgi?cmd(Retrieve&db(PubMed&dopt(Citation&list_uids(17576007.

James, C.D. et al., 2004. Extracellular recordings from patterned neuronal networks using planar microelectrode arrays. *IEEE Transactions on Biomedical Engineering*, 51(9), pp.1640–8. Available at: http://www.ncbi.nlm.nih.gov/pubmed/15376512.

Kingsley, D.M. et al., 2013. Single-step laser-based fabrication and patterning of cell-encapsulated alginate microbeads. *Biofabrication*, 5(4), p.045006. Available at: http://www.ncbi.nlm.nih.gov/pubmed/24192221 [Accessed June 13, 2014].

Köstler, W.J. et al., 2013. Epidermal growth-factor-induced transcript isoform variation drives mammary cell migration. *PloS one*, 8(12), p.e80566. Available at: http://www.pubmedcentral.nih.gov/articlerender.fcgi?artid(3855657&tool(pmcentrez&rendertype(abstract [Accessed June 13, 2014].

Kuschel, C. et al., 2006. Cell adhesion profiling using extracellular matrix protein microarrays. *BioTechniques*, 40(4), pp.523-531. Available at: http://www.biotechniques.com/article/06404RR03 [Accessed June 13, 2014].

Kuss, S. et al., 2013. Assessment of multidrug resistance on cell coculture patterns using scanning electrochemical microscopy. *Proceedings of the National Academy of Sciences of the United States of America*, 110(23), pp.9249–54. Available at: http://www.pubmedcentral.nih.gov/articlerender.fcgi?artid(3677433&tool(pmcentrez&rendertype(abstract [Accessed June 10, 2014].

Lewinski, N., Colvin, V. & Drezek, R., 2008. Cytotoxicity of nanoparticles. *Small (Weinheim an der Bergstrasse, Germany)*, 4(1), pp.26–49. Available at: http://www.ncbi.nlm.nih.gov/pubmed/18165959 [Accessed May 24, 2014].

Liang, P. et al., 2013. Drug screening using a library of human induced pluripotent stem cell-derived cardiomyocytes reveals disease-specific patterns of cardiotoxicity. *Circulation*, 127(16), pp.1677–91. Available at: http://www.pubmedcentral.nih.gov/articlerender.fcgi?artid(3870148&tool(pmcentrez&rendertype(abstract [Accessed May 27, 2014].

Liberski, A.R., Delaney, J.T., Schubert, U.S., 2011. One cell-one well: a new approach to inkjet printing single-cell microarrays. ACS Combinatorial Science 13 (2), 190–195.

Lin, Y. et al., 2010. Effect of laser fluence in laser-assisted direct writing of human colon cancer cell. *Rapid Prototyping Journal*, 16(3), pp.202–208. Available at: http://www.emeraldinsight.com/10.1108/13552541011034870 [Accessed June 26, 2014].

Liu, Y. et al., 2013. Development of a single-cell array for large-scale DNA fluorescence *in situ* hybridization. *Lab on a Chip*, 13(7), pp.1316–24. Available at: http://www.pubmedcentral.nih.gov/articlerender.fcgi?artid(35945 24&tool(pmcentrez&rendertype(abstract [Accessed June 13, 2014].

Luongo, L., Maione, S. & Di Marzo, V., 2014. Endocannabinoids and neuropathic pain: focus on neuron-glia and endocannabinoid-neurotrophin interactions. *European Journal of Neuroscience*, 39(3), pp.401–408. Available at: <Go to ISI > ://000330558300008.

Ma, Z. et al., 2013. Laser patterning for the study of MSC cardiogenic differentiation at the single-cell level. *Light, Science & Applications*, 2(January). Available at: http://www.pubmedcentral.nih.gov/articlerender.fcgi?artid(3920285&tool(pmcentrez&rendertype(abstract [Accessed June 15, 2014].

Macis, E. et al., 2007. An automated microdrop delivery system for neuronal network patterning on microelectrode arrays. *Journal of Neuroscience Methods*, 161(1), pp.88–95. Available at: http://www.sciencedirect.com/science/article/pii/S0165027006005115.

McCormick, A.M. & Leipzig, N.D., 2012. Neural regenerative strategies incorporating biomolecular axon guidance signals. *Annals of Biomedical Engineering*, 40(3), pp.578-597. Available at: <Go to ISI > ://000300770200003.

McMahon, S.B. & Malcangio, M., 2009. Current challenges in glia-pain biology. *Neuron*, 64(1), pp.46–54. Available at: <Go to ISI > ://000271454400010.

Nikkhah, M. et al., 2011. MCF10A and MDA-MB-231 human breast basal epithelial cell coculture in silicon microarrays. *Biomaterials*, 32(30), pp.7625–7632. Available at: http://www.sciencedirect.com/science/article/pii/S014296121100706X.

Paltauf, G., Schmidt-Kloiber, H. & Frenz, M., 1998. Photoacoustic waves excited in liquids by fiber-transmitted laser pulses. *The Journal of the Acoustical Society of America*, 104(2), pp.890–897. Available at: http://scitation.aip.org/content/asa/journal/jasa/104/2/10.1121/1.423334.[Accessed June 13, 2014].

Park, J. et al., 2012. Multicompartment neuron-glia coculture platform for localized CNS axon-glia interaction study. *Lab on a Chip*, 12(18), pp.3296–3304. Available at: <Go to ISI > ://000307583400010.

Patz, T.M. et al., 2006. Three-dimensional direct writing of B35 neuronal cells. *Journal of Biomedical Materials Research Part B-Applied Biomaterials*, 78B(1), pp.124–130. Available at: <Go to ISI > ://000238737300017.

Phamduy, T.B. et al., 2012. Laser direct-write of single microbeads into spatially-ordered patterns. *Biofabrication*, 4(2), p.025006. Available at: http://www.ncbi.nlm.nih.gov/pubmed/22556116 [Accessed November 12, 2013].

Phamduy, T.B., Corr, D.T. & Chrisey, D.B., 2010. Bioprinting. In *Encyclopedia of Industrial Biotechnology*. Hoboken, NJ, USA:;1; John Wiley & Sons, Inc., pp. 1–18. Available at: http://onlinelibrary.wiley.com/doi/10.1002/9780470054581.eib131/abstract;jsessionid(1D450AAE46B475ADAFEC4B6CF0D542B6.f02t0 3?deniedAccessCustomisedMessage(&userIsAuthenticated(false [Accessed June 24, 2014].

Piqué, A., Chrisey, D. & Auyeung, R., 1999. A novel laser transfer process for direct writing of electronic and sensor materials. *Applied Physics A*, 284, pp.279–284. Available at: http://link.springer.com/article/10.1007/s003390051400 [Accessed June 25, 2014].

Pirlo, R.K. et al., 2011. Laser-guided cell micropatterning system. *The Review of Scientific Instruments*, 82(1), p.013708. Available at: http://www.pubmedcentral.nih.gov/articlerender.fcgi?artid(3045411&tool(pmcentrez &rendertype(abstract [Accessed June 15, 2014].

Potter, S.M. & DeMarse, T.B., 2001. A new approach to neural cell culture for long-term studies. *Journal of Neuroscience Methods*, 110(1-2), pp.17–24. Available at: <Go to ISI > ://000171598000003.

Raspopovic, S. et al., 2014. Restoring natural sensory feedback in real-time bidirectional hand prostheses. *Science Translational Medicine*, 6(222), p.222ra19. Available at: http://www.ncbi.nlm.nih.gov/entrez/query.fcgi?cmd (Retrieve&db(PubMed&dopt(Citation&list_uids(24500407 [Accessed June 13, 2014].

Riggs, B.C. et al., 2011. Matrix-assisted pulsed laser methods for biofabrication. MRS Bulletin, 36(12), pp.1043–1050. Available at: http://www.journals.cambridge.org/abstract_S0883769411002764 [Accessed February 26, 2014].

Ringeisen, B.R., Ph, D., et al., 2004. Laser printing of pluripotent embryonal carcinoma cells. *Tissue Engineering*, 10(3).

Ringeisen, B.R., Kim, H., et al., 2004. Laser printing of pluripotent embryonal carcinoma cells. *Tissue Engineering*, 10(3-4), pp.483–91. Available at: http://www.ncbi.nlm.nih.gov/pubmed/15165465.

Sanjana, N.E. & Fuller, S.B., 2004. A fast flexible inkjet printing method for patterning dissociated neurons in culture. *Journal of Neuroscience Methods*, 136(2), pp.151-163. Available at: http://www.sciencedirect.com/science/article/pii/S016502700400024X.

Schiele, N., 2010. Gelatin-based laser direct-write technique for the precise spatial patterning of cells. *Tissue Engineering Part C: ...*, 17(3). Available at: http://online.liebertpub.com/doi/abs/10.1089/ten.tec.2010.0442 [Accessed July 9, 2014].

Schiele, N.R. et al., 2010. Laser-based direct-write techniques for cell printing. *Biofabrication*, 2(3), p.032001. Available at: http://www.ncbi.nlm.nih.gov/pubmed/20814088 [Accessed May 28, 2014].

Shackleton, M. et al., 2006. Generation of a functional mammary gland from a single stem cell. *Nature*, 439(7072), pp.84–8. Available at: http://www.ncbi.nlm.nih.gov/pubmed/16397499 [Accessed May 27, 2014].

Shahaf, G. & Marom, S., 2001. Learning in networks of cortical neurons. *Journal of Neuroscience*, 21(22), pp.8782–8788. Available at: <Go to ISI > ://000172012700012.

Shankar, G.M. et al., 2007. Natural oligomers of the Alzheimer amyloid-beta protein induce reversible synapse loss by modulating an NMDA-type glutamate receptor-dependent signaling pathway. Journal of Neuroscience, 27(11), pp.2866–2875. Available at: <Go to ISI > ://000245103600015.

Sherman, D.L. & Brophy, P.J., 2005. Mechanisms of axon ensheathment and myelin growth. *Nature Reviews Neuroscience*, 6(9), pp.683–690. Available at: <Go to ISI > ://000231591700012.

Smausz, T. et al., 2006. Study on metal microparticle content of the material transferred with absorbing film assisted laser induced forward rransfer when using silver absorbing layer. *Applied Surface Science*, 252(13), pp.4738–4742. Available at: http://linkinghub.elsevier.com/retrieve/pii/S0169433205014431 [Accessed June 25, 2014].

Sofroniew, M. V & Vinters, H. V, 2010. Astrocytes: biology and pathology. *Acta Neuropathologica*, 119(1), pp.7–35. Available at: <Go to ISI > ://000273174400003.

Spemann, H. & Mangold, H., 2001. Induction of embryonic primordia by implantation of organizers from a different species. 1923. *Int J Dev Biol*, 45(1), pp.13–38. Available at: http://www.ncbi.nlm.nih.gov/entrez/query.fcgi?cmd(Retrieve&db(PubMed&dopt(Citation&list_uids(11291841.

Staii, C. et al., 2009. Positioning and guidance of neurons on gold surfaces by directed assembly of proteins using atomic force microscopy. *Biomaterials*, 30(20), pp.3397–3404. Available at: <Go to ISI > ://000266744100005.

Sudhof, T.C., 2008. Neuroligins and neurexins link synaptic function to cognitive disease. *Nature*, 455(7215), pp.903–911. Available at: <Go to ISI > ://000260038300038.

Théry, M., 2010. Micropatterning as a tool to decipher cell morphogenesis and functions. *Journal of Cell Science*, 123(Pt 24), pp.4201–13. Available at: http://www.ncbi.nlm.nih.gov/pubmed/21123618 [Accessed May 27, 2014].

Turrigiano, G., 2012. Homeostatic synaptic plasticity: local and global mechanisms for stabilizing neuronal function. *Cold Spring Harbor Perspectives in Biology*, 4(1). Available at: <Go to ISI > ://000299165700003.

Vanneste, E. et al., 2012. New array approaches to explore single cells genomes. *Frontiers in Genetics*, 3(March), p.44. Available at: http://www.pubmedcentral.nih.gov/articlerender.fcgi?artid(3325760&tool(pmcentrez&rendertype(abstract [Accessed June 4, 2014].

Vogel, A. & Venugopalan, V., 2003. Mechanisms of pulsed laser ablation of biological tissues. *Chemical Reviews*, 103(2), pp.577–644. Available at: http://www.ncbi.nlm.nih.gov/pubmed/12580643.

Vogt, A.K. et al., 2005. Synaptic plasticity in micropatterned neuronal networks. *Biomaterials*, 26(15), pp.2549–2557. Available at: <Go to ISI > ://000226698400037.

Vukicevic, S. et al., 1992. Identification of multiple active growth factors in basement membrane matrigel suggests caution in interpretation of cellular activity related to extracellular matrix components. *Experimental Cell Research*, 202(1), pp.1–8. Available at: http://linkinghub.elsevier.com/retrieve/pii/001448279290397Q [Accessed July 24, 2014].

Wang, W. et al., 2008. Study of impact-induced mechanical effects in cell direct writing using smooth particle hydrodynamic method. *Journal of Manufacturing Science and Engineering*, 130(2), p.021012. Available at: http://manufacturingscience.asmedigitalcollection.asme.org/article.aspx?articleid(1452027.[Accessed June 13, 2014].

Wang, W., Huang, Y. & Chrisey, D., 2007. Numerical study of cell droplet and hydrogel coating impact process in cell direct writing. *Transactions of NAMRI/SME*, 35, pp.217–224. Available at: http://plaza.ufl.edu/yongh/publications/jpaper16.pdf [Accessed June 13, 2014].

Wang, W., Li, G. & Huang, Y., 2009. Modeling of bubble expansion-induced cell mechanical profile in laser-assisted cell direct writing. *Journal of Manufacturing Science and Engineering*, 131(5), p.051013. Available at: http://manufacturingscience.asmedigitalcollection.asme.org/article.aspx?articleid(1469096.[Accessed June 13, 2014].

Wang, W., Lin, Y. & Huang, Y., 2011. Modeling of thermoelastic stress wave in laser-assisted cell direct writing. *Journal of Manufacturing Science and Engineering*, 133(2), p.024502. Available at: http://manufacturingscience.asmedigitalcollection.asme.org/article.aspx?articleid(1470801 [Accessed June 13, 2014].

Wilson, J.L. et al., 2014. Single-cell analysis of embryoid body heterogeneity using microfluidic trapping array. *Biomedical Microdevices*, 16(1), pp.79–90. Available at: http://www.ncbi.nlm.nih.gov/pubmed/24085533 [Accessed May 28, 2014].

Wood, D.K. et al., 2010. Single cell trapping and DNA damage analysis using microwell arrays. *Proceedings of the National Academy of Sciences of the United States of America*, 107(22), pp.10008–13. Available at: http://www.pubmedcentral.nih.gov/articlerender.fcgi?artid(2890454&tool(pmcentrez&rendertype(abstract [Accessed May 26, 2014].

Wu, L.M.N. et al., 2012. Increasing internodal distance in myelinated nerves accelerates nerve conduction to a flat maximum. Current Biology, 22(20), pp.1957–1961. Available at: http://www.sciencedirect.com/science/article/pii/S0960982212009906.

Xu, T. et al., 2006. Viability and electrophysiology of neural cell structures generated by the inkjet printing method. *Biomaterials*, 27(19), pp.3580–3588. Available at: http://www.ncbi.nlm.nih.gov/entrez/query.fcgi?cmd(Retrieve&db(PubMed&dopt(Citation&list_uids(16516288.

Yamaguchi, S. et al., 2003. Synchronization of cellular clocks in the suprachiasmatic nucleus. *Science*, 302(5649), pp.1408–1412. Available at: <Go to ISI > ://000186683500056.

Yu, L.M.Y., Leipzig, N.D. & Shoichet, M.S., 2008. Promoting neuron adhesion and growth. *Materials Today*, 11(5), pp.36–43. Available at: <Go to ISI > ://000255691600021.

Zhang, K. et al., 2014. Block-cell-printing for live single-cell printing. *Proceedings of the National Academy of Sciences of the United States of America*, 111(8), pp.2948–53. Available at: http://www.ncbi.nlm.nih.gov/pubmed/24516129 [Accessed May 31, 2014].

Zheng, C. et al., 2012. An integrated microfluidic device for long-term culture of isolated single mammalian cells. *Science China Chemistry*, 55(4), pp.502–507. Available at: http://link.springer.com/10.1007/s11426-012-4493-1 [Accessed June 13, 2014].

Zheng, C. et al., 2012. Quantitative study of the dynamic tumor-endothelial cell interactions through an integrated microfluidic coculture system. *Analytical Chemistry*, 84(4), pp.2088–93. Available at: http://www.ncbi.nlm.nih.gov/pubmed/22263607.

ENGINEERING 2D AND 3D CELLULAR MICROENVIRONMENTS USING LASER DIRECT WRITE

Andrew D. Dias, David M. Kingsley and David T. Corr

Department of Biomedical Engineering, Rensselaer Polytechnic Institute, Troy, NY, USA

5.1 INTRODUCTION

5.1.1 SPATIAL INFLUENCES OF THE CELLULAR MICROENVIRONMENT

For tissue engineering and regenerative medicine, engineered cell microenvironments, or niches, provide a platform for directing cell fate and function. In other words, cellular behavior is influenced by the signaling of various stimuli provided by the local environment. Signals can be soluble, mechanical, and/or cellular, and the origin and transduction of each type of signal is potentially very broad and complex (Figure 5.1). The effect and/or potency of these signaling mechanisms can be influenced by distance, time, frequency, and concentration, among other variables. This concept has been reviewed extensively (Lund et al., 2009; Freytes et al., 2009; Godier et al., 2008), and can be applied to a variety of tissue engineering applications, such as creating cellular scaffolds/functional constructs, directing stem cell differentiation, and inducing desired cellular alignment, migration, proliferation, or protein production. There are numerous engineering approaches to produce microenvironments, such as introducing growth factors into cellular media, engineering substrates of desired material properties, creating 3D hydrogels, and delivering mechanical stimuli, such as flow, tension, or compression. All of these approaches take advantage of signaling mechanisms inherent in cells, such as cellular receptors for soluble signals and transduction through the cytoskeleton for mechanical signals. However, *in vivo* cellular communication is very sophisticated, and cells respond to more biochemical and mechanical signals than can typically be controlled *in vitro*. The mechanisms of cell-cell signaling are spatially dependent, where cells a short distance from each other can communicate by paracrine signaling, and cells immediately adjacent can communicate by direct cell-cell contact or juxtacrine signaling. One of the challenges facing emerging applications is how to harness this complex signaling to better direct cell fate and function *in vitro*. The cellular microenvironment is further complicated by cellular population dynamics, such as the composition and distance of a neighboring population of cells, size of populations/colonies, and of course, the types of cells within the environment. The fact that neighboring cells influence cell behaviors, and that intercellular spacing is a determining factor for the mode of

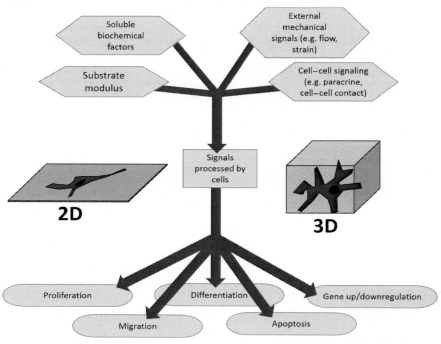

FIGURE 5.1

Cellular signaling schematic. Cells can receive a wide variety of input signals from the microenvironment, which they process to produce any number of outputs. The combination of signals can potentially be very complex.

communication, suggest that *spatial sensitivity and control* is highly desirable in many emerging tissue engineering and regenerative medicine applications.

In addition to cell placement, control over adsorbed/encapsulated proteins may have a profound effect on cellular behavior. Cell–cell and cell–extracellular matrix (ECM) interactions are spatially sensitive, and examples of how they influence cell signaling are shown in Figure 5.1. Whether the cellular microenvironments are planar or are 3D is also very important, since depending on the application, switching from 2D to 3D (or vice versa) can greatly impact cell signaling, function, and fate. In order to harness the cells' natural ability to process complex environmental signals, 2D/3D approaches for engineering cellular microenvironments must not only provide a degree of spatial control in fabrication, but must also control the density/concentration of cells, proteins, or other factors at these locations. Various printing/patterning techniques may be appropriate depending on the desired application, and we will briefly examine a few, weighing their potential advantages and disadvantages. We will pay particular attention to laser direct-write (LDW).

5.1.2 OVERVIEW OF PRINTING TECHNIQUES FOR ENGINEERING CELLULAR MICROENVIRONMENTS

A number of patterning, deposition, and printing techniques have been employed to achieve spatial control in engineered microenvironments. Various biologics, including proteins, nucleic acids, and even viable cells have been successfully deposited with spatial precision. Some of the early work for cell patterning involved controlled spatial adsorption of an adhesive protein, with subsequent seeding of cells that preferentially grow on the patterned protein. While this sort of patterning was first demonstrated using lithography-based techniques such as microcontact printing (Chen et al., 1998; Tien et al., 2002), other deposition methods like ink-jet printing (reviewed in (Calvert, 2001)) and LDW (Ringeisen et al., 2002a; Wu et al., 2003; Colina et al., 2005) have also shown successful protein or nucleic acid deposition. Moreover, ink-jet printing (Xu et al., 2005; Roth et al., 2004; Saunders et al., 2008) and LDW (Odde et al., 2000; Pirlo et al., 2006; Barron et al., 2004a; Schiele et al., 2010; Schiele et al., 2011) have been used to deposit viable mammalian cells directly to a homogeneous substrate, without requiring the prior patterning of a protein. There are other patterning techniques available, such as dip-pen nanolithography (Piner et al., 1999) or AFM-based patterning (Xie et al., 2006), that focus on patterning at submicron scales. Cells certainly sense nanoscale features in their environments, but in order to study and produce cellular microenvironments where cell- and population-level interactions are controlled, patterning at the micro- and mesoscale may be most relevant for directing cellular signaling. However, patterning at this scale can be quite challenging; it is too large to be accomplished directly by chemistry, and too small to use many traditional fabrication methods.

Some methods that have proven suitable for patterning at this scale include microcontact printing, ink-jet printing, and LDW. Each of these has its own specific advantages and disadvantages, and they are quite complementary. Microcontact printing employs a stamp with relief features attained via photolithography. The resolution of the pattern is limited only by the wavelength of light, making submicron resolution attainable with this technique. By contrast, LDW is a noncontact technique that propels material to a substrate by laser energy absorption and partial volatilization of a sacrificial layer. The spatial location of the transferred material is determined by the programmed position of a computer-aided design/computer-aided manufacturing (CAD/CAM) stage. The resolution of this technique can be under 10 μm (Zhang et al., 2003), because of the dynamics of the material transfer event and controlled stage movement. Ink-jet printing, by comparison, has a printing resolution on the order of 50 μm (Calvert, 2001), which is appreciably lower but still very good. Material deposition is achieved by one of several methods of propulsion through a nozzle (Saunders et al., 2008; Boland et al., 2006; Lee et al., 2009a; Gonzalez-Macia et al., 2010), and both the size of the nozzle and method of material ejection can influence printing resolution.

Because of their different mechanisms and printing resolutions, each of these methods is particularly well suited for a specific subset of applications, summarized in Table 5.1. Micropatterning excels at creating patterns of proteins on 2D flat, or even curved (Jackman et al., 1995) surfaces. Cells can be seeded on adhesive proteins, allowing their behavior in response to the protein or protein patterns to be studied. However, once the pattern of proteins is set, it is immobilized, and cells generally do not proliferate

Table 5.1   Patterning techniques and applications

Technique	Resolution	Application
Micropatterning	<1 μm	Patterns of adhesive proteins in 2D and cell culture on patterned adhesion islands. Study of cell behavior in controlled population or cell size and/or geometry. New mask must be fabricated for each new pattern.
Ink-jet printing	~50 μm	High-throughput or large constructs in 2D or 3D on any suitable substrate. Direct patterning of cells or less viscous materials.
LDW	~10 μm	Patterning of cells or material in 2D or 3D on any suitable flat substrate. Direct patterning of cells with high spatial resolution. More viscous materials can be patterned, but with lower throughput than ink-jet printing.

beyond the adhesion islands. Moreover, controlled coculture is difficult, because different cell types, if seeded simultaneously, will all adhere to the protein islands. Coculture patterning (reviewed in (Kaji et al., 2011)) requires either sequential seeding on micropatterns, masking the substrate, switching regions of the substrate to be favorable to cell binding (Yamato et al., 2001; Yousaf et al., 2001), or combining multiple substrates (Hui and Bhatia S, 2007).

On the other hand, ink-jet printing and LDW do not require patterning of adhesive proteins to control cellular placement. Therefore, cells can be directly deposited to a homogeneous substrate, allowing evolution of a printed structure from a prescribed initial condition. Furthermore, both ink-jet printing and LDW have also been extended to print in 3D, allowing 3D spatial control over the microenvironment. In ink-jet printing, the biologic payload (e.g. cells, proteins, or other biomolecules) is printed through the controlled deposition of solution containing the desired payload, much like the multiple colors of ink in color printing. In this way, the amount of payload-containing solution, and its placement, can be precisely controlled to rapidly generate patterns with multiple cell types or materials. However, the location of the cells, or other biopayload, within the areas of dispensed solution is not controlled. For discrete components like cells, the payload is randomly distributed within a liquid volume, although parameters such as concentration are controllable. Therefore, ink-jet printing is particularly well suited for rapidly fabricating larger patterns, including constructs of intricate geometries and multiple cell types, in which the geometric precision is important, but spatial control of the cells or groups of cells is not.

In a traditional LDW setup, a camera is coincidently focused with the laser, allowing direct visualization of the biologic payload on the print ribbon. As a result, the specific cell or group of cells (or other biopayload) to be printed can be visualized, targeted, and transferred to the substrate with high spatial precision. In contrast to ink-jet printing, where a controlled volume of liquid is deposited, LDW allows targeting of specific cells, so the dispersal of cells within the volume transferred is *not* random. LDW is also capable of printing more viscous materials that may not be dispensed by an ink-jet nozzle. This unique ability to combine the high-resolution placement of selected biological payloads with various substrate materials, makes LDW particularly attractive to engineer *in vitro* cellular microenvironments.

5.1.3 LDW OVERVIEW

LDW is a noncontacting material deposition technique. While some differences in configuration exist, a typical LDW setup consists of two coplanar plates: a laser-transparent print ribbon, which holds the material/cells to be deposited, and a receiving substrate onto which material is printed (Figure 5.2). Both the receiving substrate and print ribbon can be independently moved with CAD-CAM-controlled stages. A charge-coupled device (CCD) camera also allows real-time visualization of the ribbon and receiving substrate. The underlying side of the mounted ribbon is coated with two thin layers, the first is a sacrificial layer, and the second a transfer layer. The sacrificial layer directly interacts with the laser, while the transfer layer consists of the actual printed material. A pulse from the laser passes through the transparent ribbon and volatilizes the sacrificial layer.

The consensus mechanism for deposition is that laser energy is absorbed by the sacrificial layer, forming a vapor pocket at the ribbon–material interface (Barron et al., 2004a). Expansion of the vapor pocket allows the printed material – the donor material – to form a droplet that is ejected from the surface of the print ribbon, on a trajectory perpendicular to the plane of the ribbon. It is also generally accepted that for a high-power laser and an appropriate sacrificial layer, mass transfer

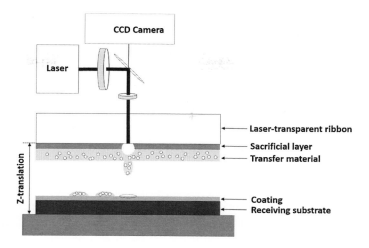

FIGURE 5.2

Schematic of typical LDW setup for material or cell deposition. A laser beam passes through a laser-transparent print ribbon to interact with material, partially volatalizing a sacrificial layer and forming a vapor pocket to eject the transfer to a receiving substrate. Independent CAD/CAM control of the ribbon and receiving stages allows programmatic deposition of material.

occurs at a much faster time scale than heat transfer (Barron et al., 2004a). Furthermore, heat shock protein expression does not appear to be elevated with LDW (Chen et al., 2006), meaning that thermal damage to the transferred material is negligible. Moreover, the laser-material interaction occurs at the surface of the material (Barron et al., 2004a; Barron et al., 2004b), so the bulk of the material that is transferred using LDW never directly interacts with the laser. After deposition, the receiving substrate is moved to the next programmed position, and a new spot is used for volatilization on the ribbon; serial deposition creates a desired structure or pattern according to the programmed stage positions. Typically, a receiving substrate is a Petri dish, cover slip, or glass slide, and is often coated with a thin layer of material that serves as the matrix material.

5.2 MATERIALS IN LDW

5.2.1 MATERIAL PROPERTIES INFLUENCING CELLULAR MICROENVIRONMENTS

In this section, we will focus on how the materials used in LDW can affect cell response and thus be used to engineer desired microenvironments. Functional and mechanical properties of the material will determine not only how the cells interact with the substrate, but also how they interact with other nearby cells.

Mechanical properties of a substrate play a critical role in the gene expression and determination of the functional role a cell performs. As cells adhere to a substrate, they form focal complexes that become points of force transfer between the substrate and the cell. On a softer substrate, in response to a stress initiated either externally or by the cell itself, the substrate-cell attachment region deforms more than the cell. On stiffer surfaces, less deformation of the substrate interacting with the focal contact point takes place, and there is a greater stress on the cytoskeleton elements causing cell changes. Significant differences in substrate stiffness manifest in different cell morphologies, cytoskeletal reorganization, gene expression, and fate decisions. The pioneering work of Dennis Discher's group (Engler et al., 2006) showed that mesenchymal stem cells (MSCs), when cultured on substrates of different elastic moduli, exhibited different morphologies and gene profiles; substrate stiffness could be used to direct their differentiation (Engler et al., 2006). Thus, the substrate stiffness is an important factor to consider when engineering the cellular microenvironment.

Functional groups present and their relative concentration in the ECM also play a pivotal role in the overall behavior of the cell. The presentation and binding of ligands is necessary to induce adhesion, motility, survival, differentiation, and other specific cell functions. For example, human embryonic stem cell (hESC) culture differentiation typically requires the use of a Matrigel® substrate, a hydrogel derived from mouse tumor basement membrane. However, synthetic substrates can be functionalized with the integrin ligands found within Matrigel® to support hESC attachment and growth (Liu et al., 2011). In addition to the presence of the specific functional groups, their relative concentrations also play a key role in cell behavior. Substrates modified with different surface concentrations of the arginine–glycine–aspartic acid (RGD) motif exhibit differing degrees of cellular attachment, affecting the cytoskeletal structure, its organization, and the cellular morphology (Massia et al., 1991). Another feature of engineered motifs is that binding affinity is different from that of the same motif within a natural protein, such as collagen or fibronectin. As a result, engineered substrates may have different binding kinetics than natural substrates. However, the surface-coated RGD can be engineered to contain additional peptide sequences that enhance its specificity or binding kinetics to a ligand (Petrie et al., 2006). Overall, mechanical properties of the material, such as elastic modulus, and biochemical properties of the material, such as functionalization, can have a profound influence on cells within the microenvironment.

5.2.2 MATRIGEL-BASED LDW

Initial studies of cell printing by LDW utilized Matrigel® as a coating on the receiving substrate, as well as on the ribbon as a sacrificial layer/transfer material (Figure 5.3). Matrigel® is a soluble tumor extract comprised of an assortment of proteins, and it undergoes thermal gelation at 37°C, creating a 3D gel (Kleinman et al., 1986). There are several features that make Matrigel® a good candidate material for the receiving substrate and as a transfer material for LDW studies. Matrigel® contains large quantities of all the essential structural proteins along with many of the other proteins, proteases, and growth factors found in the basement membrane (Kleinman et al., 2005). Matrigel® has been used widely for *in vitro* cell culture to differentiate a variety of cells and is currently used almost exclusively as the scaffolding material for maintaining undifferentiated human embryonic stem cell culture (Xu et al., 2001). Additionally, Matrigel® provides a permanent matrix to immobilize printed material on the receiving substrate, contributing to the high pattern fidelity.

However, despite these many benefits, Matrigel® has a number of shortcomings that may preclude or limit its use in certain applications, causing some researchers to seek alternative ribbon and substrate materials. As a transfer material, Matrigel® requires a cell-based biopayload to loosely attach to the Matrigel® matrix. Cellular attachment times vary depending on the cell type, making coculture transfers difficult and limited to adherent cell types. As an ECM-mimetic substrate material, Matrigel® has an inherent batch-to-batch variability because it is grown and extracted from a mouse tumor. This

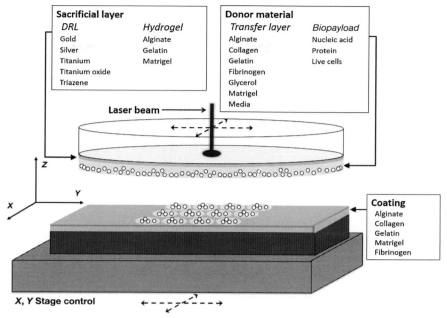

FIGURE 5.3

Materials utilized in LDW. The laser first interacts with a thin sacrificial layer, which can be either a hydrogel, or a dynamic release layer (DRL), such as triazene, metals, or metal oxides. The biologic payload is typically suspended in a transfer layer (e.g. glycerol for printing nucleic acids or proteins, hydrogel, or media for printing live cells). When printing live cells, the cells can either be attached to the material to be transferred (e.g. when using Matrigel®), or suspended in hydrogel or media in their trypisinized state. The receiving substrate typically utilizes a thin hydrogel coating as well, to provide viscous dissipation of energy.

variability may be a source of inconsistencies among researchers, and can make it difficult to reproduce previous findings. Further, not all Matrigel® components are defined, and some concentrations are unknown. Without knowing Matrigel®'s exact content and concentration of proteases, growth factors, and other proteins, the amount of information that may be inferred from an experiment is largely limited. In an experiment using Matrigel®, it is impossible to know what factor or factor combination is responsible for the observed cellular behavior, since any potential influence by Matrigel®'s numerous unspecified constituents cannot be ruled out. Although Matrigel® provides a cost-effective basement membrane substitute because it is derived from mouse sarcoma cells, it could elicit an extreme immune response if implanted into the body. Moreover, extrapolating cell behavior observed on a mouse tumor extract to a human condition could be inaccurate.

5.2.3 GELATIN-BASED LDW

Among the hydrogel materials explored for use with LDW, gelatin has proven quite promising. Gelatin is a natural polymer that is a reduced and degraded form of collagen. At room temperature, gelatin is a stiff gel, but at incubation temperatures (37°C), the polymer network becomes soluble. In LDW applications, gelatin is currently used both as a sacrificial and transfer material, as well as a receiving substrate (Schiele et al., 2011). As previously described, a laser pulse vaporizes the sacrificial gelatin, ejecting a droplet of cells. What is truly powerful about this approach is its exploitation of the thermally reversible properties of gelatin. On the print ribbon, exploiting gelatin's thermal properties allows for partial encapsulation of the biopayload, giving the opportunity for a wide variety of cells and other payload to be transferred. Further, this technique does not require cells to attach to the matrix on the ribbon (as with Matrigel®), and thus allows cells to be printed in their trypsinized state. This is less traumatic to the cells because it does not disrupt the focal adhesions, thereby allowing even nonadherent cells to be printed. In addition to gelatin, it is possible to use other materials with temperature-dependent gelation or viscosity, such as glycerol (Guillemot et al., 2010), to encapsulate the biopayload for transfer. Gelatin's thermal properties also offer unique benefits when used on the receiving substrate. Following transfer, after a brief incubation period, the gelatin on the receiving substrate is liquefied, leaving only the LDW-patterned cells. This further takes advantage of gelatin's thermal reversibility to minimize the influence of potentially unwanted matrix factors in simple cell studies.

5.2.4 DYNAMIC RELEASE LAYERS

A dynamic release layer on the ribbon is often used to control the material interacting directly with the laser, decoupling the laser from the transfer material. Dynamic release layers are thin, sacrificial layers of material that interact with the laser at its operational wavelength, and can amplify energy from the laser (Figure 5.3). Some example materials for this layer are triazene or a metal/metal oxide (Schiele et al., 2009; Ringeisen et al., 2008). With LDW, the size of the transferred material droplet is directly related to laser fluence. However, inconsistencies of the transfer material on the ribbon, such as uneven coating, could produce unwanted spot-to-spot variation in the pattern. The use of a dynamic release layer has been shown to help minimize the energy threshold required for transfer, and decrease the thermal impact on the transfer material (Banks et al., 2008). The other benefit of using such a dynamic release layer is that it provides a consistent material at the ribbon interface, thereby granting a more predictable laser–material interaction when utilizing a variety of transfer materials. Without a dynamic release layer, each transferred material will exhibit different transfer dynamics due to specific laser–material interactions. The use of a dynamic release layer will result in a consistent vapor pocket, independent of the transfer material. However, utilizing a dynamic release layer may impair or preclude ribbon and substrate visualization.

5.2.5 ADDITIONAL HYDROGELS AND HYDROGEL PROCESSING USED IN LDW

An extensive amount of hydrogels and hydrogel blends have been used for LDW applications. The use of a dynamic release layer has made it simple to change between different hydrogel materials on the ribbon as a transfer material for cells or other biological contents. To date, the most commonly used hydrogels as a transfer material are gelatin, Matrigel®, and alginate (Barron et al., 2004a; Koch et al., 2010; Koch et al., 2012). The same hydrogels are also commonly used to coat the receiving substrate during printing, to cushion the impact of printed cells, and to immobilize the printed spots to increase pattern fidelity. The feature that these hydrogels have in common is that they have rapid and controllable gelation properties that benefit immobilization of the biopayload. There has also been use of fibrinogen/hyaluronic acid and collagen both as transfer material and for coating the receiving substrate (Koch et al., 2012; Gruene et al., 2011a). Current research in LDW has not yet focused on customizing receiving substrate properties, which can be achieved using a combination of natural and synthetic hydrogels.

Hydrogels as substrate coating materials allow for the manipulation of both the mechanical and biochemical properties of the environment. The biochemical properties of the surface can be engineered either by using protein-based gels or grafting functional moieties into the polymer that makes up the gel. Additionally, the mechanical properties of the hydrogel are tunable by manipulating the amount of cross-linking and the relative polymer concentration. LDW applications have not yet explored this parameter space. However, nonpatterning applications have begun customizing both of these properties in hydrogels for a wide range of applications. One example of manipulation of hydrogel biochemical and mechanical properties is within alginate gels, where the elastic modulus and RGD grafting density were independently manipulated to optimize the environment for stem cell differentiation (Huebsch et al., 2010). Printed material can also be manipulated, through *in situ* cross-linking on the receiving substrate. Example of this include the printing of cells suspended in fibrinogen to substrates of thrombin, which after a period of incubation, forms fibrin (Gruene et al., 2011a), and the printing of alginate into a substrate of calcium, to fabricate and localize microbeads (Kingsley et al., 2013).

5.2.6 CUSTOMIZABLE TOPOGRAPHY OF NONHYDROGEL RECEIVING SUBSTRATES

Another means to control the receiving substrate surface properties is through the use of engineered nonhydrogel materials or scaffolds. One such example of this is electrospun nanofibrous structures. As a surface coating material, electrospun fibers can be fabricated with a variety of natural and synthetic polymers, where fabrication parameters allow for the tuning of stiffness, topography, degradation, and fiber size (Pham et al., 2006). Fibers can also be chemically treated after fabrication with desired functional units, similar to protein grafting in hydrogels. Additionally, electrospun fibers can be fabricated in such a way that they are aligned, giving cells directional cues (Schaub et al., 2013). Utilizing electrospun fiber substrates with cell printing is another way to achieve idealized microenvironments, with the addition of directing cell growth through properties of the fibers and their structural alignment. A 1-day time course of fibroblasts printed on electrospun fibers (Figure 5.4) indicates that cells maintain registry to the printed pattern. Aligned fibers appear to direct the cell elongation and migration after LDW. Printing onto substrates with topographical features enables the fabrication of unique constructs or cell studies that are difficult to perform using other techniques.

5.3 LDW APPLICATIONS IN 2D

LDW was first used in 2D for electronics applications (Chrisey et al., 2000). Once it was shown that LDW could be adapted for use in soft materials transfer and deposition of biologics, 2D printing of nucleic acids (Colina et al., 2005; Fernández-Pradas et al., 2004), proteins (Dinca et al., 2008), and even live cells (Wu et al., 2001) was demonstrated. These bioprinting approaches hold many promising applications, ranging from biosensors fabrication, to the creation of small grafts or biological constructs, to building spatially precise cultures for *in vitro* diagnostics and cellular signaling studies.

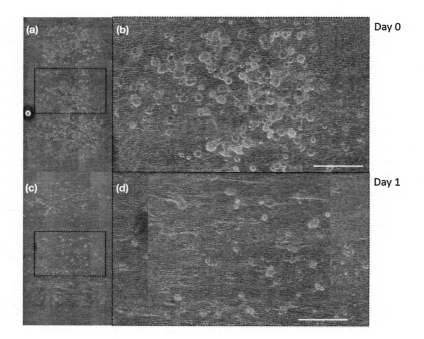

FIGURE 5.4

RFP normal human lung fibroblasts printed by LDW onto PLLA electrospun nanofiber substrates. Nanofibers are aligned on the substrate, oriented left-to-right in images. Fibroblasts (a–b) immediately after printing appear rounded in their trypsinized state, but (c–d) after one day begin to align in the direction of the electrospun fibers. Scale bars are 200 μm.

Various methods for laser-based deposition have been utilized, each quite similar, but with some distinct (although often subtle) differences. In this chapter, we will group all of these methods under "LDW," although there are different preferences in the field about the most appropriate term to use. LDW-based methods operate on the same general principle, illustrated in Figure 5.2. The major difference among the various LDW methods is their choice of energy-absorbing layers that can be used to amplify laser energy.

The purpose of using a sacrificial material is so that biologics or cells are not themselves sacrificed during a deposition event. Laser-induced forward transfer (LIFT) typically uses a metal or foil as a sacrificial layer, while matrix-assisted pulsed laser evaporation direct-write (MAPLE-DW) typically uses a biologic matrix, such as Matrigel® or gelatin. Both methods have shown success depositing multiple mammalian cell types in controlled patterns (Table 5.2). Moreover, custom configurations of cells, such as grids, lines, and sheets have been demonstrated, with unrestricted cell growth from the initial printed pattern (Figure 5.5).

Because the substrate is generally homogeneous, following LDW, cells are free to migrate, cluster, or form structures uninhibited by geometric or biochemical restrictions on the substrate. This property of LDW allows cellular migration and migration-based behavior to be studied. By observing structural evolution, LDW enables different types of studies than what can be explored using micropatterned proteins,

Table 5.2   Examples of cell types and materials successfully deposited by LDW, illustrating the applicability and broad range of research

Cell type	Ribbon material (not including laser-transparent ribbon or cell suspension)	Receiving substrate material	References
Rat Schwann	Silver	Gelatin	Hopp et al., 2005
Rat astroglial	Silver	Gelatin	Hopp et al., 2005
Pig lens epithelial	Silver	Gelatin	Hopp et al., 2005
Bovine aortic endothelial cells (BAEC)	Metal or metal oxide	Matrigel®or media	Chen et al., 2006
Human Osteosarcoma (MG 63)	Au,Ti, or TiO$_2$ Matrigel or glycerol Hydroxyapatite or zirconia/ glycerol Extracellular Matrix Solution	Matrigel® Extracellular Matrix Solution	Barron et al., 2004a; Barron et al., 2004b; Barron et al., 2005; Doraiswamy et al., 2006a
Rat cardiac	Au,Ti, or TiO$_2$	Matrigel®	Barron et al., 2004a
Mouse endothelial	Au,Ti, or TiO$_2$	Matrigel®	Barron et al., 2004b
Human endothelial (EA.hy926)	Gold, alginate, glycerol, nano-hydroxyapatite	no material	Guillemot et al., 2010
Olfactory ensheathing cells	TiO$_2$	Matrigel®	Othon et al., 2008
Human umbilical vein endothelial cells (HUVEC)	Au,Ti, or TiO$_2$ Gold, alginate	Matrigel®	Wu et al., 2010; Gaebel et al., 2011
Endothelial colony forming cells (ECFC)	Gold, hydrogel precursor	Hydroxyapatite, fibrinogen, thrombin	Gruene et al., 2011a
Mouse embryonic stem cells (mESC)	Cured polyimide Gelatin	Matrigel® Gelatin	Kattamis et al., 2007, Raof et al., 2011; Dias et al., 2014
Human dermal fibroblasts	Gelatin	Gelatin	Schiele et al., 2011; Schiele et al., 2009
3T3 fibroblasts	Gold, alginate	Matrigel®	Koch et al., 2010
Human mesenchymal stem cells (hMSC)	Gold, alginate/ Matrigel®	Matrigel®	Koch et al., 2010; Gaebel et al., 2011; Gruene et al., 2011b
Human adipose-derived stem cells (hASC)	Gold, alginate	Matrigel®	Gruene et al., 2011c
Mouse embryonal carcinoma (P19)	Matrigel®	Matrigel®	Ringeisen et al., 2004
Chinese hamster ovary (CHO) cells	Matrigel®	Matrigel®	Ringeisen et al., 2002b
Human osteoblast	Matrigel®	Matrigel®	Ringeisen et al., 2002b

(Continued)

Table 5.2   Examples of cell types and materials successfully deposited by LDW, illustrating the applicability and broad range of research (cont.)

Cell type	Ribbon material (not including laser-transparent ribbon or cell suspension)	Receiving substrate material	References
Rat B35 neuronal neuroblast cells	Matrigel® Triazene, Extracellular Matrix Solution	Matrigel® Extracellular Matrix Solution	Patz et al., 2005; Doraiswamy et al., 2006b
Ovine endothelial cells	Gold	PEG scaffold	Ovsianikov et al., 2010
Ovine vascular smooth muscle cells (vSMC)	Gold	PEG scaffold	Ovsianikov et al., 2010
Human keratinocytes	Gold, alginate	Matrigel®	Koch et al., 2010
Human breast cancer	Alginate, gelatin	Gelatin, calcium chloride	Kingsley et al., 2013
Bovine Pulmonary Artery Endothelial Cells (BPAEC)	Matrigel®	Matrigel®	Schiele et al., 2009
Rat neural stem cells	Matrigel®	Matrigel®	Schiele et al., 2009
Mouse myoblast	Matrigel®	Matrigel®	Schiele et al., 2009

because patterning proteins restricts cell migration to adhesion islands. Unrestricted cell growth following LDW has been demonstrated with a Matrigel® coating on the substrate in earlier work (Wu et al., 2003; Schiele et al., 2009), and more recently with a gelatin substrate coating (Schiele et al., 2011). An advantage of the gelatin substrate is that the gelatin coating is temporary (as mentioned in Section 7.2.3), and does not provide cells with a permanent scaffold or a large assortment of unknown complex signaling factors. Thus, after the gelatin has liquefied during incubation, and is removed with the first media exchange, the cells remain attached to the underlying substrate. This allows various substrates and materials to be utilized, provided the thin temporary gelatin layer is applied to enable viable transfers. Despite the temporary nature of the gelatin layer, cells printed to the substrate maintain excellent pattern registry, moving on average, less than 6 μm from their initial location within half an hour of printing (Schiele et al., 2011). One restriction of this technique is the limitation to a flat surface parallel to the ribbon, but otherwise potentially any cell-adhesive material can be used with LDW, such as hydrogels or scaffold materials with engineered topography.

Although LDW can potentially be used to study the effects of cell location and cell-cell interaction, exploration of cellular phenomena after LDW patterning has been examined in only a few instances. One study of embryonic stem cell (ESC) behavior examined the clustering of ESCs after LDW. ESCs are pluripotent, meaning they can differentiate into any of the three primitive germ layers. It was shown that following LDW, mouse ESCs (mESCs) maintained their pluripotency, as evidenced by their ability to express markers of all three primitive germ layers (Raof et al., 2011). An additional indication of pluripotency is the formation of 3D cell clusters called embryoid bodies (EBs), which contain cells of all three germ layers. The cell–cell interactions within the EB are greatly important. The size of the EB, which may impart complex signals to cells within the structure, can influence differentiation, as can the size of a stem cell colony (Peerani et al., 2007; Bauwens et al., 2008; Lee et al., 2009b). The

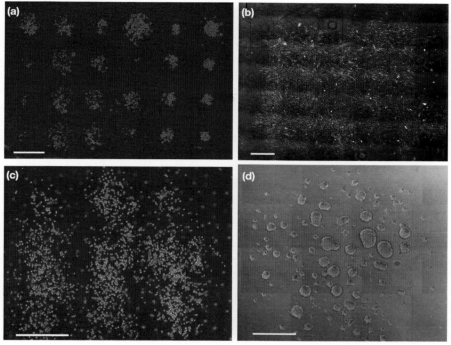

FIGURE 5.5

Examples of cells printed by LDW, and evolution of structure over time. Array of human dermal fibroblasts
(a) immediately after printing and (b) 24 h after printing, showing the evolution of cellular network structure
from the initial printing positions. Contrast is adjusted to show detail. Lines of mouse embryonic stem cells
(c) immediately after printing and (d) 48 h after printing, demonstrating the formation of embryoid bodies (EBs) on
an unrestricted uniform substrate due to collective cellular behavior, rather than constrained growth. Bubbles in the
background are artifacts of securing the substrate. Scale bars are 500 μm.

micropatterning approach previously used to study this phenomenon only allows the combined effect
of the patterned protein and colony size to be studied. However, it would be ideal to study the effect of
colony size and the surface-coated protein independently.

Controlling EB size can potentially be very useful to direct differentiation, and LDW has been
used to control EB size, via the density of printed cells, independent of the stem cell colony diameter (Dias et al., 2014). While colony size did not influence the size of the EBs that formed, the
effect of colony size on stem cell differentiation based on cellular patterning on a homogeneous
substrate still needs to be determined. Prescribing these factors in engineered microenvironments
could allow more efficient directed differentiation of stem cells. Additionally, differentiation can
also be influenced by printing protein gradients, as reviewed (Tasoglu et al., 2013). The versatility afforded by LDW for printing cells, biomaterials, and proteins, enables complex studies to
differentiate stem cells, influence migration, and answer many other questions using engineered
microenvironments.

5.4 LDW APPLICATIONS IN 3D
5.4.1 MICROENVIRONMENTS IN 3D

Planar, 2D cell culture has long been a paradigm for studying mammalian biological phenomena *in vitro*, ranging from stem cell differentiation and tissue development to drug testing. However, fundamental differences exist in the way cells behave between 2D and 3D microenvironments. Cells with 3D microenvironment interactions show differences in their cytoskeletal structure, morphology, membrane protein distribution, and interaction with soluble factors (Pampaloni et al., 2007). Additionally, cells cultured in a 3D ECM experience limitations for cell migration, while those grown on a 2D substrate can migrate and proliferate without the same restrictions.

In a 2D environment, cells are only able to interact and attach to the ECM on the substrate. This produces a difference in the receptor density and orientation along the surface of the cell compared with receptor orientation in 3D (Meshel et al., 2005). Further, the composition and strength of the complexes forming the adhesions differ between 2D and 3D microenvironments. As an example, on a 2D substrate, fibroblasts form adhesions along only the ventral surface (Berrier et al., 2007). The localization of cell traction forces results in a morphological polarity that does not exist in 3D. In a 3D environment, adhesion is formed via focal complexes (as opposed to 2D focal adhesions), all along the cell membrane, and has a different composition. The difference in cell distribution and type of adhesion site in 3D affects the organization and generation of tension in the cytoskeleton (Pedersen et al., 2005). The compounding discrepancy between 2D and 3D cell-matrix interactions can yield very different behaviors and response to stimuli, when explored experimentally.

In a 3D ECM, cell migration occurs either by moving through the material's pores, or by breaking down the surrounding ECM with proteases. Highly porous materials, such as sponges and foams, typically have pore sizes greater than the cell diameter, allowing for nonproteolytic migration. In hydrogels, the pores are on the nano scale (much smaller than the actual cell). For infiltration to occur, the cell must produce proteases to break up the restrictive matrix, assuming the matrix is made from a peptide-based gel. The ease of infiltration will determine the migratory rate of the cells within the material. Additionally, the pore size will be a factor in determining the rate that nutrients can diffuse into and waste products can flow out of the bulk material. If the bulk material is too thick and/or pore size too small, toxins can build up in the environment, or cells can die due to ischemic effects.

Fundamental biological questions have been, and will continue to be solved using 2D culture models. However, 3D alternatives are necessary to overcome 2D matrix interactions that fundamentally change cell behavior (e.g. cytoskeletal structure, morphology, membrane protein distribution and interaction with soluble factors (Pampaloni et al., 2007), and focal adhesions). Additionally, the creation of large tissue-engineered constructs, for fundamental research or *in vivo* transplantation, will require a means of 3D fabrication with precise spatial arrangement of the contents. To overcome the limitations set by 2D environments, LDW has been adapted to build more physiologically relevant 3D culture models, *in vitro* diagnostic tools, and tissue-engineered constructs.

5.4.2 LAYER-BY-LAYER APPROACHES

Layer-by-layer (LbL) printing is a 3D biofabrication approach adapted from industrial rapid prototyping technologies. Traditionally, LbL printing utilizes a 2D method, sequentially, to produce an overall 3D construct. After one layer is fabricated, it is often stabilized before the next layer is printed to provide a flat new printing surface. For LbL LDW, liquid hydrogel, or hydrogel precursor suspending a desired biologic, is used as a transfer material, similar to the 2D printing method. The hydrogel is deposited at preprogrammed coordinates, according to the specific layer design. Once the printed layer is completed, it is gelled by either cross-linking the polymer solution, or inducing polymerization. Gelation can be triggered by a number of different cross-linking or polymerization mechanisms (e.g. thermal, ionic, pH, or enzymatic), depending on the hydrogel or hydrogel blend. The rate of gelation for each of these mechanisms differs, and generally a fast gelation time is desired to maintain pattern fidelity. Once the layer is gelled, the printing surface is stabilized, and the process can be repeated to add additional layers until the overall construct is completed.

Another method of performing LbL fabrication with LDW utilizes a hydrogel precoated on the substrate, rather than in the transferred material. Thin-film coating mechanisms can consistently make thin layers of liquid hydrogel polymer or precursor on a substrate at a desired height. The selected material is transferred into the liquid hydrogel layer in a programmed pattern. The layer is gelled, and a new printing layer is produced by the addition of new hydrogel solution, again coated to the desired layer thickness. This technique is very similar to the previous method, but may hold advantages if each layer needs only a small amount of printed substance, relative to the overall bulk material. These techniques can be repeated for a desired number of layers. 3D resolution is determined by the hydrogel coating mechanism's control over the height of the newly laid hydrogel layer. The average height for individual layers with one such coating technique, blade-coating, approaches approximately 40 μm (Gruene et al., 2011b). Combinations of hydrogel materials previously used in LbL LDW, as well as other candidate materials, have been listed in Figure 5.3. LbL LDW has been used for *in vitro* and *in vivo* skin tissue, an osteosarcoma model, and cardiac regeneration (Koch et al., 2012; Gaebel et al., 2011; Catros et al., 2011). Beyond creating tissue models, 3D LDW can be used to study cell-to-cell signaling. One powerful example of 3D LDW studied the coculture signaling and migration between adipose-derived stem cells and endothelial colony forming cells in hyaluronic acid and fibrin gelsu (Gruene et al., 2011a).

The LbL LDW printing technique appears to be very similar to another LbL technique, ink-jet printing, which has also been adapted for biological applications. Ink-jet printing deposits hydrogel material, which suspends cells, directly from a nozzle to a substrate, either spot by spot or by continuous flow. Discrete layers of printed material are often gelled prior to printing additional layers. The advantages of ink-jet printing include that it is generally less expensive than LDW, and offers higher throughput.

5.4.3 LDW MICROBEADS

Microbeads are spheroidal microstructures that can encapsulate a desired biologic, and have been investigated for applications ranging from drug delivery to cell culture (Xie et al., 2009; Amsden et al., 1997). For cell-based applications, microbeads are fabricated from cross-linkable polymers, or precursor as used for hydrogels. Like bulk hydrogels, microbeads provide cells with a 3D environment, but there are distinct advantages to the microbead structure. Nutrient diffusion into bulk hydrogels occurs slowly, and large gels may even require a bioreactor to prevent encapsulated cells from suffering from ischemic effects or buildup of metabolic waste. Microbeads, on the other hand, have a high surface area-to-volume ratio, allowing for more rapid exchange of nutrients and waste products. Further, microbeads provide greater control of the cell microenvironment, especially in studies involving multiple cell types. Properties of the individual beads can be tailored for the specific cell type, whereas bulk gels are generally isotropic.

Current microbead fabrication techniques include electrostatic bead generators, microfluidic devices, and other emulsion-based techniques (Amsden et al., 1997; Desmarais et al., 2012). These technologies operate under the same general principles to fabricate beads: cells are suspended in a polymer solution, and a single droplet is extruded into a cross-linking bath, where rapid gelation occurs to encapsulate the suspended contents. For example, in a solution of alginate with suspended cells, a droplet can be extruded into a bath containing calcium. The divalent Ca^{2+} cation ionically cross-links the alginate solution, causing rapid gelation, and encapsulating the suspended cells. The current fabrication technologies used to produce microbeads have certain limitations in their uses. Electrostatic bead generators limit bead fabrication to only polyelectrolyte materials, whereas other pressure- and flow-based techniques appear to have relatively consistent control of bead size, but only over a limited range. None of these technologies are able to control microbead placement.

A recently developed method for bead fabrication uses LDW to both fabricate and pattern cell-containing alginate microbeads in a single step (Kingsley et al., 2013). This LDW setup consists of a ribbon with an alginate–gelatin sacrificial layer and cell-suspending alginate transfer material. The receiving substrate consists of a thinly coated layer of gelatin/calcium chloride. A pulse from a laser ejects cell-containing alginate droplets to the substrate below in a prescribed location, where *in situ* gelation occurs via calcium chloride/alginate cross-linking. This technique manipulates microbead size by adjusting the diameter of the laser beam used to eject the material. Beads fabricated by this technique in our lab were consistently produced from as large as 500 μm to as small as 50 μm in diameter (Figure 5.6). Further, bead spacing in printed arrays was accurate within 2% of the desired spacing. The viability of cells encapsulated within beads remained high, very close to what is found using gelatin-based planar LDW. This work leads into ongoing research that utilizes single microbeads as "voxels" to form larger 3D structures. Microbeads are printed consecutively, layer-by-layer, on top of one another, to produce an overall 3D construct in which the architecture and composition are prescribed with microbead-level fidelity.

5.4.4 FABRICATION OF SHELLED MICROENVIRONMENTS

Microbeads have many uses for 3D cell-based applications. However, in certain circumstances where one would want the cells to produce their own scaffold material, a microcapsule may be a more ideal structure. A microcapsule is a hollow shelled spherical microenvironment, made from processing a microbead (Orive et al., 2006). In the case of an alginate microbead (negatively charged polymer), a microcapsule is formed by the addition of an oppositely charged polyelectrolyte, such as poly-L-lysine or chitosan (positively charged polymers). The oppositely charged polymer complexes with the outside of the alginate microbead, forming a shell surrounding the bead. The bulk alginate in the microbead is then liquefied by the addition of a chelating agent to remove the divalent cation that cross-links the alginate. This results in a hollow shell composed of the two oppositely charged polymers, while still retaining the encapsulated cells.

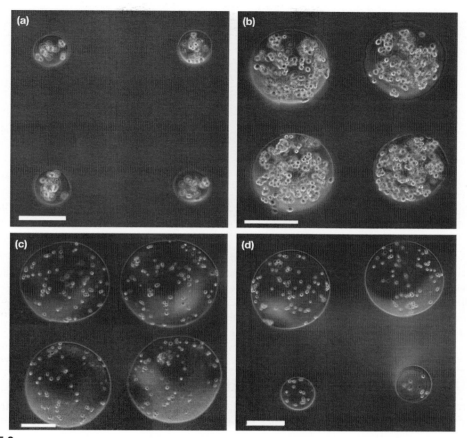

FIGURE 5.6

Examples of microbeads fabricated and patterned with LDW in a single step. Patterns are shown with microbead diameter sizes of (a) 150 μm, (b) 350 μm with human breast cancer cells, (c) 500 μm, and (d) multiple bead sizes in a single pattern with mouse embryonic stem cells. Scale bars are 200 μm. Bead size is controlled by adjusting the laser beam diameter, and cell density is adjusted via the cell density on the print ribbon. Microbeads can be patterned into custom configurations to study spatially sensitive aspects of the microenvironment.

Processing of LDW patterned and fabricated microbeads to microcapsules is current and ongoing work. Figure 5.7 shows preliminary results of processed LDW microbeads into chitosan-alginate microcapsules. It appears that processed microcapsules retain initial bead position on the patterned substrate. The advantage of these capsules fabricated by LDW is that they can be placed accurately (within 2% of their target spacing). By exchanging the ribbon during printing, coculture studies can examine how signaling occurs from capsule to capsule. The spatial precision afforded by LDW allows fabrication of capsules or beads that are approximately touching. This suggests future applications where beads could be placed overlapping and processed to create hollow complex structures, or any planar geometry.

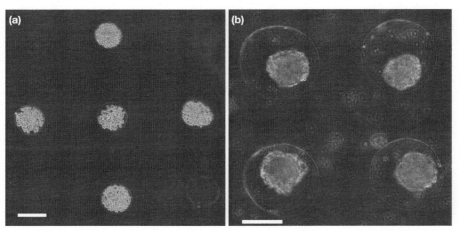

FIGURE 5.7

Microcapsules created by processing microbeads fabricated and patterned by LDW. (a) Human breast cancer cells in poly-L-lysine capsules after 7 days, and (b) mouse embryonic stem cells in chitosan capsules after 3 days, illustrate that the hollow microcapsules allow cell growth over time within a constrained geometry. The breast cancer cells grow to fill the hollow capsule, and the stem cells form an embryoid body (EB) contained with the microcapsule. The many differences between these images illustrate that the material, cell type, and other features of the microenvironment can greatly influence cell behavior. Scale bars are 500 μm.

5.5 CONCLUSIONS AND FUTURE DIRECTIONS

There are some aspects of cellular behavior, such as migration, differentiation, and certain types of gene expression, that are heavily influenced by the cellular microenvironment. Both mechanical signals from the substrate and biochemical signals, either soluble or insoluble, have profound effects on cell fate and function, and are being widely studied. However, cell–cell signaling is also an important influential factor in cell behavior, yet this is rarely studied, due, in part, to the complexity and technical challenges in doing so. Therefore, it may be prudent to control how cells signal within the microenvironment. Factors that can be manipulated include whether the cellular signaling is homotypic or heterotypic, the strength of the signal based on the number/density and placement of cells, and the signaling dynamics. Herein, we have discussed patterning approaches, in particular LDW, to control cell placement in engineered microenvironments. Although it is all but impossible to mimic the sheer complexity of multiple cell types and signals *in vivo*, spatially precise patterning approaches onto engineered substrates offers a powerful tool to prescribe and control cellular signaling for *in vitro* experiments.

LDW has been used to print multiple types of cells in custom patterns, which allows the fabrication of microenvironments that can maximize (or minimize) a desired behavior based solely on cellular arrangement. Micropatterning has been used to show that cell size/shape or colony size can influence cell fates, but it generally does not allow evolution of structure. Moreover, the adhesive proteins that are patterned confound cell signaling, making it difficult to decouple the effect of the adhesive protein and the effect of cellular signaling.

While LDW offers capabilities for generating spatially precise 2D cellular microenvironments, it has even more power beyond micropatterning approaches in fabricating 3D microenvironments. Similar to ink-jet printing, LDW has demonstrated the ability to print cells and biomaterials in a layer-by-layer fashion. While the throughput of LDW is lower than that of ink-jet, it offers higher resolution, potentially allowing controlled studies of 3D cell-cell interactions in complex geometries. For 3D bioprinting approaches, in order to get sufficient height to the structure, fairly robust materials such as hydrogels are often used to ensure both cell viability within the printed construct and 3D structural integrity. Although this restriction on material configuration can help maintain construct geometry, it is also a limitation from a cellular perspective; in order for cells to proliferate and migrate, they must break down the nanoporous hydrogel, and replace it with ECM. 3D constructs have been realized using layer-by-layer approaches, but they take weeks, or even months to become fully cellularized because of the geometric restriction of nanoporous hydrogels. Despite this potential limitation, future applications could include complex *in vitro* tumor models, vessels, lamina, and other structures where the small attainable size is not a restriction.

Another recent method for creating 3D microenvironments using LDW is the fabrication of 3D microbeads of controlled size and placement (Kingsley et al., 2013). In contrast to traditional methods for fabricating microbeads, LDW allows fabrication and placement in a single step, via *in situ* crosslinking of a hydrogel. This enables cells encapsulated in the microbead to be precisely placed in 3D microenvironments. What follows is that this method allows the study of cells within a 3D microenvironment, but on a 2D substrate, which, in turn, permits high-quality imaging and analysis. This feature also makes microbead printing compatible with planar (2D) LDW, so hybrid 2D/microbead constructs can be fabricated (Kingsley et al., 2013). Hybrid constructs allow 2D cellular studies based on point sources of material or factors delivered by beads. Encapsulated cells within beads can potentially deliver factors continuously, or beads themselves could be used for delivery. Beads can serve also as nodes at precise spatial locations to direct 2D spatial migration. The true power of this technique is realized when additional processing of microbeads with a cationic polymer allows them to be shelled, and the hydrogel liquefied, leaving a macroporous capsule that allows cellular migration and proliferation within the boundary of the capsule. The macroporous structure afforded by microbead printing and capsule formation may also allow a highly cellular structure to be realized much more rapidly because cells do not have to break down matrix in order to proliferate.

While the LDW field seems to be moving in the direction of 3D patterning, the potential of 2D LDW has not yet been fully realized. As discussed, micropatterning, ink-jet printing, and LDW are complimentary cell printing techniques that offer unique advantages for particular applications. LDW is particularly well suited for applications that require spatial precision on homogeneous substrate and/or evolution of the printed structure are desired. 3D layer-by-layer and microbead printing approaches both hold promise for studying 3D cellular microenvironments, and allow a wide range of applications based on the same technology. The coming decade holds great promise for the advancement of LDW and cellular studies for tissue engineering and regenerative medicine based on 2D and 3D control of the microenvironment.

ACKNOWLEDGMENTS

We would like to thank Nick Schaub and Dr. Ryan Gilbert (RPI) for providing electrospun fiber substrates, Dr. Yubing Xie (SUNY CNSE) for providing human breast cancer cells, and Dr. Guohao Dai (RPI) for providing normal human lung fibroblasts. This work was also supported, in part, by NIH R56-DK088217 (DTC) and DoD, Air Force Office of Scientific Research, National Defense Science and Engineering Graduate (NDSEG) Fellowship, 32 CFR 168a (ADD).

REFERENCES

Amsden, B.G., Goosen, M.F.A., 1997. An examination of factors affecting the size, distribution, and release characteristics of polymer microbeads made using electrostatics. Control. Release 43, 183–196.

Banks, D.P., Kaur, K., Gazia, R., Fardel, R., Nagel, M., Lippert, T., Eason, R.W., 2008. Triazene photopolymer dynamic release layer-assisted femtosecond laser-induced forward transfer with an active carrier substrate. EPL (Europhysics Letter) 83, 38003.

Barron, J.A., Krizman, D.B., Ringeisen, B.R., 2005. Laser printing of single cells: statistical analysis, cell viability, and stress. Biomed. Eng 33, 121–130.

Barron, J.A., Ringeisen, B.R., Kim, H., Spargo, B.J., Chrisey, D.B., 2004a. Application of laser printing to mammalian cells. Thin Solid Films 453, 383–387.

Barron, J.A., Wu, P., Ladouceur, H.D., Ringeisen, B.R., 2004b. Biological laser printing: a novel technique for creating heterogeneous 3-dimensional cell patterns. Biomed. Microdevices 6, 139–147.

Bauwens, C.L., Peerani, R., Niebruegge, S., Woodhouse, K.A., Kumacheva, E., Husain, M., Zandstra, P.W., 2008. Control of human embryonic stem cell colony and aggregate size heterogeneity influences differentiation trajectories. Stem Cells 26, 2300–2310.

Berrier, A., Yamada, K., 2007. Cell–matrix adhesion *J*. Cell. Physiol, 565–573.

Boland, T., Xu, T., Damon, B., Cui, X., 2006. Application of inkjet printing to tissue engineering. Biotechnol. J 1, 910–917.

Calvert, P., 2001. Inkjet printing for materials and devices. Mater 13, 3299–3305.

Catros, S., Guillemot, F., Nandakumar, A., Ziane, S., Moroni, L., Habibovic, P., van Blitterswijk, C., Rosseau, B., Chassande, O., Amedee, J., Fricain, J.-C., 2011. Layer-by-layer tissue microfabrication supports cell proliferation *in vitro* and *in vivo*. Tissue Eng Part C: Methods 18, 62–70.

Chen, C., Barron, J., Ringeisen, B., 2006. Cell patterning without chemical surface modification: cell-cell interactions between printed bovine aortic endothelial cells (BAEC) on a homogeneous cell-adherent hydrogel. Appl. Surface Sci. 252, 8641–8645.

Chen, C.S., Mrksich, M., Huang, S., Whitesides, G.M., Ingber, D.E., 1998. Micropatterned surfaces for control of cell shape, position, and function. Biotechnol. Prog 14, 356–363.

Chrisey, D.B., Pique, A., Fitz-Gerald, J., Auyeung, R.C.Y., McGill, R.A., Wu, H.D., Duignan, M., 2000. New approach to laser direct writing active and passive mesoscopic circuit elements. Appl. Surface Sci. 154–155, 593–600.

Colina, M., Serra, P., Fernández-Pradas, J.M., Sevilla, L., Morenza, J.L., 2005. DNA deposition through laser induced forward transfer. Biosens. Bioelectron 20, 1638–1642.

Desmarais, S.M., Haagsman, H.P., Barron, A.E., 2012. Microfabricated devices for biomolecule encapsulation. Electrophoresis 33, 2639–2649.

Dias, A.D., Unser, A.M., Xie, Y., Chrisey, D.B., Corr, D.T., 2014. Generating size-controlled embryoid bodies using laser direct-write. Biofabrication 6, 025007.

Dinca, V., Farsari, M., Kafetzopoulos, D., Popescu, A., Dinescu, M., Fotakis C, 2008. Patterning parameters for biomolecules microarrays constructed with nanosecond and femtosecond UV lasers. Thin Solid Films 516, 6504–6511.

Doraiswamy, A., Narayan, R., Lippert, T., Urech, L., Wokaun, A., Nagel, M., Hopp, B., Dinescu, M., Modi, R., Auyeung, R., 2006b. Excimer laser forward transfer of mammalian cells using a novel triazene absorbing layer. Appl. Surface Sci. 252, 4743–4747.

Doraiswamy, A., Narayan, R.J., Harris, M.L., Qadri, S.B., Modi, R., Chrisey, D.B., Hill, C., Carolina, N., 2006a. Laser microfabrication of hydroxyapatite-osteoblast-like cell composites. Biomed. Mater. Res. Part A 80, 635–643.

Engler, A., Sen, S., Sweeney, H., Discher D, 2006. Matrix elasticity directs stem cell lineage specification. Cell 126, 677–689.

Fernández-Pradas, J.M., Colina, M., Serra, P., Domınguez, J., Morenza, J.L., 2004. Laser-induced forward transfer of biomolecules. Thin Solid Films 453, 27–30.

Freytes, D.O., Wan, L.Q., Vunjak-Novakovic, G., 2009. Geometry and force control of cell function. J. Cell. Biochem 108, 1047–1058.

Gaebel, R., Ma, N., Liu, J., Guan, J., Koch, L., Klopsch, C., Gruene, M., Toelk, A., Wang, W., Mark, P., Wang, F., Chichkov, B., Li, W., Steinhoff, G., 2011. Patterning human stem cells and endothelial cells with laser printing for cardiac regeneration. Biomaterials 32, 9218–9230.

Godier, A.F.G., Marolt, D., Gerecht, S., Tajnsek, U., Martens, T.P., Vunjak-Novakovic, G., 2008. Engineered microenvironments for human stem cells. Birth Defects Res. C. Embryo Today 84, 335–347.

Gonzalez-Macia, L., Morrin, A., Smyth, M.R., Killard, A.J., 2010. Advanced printing and deposition methodologies for the fabrication of biosensors and biodevices. Analyst 135, 845–867.

Gruene, M., Deiwick, A., Koch, L., Ph, D., Schlie, S., Unger, C., Hofmann, N., Bernemann, I., Glasmacher, B., Chichkov, B., 2011b. Laser printing of stem cells for biofabrication scaffold-free autologous grafts. Tissue Eng Part C: Methods 17, 79–87.

Gruene, M., Pflaum, M., Deiwick, A., Koch, L., Schlie, S., Unger, C., Wilhelmi, M., Haverich, A., Chichkov, B., 2011c. Adipogenic differentiation of laser-printed 3D tissue grafts consisting of human adipose-derived stem cells. Biofabrication 3, 015005.

Gruene, M., Pflaum, M., Hess, C., Diamantouros, S., Schlie, S., Deiwick, A., Koch, L., Wilhelmi, M., Jockenhoevel, S., Haverich, A., Chichkov, B., 2011a. Laser printing of three-dimensional multicellular arrays for studies of cell-cell and cell-environment interactions. Tissue Eng Part C: Methods 17, 973–982.

Guillemot, F., Souquet, A., Catros, S., Guillotin, B., Lopez, J., Faucon, M., Pippenger, B., Bareille, R., Rémy, M., Bellance, S., Chabassier, P., Fricain, J.C., Amédée, J., 2010. High-throughput laser printing of cells and biomaterials for tissue engineering. Acta Biomater 6, 2494–2500.

Hopp, B., Smausz, T., Kresz, N., Barna, N., Bor, Z., Kolozsvári, L., Chrisey, D.B., Szabó, A., Nógrádi, A., 2005. Survival and proliferative ability of various living cell types after laser-induced forward transfer. Tissue Eng 11, 1817–1823.

Huebsch, N., Arany, P.R., Mao, A.S., Shvartsman, D., Ali, O.A., Bencherif, S.A., Rivera-Feliciano, J., Mooney, D.J., 2010. Harnessing traction-mediated manipulation of the cell/matrix interface to control stem-cell fate. Nat. Mater 9, 518–526.

Hui, E., Bhatia S, 2007. Micromechanical control of cell-cell interactions. Proceedings of the National Academy of Sciences 104, 5722–5726.

Jackman, R.J., Wilbur, J.L., Whitesides, G.M., 1995. Fabrication of submicrometer features on curved substrates by microcontact printing. Science 269, 664–666.

Kaji, H., Camci-Unal, G., Langer, R., Khademhosseini, A., 2011. Engineering systems for the generation of patterned cocultures for controlling cell-cell interactions. Biochim. Biophys. Acta 1810, 239–250.

Kattamis, N.T., Purnick, P.E., Weiss, R., Arnold, C.B., 2007. Thick film laser induced forward transfer for deposition of thermally and mechanically sensitive materials. Appl Phys. Lett 91, 171120.

Kingsley, D.M., Dias, A.D., Chrisey, D.B., Corr, D.T., 2013. Single-step laser-based fabrication and patterning of cell-encapsulated alginate microbeads. Biofabrication 5, 045006.

Kleinman, H.K., Martin, G.R., 2005. Matrigel: basement membrane matrix with biological activity. Semin. Cancer Biol 15, 378–386.

Kleinman, H.K., McGarvey, M.L., Hassell, J.R., Star, V.L., Cannon, F.B., Laurie, G.W., Martin, G.R., 1986. Basement membrane complexes with biological activity. Biochemistry 25, 312–318.

Koch, L., Deiwick, A., Schlie, S., Michael, S., Gruene, M., Coger, V., Zychlinski, D., Schambach, A., Reimers, K., Vogt, P.M., Chichkov, B., 2012. Skin tissue generation by laser cell printing. Biotechnol. Bioeng 109, 1855–1863.

Koch, L., Kuhn, S., Sorg, H., Gruene, M., Schlie, S., Gaebel, R., Polchow, B., Reimers, K., Stoelting, S., Ma, N., Vogt, P.M., Steinhoff, G., Chichkov, B., 2010. Laser printing of skin cells and human stem cells. Tissue Eng. Part C: Methods 16, 847–854.

Lee, L.H., Peerani, R., Ungrin, M., Joshi, C., Kumacheva, E., Zandstra, P.W., 2009b. Micropatterning of human embryonic stem cells dissects the mesoderm and endoderm lineages. Stem Cell Res 2, 155–162.

Lee, W., Debasitis, J.C., Lee, V.K., Lee, J.-H., Fischer, K., Edminster, K., Park, J.-K., Yoo, S.-S., 2009a. Multi-layered culture of human skin fibroblasts and keratinocytes through three-dimensional freeform fabrication. Biomaterials 30, 1587–1595.

Liu, Y., Wang, X., Kaufman, D.S., Shen, W., 2011. A synthetic substrate to support early mesodermal differentiation of human embryonic stem cells. Biomaterials 32, 8058–8066.

Lund, A.W., Yener, B., Stegemann, J.P., Plopper, G.E., 2009. The natural and engineered 3D microenvironment as a regulatory cue during stem cell fate determination. Tissue Eng. Part B: Rev 15, 371–380.

Massia, S.P., Hubbell, J.A., 1991. An RGD spacing of 440 nm is sufficient for integrin alpha V beta 3-mediated fibroblast spreading and 140 nm for focal contact and stress fiber formation. J. Cell Biol 114, 1089–1100.

Meshel, A.S., Wei, Q., Adelstein, R.S., Sheetz, M.P., 2005. Basic mechanism of three-dimensional collagen fibre transport by fibroblasts. Nat. Cell Biol 7, 157–164.

Odde, D.J., Renn, M.J., 2000. Laser-guided direct writing of living cells. Biotechnol. Bioeng 67, 312–318.

Orive, G., Tam, S.K., Pedraz, J.L., Hallé, J.-P., 2006. Biocompatibility of alginate-poly-L-lysine microcapsules for cell therapy. Biomaterials 27, 3691–3700.

Othon, C.M., Wu, X., Anders, J.J., Ringeisen, B.R., 2008. Single-cell printing to form three-dimensional lines of olfactory ensheathing cells. Biomed. Mater 3, 034101.

Ovsianikov, A., Gruene, M., Pflaum, M., Koch, L., Maiorana, F., Wilhelmi, M., Haverich, A., Chichkov B, 2010. Laser printing of cells into 3D scaffolds. Biofabrication 2, 014104.

Pampaloni, F., Reynaud, E.G., Stelzer, E.H.K., 2007. The third dimension bridges the gap between cell culture and live tissue. Nat. Rev. Mol. Cell Biol 8, 839–845.

Patz, T.M., Doraiswamy, A., Narayan, R.J., He, W., Zhong, Y., Bellamkonda, R., Modi, R., Chrisey, D.B., 2005. Three-dimensional direct writing of B35 neuronal cells. Biomed. Mater. Res 78, 124–130.

Pedersen, J., Swartz, M., 2005. Mechanobiology in the third dimension. Ann Biomed. Eng 33, 1469–1490.

Peerani, R., Rao, B.M., Bauwens, C., Yin, T., Wood, G.A., Nagy, A., Kumacheva, E., Zandstra, P.W., 2007. Niche-mediated control of human embryonic stem cell self-renewal and differentiation. EMBO J 26, 4744–4755.

Petrie, T.A., Capadona, J.R., Reyes, C.D., García, A.J., 2006. Integrin specificity and enhanced cellular activities associated with surfaces presenting a recombinant fibronectin fragment compared to RGD supports. Biomaterials 27, 5459–5470.

Pham, Q.P., Sharma, U., Mikos, A.G., 2006. Electrospinning of polymeric nanofibers for tissue engineering applications: a review. Tissue Eng 12, 1197–1211.

Piner, R., Zhu, J., Xu, F., Hong, S., Mirkin C, 1999. "Dip-pen" nanolithography. Science 283, 661–663.

Pirlo, R.K., Dean, D.M.D., Knapp, D.R., Gao, B.Z., 2006. Cell deposition system based on laser guidance. Biotechnol. J 1, 1007–1013.

Raof, N.A., Schiele, N.R., Xie, Y., Chrisey, D.B., Corr, D.T., 2011. The maintenance of pluripotency following laser direct-write of mouse embryonic stem cells. Biomaterials 32, 1802–1808.

Ringeisen, B., Chrisey, D., Krizman, D., Kim, H., Young, P., Spargo B, 2002b. Cell-by-cell construction of living tissue by ambient laser transfer. 2nd Annual International IEEE-EMBS Special Topic Conference on Microtechnologies in Medicine & Biology, 120–125.

Ringeisen, B.R., Barron, J.A., Young, D., Othon, C.M., Wu, P.K., Ladoucuer, D., Spargo, B.J., 2008. Laser Printing Cells. Virtual Prototyping & Bio Manufacturing in Medical Applications (Springer), 207-28.

Ringeisen, B.R., Kim, H., Barron, J.A., Krizman, D.B., Chrisey, D.B., Jackman, S., Auyeung, R.Y.C., Spargo, B.J., 2004. Laser printing of pluripotent embryonal carcinoma cells. Tissue Eng 10, 483–491.

Ringeisen, B.R., Wu, P.K., Kim, H., Piqué, A., Auyeung, R.Y.C., Young, H.D., Chrisey, D.B., Krizman, D.B., 2002a. Picoliter-scale protein microarrays by laser direct write Chem. Biotechnol. Prog 18, 1126–1129.

Roth, E.A., Xu, T., Das, M., Gregory, C., Hickman, J.J., Boland, T., 2004. Inkjet printing for high-throughput cell patterning. Biomaterials 25, 3707–3715.

Saunders, R.E., Gough, J.E., Derby, B., 2008. Delivery of human fibroblast cells by piezoelectric drop-on-demand inkjet printing. Biomaterials 29, 193–203.

Schaub, N.J., Britton, T., Rajachar, R., Gilbert, R.J., 2013. Engineered Nanotopography on Electrospun PLLA Micro fibers Modifies RAW 264. 7 Cell Response ACS Appl. Mater. Interfaces 5, 10173–10184.

Schiele, N.R., Chrisey, D.B., Corr, D.T., 2011. Gelatin-based laser direct-write technique for the precise spatial patterning of cells. *Tissue Eng* Part C: Methods 17, 289–298.

Schiele, N.R., Corr, D.T., Huang, Y., Raof, N.A., Xie, Y., Chrisey, D.B., 2010. Laser-based direct-write techniques for cell printing. Biofabrication 2, 032001.

Schiele, N.R., Koppes, R.A., Corr, D.T., Ellison, K.S., Thompson, D.M., Ligon, L.A., Lippert, T.K.M., Chrisey, D.B., 2009. Laser direct writing of combinatorial libraries of idealized cellular constructs: biomedical applications. Appl. Surface Sci. 255, 5444–5447.

Tasoglu, S., Demirci, U., 2013. Bioprinting for stem cell research. Trends Biotechnol 31, 10–19.

Tien, J., Nelson, C.M., Chen, C.S., 2002. Fabrication of aligned microstructures with a single elastomeric stamp. Proceedings of the National Academy of Sciences 99, 1758–1762.

Wu, P.K., Ringeisen, B.R., 2010. Development of human umbilical vein endothelial cell (HUVEC) and human umbilical vein smooth muscle cell (HUVSMC) branch/stem structures on hydrogel layers via biological laser printing (BioLP). Biofabrication 2, 014111.

Wu, P.K., Ringeisen, B.R., Callahan, J., Brooks, M., Bubb, D.M., Wu, H.D., Pique, A., Spargo, B., McGill, R.A., Chrisey, D.B., 2001. deposition, structure, pattern deposition, and activity of biomaterial thin-films by matrix-assisted pulsed-laser evaporation (MAPLE) and MAPLE direct write. Thin Solid Films, 398–399, 607-14.

Wu, P.K., Ringeisen, B.R., Krizman, D.B., Frondoza, C.G., Brooks, M., Bubb, D.M., Auyeung, R.C.Y., Piqué, A., Spargo, B., McGill, R.A., Chrisey, D.B., 2003. Laser transfer of biomaterials: matrix-assisted pulsed laser evaporation (MAPLE) and MAPLE direct write *Rev.* Sci. Instrum 74, 2546.

Xie, X.N., Chung, H.J., Sow, C.H., Wee, A.T.S., 2006. Nanoscale materials patterning and engineering by atomic force microscopy nanolithography *Mater.* Sci. Eng. R Reports 54, 1–48.

Xie, Y., Castracane, J., 2009. High-voltage, electric field-driven micro/nanofabrication for polymeric drug delivery systems. Eng Med. Biol, 23–30.

Xu, C., Inokuma, M.S., Denham, J., Golds, K., Kundu, P., Gold, J.D., Carpenter, M.K., 2001. Feeder-free growth of undifferentiated human embryonic stem cells. Nat. Biotechnol 19, 971–974.

Xu, T., Jin, J., Gregory, C., Hickman, J.J., Boland T, 2005. Inkjet printing of viable mammalian cells. Biomaterials 26, 93–99.

Yamato, M., Kwon, O., Hirose, M., Kikuchi, A., Okano, T., 2001. Novel patterned cell coculture utilizing thermally responsive grafted polymer surfaces *J.* Biomed. Mater. Res 55, 137–140.

Yousaf, M.N., Houseman, B.T., Mrksich, M., 2001. Using electroactive substrates to pattern the attachment of two different cell populations. Proceedings of the National Academy of Sciences 98, 5992–5996.

Zhang, C., Liu, D., Mathews, S.A., Graves, J., Schaefer, T.M., Gilbert, B.K., Modi, R., Wu, H., Chrisey, D.B., 2003. Laser direct-write and its application in low temperature Co-fired ceramic (LTCC) technology. Microelectron. Eng 70, 41–49.

BIOMATERIALS FOR BIOPRINTING

6

Lee Jia Min, Tan Yong Sheng Edgar, Zhu Zicheng and Yeong Wai Yee

School of Mechanical & Aerospace Engineering, Nanyang Technological University, Singapore

6.1 INTRODUCTION

Modern medicine and research rely heavily on biomaterials. Biomaterials are used in small objects, such as early diagnosis devices and sustained drug release, and in larger objects, such as permanent implants. In the case of bioprinting, such biomaterials serve as an artificial extracellular matrix (ECM) for the cells. They provide structural support and enable cellular attachment, help isolate tissues and cells, and regulate cellular functions and behaviors (Chan and Leong, 2008). For biomaterials to be considered for bioprinting, they need to possess certain characteristics, just as the process used to fabricate them must meet its own set of requirements.

Hydrogels make up the bulk of ink used in bioprinting. They are generally used as they closely mimic the natural ECM in our body. Additionally, due to their high water content, hydrogels are able to protect cells and fragile drugs, such as peptides and proteins. These polymers can be modified with ligands for better cell adhesion.

This chapter will first discuss the important characteristics of "ink" used in bioprinting. It will then explore different cross-linking mechanisms of hydrogel and describe examples of hydrogel used in bioprinting. Finally, potential developments and applications as well as research directions will be discussed.

6.2 PARAMETERS FOR BIOPRINTING
6.2.1 BIOCOMPATIBILITY

The biocompatibility of biomaterials is the most important prerequisite for their use as a scaffold. Biomaterials have to support cell attachment, migration, and other basic cellular functions (Chan and Leong, 2008). Chemicals involved in cross-linking of hydrogels should not affect cell viability (Bryant, 2008). Some photo-initiators and monomers used during cross-linking of hydrogels have been found to be toxic if left unreacted (Bryant et al., 2000).

To evaluate biocompatibility, live/dead staining is usually employed after 24 h to determine the cytocompatibility of the hydrogel and cross-linking process used with the cells. The results can be quantified using a microplate reader to measure the fluorescence intensity through which the percentage of viable cells can be determined (Klouda et al., 2009). MTS proliferative assay can be performed on alternate days to assess cytocompatibility and to determine the metabolic activity of the cells (Park et al., 2009).

6.2.2 SHEAR THINNING AND THIXOTROPY

Non-Newtonian behavior, which shows the viscosity of the biomaterials decreasing as shear rate increases, is one of the fundamental requirements for a bioprintable material. During dispensing, the shear force reorganizes the random polymer chains into an aligned conformation that reduces hydrogel viscosity during the process (Yucel et al., 2009). The effect of shear thinning varies among biomaterials with high molecular weight. This property is important for bioprinting as it enables easy dispensing of the fluid when pressure is applied, and allows the fluid to return to its gel state by removing its stress. However, the hydrogel must be given time to return to its viscous state. The degree of this behavior can be determined through a hysteresis loop test (Barbucci et al., 2008). During this test, the sample is subjected to increasing and decreasing shear. By determining the area between the two curves, the degree of thixotropy can be known.

6.2.3 YIELD STRESS

Yield stress is the instantaneous pressure required to initiate the flow of material. For bioprinting, this property affects the required minimum stress applied on the material before it is dispensed This property is important as the stress applied on the gel will in turn affect the shear stress induced on the cells during printing. Physically cross-linked soft polymers form a fragile network when at rest. When shear force above the yield stress is applied, the network breaks. It slowly forms back when shear force is detached. Such stress is responsible for delaying the flow and collapse of the gel structure. Yield stress also helps keep the cells homogenous in the hydrogel reservoir (Malda et al., 2013). To determine the yield stress required for these materials, an extrapolation of the flow curve at low shear rate to zero shear rate can be done (Malana et al., 2012). However, researchers have suggested the use of the Bingham equation ($\tau = \tau B + \mu B \gamma$) and modified Bingham equation ($\tau = \tau MB + \mu MB \gamma + C \gamma^2$) at low shear rates to determine these values, as doing so minimizes errors during extrapolation of the flow curve.

6.2.4 WATER CONTENT

The amount of water in hydrogel determines the adsorption rate and the diffusion of the solutes through the material. Hydrogel's water content must be controlled; if it is too high, cell proliferation rate will deteriorate (Ogawa et al., 2010). Water can be bound to the gel by two interactions. First, the polar hydrophilic groups are hydrated and the network of polymer swells. Second, hydrophobic groups in the gels are exposed and water molecules are bound to the hydrophobic groups (Hoffman, 2002). Additional water is absorbed as free water, which fills the space such as macropores and voids formed between the chains. Hydrogel's water content can easily be determined by either measuring the percentage change in the mass between the hydrogel and its dry form or determining the gel's light absorbance (Huglin and Yip, 1987; Nagaoka et al., 1990).

6.2.5 SWELLING BEHAVIOR

Swelling and contraction of the bioprinted construct have to be considered and accounted for during the design of the tissue construct. Differences in the swelling properties among different hydrogels will cause complications, such as restricted grafting of the layers and overall uneven deformation of the construct (Murphy et al., 2013). Swelling behavior is also strongly dependent on the local environment, such as pH, ionic strength, and solvent conditions. Swelling rates also affect the surface properties of the gel, the solute diffusion coefficient through hydrogels, and the optical properties. The dynamic swelling and equilibrium in solutions of hydrogel governs the mechanical integrity of hydrogels. As described earlier, most water molecules within hydrogel are bound by either hydrophilic or hydrophobic groups. Most solutes can only diffuse into and through hydrogel from within the unbound regions of the macropores and voids. Solutes that are chaotropic can diffuse through hydrogel by disturbing the interaction of the bound water layers around the polymer chain.

The swelling behavior of hydrogel can be thermodynamically described by the Flory–Huggins theory (Flory and Rehner, 1943). However, this theory does not account for the imperfections of the network or the finite volume of network chains and cross-links in the gel. In this model, the equilibrium of cross-linked polymer network swelling was described by the elastic forces of the polymer chains and the thermodynamic compatibility of the hydrogel polymer and the solvent, as shown in the Gibbs Free energy equation shown here. Swelling rates can be controlled by copolymerizing monomers of varying hydrophobicity.

$$\Delta G_{total} = \Delta G_{elastic} + \Delta G_{mixing}$$

However, for ionic gels, the total free energy contribution by hydrogel would require the involvement of the ionic properties of the network.

$$\Delta G_{total} = \Delta G_{elastic} + \Delta G_{mixing} + \Delta G_{ionic}$$

At equilibrium potential, the net chemical potential between the solvent and the surrounding solution is zero. This zero net chemical potential balances the elastic and mixing potential. The Flory–Rhener theory evolves the expression that the hydrogel prepared in the absence of solvent should contain the following relationship:

$$\frac{1}{M_c} = \frac{1}{M_N} - \frac{\left(\dfrac{\bar{v}}{V_1}\right)\left[\ln(1 - v_2) + v_{2,s} + x_1 v_2^2\right]}{v_2^{\frac{1}{3}} - \dfrac{v_2}{2}}$$

where $\bar{v}$ is the specific volume of the hydrophilic polymer, M_N is the primary molecular mass, M_c is the average molecular mass between the cross-links, v_2 is the volume fraction of the polymer in the swollen mass, and V_1 is the molar volume of the solvent. The mixing term ΔG_{mixing} depends on the compatibility of the hydrophilic polymer and the solvent, and is expressed as the polymer-solvent interaction parameter, χ_1.

The equation is then modified by Peppas and Merrill for gels prepared in the presence of a solvent by including the changes in the elastic potential due to the solvent (Peppas and Merrill, 1977).

$$\frac{1}{M_c} = \frac{1}{M_N} - \frac{\left(\frac{\bar{v}}{V_1}\right)\left[\ln(1-v_{2,s})+v_{2,s}+x_1 v_{2,s}^2\right]}{v_{2,r}\left[\left(\frac{v_{2,s}}{v_{2,r}}\right)^{\frac{1}{3}} - \frac{v_{2,s}}{2v_{2,r}}\right]}$$

Here, the term $v_{2,r}$ and $v_{2,s}$ show the volume fraction of the polymer in the relaxed and swollen state. The relaxed state refers to the cross-linked polymer without the occurrence of swelling. Thus, in the absence of a solvent, the volume fraction of the relaxed state is 1, simplifying the equation back to the Flory–Rhener equation.

For ionic gels, two separate but equivalent expressions were derived for anionic and cationic gels (Brannon-Peppas and Peppas, 1991). These equations for average molecular weight between cross-links require ionic strength I and dissociation constants K_a and K_b.

$$\left(\frac{V_1}{4I}\right)\left(\frac{v_{2,s}^2}{\bar{v}}\right)\left(\frac{K_a}{10^{-pH}-K_a}\right)^2$$

$$= [\ln(1-v_{2,s})+v_{2,s}+x_1 v_{2,s}^2]+\left(\frac{V_1}{\bar{v}M_c}\right)\left(1-\frac{2M_c}{M_N}\right)v_{2,r}\left[\left(\frac{v_{2,s}}{v_{2,r}}\right)^{\frac{1}{3}} - \frac{v_{2,s}}{2v_{2,r}}\right]$$

$$\left(\frac{V_1}{4I}\right)\left(\frac{v_{2,s}^2}{\bar{v}}\right)\left(\frac{K_a}{10^{pH-14}-K_a}\right)^2$$

$$= [\ln(1-v_{2,s})+v_{2,s}+x_1 v_{2,s}^2]+\left(\frac{V_1}{\bar{v}M_c}\right)\left(1-\frac{2M_c}{M_N}\right)v_{2,r}\left[\left(\frac{v_{2,s}}{v_{2,r}}\right)^{\frac{1}{3}} - \frac{v_{2,s}}{2v_{2,r}}\right]$$

6.2.6 ELASTICITY

When pressure is applied on the hydrogel, it displays a variety of responses, from rapid recovery to slow time-dependent approach. This characteristic is dependent on the hydrogel's glass transition state (Tg). At temperatures below Tg, the rearrangement of the polymer segments is slowed down. Thus, the gel becomes more viscoelastic. During this system, the mechanical properties of the gel, such as creep, stress relaxation, and dynamic loading, are most important. However, during tissue culturing, the hydrogel is usually water swollen to allow cell to proliferate. The excess water and solutes entering the hydrogel plasticize it and induce a reduction in the polymer glass transition due to the swelling. This decreases the Tg to below 37 °C and subsequently causes the hydrogel to move toward a rubbery elastic region.

6.2.7 SOLUTE TRANSPORTATION

Another important aspect of designing hydrogel for tissue engineering is the development of an effective solute transport. This determines how cellular products, nutrients, and waste are exchanged within the scaffold. The driving force of solute transportation is diffusion since the pores within the gels are too small for convection to play a significant role. In ionic gels, research has revealed that hydrogel mesh size and culture conditions such as pH and temperature play an important role in diffusion (am Ende et al., 1995). In biological systems, the hydrophilic polymer and their solutes frequently interact and ionize with their surroundings, thus becoming an important factor in determining transport behavior. Further investigation has shown that this interaction tends to decrease transport rate of solutes into the hydrogel (Collins and Ramirez, 1979). Another study on the permeability of the hydrogel has also confirmed the significance of this polymer-solute interaction and shown that it can be controlled by size exclusion (Gudeman and Peppas, 1995).

For nutrient transportation, an important measure for determining the maximum size of solute that can diffuse into the gel is its porosity. This can be predicted by the mesh size, $\in$, and the average linear distance between the cross-links. To predict the mesh size, we use an equation containing the bond length of the polymer backbone, l (which is often 1.54 due to the carbon-carbon bone), the characteristic ratio, C_N, the average molecular weight between the cross-links, M_c, the swollen polymer volume fraction, $v_{2,s}$ and the molecular weight of the monomer, M_r (Peppas et al., 2006; Tesoro, 1984).

$$\in = (v_{2,s})^{\frac{1}{3}} \left(\frac{2 C_N M_c}{M_r} \right)^{1/2}$$

In tissue-engineered hydrogel, the mesh size of hydrogel has to include the effects of the solution containing salts, ions, and nutrients. Thus the swollen polymer fraction $v_{2,s}$ has to be revised to account for the solution used.

All solute transport models involving hydrogel are based on diffusion. This enables the transport rate of solutes to be described using Fick's Law (Crank, 1979; Park and Crank, 1968). Fickian diffusion is however only applicable if the gel is amorphous. If the gel is significantly heterogenous, this law is no longer sufficient.

$$\frac{\partial c_i}{\partial t} = \nabla.(D_{ig}(C_i)\nabla C_i)$$

where c_i is the concentration of the species i and D_{ig} is the concentration-dependent coefficient of species i in the hydrogel.

Another model based on the free volume theory for water solute and polymer was developed by Peppas et al. to predict the dependency of the solute diffusion coefficient on solute size, mesh size, degree of swelling, and other structural properties of the hydrogel (Peppas and Reinhart, 1983).

$$\frac{D_{SM}}{D_{Sw}} = \frac{k_1 \left(M_C - M_C^* \right)}{M_N - M_C^*} \exp\left(-\frac{K_2 r_s^2}{Q_M - 1}\right)$$

where D_{SM} and D_{Sw} are the diffusion coefficients of the solutes in the membrane and solvent. This ratio is also known as the normalized diffusion coefficient. k_1 and k_2 are the structural parameters of the polymer–water complex. M_C^* is the average critical molecular weight between the cross-links at which diffusion is excluded. r_s is the Stokes hydrodynamic radius of the solute and Q_m is the degree of swelling of the hydrogel. This experiment was validated using PVA membrane.

6.2.8 **DEGRADATION**

The degradation rate of a material is usually matched with the rate of tissue regeneration (Dhanda-yuthapani, 2011). This is to ensure that the scaffold can still provide the necessary support for cells to regenerate completely. If the degradation rate is too short, the hydrogel support may be encapsulated by fibrous tissue when implanted into the body, blocking the overall cellular regeneration. If the degradation rate is too long, chronic inflammatory response may set in, which in turn will cause greater damage to the body.

Unlike solid polymers, hydrogels undergo purely bulk degradation since they are hydrated within the structures. To control the degradation of these polymers, researchers have come up with hydrogels that have been copolymerized with peptides sensitive to enzymatic degradation (Drury and Mooney, 2003). One such research by Lei controlled the degradation with the use of polyethylene glycol (PEG) hydrogel that incorporates with matrix metalloproteinase (MMP) sensitive peptides (Lei et al., 2010). These peptides respond to local protease activity on the cell surface and degrade accordingly.

Determination of these changes in degradation can be characterized by observing changes in their weight, elastic modulus, and swelling degree with respect to their initial condition. The hydrolysis and enzymatic degradation of the hydrogel polymers reduce the efficiency of the cross-linking mechanism in the gel thus reducing the above properties of the gel (Lee *et al.*, 2004).

6.3 **HYDROGEL CROSS-LINKING MECHANISM**

In general, all biomaterials used for bioprinting are made of cross-linking mechanisms that can be categorized as either physical or chemical. The type of cross-linking initiated on the hydrogel will yield either a reversible or a permanent gel, which in turn will affect the printing process. Permanent hydrogels created by chemical cross-linking require postcuring, but have higher mechanical strength compared to their physically cross-linked counterparts. The different cross-linking mechanisms also determine the mechanical strength of the scaffold and the compatibility of the cells with the hydrogel.

In the cross-linked state, hydrogels can attain an equilibrium swelling in aqueous solutions depending on the cross-linking density. These hydrogels are also not homogenous (Drumheller and Hubbell, 1995). They have regions of low water swelling clusters that are dispersed within the regions of high swelling clusters. This may be caused by hydrophobic aggregation of cross-linking molecules leading to higher cross-linking density clusters within the region. Phase separation and formation of water-filled voids can occur in some circumstances during gel formation, depending on the solvent composition, temperature, and solid concentration used.

6.3.1 **PHYSICAL**

Physically cross-linked hydrogels are formed by molecular entanglements, and other secondary forces such as ionic, hydrogen bonding, and hydrophobic forces (Campoccia et al., 1998). They are mostly reversible; however, they cannot easily become homogenous as the clusters of molecular entanglements, or hydrophobic and ionic domains found within the polymer chains, are nonhomogenous (Hoffman, 2002). These free chain ends or chain loops create a temporary network defect within the gels.

6.3.1.1 Ionic Cross-linking

Ionic hydrogel is one of the most commonly used physical hydrogel. It has a reversible characteristic that helps to ensure consistent viscosity during printing. Moreover, it has excellent cytocompatibility since no other chemicals are used. These interactions can be between oppositely charged polyelectrolytes or between polyelectrolytes and charged molecules. The length of the polymers and their charges determine the strength of the ionic interactions. However, due to its simplicity in cross-linking interactions, its construct is also mechanically weak.

6.3.1.2 Thermal Cross-linking

Thermo-sensitive gelation and cross-linking is triggered by interactions between polymer chains in the solution. These interactions are usually reversible and form cross-linked polymers at low temperatures, but viscous solutions above a specific temperature. This behavior can be commonly seen in many naturally occurring, water-soluble polymers, such as gelatin and carrageenan.

However, researchers have developed a reverse thermal behavior hydrogel that forms a gel at higher temperatures but melts at lower temperatures (Yoshioka et al., 1994). This mechanism of gelation is related to the temperature dependency of polymer–solvent interaction and a heterogenous polymer microstructure. Part of the chain becomes more hydrophobic at higher temperatures, inducing a local cross-linking, and consequently, a gelation process. Some of the common hydrogels that participate in such behaviors are methylcellulose and Poly(N-isopropylacrylamide) hydrogels.

6.3.1.3 Stereocomplex Cross-linking

Stereocomplex formation of hydrogel occurs between polymers of opposite chirality, interactive with one another (Hennink et al., 2004). They have similar characteristics in terms of properties and advantages as ionic cross-linked hydrogel. These hydrogels usually take longer to stabilize.

6.3.2 CHEMICAL

Chemically cross-linked hydrogels are formed by chemically binding the chains of polymer with one another. These polymers are permanent and can be generated by covalently cross-linking hydrophilic polymers or converting hydrophobic polymers into hydrophilic polymers. Since these hydrogels' formation is permanent, dispensing of such materials is usually done before cross-linking reaction has occurred.

6.3.2.1 Free Radical Polymerization

Free radical polymerization is one of the most frequently used techniques in bioprinting to create chemically cross-linked hydrogels. Polymer with vinyl groups are usually polymerized to form hydrogel using redox reaction or through the use of thermal or photo-initiator. The photo-initiators used are wavelength-specific and must meet a minimum energy level before reaction can occur. These reactions are usually fast and the gels are usually permanently cross-linked. However, UV light exposed to cells has to be controlled as the UV radiation may cause cellular necrosis due to the heat released during curing. Long exposure time may also cause damage and mutations in the DNA of the cells, leading to long-term damage to cellular metabolic activity. Thus, the intensity of the UV light is usually limited to between 5 and 10 mW/cm^2 to prevent such occurrence (Cho et al., 2009).

6.3.2.2 Functional Group Cross-linking
6.3.2.2.1 Schiff-base Formation

During a Schiff-base formation, the polymer with an amine side chain attaches itself to a carbonyl group via a nucleophilic addition to form a functional group containing a carbon–nitrogen double bond with the nitrogen atom connecting to another carbon chain. Such chemical methods can be done to cross-link natural polymers such as collagen and other proteins using glutaraldehyde. However, the excessive use of aldehyde may be toxic to the body and induce reactions within the body (Tan et al., 2009). These linkages are also unstable in low pH, thus these chemical bonds may break if foreign body response is triggered at the site of transplantation (Walt and Agayn, 1994).

6.3.2.2.2 Michael Type Addition

Michael addition is another nucleophilic addition reaction to produce hydrogels. In this reaction, a carbanion is attached to an unsaturated carbonyl compound. The nucleophile or carbanion usually contains an electron-withdrawing group that makes the hydrogen more reactive. The acceptor usually contains a ketone or an amine. These reactions usually require a milder reaction condition and allow more tunable properties, thus making them more suitable for cell encapsulation purposes (Fairbanks et al., 2009; Metters and Hubbell, 2005). However, excess or unreacted thiol polymers in the group may be cytotoxic to the cells (Di Monte et al., 1984).

6.3.2.2.3 Native Chemical Ligation

Native chemical ligation of thioesters is one of the naturally occurring chemical reactions. They help in the synthesis of numerous cellular components, such as peptides (Paramonov et al., 2005) and lipids (Reulen et al., 2007), and can easily be synthesized. Some of the typical functional groups involve the use of N-terminal cysteine and aldehydes (Yeo et al., 2004). They are relatively nonreactive to aminolysis, but reacts readily to a thiol group through transesterification to form a new thioester. These reactions are usually highly substrate- specific, efficient in cross-linking, and occur in natural conditions. However, they have complicated synthesis procedures as they involve protecting and deprotecting the peptides.

6.3.2.2.4 Click Chemistry

Click Chemistry is not a specific chemical reaction but one of the reactions created to mimic naturally occurring polymerization (Crescenzi et al., 2007). Developed to aid in the discovery of new pharmaceuticals (Hein et al., 2008), this reaction has been described as modular, wide in scope, high-yield, stereo-specific, and unable to generate offensive byproducts. However, the use of such chemistry may prove toxic as copper ions are used as catalysts in the process. The use of copper causes the formation of reactive oxygen species in the body (Li and Trush, 1993). Research efforts are underway to develop a copper-free method for click chemistry (Baskin et al., 2007; Orski et al., 2010).

6.4 HYDROGELS IN BIOPRINTING

Hydrogels have drawn significant attention in bioprinting and tissue engineering due to their high water content. Networks of hydrophilic polymer chains are cross-linked through chemical or physical interaction. Hydrogels formed through chemical bonds, such as covalent bonds, are permanent and irreversible. On the other hand, hydrogels formed through physical interactions (e.g. hydrogen

Table 6.1   Classifications of hydrogel
Classification of hydrogels (Ratner et al., 2013)

Method of preparation	Ionic hydrogel	Physiochemical structural features	Complexation hydrogel

i) Homopolymer
  – One type of hydrophilic monomer unit
ii) Copolymer
  – Crosslinking two comonomer units with at least on hydrophilic monomer
iii) Multi polymer
  – Three of more comonomers
iv) Interpentrating Network
  – Intermeshed network of monomer with cross-linked hydrogel of different mechanism
	i) Neutral
	  – Uncharged
	ii) Anionic
	  – Negative charges only
	iii) Cationic
	  – Positive charges only
	iv) Ampholytic
	  – Positive and negative charges
		i) Amorphous
		  – Covalent crosslinks
		  – Random arrangement
		ii) Semicrystalline
		  – May or may not have covalent crosslinks
		  – Self-assembled, ordered macromolecular chains
			i) Hydrogen bonds
			ii) Hydrophobic group associations
			iii) Affinity Complexes
			  – Heterodimers
			  – biotin/streptavidin
			  – antibody/antigen
			  – conA/glucose
			  – PDLA/PLLA
			  – stereocomplexes
			  – Cyclodextrin Inclusion complexes

bonds, ionic forces, physical entanglements of individual polymer chains, polymer crystallites, and hydrophobic) are reversible under certain conditions. Alternatively, a hydrogel can be formed through a combination of two or more of the mentioned interactions. There are a number of ways to categorize hydrogels, based on preparation methods, ionic charge, or physico-chemical structural features, and these are summarized in Table 6.1 (Ratner et al., 2013). In this section, examples of hydrogels used are discussed based on their origin, for example, natural vs. synthetic.

6.4.1 NATURAL HYDROGEL

Natural hydrogels have been heavily utilized for tissue engineering because they usually already contain specific bioactive regions that give them good cellular compatibility with the cells of interest. Since they are usually produced through biological methods, they also tend to have a more defined structure and a monodispersed molecular weight. They are usually biodegradable and possess mechanical properties similar to that of the cell's natural ECM. However, they have issues concerning immunogenicity and are relatively unstable compared to their synthetic counterparts. They also vary in terms of properties between species, tissue, and batch of production. In addition, characterizing their innate properties can be difficult due to their complex composition.

6.4.1.1 Collagen

Collagen, a well-known protein, has been extensively used in biomedical applications as it is the main component of natural ECM and the most abundant protein in mammalian tissues (Lee et al., 2001). Although a large number of natural and synthetic polymers are used as biomaterials, the characteristics of collagen are distinct from those of synthetic polymers, specifically in its mode of interaction in the body. More than 20 different types of collagen have been found, but their basic structures are the same, namely, consisting of three polypeptide chains. These three chains wrap around one another, thereby creating a three-stranded rope structure. The strands are held together by covalent bonds and hydrogen. Stable collagen fibers can immediately form as a result of strands self-aggregating. Furthermore, the mechanical properties of collagen fibers can be enhanced by introducing chemical cross-linkers such as carbodiimide and glutaraldehyde (Lee et al., 2002). Collagen is naturally degraded by metalloproteases, hence, the degradation process can be locally controlled by cells in the engineered tissue.

6.4.1.2 Gelatin

Gelatin, a protein-based polymer, is derived through partial hydrolysis of collagen. Chemical pretreatment followed by heat treatment disorganizes collagen protein structure, resulting in helix-to-coil transition and conversion into soluble gelatin. Gelatin has a lower antigenicity compared to collagen (Ratner et al., 2013). It undergoes gelation during a change in temperature. However, gelatin has been modified to photopolymerizable hydrogel by the addition of the methylacrylate group (GelMA). Several studies have used GelMA for bioprinting (Billiet et al., 2014; Duan et al., 2013; Soman et al., 2013; Visser et al., 2013).

6.4.1.3 Fibrin

Fibrin is a biopolymer formed through thrombin-mediated cleavage of fibrinogen, which resulted in self-associated, insoluble fibrin monomer (Ahmed et al., 2008). This naturally occurring material has been used as injectable scaffolds and cell delivery vehicles (Dare et al., 2007). The major advantage is that fibrinogen can be autologously obtained from the plasma, which reduces the risks of foreign body reaction. Moreover, fibrin has good adhesion capabilities. Generally, fibrin is used as a glue to control bleeding and adhere tissues in surgery. In addition, it has also shown promise in skin grafting and in the delivery of exogenous growth factors to reduce wound healing time.

6.4.1.4 Alginate

Alginate, a natural polymeric material, is derived from brown seaweed and bacteria. It has a wide range of medical applications, particularly in drug stabilization and delivery, and cell encapsulation. It is a linear polysaccharide copolymer of α-L-guluronic acid (G) and (1–4)-linked β-D-mannuronic acid (M) monomers (see Figure 6.1). The G and M monomers are distributed sequentially in either alternating or repeating blocks (Donati and Paoletti, 2009; Johnson et al., 1997). Under neutral pH, alginate is a polyanion. It has low toxicity and gels under gentle conditions. The species, location, and age of seaweed are the influential factors that determine the amount and distribution of each monomer. Alginate gels are formed as a result of divalent cations, such as Ba^{2+}, Ca^{2+}, or Sr^{2+}, cooperatively interacting with G monomer blocks to generate ionic bridges between different polymer chains. By changing and adjusting the G and M ratio together with the molecular weight of the polymer chain, the cross-linking density and the resulting mechanical properties can be easily manipulated. It is noted that ionically cross-linked alginate hydrogels undergo slow and uncontrolled dissolution rather than following a specific degradation trend. Mass is gradually lost through ion exchange of calcium and, consequently, individual chains are dissociated, leading to the loss of mechanical stiffness over time.

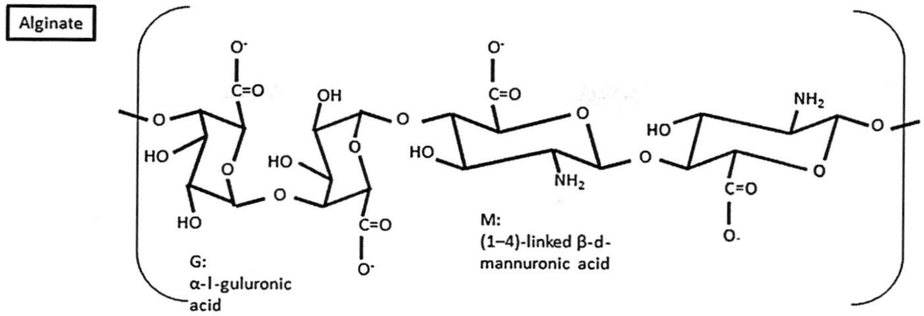

FIGURE 6.1

Chemical structure of alginate.

6.4.1.5 Chitosan and Chitin

Chitin is a copolymer of *N*-acetyl-glucosamine and *N*-glucosamine units randomly or block distributed throughout the biopolymer chain, depending on the processing method used to derive the biopolymer (Khor and Lim, 2003). The biopolymer is termed chitin if the number of *N*-acetyl-glucosamine units is higher than 50%. Otherwise, the biopolymer is termed chitosan. Chitin and chitosan are commercially obtained from shellfish sources such as shrimps and crabs.

Chitosan is a variant of chitin, a partially deacetylated counterpart. Figure 6.2 depicts the difference in molecular structure of chitin and chitosan. Chitosan is degradable by enzymes in humans and is structurally similar to naturally occurring glycosaminolglycans. Chitosan is soluble in dilute acids and once dissolved, it can be gelled by extruding the solution into a nonsolvent or increasing the pH value. Chitosan is degraded by lysozyme and the degradation kinetics is

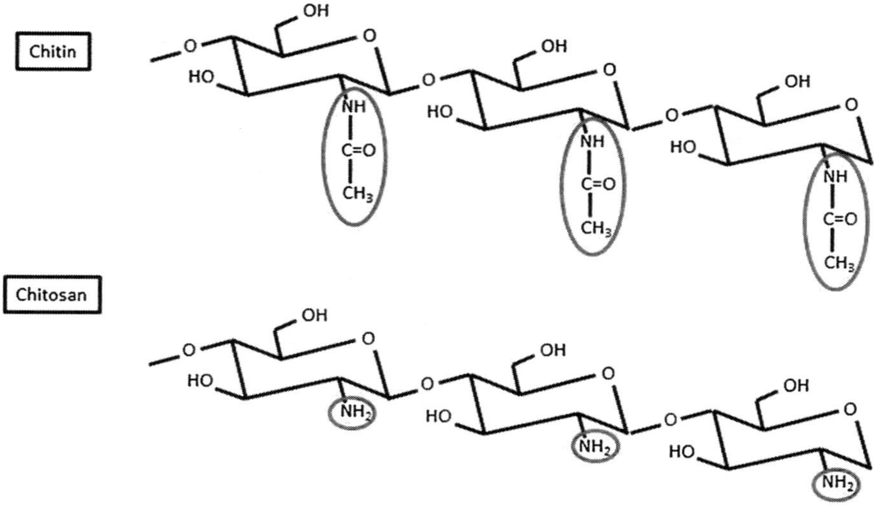

FIGURE 6.2

Difference between the chemical structure of chitin and chitosan.

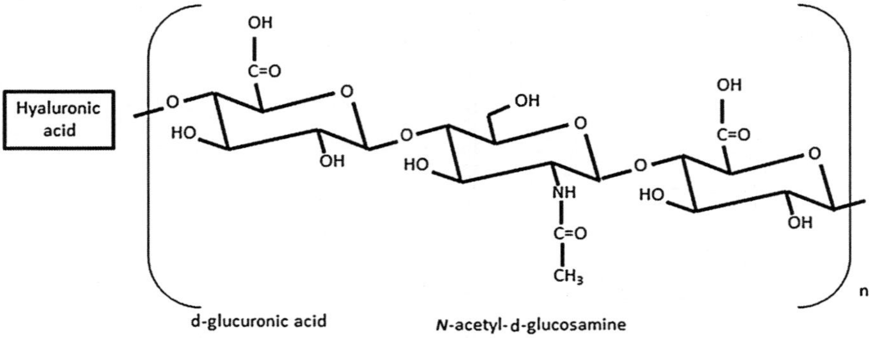

FIGURE 6.3

Chemical structure of hyaluronic acid.

inversely proportional to the degree of crystallinity. Its degraded product includes chitosan with lower molecular weight, N-acetyl-D-glucosamine residues, and chitogligomers (Ren et al., 2005).

6.4.1.6 Hyaluronic acid

Hyaluronic acid (HA) is a linear polysaccharide consisting of a repeating disaccharide of (1–3) and (1–4)-linked β-D-glucuronic acid, and N-acetyl-β-D-glucosamine units (Alberts et al., 1994). The molecular structure is shown in Figure 6.3. It is a highly hydrated polyanionic macromolecule (Khan and Ahmad, 2013) and can be found in almost every mammalian tissue and fluid. It has been widely used in wound healing and the synovial fluid of joints. A number of methods can be used to form HA hydrogels, such as covalent cross-linking with hydrazide derivatives, polyfunctional epoxides, carbodiimides, self-crosslink, and esterification (Collins and Birkinshaw, 2013). The reactive groups of HA (–OH and –NHCOCH$_3$) provides cross-linking sites for hydrogel formation or can be modified for enhanced properties (Khan and Ahmad, 2013). HA is naturally degraded by hyaluronidase, which allows cells to regulate the clearance of the material locally.

6.4.2 SYNTHETIC HYDROGELS

Synthetic hydrogels are some of the earliest biomaterials used as scaffolds in tissue engineering. The natural hydrogels presented in Section 4.1.1 are becoming increasingly popular due to their inherent biocompatibility. Comparatively, synthetic hydrogels have shown their benefits (e.g. highly tunable and consistent properties, and large-scale production capacity) in the field of regenerative medicine. This subsection presents three of the most commonly used synthetic hydrogels.

6.4.2.1 Poly(2-hydroxethyl methacrylate)

Poly(2-hydroxethyl methacrylate) (PHEMA) hydrogels have been used as implant materials since the late 1960s. They are synthesized using free radical precipitation polymerization of 2-hydroxyethyl methacrylate (Slaughter et al., 2009). The resultant hydrogel is biologically inert but relatively weak. It is also a highly resistant material to protein adsorption and cell adhesion. However, it has been found that PHEMA implants undergo delayed, episodic calcification *in vivo* (Belkas et al., 2005). PHEMA is a neutrally charged gel and the molecular structure of neutrally charged PHEMA repeat units is shown in Figure 6.4.

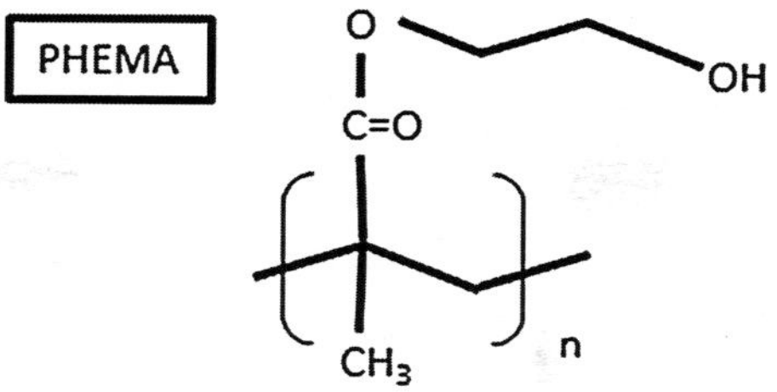

FIGURE 6.4

Neutrally charged PHEMA monomer.

6.4.2.2 Poly(vinyl alcohol)

Poly(vinyl alcohol) (PVA) (Figure 6.5) is prepared from the partial hydrolysis of poly(vinyl acetate). It can be cross-linked into a gel through chemical or physical methods (e.g. via treatment with monoaldehydes). PVA can also be photocured to fabricate hydrogels as an alternative to chemical cross-linking. It is similar to PHEMA in terms of having available pendant alcohol groups that function as attachment sites for biological molecules. PVA is an elastic material, which means it can induce matrix synthesis or cell orientation by enhancing the transmission of mechanical stimuli to seeded cells (Schmedlen et al., 2002). PVA hydrogels are neutral and nonadhesive to proteins and cells. They have a low friction coefficient and their structural properties are similar to natural cartilage. More importantly, they are generally stronger than most other synthetic gels, making them successful for avascular tissue. Furthermore, PVA can be copolymerized with Poly(ethylene glycol) (PEG) to produce a biodegradable hydrogel that has a degradation rate faster than that of PEG hydrogels but slower than that of PVA homopolymer hydrogels.

6.4.2.3 Poly(ethylene glycol)

PEG (Figure 6.6) is a biocompatible and hydrophilic material, with properties that limit antigenicity, immunogenicity, cell adhesion, and protein binding (Alcantar et al., 2000). PEG homopolymer is a polyether that can be polymerized from ethylene oxide by condensation. It is nonadsorptive due to the lack of protein binding sites on the polymer chain. Compared to PHEMA and PVA, PEG does not have hydrogen bond donating groups, a feature that is critical in reducing protein binding.

PEG hydrogels have been considered as one of the most successful synthetic hydrogels for tissue engineering applications. The ends of a PEG polymer can be modified with either acrylates or methacrylate to form photo-cross-linkable polyethylene glycol dimethacrylate (PEGDA) (Drury and Mooney, 2003). PEGDA is extensively used to encapsulate cells into scaffolds. Furthermore, PEG can act as a mediator for immobilizing the RGD sequence.

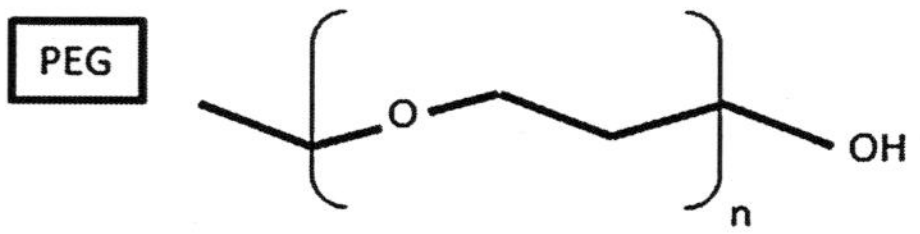

FIGURE 6.5

Neutrally charged PVA monomer.

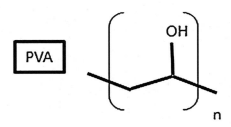

FIGURE 6.6

Neutrally charged PEG monomer.

6.5 INTEGRATIVE SUPPORT MATERIALS

The role of hydrogel in bioprinting is recognized for its biocompatibility. However, hydrogels usually lack mechanical strength. Some research groups have integrated the use of hard thermoplastic material such as polycaprolactone (PCL) and poly(lactic-co-glycolic acid) (PLGA) with soft materials (hydrogel) to enhance the mechanical strength of constructs and ensure better shape fidelity. The solid scaffold provides a suitable macrostructure and creates open pore geometry with highly porous surface and microstructure. Since these materials are synthetic, their degradation rate and cytotoxic effects can be controlled. The use of soft and hard polymers also helps to enhance cell adhesion and proliferation on the scaffold.

The conditions required to process thermoplastic PCL and PLGA are unfavorable for cells. Hence, most hybrid bioprinting systems comprise two components: (i) dispensing thermoplastic–melt plotting system and (ii) dispensing hydrogel/cell materials. Table 6.2 summarizes some of the current biomaterials used in bioprinting along with the different printing techniques used.

Table 6.2   Hydrogel used for bioprinting classified according to the type of material used

Materials	Printing technique	References
Soft		
Natural Hydrogel		
Alginate	Ink-jet Printing	(Xu et al., 2013b)
	Ink-jet Printing	(Xu et al., 2012)
	Pneumatic Extrusion	(Ozbolat et al., 2014)
	Pneumatic Extrusion	(Ahn et al., 2013)
	Pneumatic Extrusion	(Huang et al., 2013)
	Pneumatic Extrusion	(Lee et al., 2013)
	Pneumatic Extrusion	(Fedorovich and Wouter, 2011)
	Pneumatic Extrusion	(Shim et al., 2012)
	BioLP / AFA-LIFT / MAPLE-DW	(Guillotin et al., 2010)
	Positive Displacement Extrusion	(Sawkins et al., 2012)

Table 6.2   Hydrogel used for bioprinting classified according to the type of material used *(cont.)*

Materials	Printing technique	References
Soft		
Natural Hydrogel		
Matrigel	BioLP / AFA-LIFT / MAPLE-DW	(Guillotin et al., 2010)
	Pneumatic Extrusion	(Snyder et al., 2011)
Collagen	Ink-jet Printing	(Xu et al., 2013b)
	Pneumatic Extrusion	(Lee, et al., 2009)
	BioLP / AFA-LIFT / MAPLE-DW	(Michael et al., 2013)
Gelatin	Pneumatic Extrusion	(Wang et al., 2006)
	Pneumatic Extrusion	(Huang et al., 2013)
	BioLP / AFA-LIFT / MAPLE-DW	(Schiele et al., 2011)
	Positive Displacement Extrusion	(Visser et al., 2013)
Gel-MA	Positive Displacement Extrusion	(Visser et al., 2013)
	Dynamic Optical Projection Stereolithography (DOPsL)	(Soman et al., 2013)
Fibrinogen	Pneumatic Extrusion	(Huang et al., 2013)
Thrombin	BioLP / AFA-LIFT / MAPLE-DW	(Guillotin et al., 2010)
Hyaluronan	Positive Displacement Extrusion	(Skardal et al., 2010)
Synthetic Hydrogel		
PEG	Accoustic Droplet Ejection	(Fang et al., 2012)
	Positive Displacement Extrusion	(Skardal et al., 2010)
	Stereolithography (SLA)	(Chan et al., 2010)
	Dynamic Optical Projection Stereolithography (DOPsL)	(Lu et al., 2006)
PLGA-PEG PLGA	Positive Displacement Extrusion	(Sawkins et al., 2012)
Hybrid		
PCL, Alginate	Melt-Plotting System, Pneumatic Extrusion	(H. Lee et al., 2013) (Schuurman et al., 2011)
PCL, PLGA, HA, gelatin, collagen	Melt-Plotting System, Pneumatic Extrusion	(Shim et al., 2011)
PCL, Fibrinogen, Collagen	Electrospinning Apparatus, Inkjet Printing	(Xu et al., 2013a)

6.6 CONCLUSION

The development of materials specific for bioprinting is still in its infancy, as much of the biomaterials used for this technology are mainly based on the current materials used in tissue engineering. Most of these materials usually lack processing capability in terms of fast gelation time and good mechanical strength, thus impacting the overall fidelity of the construct. The ideal hydrogel must have features such as good porosity and particle size, low amount of residual monomer and soluble content, pH neutrality, nontoxicity, photostability, high durability, stability during cell culturing and storage, and high degradability without forming toxic chemicals. The combined use of solid and soft scaffold may potentially lead to the production of scaffolds possessing both mechanical strength and ECM mimicry features. However, this method of integrating these materials may be viable only for bone tissue regeneration due to their overall mechanical properties. Hence, in the future, smarter and more mechanically stable hydrogels must be developed to ultimately serve as the "bioink" for bioprinting tissue constructs.

REFERENCES

Ahmed, T.A., Dare, E.V., Hincke, M., 2008. Fibrin: a versatile scaffold for tissue engineering applications. Tissue Eng Part B Rev 14 (2), 199–215, doi: 10.1089/ten.teb.2007.0435.

Ahn, S., Lee, H., Kim, G., 2013. Functional cell-laden alginate scaffolds consisting of core/shell struts for tissue regeneration. Carbohydrate Polymers 98 (1), 936–942, doi: 10.1016/j.carbpol.2013.07.008.

Alberts, B., Bray, D., Lewis, J., Raff, M., Roberts, K., Watson, J.D., 1994. Molecular biology of the cell, 3rd ed. Garland Science, New York, US.

Alcantar, N.A., Aydil, E.S., Israelachvili, J.N., 2000. Polyethylene glycol–coated biocompatible surfaces. Journal of biomedical materials research 51 (3), 343–351.

am Ende, M. T., Hariharan, D., & Peppas, N. A., 1995. Factors influencing drug and protein transport and release from ionic hydrogels. *Reactive Polymers, 25*(2-3), 127-137. doi: http://dx.doi.org/10.1016/0923-1137(94)00040-C.

Barbucci, R., Pasqui, D., Favaloro, R., & Panariello, G., 2008. A thixotropic hydrogel from chemically cross-linked guar gum: synthesis, characterization and rheological behaviour. *Carbohydrate Research, 343*(18), 3058-3065. doi: http://dx.doi.org/10.1016/j.carres.2008.08.029.

Baskin, J.M., Prescher, J.A., Laughlin, S.T., Agard, N.J., Chang, P.V., Miller, I.A., Bertozzi, C.R., 2007. Copper-free click chemistry for dynamic in vivo imaging. Proceedings of the National Academy of Sciences 104 (43), 16793–16797.

Belkas, J.S., Munro, C.A., Shoichet, M.S., Johnston, M., Midha, R., 2005. Long-term in vivo biomechanical properties and biocompatibility of poly (2-hydroxyethyl methacrylate-co-methyl methacrylate) nerve conduits. Biomaterials 26 (14), 1741–1749.

Billiet, T., Gevaert, E., De Schryver, T., Cornelissen, M., Dubruel, P., 2014. The 3D printing of gelatin methacrylamide cell-laden tissue-engineered constructs with high cell viability. Biomaterials 35 (1), 49–62, doi: 10.1016/j.biomaterials.2013.09.078.

Brahatheeswaran Dhandayuthapani, Y. Y., Toru Maekawa, and D. Sakthi Kumar., 2011. Polymeric Scaffolds in Tissue Engineering Application: A Review. International Journal of Polymer Science, *2011*. doi: 10.1155/2011/290602.

Brannon-Peppas, L., & Peppas, N.A., 1991. Equilibrium swelling behavior of pH-sensitive hydrogels. *Chemical Engineering Science, 46*(3), 715-722. doi: http://dx.doi.org/10.1016/0009-2509(91)80177-Z.

Bryant, S.J., Nuttelman, C.R., Anseth, K.S., 2000. Cytocompatibility of UV and visible light photoinitiating systems on cultured NIH/3T3 fibroblasts in vitro. Journal of Biomaterials Science -- Polymer Edition 11 (5), 439–457.

Bryant, G.D.N. a. S.J., 2008. Cell Encapsulation in Biodegradable Hydrogels for Tissue Engineering Applications. Tissue Engineering Part B: Reviews 14 (2), 149–165.

Campoccia, D., Doherty, P., Radice, M., Brun, P., Abatangelo, G., Williams, D.F., 1998. Semisynthetic resorbable materials from hyaluronan esterification. Biomaterials 19 (23), 2101–2127.

Chan, B.P., Leong, K.W., 2008. Scaffolding in tissue engineering: general approaches and tissue-specific considerations. European Spine Journal 17 (4), 467–479, doi: 10.1007/s00586-008-0745-3.

Chan, V., Zorlutuna, P., Jeong, J.H., Kong, H., Bashir, R., 2010. Three-dimensional photopatterning of hydrogels using stereolithography for long-term cell encapsulation. Lab on a chip 10 (16), 2062–2070, doi: 10.1039/c004285d.

Cho, N.-J., Elazar, M., Xiong, A., Lee, W., Chiao, E., Baker, J., Glenn, J.S., 2009. Viral infection of human progenitor and liver-derived cells encapsulated in three-dimensional PEG-based hydrogel. Biomedical Materials 4 (1), 011001.

Collins, M.C., Ramirez, W.F., 1979. Transport through polymeric membranes. The Journal of Physical Chemistry 83 (17), 2294–2301, doi: 10.1021/j100480a022.

Collins, M.N., Birkinshaw, C., 2013. Hyaluronic acid based scaffolds for tissue engineering-A review. Carbohydrate polymers 92 (2), 1262–1279, doi: DOI 10.1016/j.carbpol.2012.10.028.

Crank, J., 1979. The mathematics of diffusion. Oxford university press.

Crescenzi, V., Cornelio, L., Di Meo, C., Nardecchia, S., Lamanna, R., 2007. Novel hydrogels via click chemistry: synthesis and potential biomedical applications. Biomacromolecules 8 (6), 1844–1850.

Dare, E., Vascotto, S., Carlsson, D., Hincke, M., Griffith, M., 2007. Differentiation of a fibrin gel encapsulated chondrogenic cell line. The International journal of artificial organs 30 (7), 619.

Di Monte, D., Bellomo, G., Thor, H., Nicotera, P., Orrenius, S., 1984. Menadione-induced cytotoxicity is associated with protein thiol oxidation and alteration in intracellular Ca^{2+} homeostasis. Archives of biochemistry and biophysics 235 (2), 343–350.

Donati, I., & Paoletti, S., 2009. Material Properties of Alginates. *13*, 1-53. doi: 10.1007/978-3-540-92679-5_1.

Drumheller, P.D., Hubbell, J.A., 1995. Densely crosslinked polymer networks of poly (ethylene glycol) in trimethylolpropane triacrylate for cell-adhesion-resistant surfaces. Journal of biomedical materials research 29 (2), 207–215.

Drury, J.L., Mooney, D.J., 2003. Hydrogels for tissue engineering: scaffold design variables and applications. Biomaterials 24 (24), 4337–4351, doi: Doi 10.1016/S0142-9612(03)00340-5.

Duan, B., Kapetanovic, E., Hockaday, L. a., Butcher, J.T., 2013. Three-dimensional printed trileaflet valve conduits using biological hydrogels and human valve interstitial cells. Acta biomaterialia, doi: 10.1016/j.actbio.2013.12.005.

Fairbanks, B.D., Schwartz, M.P., Halevi, A.E., Nuttelman, C.R., Bowman, C.N., Anseth, K.S., 2009. A Versatile Synthetic Extracellular Matrix Mimic via Thiol-Norbornene Photopolymerization. Advanced Materials 21 (48), 5005–5010.

Fang, Y., Frampton, J.P., Raghavan, S., Sabahi-Kaviani, R., Luker, G., Deng, C.X., Takayama, S., 2012. Rapid Generation of Multiplexed Cell Cocultures Using Acoustic Droplet Ejection Followed by Aqueous Two-Phase Exclusion Patterning. Tissue Engineering Part C: Methods 18 (9), 647–657, doi: 10.1089/ten.tec.2011.0709.

Fedorovich, N.E.W.S.H.M.W.H.-J.P.P.R.v.W.J.M.J.A., Wouter, J.A.D., 2011. Biofabrication of osteochondral tissue equivalents by printing topologically defined, cell-laden hydrogel scaffolds. Tissue Engineering … 18 (1), 33–44, doi: 10.1089/ten.tec.2011.0060.

Flory, P.J., & Rehner, J., 1943. Statistical Mechanics of Cross Linked.

Polymer Networks II. Swelling. *The Journal of Chemical Physics, 11*(11), 521-526. doi: doi:http://dx.doi.org/10.1063/1.1723792.

Gudeman, L.F., & Peppas, N.A., 1995. pH-sensitive membranes from poly(vinyl alcohol)/poly(acrylic acid) interpenetrating networks. *Journal of Membrane Science, 107*(3), 239-248. doi: http://dx.doi.org/10.1016/0376-7388(95)00120-7.

Guillotin, B., Souquet, A., Catros, S., Duocastella, M., Pippenger, B., Bellance, S., Guillemot, F., 2010. Laser assisted bioprinting of engineered tissue with high cell density and microscale organization. Biomaterials 31 (28), 7250–7256, doi: 10.1016/j.biomaterials.2010.05.055.

Hein, C.D., Liu, X.-M., Wang, D., 2008. Click chemistry, a powerful tool for pharmaceutical sciences. Pharmaceutical research 25 (10), 2216–2230.

Hennink, W.E., De Jong, S.J., Bos, G.W., Veldhuis, T.F. J., & van Nostrum, C.F., 2004. Biodegradable dextran hydrogels crosslinked by stereocomplex formation for the controlled release of pharmaceutical proteins. *International journal of pharmaceutics, 277*(1-2), 99-104. doi: http://dx.doi.org/10.1016/j.ijpharm.2003.02.002.

Hoffman, A.S., 2002. Hydrogels for biomedical applications. *Advanced Drug Delivery Reviews, 54*(1), 3-12. doi: http://dx.doi.org/10.1016/S0169-409X(01)00239-3.

Huang, Y., He, K., Wang, X., 2013. Rapid prototyping of a hybrid hierarchical polyurethane-cell/hydrogel construct for regenerative medicine. Materials science & engineering. C, Materials for biological applications 33 (6), 3220–3229, doi: 10.1016/j.msec.2013.03.048.

Huglin, M.B., Yip, D.C.F., 1987. An alternative method of determining the water content of hydrogels. Die Makromolekulare Chemie, Rapid Communications 8 (5), 237–242, doi: 10.1002/marc.1987.030080506.

JOHNSON, F.A., CRAIG, D.Q., MERCER, A.D., 1997. Characterization of the block structure and molecular weight of sodium alginates. Journal of pharmacy and pharmacology 49 (7), 639–643.

Khan, F., Ahmad, S.R., 2013. Polysaccharides and Their Derivatives for Versatile Tissue Engineering Application. Macromolecular Bioscience 13 (4), 395–421, doi: DOI 10.1002/mabi.201200409.

Khor, E., Lim, L.Y., 2003. Implantable applications of chitin and chitosan. Biomaterials 24 (13), 2339–2349.

Klouda, L., Hacker, M.C., Kretlow, J.D., Mikos, A.G., 2009. Cytocompatibility evaluation of amphiphilic, thermally responsive and chemically crosslinkable macromers for in situ forming hydrogels. Biomaterials 30 (27), 4558–4566.

Lee, C., Grodzinsky, A., Spector, M., 2001. The effects of cross-linking of collagen-glycosaminoglycan scaffolds on compressive stiffness, chondrocyte-mediated contraction, proliferation and biosynthesis. Biomaterials 22 (23), 3145–3154.

Lee, C.H., Singla, A., Lee, Y., 2001. Biomedical applications of collagen. International journal of pharmaceutics 221 (1), 1–22.

Lee, H., Ahn, S., Bonassar, L.J., Chun, W., Kim, G., 2013. Cell-laden poly(varepsilon-caprolactone)/alginate hybrid scaffolds fabricated by an aerosol cross-linking process for obtaining homogeneous cell distribution: fabrication, seeding efficiency, and cell proliferation and distribution. Tissue Eng Part C Methods 19 (10), 784–793, doi: 10.1089/ten.TEC. 2012.0651.

Lee, K.Y., Bouhadir, K.H., & Mooney, D.J., 2004. Controlled degradation of hydrogels using multi-functional cross-linking molecules. *Biomaterials, 25*(13), 2461-2466. doi: http://dx.doi.org/10.1016/j.biomaterials.2003.09.030.

Lee, W., Lee, V.K., Polio, S., 2009. Three-dimensional cell-hydrogel printer using electromechanical microvalve for tissue engineering. Solid-State Sensors, …, 2230–2233.

Lei, Y., Ng, Q.K., Segura, T., 2010. Two and three-dimensional gene transfer from enzymatically degradable hydrogel scaffolds. Microscopy research and technique 73 (9), 910–917.

Li, Y., Trush, M.A., 1993. DNA damage resulting from the oxidation of hydroquinone by copper: role for a Cu (II)/Cu (I) redox cycle and reactive oxygen generation. Carcinogenesis 14 (7), 1303–1311.

Lu, Y., Mapili, G., Suhali, G., Chen, S., Roy, K., 2006. A digital micro-mirror device-based system for the microfabrication of complex, spatially patterned tissue engineering scaffolds. Journal of biomedical materials research. Part A 77 (2), 396–405, doi: 10.1002/jbm.a.30601.

Malana, M., Zohra, R., Khan, M., 2012. Rheological characterization of novel physically crosslinked terpolymeric hydrogels at different temperatures. Korea-Australia Rheology Journal 24 (3), 155–162, doi: 10.1007/s13367-012-0019-9.

Malda, J., Visser, J., Melchels, F.P., Jüngst, T., Hennink, W.E., Dhert, W.J.A., Hutmacher, D.W., 2013. 25th Anniversary Article: Engineering Hydrogels for Biofabrication. Advanced Materials 25 (36), 5011–5028, doi: 10.1002/adma.201302042.

Metters, A., Hubbell, J., 2005. Network formation and degradation behavior of hydrogels formed by Michael-type addition reactions. Biomacromolecules 6 (1), 290–301.

Michael, S., Sorg, H., Peck, C.-T., Koch, L., Deiwick, A., Chichkov, B., Reimers, K., 2013. Tissue engineered skin substitutes created by laser-assisted bioprinting form skin-like structures in the dorsal skin fold chamber in mice. PloS one 8 (3), e57741–e157741, doi: 10.1371/journal.pone.0057741.

Murphy, S.V., Skardal, A., Atala, A., 2013. Evaluation of hydrogels for bio-printing applications. Journal of Biomedical Materials Research Part A 101A (1), 272–284, doi: 10.1002/jbm.a.34326.

Nagaoka, S., Tanzawa, H., Suzuki, J., 1990. Cell proliferation on hydrogels. In Vitro Cellular & Developmental Biology 26 (1), 51–56, doi: 10.1007/BF02624154.

Ogawa, T., Akazawa, T., Tabata, Y., 2010. In Vitro Proliferation and Chondrogenic Differentiation of Rat Bone Marrow Stem Cells Cultured with Gelatin Hydrogel Microspheres for TGF-β1 Release. Journal of Biomaterials Science -- Polymer Edition 21 (5), 609–621, doi: 10.1163/156856209X434638.

Orski, S.V., Poloukhtine, A.A., Arumugam, S., Mao, L., Popik, V.V., Locklin, J., 2010. High density orthogonal surface immobilization via photoactivated copper-free click chemistry. Journal of the American Chemical Society 132 (32), 11024–11026.

Ozbolat, I.T., Chen, H., Yu, Y., 2014. Development of 'Multi-arm Bioprinter' for hybrid biofabrication of tissue engineering constructs. Robotics and Computer-Integrated Manufacturing 30 (3), 295–304, doi: 10.1016/j.rcim.2013.10.005.

Paramonov, S.E., Gauba, V., Hartgerink, J.D., 2005. Synthesis of collagen-like peptide polymers by native chemical ligation. Macromolecules 38 (18), 7555–7561.

Park, G.S., & Crank, J. (1968). Diffusion in polymers.

Park, K.M., Lee, S.Y., Joung, Y.K., Na, J.S., Lee, M.C., & Park, K.D., 2009. Thermosensitive chitosan–Pluronic hydrogel as an injectable cell delivery carrier for cartilage regeneration. Acta biomaterialia, 5(6), 1956-1965. doi: http://dx.doi.org/10.1016/j.actbio.2009.01.040.

Park, S.-N., Park, J.-C., Kim, H.O., Song, M.J., Suh, H., 2002. Characterization of porous collagen/hyaluronic acid scaffold modified by 1-ethyl-3-(3-dimethylaminopropyl) carbodiimide cross-linking. Biomaterials 23 (4), 1205–1212.

Peppas, N.A., Hilt, J.Z., Khademhosseini, A., Langer, R., 2006. Hydrogels in Biology and Medicine: From Molecular Principles to Bionanotechnology. Advanced Materials 18 (11), 1345–1360, doi: 10.1002/adma.200501612.

Peppas, N.A., Merrill, E.W., 1977. Crosslinked poly(vinyl alcohol) hydrogels as swollen elastic networks. Journal of Applied Polymer Science 21 (7), 1763–1770, doi: 10.1002/app.1977.070210704.

Peppas, N.A., & Reinhart, C.T., 1983. Solute diffusion in swollen membranes. Part I. A new theory. *Journal of Membrane Science, 15*(3), 275-287. doi: http://dx.doi.org/10.1016/S0376-7388(00)82304-2.

Ratner, B.D., Hoffman, A.S., Schoen, F.J., Lemons, J.E., 2013. Biomaterials science: an introduction to materials in medicine, 3rd ed. Elsevier, London, United Kingdom.

Ren, D.W., Yi, H.F., Wang, W., Ma, X.J., 2005. The enzymatic degradation and swelling properties of chitosan matrices with different degrees of N-acetylation. Carbohydrate Research 340 (15), 2403–2410, doi: 10.1016/j.carres.2005. 07. 022.

Reulen, S.W., Brusselaars, W.W., Langereis, S., Mulder, W.J., Breurken, M., Merkx, M., 2007. Protein-liposome conjugates using cysteine-lipids and native chemical ligation. Bioconjugate chemistry 18 (2), 590–596.

Sawkins, M. J., Brown, B. N., Bonassar, L. J., Rose, F., & Shakesheff, K. M., 2012. Bioplotting of novel scaffold materials and complex constructs for osteochondral tissue engineering. *23*(page 36), 2262-2262.

Schiele, N. R., Chrisey, D. B., Ph, D., & Corr, D. T., 2011. Gelatin-Based Laser Direct-Write Technique for the Precise Spatial Patterning of Cells. *17*(3). doi: 10.1089/ten.tec.2010.0442.

Schmedlen, R.H., Masters, K.S., West, J.L., 2002. Photocrosslinkable polyvinyl alcohol hydrogels that can be modified with cell adhesion peptides for use in tissue engineering. Biomaterials 23 (22), 4325–4332.

Schuurman, W., Khristov, V., Pot, M.W., van Weeren, P.R., Dhert, W.J., Malda, J., 2011. Bioprinting of hybrid tissue constructs with tailorable mechanical properties. Biofabrication 3 (2), 021001, doi: 10.1088/1758-5082/3/2/021001.

Shim, J.-H., Lee, J.-S., Kim, J.Y., Cho, D.-W., 2012. Bioprinting of a mechanically enhanced three-dimensional dual cell-laden construct for osteochondral tissue engineering using a multi-head tissue/organ building system. Journal of Micromechanics and Microengineering 22 (8), 085014–185014, doi: 10.1088/0960-1317/22/8/085014.

Shim, J.H., Kim, J.Y., Park, M., Park, J., Cho, D.W., 2011. Development of a hybrid scaffold with synthetic biomaterials and hydrogel using solid freeform fabrication technology. Biofabrication 3 (3), 034102, doi: 10.1088/1758-5082/3/3/034102.

Skardal, A., Zhang, J., Prestwich, G.D., 2010. Bioprinting vessel-like constructs using hyaluronan hydrogels crosslinked with tetrahedral polyethylene glycol tetracrylates. Biomaterials 31 (24), 6173–6181, doi: 10.1016/j.biomaterials.2010.04.045.

Slaughter, B.V., Khurshid, S.S., Fisher, O.Z., Khademhosseini, A., Peppas, N.A., 2009. Hydrogels in regenerative medicine. Advanced Materials 21 (32-33), 3307–3329.

Snyder, J.E., Hamid, Q., Wang, C., Chang, R., Emami, K., Wu, H., Sun, W., 2011. Bioprinting cell-laden matrigel for radioprotection study of liver by pro-drug conversion in a dual-tissue microfluidic chip. Biofabrication 3 (3), 034112–134112, doi: 10.1088/1758-5082/3/3/034112.

Soman, P., Chung, P.H., Zhang, a.P., Chen, S., 2013. Digital microfabrication of user-defined 3D microstructures in cell-laden hydrogels. Biotechnology and bioengineering 110 (11), 3038–3047, doi: 10.1002/bit.24957.

Tan, H., Chu, C.R., Payne, K.A., & Marra, K.G. (2009). Injectable in situ forming biodegradable chitosan–hyaluronic acid based hydrogels for cartilage tissue engineering. *Biomaterials, 30*(13), 2499-2506. doi: http://dx.doi.org/10.1016/j.biomaterials.2008.12.080.

Tesoro, G. (1984). Textbook of polymer science, 3rd ed., Fred W. Billmeyer, Jr., Wiley-Interscience, New York, 1984, 578 pp. No price given. *Journal of Polymer Science: Polymer Letters Edition, 22*(12), 674-674. doi: 10.1002/pol.1984.130221210.

Visser, J., Peters, B., Burger, T.J., Boomstra, J., Dhert, W.J.a., Melchels, F.P.W., Malda, J., 2013. Biofabrication of multi-material anatomically shaped tissue constructs. Biofabrication 5 (3), 035007–135007, doi: 10.1088/1758-5082/5/3/035007.

Walt, D.R., Agayn, V.I., 1994. The chemistry of enzyme and protein immobilization with glutaraldehyde. TrAC Trends in Analytical Chemistry 13 (10), 425–430.

Wang, X., Yan, Y., Pan, Y., Xiong, Z., Liu, H., Cheng, J., Lu, Q., 2006. Generation of three-dimensional hepatocyte/gelatin structures with rapid prototyping system. Tissue engineering 12 (1), 83–90, doi: 10.1089/ten.2006.12.83.

Xu, C., Chai, W., Huang, Y., Markwald, R.R., 2012. Scaffold-free inkjet printing of three-dimensional zigzag cellular tubes. Biotechnology and bioengineering 109 (12), 3152–3160, doi: 10.1002/bit.24591.

Xu, T., Binder, K.W., Albanna, M.Z., Dice, D., Zhao, W., Yoo, J.J., Atala, A., 2013a. Hybrid printing of mechanically and biologically improved constructs for cartilage tissue engineering applications. Biofabrication 5 (1), 015001, doi: 10.1088/1758-5082/5/1/015001.

Xu, T., Zhao, W., Zhu, J.-M., Albanna, M.Z., Yoo, J.J., Atala, A., 2013b. Complex heterogeneous tissue constructs containing multiple cell types prepared by inkjet printing technology. Biomaterials 34 (1), 130–139, doi: 10.1016/j.biomaterials.2012.09.035.

Yeo, D.S., Srinivasan, R., Chen, G.Y., Yao, S.Q., 2004. Expanded utility of the native chemical ligation reaction. Chemistry-A European Journal 10 (19), 4664–4672.

Yoshioka, H., Mikami, M., Mori, Y., Tsuchida, E., 1994. A synthetic hydrogel with thermoreversible gelation. I. Preparation and rheological properties, Journal of Macromolecular Science—Pure and Applied Chemistry 31 (1), 113–120.

Yucel, T., Cebe, P., & Kaplan, D.L. (2009). Vortex-Induced Injectable Silk Fibroin Hydrogels. Biophysical Journal, 97(7), 2044-2050. doi: http://dx.doi.org/10.1016/j.bpj.2009.07.028.

BLOOD VESSEL REGENERATION

Jesse K. Placone and John P. Fisher

Fischell Department of Bioengineering, University of Maryland, College Park, Maryland, USA

7.1 INTRODUCTION

In the field of tissue engineering, one of the most rapidly advancing areas of study is blood vessel regeneration. This area focuses on the development of technologies and methods to encourage new vessel growth, angiogenesis, or the repair and replacement of native blood vessels. New vessel formation has specific applications to the survival of implanted organs as discussed in another chapter in this book. The replacement of native blood vessels has direct therapeutic applications due to the prevalence of cardiovascular diseases worldwide (Go et al., 2012; Roger et al., 2012). Cardiovascular disease and congenital heart disease have had significant advances in their treatment and care; however, the usage of synthetic materials still leaves much to be desired due to complications associated with their use. These artificial grafts typically have problems with the progressive occlusion of the implant because of a lack of growth potential and an increase in thromboebolic events (Stark, 1998; Cleveland et al., 1992); (Jonas et al., 1985; Lamers et al., 2012). As such, the field of tissue engineering has been investigating multiple routes toward the development of a vascular graft that has the potential to overcome some of the shortcomings of traditional vascular grafts and improve the quality of life for the patient.

In this chapter, we will discuss some of the recent technologies that have been developed to address problems in this field, and each approach's potential future directions to become clinically relevant for cardiovascular disease treatments and for the implantation of other tissue-engineered implants. The current approaches can be divided into cell-based and cell-free systems where man-made scaffolds or grafts are generated to provide the foundation for blood vessel regeneration. Both approaches make use of micro- to nanoscale architecture, embedded or attached growth factors or therapeutics, and tunable mechanical properties to better mimic the native tissue and its heterogeneous composition of cells and extracellular matrices.

The most basic description of blood vessels typically illustrates the blood vessel as consisting of three main layers. These layers, the tunica intima, the tunica media, and the tunica adventitia, are organized as concentric layers from the inside out and all have distinct cellular and matrix components. The innermost layer, the tunica intima, consists of an endothelial cell layer that completely covers the inside of the blood vessel. These cells are critical for the control of hemostasis and maintaining the patency of the blood vessel. Additionally, these cells provide a layer that is selectively permeable while at the same time providing a lining that regulates platelet and leukocyte adhesion and aggregation, and influences smooth muscle cell migration and proliferation. The endothelial cells in this region deposit and maintain a basement membrane of type IV collagen and laminin. There is an elastic laminin layer separating this layer from the next cell population. The middle layer, or tunica media, is characterized by the presence of smooth muscle cells. These cells are organized concentrically and control vasoconstriction and vasodilatation. The outermost layer, the tunica adventitia, is comprised primarily of fibroblasts. This layer's extracellular matrix is different from the other layers in that it is made up of collagen I, elastin, laminin, and fibronectin. The exact composition of this layer and the presence of various growth factors and cytokines greatly influence the mechanical strength and local cell response, and this composition varies based upon location and function in the body. These three layers work in tandem to control and modulate the physical and biological properties of the blood vessel. The recapitulation of these three main layers is critical to the success of a tissue-engineered vascular implant, and as such, much research has been carried out to develop manufacturing methods to address the heterogeneous nature of human vasculature.

7.1.1 ADDITIVE MANUFACTURING

In order to manufacture mimetic scaffolds for vascular applications, 3D printing is increasingly being employed as it provides the ability to tailor scaffolds with stringent control of the material on a layer-by-layer basis. Additive manufacturing has been present in industrial applications, but has only recently become an emerging technology for use in tissue engineering applications (Horn and Harrysson, 2012; Melchels et al., 2012; Marga et al., 2012; Schubert et al., 2013; Derby, 2012; Barron et al., 2004; Xu et al., 2005; Campbell and Weiss, 2007). However, other technologies exist that can mimic the native structures and architecture with a more global approach and as such will be briefly discussed in this chapter.

One critical aspect of the advancement of 3D printing, or rapid prototyping and additive manufacturing, with regards to tissue engineering is the development of vascularized tissues. Efficient vascularization would allow for incorporation into the host tissue as well as provide an avenue for necessary nutrient and waste exchange. The circulatory system is how the body naturally maintains homeostasis, transports waste and nutrients, delivers chemical signals, and provides rapid delivery of the hosts' defenses. It is therefore especially important that any implanted tissue engineered construct incorporates vascularization in its design.

Advances in computer-aided design have also had an impact on the rational design of tissue-engineered implants for vascular regeneration (Horn and Harrysson, 2012; Barron et al., 2004; Almeida Hde and Bartolo, 2010; Guillotin et al., 2010; Hirt et al., 2014; Jakab et al., 2010). As mentioned previously, the main cellular component of the inner layer of vasculature is endothelial cells. These endothelial cells maintain homeostasis through the formation of a boundary reinforced with their cell–cell junctions and their deposited ECM, and secrete various growth factors to control proliferation and cell recruitment, and induce ECM production. In this regard, CAD approaches have been beneficial because they can be used to model the shear stress that the ECs will be exposed to upon implantation in the body. It is well known that EC protein expression is modulated by shear stress, and by controlling the microenvironment that the cells are exposed to, the expression of proteins, such as platelet-derived growth factor

(PDGF), transforming growth factor-β (TGFβ), and endothelial nitric oxide synthase (eNOS), can be manipulated. These proteins and cytokines can be utilized to help with the endothelialization process as well as maintain the implant integrity after implantation.

7.1.2 IMPORTANT PROTEINS FOR VASCULATURE

Platelet-derived growth factor is known to play a role in the regulation of cell growth and proliferation. Specifically, PDGF is important for angiogenesis and has been the focus of many tissue engineering applications that wish to make use of existing vasculature to invade into a tissue-engineered implant. PDGF can act as a chemoattractant for fibroblasts and influence the division of vascular smooth muscle cells.

TGFβ is involved in the production of ECM by modulating the production of elastin (Sales et al., 2006). Elastin itself has two main functions in the ECM. First, it is critical for the elastic properties of the blood vessel (Kozel et al., 2011). The development of tissue-engineered materials or implants that mimic the elastic properties is very difficult, but is absolutely necessary to prevent complications, such as neointimal hyperplasia, which is generally caused by an elastic mismatch between the native tissue and the implanted vascular graft (Creager et al., 2012). The second function that elastin performs is the sequestration and activation of secreted growth factors (Creager et al., 2012). Therefore, the presence of elastin at varying concentrations natively controls the activity of the proteins present in the ECM and modulates their release.

Endothelial nitric oxide synthase (eNOS) is critical for the functioning of an intact endothelial layer in blood vessels. It is involved in the generation of nitric oxide (NO) by endothelial cells, which is important in the regulation of vascular resistance. NO produced by endothelial cells can cause the relaxation of the smooth muscle cells surrounding the vasculature thereby resulting in vasodilatation. However, if the endothelium is disrupted or damaged, the expression of eNOS is altered and vasoconstriction can occur due to the decreased levels of NO. Additionally, the expression of eNOS can affect vessel permeability through the levels of NO produced (Dauphinee and Karsan, 2010).

Although there are many proteins and cytokines known to influence angiogenesis and vasculogenesis, the deposition of these proteins alone cannot guide the invasion of vasculature or generate new vasculature. However, these proteins and others can be used in conjunction with materials and scaffolds designed for vascular regeneration. Traditionally, techniques such as electrospinning, molding, and vapor deposition have been used to create the materials that these growth factors and signaling molecules will be attached to for vascular applications (Roy, 2010). There has, however, been a push in the past several years to have more control over the surface topologies, the spatial organization of these surface features and modifications, as well as the direct seeding of cells on the vascular implant. As such, the goal of many tissue engineers is to develop a vascular implant that has tightly controlled features and mechanical properties while at the same time has very heterogeneous properties to mimic the native vasculature.

7.1.3 APPLICATION TO VASCULAR IMPLANTS

To accomplish this goal, many researchers have been investigating the application of additive manufacturing to create replacement vasculature. In the circulatory system there are many different components that could be either replaced or mimicked using 3D printing. The three main classifications that we will discuss here are: the arteries, which provide flow to the tissue; veins, which provide a return flow from the tissue; and capillaries, which provide for the efficient exchange of waste and nutrients in the tissue (Figure 7.1). Arteries and veins have larger inner diameters, which range from 2 to 5 mm and have a wall thickness on the order of 1 mm for arteries and 0.5 mm for veins. Additionally, there are structural differences between arteries and veins even though they have the same three basic layers. Blood vessels can be thought of as a concentric structure with each concentric ring comprised of different cells and matrices. Endothelial cells form the smallest-diameter layer (tunica intima), followed by smooth muscle cells (tunica media), and then fibroblasts and primarily collagen farther out forming the outermost layer (tunica adventitia) (Junqueira, 1995; Iaizzo, 2005).

The cells are embedded inside ECMs that are primarily comprised of proteins, glycoproteins, glycosaminoglycans, and proteoglycans. In between these layers there can be a layer of elastin. However, the thickness of each layer and the composition of those layers can vary for arteries and veins. Arteries have more smooth muscle cells surrounding the endothelial cells, whereas the veins have a decreased presence of smooth muscles cells resulting in the disparity in their wall thicknesses, as well as providing different mechanical properties and functionality. It is especially important for arteries to have a thicker layer of smooth muscle cells and elastin layers due to the active role arteries play in the circulatory system. Veins on the other hand have a diminished need for smooth muscle cells due to their role in the vascular system not requiring strong contractions. As a result, the tunica media is not the thickest layer. However, veins do have valves periodically along their length that prevent the back-flow of blood. These mechanical properties and functionality need to be taken into account when designing an implant for regenerative medicine (Junqueira, 1995; Iaizzo, 2005).

The vascular system is a very dynamic environment. Typically, the average adult human heart pumps 5 l of blood per minute throughout the body. Different organs and tissues receive varied quantities of flow depending on their nutrient requirements. These levels can also be dynamically modulated in order to account for various activities such as digestion, physical exertion, and basal metabolic activities.

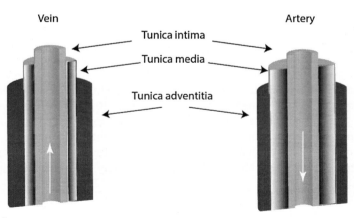

FIGURE 7.1 Vasculature.

A cross-sectional view showing the basic layers of veins and arteries. The three main layers are labeled with black arrows. The tunica media and the tunica adventitia are of different thicknesses for veins and arteries based upon their respective mechanical requirements. In between the tunica media and the inner and outer layers, there are thin layers of elastin matrix. The thickness of the elastin matrices is thicker in arteries.

This provides a unique set of environmental conditions that are continually changing, and as a result, any implant must be able to survive this range of conditions.

To address the clinical need for biocompatible and biodegradable vascular grafts, tissue-engineering approaches have become more widely utilized to improve vascular regeneration. These approaches are varied in the methods used to generate the grafts or scaffolds to promote vascularization and angiogenesis, but the main concepts behind the design of the grafts remain largely the same. The common features can be broken down into the components of the implant: (1) the scaffold, (2) functionalization, and (3) cells necessary for vascular regeneration. Some of the methods used by the tissue engineering community to fabricate these therapeutic implants are electrospinning, molding, and rapid prototyping of cell-free scaffolds and cell-based printed scaffolds. A selection of these methods, currently employed in scientific literature, will be discussed in the following sections with an emphasis on 3D printing technologies.

In the manufacture of these implants, there are currently two main approaches, consisting either of a cell-free or a cell-based system of designing and creating the implants for vascular regeneration. These two approaches can be broken down further into the underlying technologies that allow for the creation of these tissue-engineered implants. The cell-free scaffolds are typically created using technologies such as electrospinning, chemical vapor deposition, and 3D printing. These methods are discussed in more depth in the next section. Cell-based implants for vascular regeneration are produced either through the seeding of the cell-free scaffolds prior to implantation or through the direct creation of scaffolds where the cells are embedded in the scaffold during the production. The production of these cell- laden constructs is discussed in depth in Section 7.6.1 and focuses primarily on current and emerging technologies for 3D printing cells for vascular regeneration applications.

7.1.3.1 Cell-Free Scaffolds

Investigations into blood vessel regeneration and growth have focused on the use of cell-free systems. In these tissue engineering endeavors many technologies have been utilized to develop scaffolds that encourage the growth of native vasculature either *in vitro* or *in vivo*. Generally, these technologies aim to generate scaffolds that mimic the native vasculature and can be functionally modified as discussed previously to develop more bioactive scaffolds. This section aims to give an in-depth discussion of how the technologies work to generate the scaffolds and a few specific examples of how these methods are used in literature. Due to the number of methods available, this section will discuss the following for generating scaffolds: electrospinning, stereolithography, and fused deposition modeling.

7.1.4 ELECTROSPINNING

This book focuses primarily on 3D printing and its application to various tissue engineering endeavors; however, the discussion on blood vessel regeneration would not be complete without a brief discussion on electrospinning. Nanofibrous electrospun grafts have been widely utilized for vascular regeneration applications. The electrospinning process allows for the control of graft parameters in real time during the fabrication process, which is precisely why electrospun scaffolds have been investigated for multiple tissue engineering applications that are not limited to the vascular system (Fridrikh et al., 2003). There are a wide number of polymers and materials suitable for this fabrication process and the costs associated with this technology are relatively low. As a result of these two considerations, multiple studies have been carried out on a wide array of electrospun vascular grafts/scaffolds for tissue regeneration and the systems used to generate these scaffolds have been well characterized (Shapira et al., 2014; Hashi et al., 2010; Wu et al., 2010; Rayatpisheh et al., 2014; Hadjizadeh et al., 2013; Fu et al., 2014; Han et al., 2013; Lai et al., 2014). However, this system is not compatible with live cells due to the solvents typically used.

Typically, an electrospinning setup is comprised of a few main components: (1) a syringe filled with the desired polymer, (2) a syringe pump, (3) a high voltage power supply, and (4) a grounded collection plate/collector (Figure 7.2). The syringe pump allows for the control of the flow rate, and can therefore impact the size of the electrospun fibers. The syringe itself can have different types/gauges of needles to modulate the stream of the polymer being ejected from the syringe. Also, the voltage applied can be modulated to control the fiber diameter and provides the driving force due to electrostatic repulsion to encourage the polymer solution to flow toward the collection plate. There are additional external variables to consider, such as the temperature and humidity in the electrospinning chamber.

By modifying the applied voltage, the rotation velocity, and the solvent/polymer mixture, properties of the graft such as orientation and fiber diameter (hundreds of nanometers to micrometers) and density can be controlled. This allows for the control of properties such as porosity, mechanical strength, fiber alignment, and surface area (Fridrikh et al., 2003; Deitzel et al., 2001).

In the case of polycaprolactone (PCL), groups have shown that using either a high- rotation (2000 RPM) or low-rotation (20 RPM) speed resulted in either aligned fibers or randomly deposited nanofibers, respectively. Furthermore, to better mimic the native vasculature and to recapitulate the layer variation as a function of radial distance, groups have shown that by simultaneously controlling the applied voltage and rotation speed, they are able to generate vascular graft material that is comprised of distinct layers with varying fiber orientation and alignment (Wu et al., 2010). This sort of elec-trospun tissue- engineered blood vessel has the potential of impacting future designs of electrospun grafts for vascular applications because each layer's orientation can be controlled. Thus, this allows the graft to physically mimic the orientation of the tunica intima and the tunica media, which in turn can modulate the response of endothelial and vascular smooth muscle cells. By mimicking the morphology of the native tissue, these scaffolds show great promise for impacting vascular regeneration in

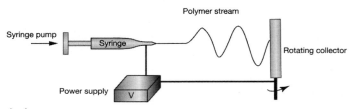

FIGURE 7.2 Electrospinning.

A typical electrospinning apparatus has four main components: (1) A high voltage power supply provides the potential difference required for the electrospinning process. The applied voltage can be modulated to vary the physical dimensions of the resulting fibers during the deposition process. (2) A syringe filled with the polymer solution. (3) A syringe pump provides the materials for the electrospinning process and controls the stream produced via the flow rate and the needle gauge/type. (4) A collection plate or cylinder is where the deposited nanofiber mesh is deposited. By depositing directly onto a cylinder, the implants' size can be directly controlled and potentially used directly for blood vessel regeneration. If a flat collection plate is used, then the resulting electrospun fiber can be rolled into a cylinder of appropriate dimensions for use in the desired application.

tissue engineering; however, recapitulation of the morphology is only one aspect critical to successful vascular regeneration.

Current studies are investigating the modification of the scaffolds by incorporation of growth factors (Han et al., 2013; Lai et al., 2014; Montero et al., 2012; Luong-Van et al., 2006; Zhang et al., 2013; He et al., 2005; Ye et al., 2012; He et al., 2012). It is important to note some of the general methods of modification of these grafts, specifically the direct incorporation of growth factors and the addition of secondary carrier molecules/material to increase the regenerative properties of these acellular scaffolds. These molecules can be added directly to the electrospinning solution or can be physically bound to the implant surface through surface modification. As mentioned previously, the growth factors or proteins selected need to be carefully chosen for the desired application and positioned in the correct locations to elicit the desired cellular response in each layer of the implant. Given the low cost, multiple materials available, ease of fabrication, and the ability to tailor the surface of these electrospun implants (Ziabari et al., 2010; Soliman et al., 2011; Subbiah et al., 2005), electrospinning will be a viable method of generating scaffolds for blood vessel regeneration in the foreseeable future.

7.1.5 STEREOLITHOGRAPHY

Another technique for the creation of cell-free scaffolds for vascular regeneration that has seen advances in the past several years is stereolithography. There are a few main technologies such as digital micromirror devices (DMD) used in parallel with digital light processing (DLP) that can be used with multiple light sources. One important aspect for this printing process is the fact that objects are rendered layer-by-layer and as a result, the process is potentially faster than other 3D printing approaches. Furthermore, in terms of reducing the cost of the 3D printer and the overall cost to manufacture parts, stereolithography-based systems have a distinct advantage over other free-form printing methods because an XY motion controller is no longer required. As a result only an accurate Z motion control system is needed, which could reduce the system cost.

There are two typical methods used for stereolithography and for the generation of solid scaffolds for tissue engineering. The basic concept is the same, but the source material reservoir and build plate behavior is different (Figure 7.3). Each layer of the print can be thought of as horizontal slices through the final object where these slices are translated into images that are projected onto the printing resin. In the first printer setup, the layer's image is projected up through the bottom of the reservoir onto the build plate. After a layer is cured, a stepping motor moves the stage up the appropriate distance and the next layer is cured. There are several advantages to this system. First the polymer solution flows back into the void left by the scaffold moving up with the stage; thus, there is no need for a pump to provide additional polymer solution, just what is in the reservoir is enough. Second, the photo-initiation of the polymerization process takes place in the solution and as a result is protected from oxygen that may potentially adversely interact with the polymerization process. In the second printer setup, the projected image comes from the top down onto the polymer solution/build plate. As each layer is cured, the build plate moves down and more printing resin is pumped in prior to the initiation of the next layer. One advantage of this system is that much larger prints can be produced without the concern of scaffold detaching from the build plate during the printing process. This can be a concern in the previous printing method due to the fact that the polymer is usually cured directly onto a glass build plate and the adhesion might not be strong enough to counteract the force of gravity, especially for a large build.

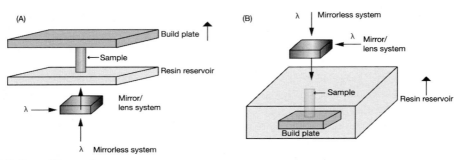

FIGURE 7.3 Stereolithography.

Two different stereolithography techniques are shown above. Each technique controls the Z-resolution by the use of a stepper motor that moves the build plate. The X- and Y-axis resolutions are controlled through the use of mirrors or direct illumination of each volumetric pixel (voxel). Both stereolithography techniques can utilize the same light source for photo-initiation and either print layer-by-layer or one voxel at a time; however, the direction of the build plate in relation to the printing resin is different. In the first case, (A) the resin reservoir remains stationary and the build plate moves upward, away from the resin, as each layer is polymerized. In the second case, (B) the resin reservoir moves up in height as each layer is built up from the stationary build plate. Both of these techniques can be used in the fabrication of scaffolds for vascular regeneration applications; however, there can be size limitations based upon the ability of the printed resin to adhere to the build plate. This can be problematic for printing method (A) where the adhered first layer must support the weight of the entire printed scaffold.

This printing method is of interest for vascular engineering due to the variety of materials that are available for the printing process and the fact that many of these materials can be biodegradable, thus, as they degrade, the native tissue can invade and resorb the printed scaffold. Since either a UV or visible light source can be used in this process, photo-initiators that are commonly used, and potentially approved by the FDA, for the polymerization of biomaterials can be utilized. Scaffolds generated using this method of fabrication can also have their mechanical properties modulated based upon the curing time for each layer of the scaffold. This allows for the generation of scaffolds that have mechanical properties very similar to the native vasculature, which is critical for the prevention of intimal hyperplasia. However, there are some drawbacks to the use of this sort of rapid fabrication process to manufacture vascular grafts. The most important drawback is the potential of the solvent to remain on or in the scaffold after the manufacturing process. This results in the need to wash the scaffold during the postprocessing steps to ensure that all of the uncross-linked monomers are removed, along with any remaining solvent. Another drawback is that support scaffolds must be generated to help hold the print in place, and as a result, these supports must be removed, which can prove to be a tedious endeavor for complex geometries and requires some additional postprocessing in order to obtain the desired surface finish on the printed vascular graft. However, as mentioned previously, one important and desirable trait of this printing process is that multiple grafts can be printed simultaneously and the printing speed is independent of the complexity of the implant design. For vascular grafts or implants, this feature can be seen as advantageous in the generation of patient-specific implants (i.e. every patient will have their own requirements that will vary in complexity based upon the native tissue architecture being replaced). Much like the other grafts/scaffolds generated for vascular regeneration, cell-free scaffolds generated in this manner can be modified postprinting to contain various bioactive molecules to encourage proper cell attachment and discourage an immune response.

One application that has seen some use in tissue engineering is the generation of artificial heart valves and conduits. Research groups have developed models based upon X-ray computed tomography and magnetic resonance imaging scans to determine the portions that need to be replaced to treat the patients' conditions. The 3D models were then printed layer-by-layer with stereolithography out of poly-4-hydroxybutyrate and polyhydroxyoctanoate, both of which allow for the generation of a biocompatible implant. For the replacement conduits and the trileaflet valves, these printed models were then tested in perfusion bioreactors (Sodian et al., 2002; Sodian et al., 2005). The application of stereolithography to the printing of biocompatible polymers for vascular regeneration has the promise of providing a method that can construct intricate implants that can have multiple components within one printed model.

7.1.6 FUSED DEPOSITION MODELING

Fused deposition modeling (FDM) has been around for several years in tissue engineering applications (Zein et al., 2002; Hutmacher et al., 2001). However, it has recently seen an increase in usage in both industrial as well as personal applications. In fact, systems based on the RepRap open source 3D printer have been utilized for tissue engineering applications (Miller et al., 2012).

In these systems, a material such as a thermoplastic or a glass is feed from a source coil into a heated extruder (Figure 7.4). The material is then deposited layer-by-layer and the scaffold is created in this stepwise fashion. The rate of extrusion, diameter of the nozzle head, temperature of the nozzle, and the speed of the nozzle head all influence the diameter of the deposited filament. Typically, as the speed of the nozzle increases for a given extrusion rate and nozzle size, the diameter of the fiber decreases. This can be used to tailor the size of the deposited filament such that it is smaller than the diameter of the extrusion nozzle. Furthermore, based upon the temperature of the extrusion process, it is possible to control the adhesion of the extruded filament to previously deposited layers. This allows for the creation of interconnected filaments that are deposited according to the 3D design.

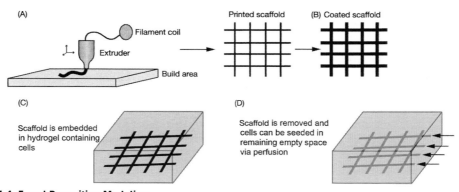

FIGURE 7.4 Fused Deposition Modeling.

This illustration shows how the extruded filament is deposited creating an interconnected 3D scaffold. (A) A scaffold is deposited in the desired pattern. (B) The scaffold is coated with a biocompatible polymer to protect it from the ECM/ hydrogel laden with cells that is (C) deposited around the scaffolding. (D) The scaffolding is then dissolved by the perfusion of water through the scaffold. Endothelial cells are injected into the interconnected network and line the walls forming vasculature/microvasculature.

However, for vascular engineering, the conditions for extruding these materials are not conducive for cell growth and, as a result, approaches using this method generally use 3D printing to deposit a scaffold that will ultimately be removed. The sacrificial scaffolds are extruded according to a 3D design and the rate of extrusion as well as temperature are tightly controlled to ensure that a proper filament diameter is being deposited as the scaffold is being printed. Since a motorized stage is being used to control the deposition in the x–y directions, the resolution in these directions is typically limited to 100 μm. This is also dependent upon the material being used so the resolution can be improved depending on the application. In the case of generating scaffolding for vasculature and microvasculature, the scaffolding material can be a carbohydrate glass that is capable of being dissolved postprinting without the use of harsh solvents (Miller et al., 2012). These glasses can be extruded at a controlled rate and result in filament fibers ranging from 150 μm in diameter up to the size of the filament.

This method allows for the unique ability to generate scaffolds that have both vasculature as well as microvasculature. After the fabrication of the sacrificial scaffold, the scaffold is then coated with a protective layer that prevents the scaffold from degrading by the deposited matrix. This matrix can be a hydrogel with embedded cells. This hydrogel can be a wide range of gels such as alginate, fibrin, or Matrigel®. Importantly, the hydrogel can be chosen based upon the desired cell type that will be encapsulated and vascularized once the scaffolding is removed. The hydrogel can be cross-linked either chemically, through the use of a photo-initiator (if the scaffold is sufficiently optically clear), or by modulating the temperature of the hydrogel plus scaffold. After the hydrogel is set with the desired cell type encapsulated, the sacrificial scaffold is then removed by perfusing water or media through the scaffold. Upon the removal of the scaffold, ECs could then be seeded onto the inner surface of the leftover porous network. This vasculature could then be placed under flow allowing for nutrient transfer from the cell culture media to the inside of the hydrogel, where typically cells would not survive due to diffusion limits and nutrient exchange.

Other applications based upon this technology have been rapidly advancing (Sodian et al., 2002; Sodian et al., 2005), and prove that cell-free methods still have the potential of providing vascularization to tissue-engineered implants. Thus, allowing for the formation of larger tissue-engineered implants while at the same time allowing for proper nutrient exchange ensures viability of the implanted vascularized construct.

7.1.6.1 Cell-Based Scaffolds

Due to the ever-changing market for commercial and research-based 3D printing systems, it is very difficult to discuss the merits of individual systems for their application to a specific research problem. As such, this section will focus mainly on the technologies and concepts behind a few cell-based printing techniques that have been directly applied to the problem of vascularization and angiogenesis. This section aims to give the reader an understanding of the principles behind the different printing processes as well as a working knowledge of how the individual technologies currently compare. 3D printing of cells or cell-laden tissue-engineered implants can be broken down into a few main technologies, many of which have been around for several decades and can be further divided based upon components utilized: (1) Ink-jet printing, (2) extrusion-based printing (fused deposition modeling), and (3) laser-assisted printing.

7.1.7 INK-JET PRINTING

As one of the main, current methods for generating 3D scaffolds for tissue engineering, ink-jet printing has several advantages (Figure 7.5) (Roth et al., 2004; Boland et al., 2006). Due to the fact that ink-jet printing technologies have been around for several decades, the cost of production of these systems is

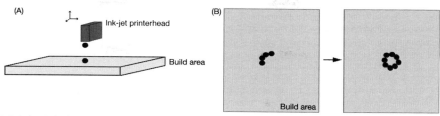

FIGURE 7.5 Ink-jet Printing.

An example of ink-jet printing of cells is shown. (A) Here the cells are embedded into a matrix that is compatible with an ink-jet printer (piezoelectric or thermal ink-jet) and printed droplet by droplet to assemble a 3D scaffold. (B) A top–down view shows how the printer head can be moved in the x–y direction laying down each droplet to form the basis of vasculature. The next layer can then be built upon these first droplets and the fabrication process continues in this iterative fashion. Although this process is capable of printing scaffolds alone, the ability to print a single cell at a time at a desired location allows for the generation of heterogeneous scaffolds required for blood vessel regeneration. Additionally, the small size of the droplets allows for the fabrication of microvessels as well as regular vessels.

relatively low and the parts necessary to build a system are readily available (Billiet et al., 2012; Cui and Boland, 2009a; Cui et al., 2012). Ink-jet printing has the ability to rapidly deposit cells and their scaffold thus enabling the rapid manufacturing of scaffolds (Nishiyama et al., 2009; Saunders et al., 2008). However, this speed also has some drawbacks since cell viability can be negatively impacted if the speeds utilized in the printing process subject the cells to very high external forces; thus, material selection, viscosity, and printing speed need to be carefully selected for a given application. Additionally, obtaining high cell densities required for the long-term stability and viability of an implant is also difficult due to the printing process. Nonetheless, this printing method has made significant advances in the past several years and provides a robust platform for developing functional tissues and can be used to generate scaffolds based off of images (Marga et al., 2012; Campbell and Weiss, 2007; Jakab et al., 2010; Arai et al., 2011).

One key aspect of developing functional 3D printed tissues is the ability to utilize multiple cell types during the printing process. Specifically, it is desirable to have heterogeneous materials and ECM deposited to fabricate the new tissue-engineered implant. Ink-jet printing is one of the technologies well suited to this application (Xu et al., 2005; Nakamura et al., 2005; Phillippi et al., 2008; Xu et al., 2006; Yamazoe and Tanabe, 2009; Cui and Boland, 2009b). This is due to the ability of this method to accurately position multiple cell types and their respective matrices in precise locations of the printed scaffold. Furthermore, it has been demonstrated that this technique can be utilized to print single cells as well as heterogeneous scaffolds (Xu et al., 2005; Cui et al., 2012; Nakamura et al., 2005; Norotte et al., 2009). In addition to printing hydrogels laden with cells, this technology can print liquids containing cells such as alginate directly into a cross-linking solution such as calcium chloride (Song et al., 2011).

Although there are commercial options available for ink-jet printing, multiple research groups have developed their own printers based off of commercial, off-the-shelf ink-jet printers. These printers need to be modified to print in the z-direction and the cartridges need to be modified and sterilized for use with cells in a sterile environment. Typically, for vascular applications, collagen-based composites are utilized due to collagen's ability to encourage vasculogenesis *in vitro* and blood vessel regeneration (Billiet et al., 2012; Nishiyama et al., 2009; Malda et al., 2013; Fedorovich et al., 2007). For microvascular applications, materials such as fibrin have been utilized for the fabrication process (Cui and Boland, 2009b).

This method has been successfully used to produce constructs that mimic native vasculature as well as in the printing of heterogeneous cell populations to encourage the vascularization of the printed biomaterial. Upon implantation into a mouse model, a group has demonstrated the ability of a printed material laden with multiple cell types to encourage vasculogenesis *in vivo*. As such, this method of producing tissue-engineered blood vessels has been steadily making progress from *in vitro* studies to *in vivo* with the end goal of translating to clinical applications.

7.1.8 EXTRUSION BIOPRINTING

Mechanical extruders provide another method of generating 3D scaffolds with embedded cells. In this approach, there are advantages due to the fact that each deposited bead can be thought of as a tissue fragment. However, this method is currently expensive due to the upfront cost of the printers and the technology required. The basic concept behind this printing method is that droplets (multicellular particles + desired scaffold material) are deposited on the template at desired locations based upon a computer-generated design.

The building blocks used in this method, much like those used for the other printing techniques, can contain either a single cell type or a heterogeneous population. For the generation of vascular tissue, however, it is important that individual cell types are used to ensure proper layering and spatial arrangement in the final printed scaffold. Another consideration that is important for the success of these implants is that the final printed construct must be as similar as possible, structurally and functionally, to the native tissue.

As such, materials such as collagen hydrogels have been successfully used as scaffolds to encapsulate the cells during the printing process. This has not been without several drawbacks. For example, the system must place the hydrogels under high pressure during the printing process. Based upon the viscosity of the printed hydrogel, the shear force exerted on the cells may be of large enough magnitude to cause cell death. Additionally this method of printing is currently restricted to a very specific set of materials, thus limiting the number of overall applications. However, by careful selection of materials and engineering, it has been possible to produce 3D printed tissues that mimic the native vasculature in terms of distinct functional layers with the ability to apply flow to the inner channel.

Figure 7.6 is an illustration of how extrusion bioprinting can be applied to the production of tissue-engineered vasculature. The layers are built in a step-wise fashion where the outermost layers are deposited with cells embedded. The inner layer, which will have the applied flow once the construct is complete, needs to be printed with an additional material that can be removed in the final printed product without compromising the seeded cell viability. One example of a secondary material that can be used for this process is agarose. The printed agarose rods fill in the void spaces during the printing process, which allow for the deposition of layers above the desired void space and prevent cell invasion and remodeling (Jakab et al., 2010). After completion of the printing process and the final gelation of the printed scaffold, the agarose rods could be removed from the center of the print. Thus, the printed scaffold has defined layers and an inner void that allows for the flow of nutrients/media as well as separate functional layers. For vascular applications, this dual print method is especially important due to the weak mechanical properties typically associated with the hydrogels compatible with this printing process.

An additional application of this printing method for vascular regeneration was in the fabrication of more complex structures such as the printing of an aortic valve. In this work, researches successfully fabricated a 3D scaffold laden with different cell types and maintained viability for one week

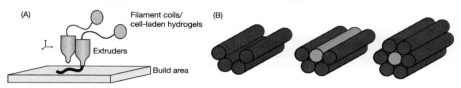

FIGURE 7.6 Extrusion Bioprinting.

(A) The basic setup of an extrusion-based 3D printer. The printer consists of a build plate and a printer head/ nozzle on a motorized axis. The printing nozzle can contain either a polymer for extrusion or a biomaterial loaded with cells and it moves in the *X*, *Y*, and *Z* directions depositing each layer in the desired location. These layers can be built off of each other much like the other 3D printing processes. However, one distinction is that the extrusion rate and the speed of the nozzle can be used to modulate the fiber size being deposited. In the case of printing vasculature, void spaces can either be included by not depositing any material or by the printing of a second material that provides structural support during the printing process. (B) The desired cell-laden scaffold is built layer-by-layer with one nozzle (black cylinders) and a second nozzle containing a secondary material (gray cylinder) that is used to deposit a support structure to be removed after the printing process is complete. A 3D vasculature can be fabricated in this manner with a controlled vasculature shape and inner diameter once the secondary matrix is removed (gray cylinder) Adapted from (Jakab et al., 2010).

(Ghista and Kabinejadian, 2013). This research demonstrates the ability to apply this technology to the various structures that comprise the vascular system. As such, this method possesses great promise in addressing the need for tissue-engineered vasculature and in aiding the blood vessel regeneration.

7.1.9 LASER-ASSISTED BIOPRINTING

One of the potential problems associated with the 3D printing of cells is due to the limited cell densities possible based upon the printing technique chosen. Laser-assisted cell printing is one method that can overcome the low viability and low cell density problem (Guillotin et al., 2010; Tasoglu and Demirci, 2013; Koch et al., 2013). The technique of laser printing is not new, as it has been used in the fabrication of circuits and biosensors. However, its application to layer-by-layer cell deposition is relatively recent and allows for easy control of cell density in the final printed product (Figure 7.7). This method overcomes limitations on cell seeding density by seeding a source gel at a high density and ensuring the cells' viability prior to the printing process (Barron et al., 2004; Guillotin et al., 2010; Barron et al., 2005). The source gel is transferred to a target or collection slide on a spot-by-spot basis through the excitation of the gel or a specialized substrate. There are currently two main modes of excitation where either the excitation energy is directly matched by the continuous wave laser source or a two-photon, pulsed excitation source is used, where the required excitation energy is approximately double the laser energy.

There are many pros and cons of using two-photon excitation; however, the most important are the ability to focus the excitation energy to very small volumes and the large upfront cost of the system especially with a tunable excitation source. Although there has been some success with Nd:YAG and Ti:sapphire tunable lasers, the costs of these systems can be prohibitive for the development of a two-photon printing system for tissue engineering applications. Even though there are differences in terms of the generation of the required excitation energy, the basic concept and procedure of using continuous wave and pulsed lasers for laser-assisted cell printing remain basically the same. Cells prior

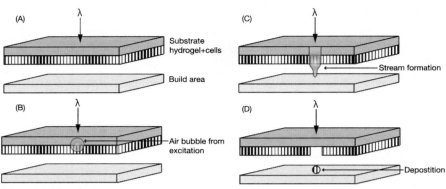

FIGURE 7.7 Laser Assisted Bioprinting.

A typical setup for laser-assisted bioprinting is shown in the illustration. (A) The excitation source is focused on a point on the glass slide containing the hydrogel with cells embedded. (B) An air bubble forms from the excitation of the substrate. (C) The air bubble then can lead to a stream formation that propels the cells + matrix toward the build area. (D) The build area is positioned micrometers to millimeters away to serve as a collection slide for the droplets produced from the rapid expansion of the radiation-absorbing layer. In the case of blood vessel regeneration, this technique has a few drawbacks in regards to the mechanical strength of the printed structure, but it excels in regards to spatial resolution as well as the ability to exchange the donor plate with another with a separate matrix and cellular composition. As such, it can print heterogeneous scaffolds that have the potential for use in the creation of vasculature as well as microvasculature. This sort of additive manufacturing has been used to generate cell based scaffolds to mimic the native vasculature.

to the printing process can be cultured normally and trypsinized to remove them from their culturing environment. Upon resuspension in a compatible hydrogel (collagen, alginate, fibrin, etc.), the hydrogel is mounted on a glass slide with or without a coating of a laser radiation absorbing layer. This glass slide is then positioned parallel to a collection slide several hundred micrometers to a few millimeters away. The basic principle of this printing process is that an excitation source is focused on a small point above the desired deposition location. The focused light causes the local expansion of gases (through the evaporation of either the laser radiation absorbing layer or the hydrogel directly) which then provides the kinetic energy required to push the hydrogel beneath the focal point toward the collection slide. The exposure time/intensity of the laser can adjust the volume ejected from hydrogel, and as such, volumes can be adjusted on the order of picoliters, allowing for very rigorous control of the ejected volume and cell seeding density of the printed construct. The spatial resolution of this technique also has some unique advantages due to the ability to precisely control the focal point of excitation, controlling the ejection of hydrogel loaded with cells from the source plate towards the collection plate.

Additionally, the properties of the ejected material can be tailored to the application at hand. Depending on the viscosity of the source hydrogel, either a droplet or a stream can be formed upon excitation by the light source. This in turn can help control the shape of the patterns deposited on the collection plate. In the case of vascular regeneration, this technique has a unique advantage because multiple source plates with different cell types embedded in different hydrogels can be used to build the vascular implant. The ECs could be deposited along an inner ring with smooth muscle cells surrounding the outside and based upon the implant location, the diameter of the scaffold and the thickness of the layers can be adjusted to mimic the relative ratios present in the native vasculature. These concentric rings can be printed one on top of another thereby building up complex architecture in a layer-by-layer approach.

One drawback to the utilization of this method for vascular regeneration is that hydrogels can have very different mechanical properties from the native tissue. Even though the end product will only have the mechanical properties of the hydrogels, which tend to be too weak on their own without additional support, the cells deposited can be precisely deposited in heterogeneous layers that very closely mimic the native heterogeneous properties of vasculature. Furthermore, these cells will be positioned close to one another such that appropriate cell–cell adhesions and communications can be established thus enabling the printed scaffold to develop into a functioning implant. Also, since this method can precisely deposit very small volumes, it may be possible to use this fabrication approach to build microvasculature that has a very small inner diameter, and very few cells are needed to create each of the concentric layers that comprise the vasculature.

Although laser-assisted 3D printing has been successfully used to create tissue- engineered blood vessels, it also has been used for patterning stem cells and ECs for cardiac regeneration (Gaebel et al., 2011). Additionally, this technology has been utilized for complex tissues such as bioprinting skin with its various ECM and cellular components (Koch et al., 2012; Pirlo et al., 2012).

7.1.9.1 Comparison of the Technologies

All of these technologies presented here represent a subset of approaches for the regeneration of vasculature. Depending on the technique used and the materials available, surface modification with proteins, cytokines, and growth factors can be included in the manufacturing process. These modifications can further enhance the cellular response to improve blood vessel regeneration. This is especially important in order to recruit endothelial cells to the tunica intima of the implant to aid in the prevention of stenosis (Hashi et al., 2010; Han et al., 2013; Zhang et al., 2013; He et al., 2005; Avci-Adali et al., 2013; Avci-Adali et al., 2010). Additionally, the mechanical properties of the vascular implant can be modified by the recruitment or incorporation of smooth muscle cells and appropriate ECM proteins such as elastin (Fu et al., 2014; Cui and Boland, 2009b; Norotte et al., 2009; Patel et al., 2006; Kasalkova et al., 2014; Greenwald and Berry, 2000; Berglund et al., 2004; Wang et al., 2013; Heydarkhan-Hagvall et al., 2006; Nerem, 2003). It is critical that these competing technologies incorporate the heterogeneous nature of each layer in the vasculature in the design and fabrication of scaffolds for therapeutic applications.

Although it has been shown that electrospinning can have control over each layer's orientation and fiber size (Shapira et al., 2014; Hashi et al., 2010; Wu et al., 2010; Zhang et al., 2013; He et al., 2005; Ziabari et al., 2010; Subbiah et al., 2005), these implants do not have as stringent control over each layer as rapid fabrication techniques. As the cost continues to decrease for 3D printing, the use of 3D printing to develop scaffolds for blood vessel regeneration will only increase. As previously mentioned, cell-free scaffolds tend to be modified in order to increase their biological activity, whereas cell-based scaffolds are already using materials that will encourage the growth of cells in substrates specific to the desired cell type. This is an important advantage and distinction because the cells are ultimately responsible for the long-term viability of the implant as well as its long-term mechanical properties. However, there is a distinct disadvantage currently with cell-based scaffolds, and that is that the base materials used have very poor mechanical properties when compared with that of native tissue or the cell-free scaffolds.

In addition, for vascular applications there are design parameters that need to be taken into account when choosing an appropriate fabrication method. For example, the complexity of the desired product in terms of surface features is very important. If only a simple cylinder is needed, then all of the above fabrication processes could be used; however, if complex overhangs or surface topologies are needed, then methods such as stereolithography and extrusion-based printing will need to be used based upon their ability to fabricate free standing structures. The desired resolution of the printing process might also be an important parameter for more complex geometries. Table 7.1 shows typical resolutions for the discussed printing techniques.

7.1.9.2 Applications to the Vascular System and Other Engineered Tissues

3D printing of the vascular system opens a wide range of applications to tissue engineering and regenerative medicine. As discussed previously, vascularization is a limiting factor in implant design and success. Additionally, complex geometries native to the vascular system could not always be replicated on the surgical table. As such, 3D printing of custom tailored implants could fulfill this need by providing surgeons a means to develop implants using off-the-shelf 3D imaging techniques such as CT and MRI. From these scans, complex architecture scaffolds could be developed and printed to meet the desired application and treat the underlying condition.

There are still challenges associated with these sorts of implants, ranging from cell source, cell viability, surface properties, mechanical properties, and a wide range of regulatory constraints that must be overcome prior to the application of these technologies to therapeutic applications.

Table 7.1 Comparison of various printing technologies commonly utilized in vascular applications

Technique	Typical resolution	Material	References
Stereolithography	25 to 50 μm	Hydrogels and polymers	(Sodian et al., 2002; Sodian et al., 2005; Billiet et al., 2012)
Fused Deposition Modeling	~ 100 μm	Polymers	(Zein et al., 2002; Miller et al., 2012)
Ink-jet Bioprinting	20 μm (picoliter droplets)	Liquids and hydrogels	(Campbell and Weiss, 2007; Boland et al., 2006; Cui and Boland, 2009a; Cui et al., 2012; Arai et al., 2011; Nakamura et al., 2005)
Extrusion Based Bioprinting	~ 100 μm	Hydrogels (natural and synthetic), polymers	(Marga et al., 2012; Jakab et al., 2010)
Laser Assisted Bioprinting	10 μm	Hydrogels	(Barron et al., 2004; Guillotin et al., 2010; Koch et al., 2013; Barron et al., 2005; Gaebel et al., 2011; Koch et al., 2012)

Investigations utilizing these methods, both cell-free scaffolds and cell-laden scaffolds, will yield the necessary insight into some of these problems. Furthermore, the developed materials and technologies must progress in order to have the desired mechanical and surface properties while at the same time steps must be taken to address the necessary regulatory requirements for clinical applications. As such, these technologies stand poised to provide insight into the direction that future research must follow to successfully develop 3D printed vasculature for therapeutic applications.

7.2 FUTURE DIRECTIONS

This research is not just limited to simple replacement of existing vascular tissue. It also has a much broader impact on the vascularization of 3D printed implants discussed throughout this book. All of these tissues require integration into the vascular system and by understanding the complex process of angiogenesis and developing mimetic tissue that can provide sufficient nutrient exchange, these artificial tissues can be further developed to make a significant impact on the field of tissue engineering and regenerative medicine.

Again, the predominant challenge that needs to be addressed for 3D printing of vasculature and its application to tissue-engineered constructs is the heterogeneous nature of the desired tissues. It is important that the printed implants not only mimic the natural shape of the organs and tissues, but that the implants recapitulate the complex, heterogeneous nature of those tissues. For vascular implants, future directions must address the need to have distinct layers with different mechanical properties to best mimic the native vasculature. Cell-laden constructs need to take this one step further by addressing the heterogeneous mechanical properties as well as scaffold materials for the different cell types being incorporated into the vascular implant.

The future of 3D printed blood vessel implants will need to address the aforementioned challenges as well as improve upon the rapid fabrication techniques such that the technologies developed in a laboratory setting can be translated into clinical settings by allowing for the scale up required for mass production. As such, the next several years will see advances in the materials used in the 3D printing application and in the ease of preparation of these materials for use with the various 3D printing technologies. Additionally, the use of stem cells, either obtained directly from the patient or induced from other cells, will have a significant impact on the design of blood vessel implants. These stem cells can be used to determine the best material environment for *in vitro* and *in vivo* population of the blood vessel implants, which will give additional insight into the critical parameters for the design of biomimetic niches within the 3D printed implants.

ACKNOWLEDGMENTS

The authors would like to thank Amanda F. Levy for help with the illustrations. This work was funded by the National Science Foundation with support from the Instrument Development for Biological Research (IDBR) Program and the Division of Chemical, Bioengineering, Environmental, and Transport Systems (CBET 1264517), and by the National Institute of Arthritis and Musculoskeletal and Skin Diseases of the National Institutes of Health (R01 AR061460).

REFERENCES

Almeida Hde, A., Bartolo, P.J., 2010. Virtual topological optimization of scaffolds for rapid prototyping. Med Eng Phys 32 (7), 775–782.

Arai, K., Iwanaga, S., Toda, H., Genci, C., Nishiyama, Y., Nakamura, M., 2011. Three-dimensional inkjet biofabrication based on designed images. Biofabrication 3 (3), 1758–5082.

Avci-Adali, M., Stoll, H., Wilhelm, N., Perle, N., Schlensak, C., Wendel, H.P., 2013. *In vivo* tissue engineering: mimicry of homing factors for self-endothelialization of blood-contacting materials. Pathobiology 80 (4), 176–181.

Avci-Adali, M., Ziemer, G., Wendel, H.P., 2010. Induction of EPC homing on biofunctionalized vascular grafts for rapid *in vivo* self-endothelialization: a review of current strategies. Biotechnol Adv 28 (1), 119–129.

Barron, J.A., Wu, P., Ladouceur, H.D., Ringeisen, B.R., 2004. Biological laser printing: a novel technique for creating heterogeneous 3-dimensional cell patterns. Biomed Microdevices 6 (2), 139–147.

Boland, T., Xu, T., Damon, B., Cui, X., 2006. Application of inkjet printing to tissue engineering. Biotechnol J 1 (9), 910–917.

Billiet, T., Vandenhaute, M., Schelfhout, J., Van Vlierberghe, S., Dubruel, P., 2012. A review of trends and limitations in hydrogel-rapid prototyping for tissue engineering. Biomaterials 33 (26), 6020–6041.

Barron, J.A., Krizman, D.B., Ringeisen, B.R., 2005. Laser printing of single cells: statistical analysis, cell viability, and stress. Ann Biomed Eng 33 (2), 121–130.

Berglund, J.D., Nerem, R.M., Sambanis, A., 2004. Incorporation of intact elastin scaffolds in tissue-engineered collagen-based vascular grafts. Tissue Eng 10 (9–10), 1526–1535.

Cleveland, D.C., Williams, W.G., Razzouk, A.J., Trusler, G.A., Rebeyka, I.M., Duffy, L., Kan, Z., Coles, J.G., Freedom, R.M., 1992. Failure of cryopreserved homograft valved conduits in the pulmonary circulation. Circulation 86 (5), 150–153.

Campbell, P.G., Weiss, L.E., 2007. Tissue engineering with the aid of inkjet printers. Expert Opin Biol Ther 7 (8), 1123–1127.

Creager, M., Loscalzo, J., Beckman, J.A., 2012. Vascular Medicine: A Companion to Braunwald's Heart Disease. Elsevier Health Sciences.

Cui, X., Boland, T., 2009a. Human microvasculature fabrication using thermal inkjet printing technology. Biomaterials 30 (31), 6221–6227.

Cui, X., Boland, T., D'Lima, D.D., Lotz, M.K., 2012. Thermal inkjet printing in tissue engineering and regenerative medicine. Recent Pat Drug Deliv Formul 6 (2), 149–155.

Cui, X., Boland, T., 2009b. Human microvasculature fabrication using thermal inkjet printing technology. Biomaterials 30 (31), 6221–6227.

Derby, B., 2012. Printing and prototyping of tissues and scaffolds. Science 338 (6109), 921–926.

Dauphinee, S., Karsan, A., 2010. Endothelial Dysfunction and Inflammation. Springer Basel AG.

Deitzel, J.M., Kleinmeyer, J., Harris, D., Tan, N.C. Beck, 2001. The effect of processing variables on the morphology of electrospun nanofibers and textiles. Polymer 42 (1), 261–272.

Fridrikh, S.V., Yu, J.H., Brenner, M.P., Rutledge, G.C., 2003. Controlling the fiber diameter during electrospinning. Physical Review Letters 90 (14), 144502.

Fu, W., Liu, Z., Feng, B., Hu, R., He, X., Wang, H., Yin, M., Huang, H., Zhang, H., Wang, W., 2014. Electrospun gelatin/PCL and collagen/PLCL scaffolds for vascular tissue engineering. Int J Nanomedicine 9, 2335–2344.

Fedorovich, N.E., Alblas, J., de Wijn, J.R., Hennink, W.E., Verbout, A.J., Dhert, W.J.A., 2007. Hydrogels as extracellular matrices for skeletal tissue engineering: state-of-the-art and novel application in organ printing. Tissue Engineering 13 (8), 1905–1925.

Go, A.S., Mozaffarian, D., Roger, V.L., Benjamin, E.J., Berry, J.D., Borden, W.B., Bravata, D.M., Dai, S., Ford, E.S., Fox, C.S., Franco, S., Fullerton, H.J., Gillespie, C., Hailpern, S.M., Heit, J.A., Howard, V.J., Huffman, M.D., Kissela, B.M., Kittner, S.J., Lackland, D.T., Lichtman, J.H., Lisabeth, L.D., Magid, D., Marcus, G.M.,

Marelli, A., Matchar, D.B., McGuire, D.K., Mohler, E.R., Moy, C.S., Mussolino, M.E., Nichol, G., Paynter, N.P., Schreiner, P.J., Sorlie, P.D., Stein, J., Turan, T.N., Virani, S.S., Wong, N.D., Woo, D., Turner, M.B., 2012. Heart disease and stroke statistics–2013 update: a report from the American Heart Association. Circulation 127 (1), e6–e245.

Guillotin, B., Souquet, A., Catros, S., Duocastella, M., Pippenger, B., Bellance, S., Bareille, R., Remy, M., Bordenave, L., Amedee, J., Guillemot, F., 2010. Laser-assisted bioprinting of engineered tissue with high cell density and microscale organization. Biomaterials 31 (28), 7250–7256.

Ghista, D.N., Kabinejadian, F., 2013. Coronary artery bypass grafting hemodynamics and anastomosis design: a biomedical engineering review. Biomed Eng Online 12 (129), 12–129.

Gaebel, R., Ma, N., Liu, J., Guan, J., Koch, L., Klopsch, C., Gruene, M., Toelk, A., Wang, W., Mark, P., Wang, F., Chichkov, B., Li, W., Steinhoff, G., 2011. Patterning human stem cells and endothelial cells with laser printing for cardiac regeneration. Biomaterials 32 (35), 9218–9230.

Greenwald, S.E., Berry, C.L., 2000. Improving vascular grafts: the importance of mechanical and haemodynamic properties. The Journal of Pathology 190 (3), 292–299.

Horn, T.J., Harrysson, O.L., 2012. Overview of current additive manufacturing technologies and selected applications. Sci Prog 95 (Pt 3), 255–282.

Hirt, M.N., Hansen, A., Eschenhagen, T., 2014. Cardiac tissue engineering: state of the art. Circ Res 114 (2), 354–367.

Hashi, C.K., Derugin, N., Janairo, R.R.R., Lee, R., Schultz, D., Lotz, J., Li, S., 2010. Antithrombogenic modification of small-diameter microfibrous vascular grafts. Arteriosclerosis, Thrombosis, and Vascular Biology 30 (8), 1621–1627.

Hadjizadeh, A., Ajji, A., Jolicoeur, M., Liberelle, B., G. De, Crescenzo, 2013. Effects of electrospun nanostructure versus microstructure on human aortic endothelial cell behavior. J Biomed Nanotechnol 9 (7), 1195–1209.

Han, F., Jia, X., Dai, D., Yang, X., Zhao, J., Zhao, Y., Fan, Y., Yuan, X., 2013. Performance of a multilayered small-diameter vascular scaffold dual-loaded with VEGF and PDGF. Biomaterials 34 (30), 7302–7313.

He, W., Yong, T., Teo, W.E., Ma, Z., Ramakrishna, S., 2005. Fabrication and endothelialization of collagen-blended biodegradable polymer nanofibers: potential vascular graft for blood vessel tissue engineering. Tissue Eng 11 (9–10), 1574–1588.

He, S., Xia, T., Wang, H., Wei, L., Luo, X., Li, X., 2012. Multiple release of polyplexes of plasmids VEGF and bFGF from electrospun fibrous scaffolds toward regeneration of mature blood vessels. Acta Biomater 8 (7), 2659–2669.

Hutmacher, D.W., Schantz, T., Zein, I., Ng, K.W., Teoh, S.H., Tan, K.C., 2001. Mechanical properties and cell cultural response of polycaprolactone scaffolds designed and fabricated via fused deposition modeling. Journal of Biomedical Materials Research 55 (2), 203–216.

Heydarkhan-Hagvall, S., Esguerra, M., Helenius, G., Soderberg, R., Johansson, B.R., Risberg, B., 2006. Production of extracellular matrix components in tissue-engineered blood vessels. Tissue Eng 12 (4), 831–842.

Iaizzo, P.A. *Handbook of cardiac anatomy, physiology, and devices*. 2005; Available from: http://public.eblib.com/EBLPublic/PublicView.do?ptiID=338359.

Jonas, R.A., Freed, M.D., Mayer, J.E., Castaneda, A.R., 1985. Long-term follow-up of patients with wynthetic right heart conduits. Circulation 72 (3), 77–83.

Jakab, K., Norotte, C., Marga, F., Murphy, K., Vunjak-Novakovic, G., Forgacs, G., 2010. Tissue engineering by self-assembly and bioprinting of living cells. Biofabrication 2 (2), 1758–5082.

Junqueira, L.C.U.C.J.K.R.O., 1995. Basic histology. Appleton & Lange, Norwalk, Conn.

Kozel, B.A., Mecham, R.P., Rosenbloom, J., 2011. *Elastin*. Extracellular Matrix: An Overview, ed. R. P. Mecham. Springer-Verlag Berlin, Berlin, 267-301.

Koch, L., Gruene, M., Unger, C., Chichkov, B., 2013. Laser-assisted cell printing. Curr Pharm Biotechnol 14 (1), 91–97.

Koch, L., Deiwick, A., Schlie, S., Michael, S., Gruene, M., Coger, V., Zychlinski, D., Schambach, A., Reimers, K., Vogt, P.M., Chichkov, B., 2012. Skin tissue generation by laser cell printing. Biotechnology and Bioengineering 109 (7), 1855–1863.

Kasalkova, N.S., Slepi Ka, P., Kolska, Z.K., Hoda Ova, P., Ku Kova, T.P., Ik, V. Vor, 2014. Grafting of bovine serum albumin proteins on plasma-modified polymers for potential application in tissue engineering. Nanoscale Res Lett 9 (1), 161.

Lamers, L.J., Frommelt, P.C., Mussatto, K.A., Jaquiss, R.D.B., Mitchell, M.E., Tweddell, J.S., 2012. Coarctectomy combined with an interdigitating arch reconstruction results in a lower incidence of recurrent arch obstruction after the Norwood procedure than coarctectomy alone. Journal of Thoracic and Cardiovascular Surgery 143 (5), 1098–1102.

Lai, H.J., Kuan, C.H., Wu, H.C., Tsai, J.C., Chen, T.M., Hsieh, D.J., Wang, T.W., 2014. Tailored design of electrospun composite nanofibers with staged release of multiple angiogenic growth factors for chronic wound healing. Acta Biomater 9 (14), 00204–209.

Luong-Van, E., Grondahl, L., Chua, K.N., Leong, K.W., Nurcombe, V., Cool, S.M., 2006. Controlled release of heparin from poly(epsilon-caprolactone) electrospun fibers. Biomaterials 27 (9), 2042–2050.

Melchels, F.P.W., Domingos, M.A.N., Klein, T.J., Malda, J., Bartolo, P.J., Hutmacher, D.W., 2012. Additive manufacturing of tissues and organs. Progress in Polymer Science 37 (8), 1079–1104.

Marga, F., Jakab, K., Khatiwala, C., Shepherd, B., Dorfman, S., Hubbard, B., Colbert, S., Gabor, F., 2012. Toward engineering functional organ modules by additive manufacturing. Biofabrication 4 (2), 1758–5082.

Montero, R.B., Vial, X., Nguyen, D.T., Farhand, S., Reardon, M., Pham, S.M., Tsechpenakis, G., Andreopoulos, F.M., 2012. bFGF-containing electrospun gelatin scaffolds with controlled nano-architectural features for directed angiogenesis. Acta Biomater 8 (5), 1778–1791.

Miller, J.S., Stevens, K.R., Yang, M.T., Baker, B.M., Nguyen, D.H., Cohen, D.M., Toro, E., Chen, A.A., Galie, P.A., Yu, X., Chaturvedi, R., Bhatia, S.N., Chen, C.S., 2012. Rapid casting of patterned vascular networks for perfusable engineered three-dimensional tissues. Nat Mater 11 (9), 768–774.

Malda, J., Visser, J., Melchels, F.P., Jüngst, T., Hennink, W.E., Dhert, W.J.A., Groll, J., Hutmacher, D.W., 2013. 25th anniversary article: engineering hydrogels for biofabrication. Advanced Materials 25 (36), 5011–5028.

Nishiyama, Y., Nakamura, M., Henmi, C., Yamaguchi, K., Mochizuki, S., Nakagawa, H., Takiura, K., 2009. Development of a three-dimensional bioprinter: construction of cell supporting structures using hydrogel and state-of-the-art inkjet technology. J Biomech Eng 131 (3), 3002759.

Nakamura, M., Kobayashi, A., Takagi, F., Watanabe, A., Hiruma, Y., Ohuchi, K., Iwasaki, Y., Horie, M., Morita, I., Takatani, S., 2005. Biocompatible inkjet printing technique for designed seeding of individual living cells. Tissue Eng 11 (11-12), 1658–1666.

Norotte, C., Marga, F.S., Niklason, L.E., Forgacs, G., 2009. Scaffold-free vascular tissue engineering using bioprinting. Biomaterials 30 (30), 5910–5917.

Nerem, R.M., 2003. Role of mechanics in vascular tissue engineering. Biorheology 40 (1–3), 281–287.

Phillippi, J.A., Miller, E., Weiss, L., Huard, J., Waggoner, A., Campbell, P., 2008. Microenvironments engineered by inkjet bioprinting spatially direct adult stem cells toward muscle- and bone-like subpopulations. Stem Cells 26 (1), 127–134.

Pirlo, R.K., Wu, P., Liu, J., Ringeisen, B., 2012. PLGA/hydrogel biopapers as a stackable substrate for printing HUVEC networks via BioLP. Biotechnol Bioeng 109 (1), 262–273.

Patel, A., Fine, B., Sandig, M., Mequanint, K., 2006. Elastin biosynthesis: the missing link in tissue-engineered blood vessels. Cardiovasc Res 71 (1), 40–49.

Roger, V.L., Go, A.S., Lloyd-Jones, D.M., Benjamin, E.J., Berry, J.D., Borden, W.B., Bravata, D.M., Dai, S., Ford, E.S., Fox, C.S., Fullerton, H.J., Gillespie, C., Hailpern, S.M., Heit, J.A., Howard, V.J., Kissela, B.M., Kittner, S.J., Lackland, D.T., Lichtman, J.H., Lisabeth, L.D., Makuc, D.M., Marcus, G.M., Marelli, A., Matchar, D.B., Moy, C.S., Mozaffarian, D., Mussolino, M.E., Nichol, G., Paynter, N.P., Soliman, E.Z., Sorlie, P.D., Sotoodehnia, N., Turan, T.N., Virani, S.S., Wong, N.D., Woo, D., Turner, M.B., Comm, A.H.A.S., Subcomm, S.S., 2012. Heart disease and stroke statistics–2012 Update: a report from the American Heart Association. Circulation 125 (1), E2–E220.

Roy, K., 2010. Biomaterials as Stem Cell Niche. Springer.

Rayatpisheh, S., Heath, D.E., Shakouri, A., Rujitanaroj, P.O., Chew, S.Y., Chan-Park, M.B., 2014. Combining cell sheet technology and electrospun scaffolding for engineered tubular, aligned, and contractile blood vessels. Biomaterials 35 (9), 2713–2719.

Roth, E.A., Xu, T., Das, M., Gregory, C., Hickman, J.J., Boland, T., 2004. Inkjet printing for high-throughput cell patterning. Biomaterials 25 (17), 3707–3715.

Stark, J., 1998. The use of valved conduits in pediatric cardiac surgery. Pediatric Cardiology 19 (4), 282–288.

Schubert, C., van Langeveld, M.C., Donoso, L.A., 2013. Innovations in 3D printing: a 3D overview from optics to organs. British Journal of Ophthalmology.

Sales, V.L., Engelmayr, G.C., Mettler, B.A., Johnson, J.A., Sacks, M.S., Mayer, J.E., 2006. Transforming growth factor-beta 1 modulates extracellular matrix production, proliferation, and apoptosis of endothelial progenitor cells in tissue-engineering scaffolds. Circulation 114, I193–I199.

Shapira, A., Kim, D.H., Dvir, T., 2014. Advanced micro- and nanofabrication technologies for tissue engineering. Biofabrication 6 (2), 020301.

Soliman, S., Sant, S., Nichol, J.W., Khabiry, M., Traversa, E., Khademhosseini, A., 2011. Controlling the porosity of fibrous scaffolds by modulating the fiber diameter and packing density. J Biomed Mater Res A 96 (3), 566–574.

Subbiah, T., Bhat, G.S., Tock, R.W., Parameswaran, S., Ramkumar, S.S., 2005. Electrospinning of nanofibers. Journal of Applied Polymer Science 96 (2), 557–569.

Sodian, R., Loebe, M., Hein, A., Martin, D.P., Hoerstrup, S.P., Potapov, E.V., Hausmann, H., Lueth, T., Hetzer, R., 2002. Application of stereolithography for scaffold fabrication for tissue engineered heart valves. Asaio J 48 (1), 12–16.

Sodian, R., Fu, P., Lueders, C., Szymanski, D., Fritsche, C., Gutberlet, M., Hoerstrup, S.P., Hausmann, H., Lueth, T., Hetzer, R., 2005. Tissue engineering of vascular conduits: fabrication of custom-made scaffolds using rapid prototyping techniques. Thorac Cardiovasc Surg 53 (3), 144–149.

Saunders, R.E., Gough, J.E., Derby, B., 2008. Delivery of human fibroblast cells by piezoelectric drop-on-demand inkjet printing. Biomaterials 29 (2), 193–203.

Song, S.J., Choi, J., Park, Y.D., Hong, S., Lee, J.J., Ahn, C.B., Choi, H., Sun, K., 2011. Sodium Alginate Hydrogel-Based Bioprinting Using a Novel Multinozzle Bioprinting System. Artificial Organs 35 (11), 1132–1136.

Tasoglu, S., Demirci, U., 2013. Bioprinting for stem cell research. Trends Biotechnol 31 (1), 10–19.

Wu, H., Fan, J., Chu, C.C., Wu, J., 2010. Electrospinning of small diameter 3D nanofibrous tubular scaffolds with controllable nanofiber orientations for vascular grafts. J Mater Sci Mater Med 21 (12), 3207–3215.

Wang, Y., Kibbe, M.R., Ameer, G.A., 2013. Photo-crosslinked biodegradable elastomers for controlled nitric oxide delivery. Biomater Sci 1 (6), 625–632.

Xu, T., Jin, J., Gregory, C., Hickman, J.J., Boland, T., 2005. Inkjet printing of viable mammalian cells. Biomaterials 26 (1), 93–99.

Xu, T., Gregory, C.A., Molnar, P., Cui, X., Jalota, S., Bhaduri, S.B., Boland, T., 2006. Viability and electrophysiology of neural cell structures generated by the inkjet printing method. Biomaterials 27 (19), 3580–3588.

Ye, L., Wu, X., Duan, H.Y., Geng, X., Chen, B., Gu, Y.Q., Zhang, A.Y., Zhang, J., Feng, Z.G., 2012. The *in vitro* and *in vivo* biocompatibility evaluation of heparin-poly(epsilon-caprolactone) conjugate for vascular tissue engineering scaffold. Biomed Mater Res A 100 (12), 3251–3258.

Yamazoe, H., Tanabe, T., 2009. Cell micropatterning on an albumin-based substrate using an inkjet printing technique. J Biomed Mater Res A 91 (4), 1202–1209.

Zhang, H., Jia, X., Han, F., Zhao, J., Zhao, Y., Fan, Y., Yuan, X., 2013. Dual-delivery of VEGF and PDGF by double-layered electrospun membranes for blood vessel regeneration. Biomaterials 34 (9), 2202–2212.

Ziabari, M., Mottaghitalab, V., Haghi, A., 2010. A new approach for optimization of electrospun nanofiber formation process. Korean Journal of Chemical Engineering 27 (1), 340–354.

Zein, I., Hutmacher, D.W., Tan, K.C., Teoh, S.H., 2002. Fused deposition modeling of novel scaffold architectures for tissue engineering applications. Biomaterials 23 (4), 1169–1185.

3D PRINTING AND PATTERNING VASCULATURE IN ENGINEERED TISSUES

Bagrat Grigoryan and Jordan S. Miller

Department of Bioengineering, Rice University, Houston, TX, USA

8.1 INTRODUCTION

Over the past half-century, researchers have made significant progress in the isolation and growth of human cells outside the body by deducing the characteristics of the extracellular environment (both soluble and insoluble) that contribute to the survival and growth of cells (Albrecht-Buehler, 1976). Moreover, the detailed mechanistic understanding of cellular biochemical activity has progressed at a rapid pace, powered by advanced genetic reporters (Shaner et al., 2005) and imaging modalities (Kanchanawong et al., 2010). However, efforts to adapt the findings from cell monolayer culture to the level of large tissues and organs have been hampered by technological limitations in keeping large masses of cells alive. While millions or perhaps even billions of human cells can now be grown and expanded as monolayers in Petri dishes, the field of bioengineering has no generic set of technologies for standardized assembly of cells into functional tissues or organ structures. The remaining major technological challenges are rooted in questions of tissue architecture and mass transport—how do we get nutrients and oxygen in and metabolic waste products out of tissues at rates akin to that seen in the human body (Miller, 2014; Hasan et al., 2014)? Here, we highlight recent efforts to fabricate vascular networks in engineered tissues to address questions of mass transport and blood perfusion. We also discuss some of the technical hurdles and conceptual targets on the horizon.

8.1.1 MACROPOROUS CONSTRUCTS AS TISSUE TEMPLATES

To enable convective transport within biocompatible materials, common approaches have utilized various material processing steps such as critical point drying (Dagalakis et al., 1980), gas foaming and salt leaching (Jun and West, 2005), or electrospinning (Pham et al., 2006) to create macroporous structures that can be perfused *in vitro* for tissue culture. Indeed, these types of porous foams have shown great utility serving as templates for the construction of functional tissue extensions because they match the mechanical compliance of native tissue while also having high surface area in which nutrients and oxygen can be delivered (Yannas et al., 1982). However, precise spatial control of scaffold architecture is not easily achieved with these methodologies (Harley et al., 2010).

Soft lithography, a suite of techniques adapted from the microprocessor industry, offered spatial control of scaffold architecture through photopatterning with photolithography. With these technologies of patterning and then replica molding patterns into silicone substrates, detailed studies could be conducted at the cell-material interface (Whitesides et al., 2001; Kane et al., 1999). Early soft lithography and photolithography work involved fabricating substrates with grooves and microbeds surfaces, and chemical modifications to the substrate surface, to study cell adhesion, migration, and proliferation (Truskett and Watts, 2006). These approaches generally resulted in planar scaffolds to elucidate topological effects on cell behavior.

To control the cellular microenvironment in a more 3D fashion, Vozzi et al. developed macroporous scaffolds out of poly(dimethylsiloxane) (PDMS) templates which were cast into poly(L-lactic-co-glycolic acid) (PLGA) multilayer scaffolds (Figure 8.1) (Vozzi et al., 2003). Fabrication of macroporous scaffolds containing living cells was investigated by Liu et al. by combining tissue engineering with photolithographic methods (Liu and Bhatia, 2002). An apparatus was designed in which the prepolymer solution was injected into a chamber, with specified height controlled by spacer thickness, and then a mask was placed on top of the chamber and exposed to UV light (Figure 8.1). Multilayered hydrogel

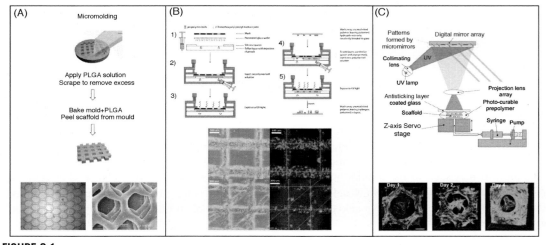

FIGURE 8.1

Patterned macroporous scaffolds can be fabricated by various methods. (A) Top: Micromachined masters can be replica-molded into multilayered microfluidic devices. Bottom: Single-layered and multilayered 3D PLGA scaffolds. Scale bars = 200 μm; figure adapted from Vozzi et al. (2003). (B) Top: Methodology for additive photopatterning with physical photomasks to fabricate cell-laden hydrogels with macroscale features. Bottom: Fluorescent images of photopatterned cell-laden hydrogels with various macroscale patterns. Scale bar = 500 μm; figure adapted from Liu Tsang et al. (2007). (C) Top: Dynamic photomasking with a computer-controlled digital mirror projection stereolithography system can be used to engineer multilayered structures in a highly automated fashion. Bottom: Cell infiltration of a 3D hexagonal printed scaffold over the course of 4 days. Scale bar = 100 μm; figure adapted from Gauvin et al. (2012).

structures were formed after rinsing the first layer with buffer, changing the mask, using a thicker spacer, and injecting prepolymer solution into the chamber followed by UV exposure. Multilayered hydrogel microstructures containing encapsulated living cells were developed by adding cells to the prepolymer solution and following the same procedure. Using this approach, cellular tissue constructs were created in which placement of cells could be spatially controlled in 3D configurations throughout a thick construct. The authors were able to demonstrate fidelity of the patterns and cell viability in the fabricated hydrogel microstructures.

Further efforts from the same group resulted in a more refined apparatus to fabricate structures with macroporous architecture (Liu Tsang et al., 2007). Due to their interest in fabricating hepatic tissues, the authors tailored the chemistry and architecture of hydrogels to support hepatocyte survival and liver-specific function. The authors observed that coculture of hepatocytes with fibroblasts stabilized the hepatocytes, allowing for better incorporation into biomaterials by increasing their affinity to ligands. Multilayered cellular hydrogels fabricated with controlled 3D microarchitecture facilitated the transport of oxygen and nutrients, resulting in higher cell viability due to the high metabolic demands of encapsulated hepatocytes. Additionally, the authors demonstrated the importance of incorporating macroporous structures within the fabricated multilayered constructs after the constructs were transferred to a perfused bioreactor, resulting in higher cellular metabolic activity for 2 weeks compared to static constructs. Although the system utilized was able to achieve 3D microarchitectures that were reminiscent of native tissue, the approach involved manually switching out spacers, masks, and adding prepolymer solution which can be time consuming and increase the chances of problems with architectural fidelity. Similar 3D constructs were automatically fabricated in a layer-by-layer fashion by using 3D projection stereolithography to fabricate an entire layer under one single UV exposure (Gauvin et al., 2012). This approach resulted in rapid fabrication of tissue constructs with controlled microarchitecture. The custom designed projection stereolithography (PSL) system developed consists of a digital light processing (DLP) chip to create dynamic photomasks from computer-aided design (CAD), which are projected onto the prepolymer solution, and a servo stage which can increment the z-axis for patterning of subsequent layers (Figure 8.1). Additionally, a syringe which dispensed prepolymer solution was connected to the system. The authors demonstrated that the physical characteristics of the gelatin methacrylate (GelMA) scaffold, such as porosity and interconnectivity, can be tailored by controlling pore geometry and architecture. The authors also demonstrated that dynamic seeding of cells on the hydrogel structure remained viable, spread, and proliferated for an extended period of incubation, indicating biocompatibility of the fabrication process. Recently, similar liver-mimetic 3D structures were fabricated by 3D printing as a potential detoxification device. In this work, poly(diacetylene) nanoparticles were incorporated into hydrogels with precise microstructures and were shown to attract, capture, and sense toxins while the 3D matrix trapped the toxins (Gou et al., 2014).

8.1.2 FABRICATING FLUIDIC NETWORKS WITHIN BIOMATERIALS

The fabrication of macroporous scaffolds, such as those described earlier, provided early insights into the relationship between structure and function in native tissues, though they failed to mimic the cardiovascular system responsible for the bulk of convective mass transport in the body. Recent efforts have turned toward the introduction of vascular networks into engineered tissues, able to support larger cell populations, and able to sustain the flow of complex fluids such as whole blood. Importantly, blood is known to be extremely sensitive to small perturbations in fluid dynamics and microenvironment, rapidly leading to clotting (McGuigan and Sefton, 2007). Here we discuss approaches to pattern perfusable channels and networks in engineered tissues, motivated by limitations in controlling individual pore size, shape, arrangement, and interconnectivity of microdevices.

Photolithography can typically pattern only a single layer of material at a time. Thus, strategies combining photolithographic processes with layer-by-layer lamination methods represent an important approach to achieve 3D scaffolds with high-resolution microstructures. To this end, Bettinger et al. developed a vasculature construct out of poly(glycerol-sebacate) (PGS), a tough, biodegradable elastomer with superior mechanical properties (Bettinger et al., 2005). Multiple layers of PGS vasculature construct were obtained by stacking and aligning layers, followed by a thermal bonding process. Although the authors demonstrated that cells seeded on the fabricated scaffold resulted in normal metabolic activity, the ability to encapsulate cells in these constructs was limited by the harsh fabrication process. Furthermore, the ability to precisely form microstructures with encapsulated cells in complex 3D structures cannot easily be obtained using this approach.

King et al. developed a novel technique for fabrication of high-resolution, high-precision features by combining a thermal fusion bonding process after fabrication of a single layer of microfluidic network composed of thermoplastic PLGA (King et al., 2004). This optimized approach involves conventional photolithography and silicon micromachining of single layers, followed by precise alignment and stacking of interconnected micropatterned films and thermal sealing, to obtain highly branched, multilayer PLGA microfluidic networks. Moreover, the authors were able to demonstrate the formation of 2 μm wide channels, close to the physical limits of traditional photolithography, all the way up to channels in the millimeter range. When perfused, single-layer networks showed no sign of leaks, occlusions, or channel-to-channel cross-talk, and demonstrated linear pressure-flow characteristics. Lactide has also been incorporated into a copolymer with PEG to form microchannels by selective degradation of bulk photopolymerized hydrogel substrates (Chiu et al., 2009). While the bulk of the scaffold was made from poly(ethylene glycol) diacrylate (PEGDA) by exposing light without a photomask, a photomask was used when PEG-PLLA-DA was injected into the system to create smaller patterns on top of the previous layer. Then, the spaces between the patterned strips were injected with PEGDA and cured. Channels were obtained in scaffolds by repeating the steps to create multilayered constructs and then incubating the construct in high pH environment to accelerate degradation of the PEG-PLLA-DA copolymer. This route for microchannel fabrication offers a 3D architecture with interconnected microchannels that can be obtained from biocompatible materials.

Inspired by the complex, multiscale microvascular network of leaf venation, He et al. fabricated nature-inspired microfluidic networks for perfusable tissue constructs (He et al., 2013). Their technique involved a microreplication method to mimic the microvascular network of natural leaf in synthetic substrates by digesting leafs to expose venation networks, sputtering with a thin layer of chrome, and then using the chrome-coated leaf venation as photomask to obtain a silicon mold with leaf venation to fabricate biomaterial hydrogels with multiscale vascular channels ranging from 30 μm to 1 mm in diameter. Seeding cells in this system resulted in high viability of endothelial cells, proliferation, and spreading along the microfluidic channels during the short culture period. The channels were perfused by mounting the hydrogels on an incline plate and introducing fluid from an inlet in a drop-wise manner, whose flow was induced by gravity. Further work by Wu et al. studied the transport efficiency of microvascular networks inspired by the hierarchical, bifurcating network of leaf venation (Wu et al., 2010). In this approach, a CAD image of leaf venation was obtained and then a modified direct-write assembly process utilizing fugitive organic ink was used to create biomimetic microvascular networks of varying design. The authors explored the effects of network design on fluid transport efficiency and found that fluid transport efficiency is maximized when the network architecture obeys Murray's law, which relates the radii of branching vasculature to volumetric flow and velocity profiles showing maximal efficiency of mass transport (Murray, 1926; Sherman, 1981).

8.1.3 APPROACHES TO FABRICATE ENDOTHELIALIZED AND CELL-LADEN TISSUE CONSTRUCTS

Initial efforts to incorporate cells into constructs that resemble vasculature involved seeding of endothelial cells (ECs) inside channels after fabrication of 3D constructs. Lining channels with endothelial cells is critical for control of vascular functions such as barrier function, blood vessel formation, coagulation, and inflammatory response. To this end, Golden et al. have fabricated perfusable microfluidic extracellular matrix (ECM) hydrogels by using molded gelatin as a sacrificial element for the transport of materials in a tissue analogue (Golden and Tien, 2007). By encapsulating micromolded meshes of gelatin inside a gel followed by removal of gelatin by heating and flushing, interconnected channels as narrow as 6 μm were obtained. These gels were then seeded with microvascular endothelial cells (MECs) to demonstrate attachment, spreading, and proliferation of cells on the lumen of the channels, indicating normal function of seeded cells, which could later be perfused. To produce a multiplanar network that can be used to transport and exchange materials in a 3D manner, the authors coencapsulated and melted multiple gelatin meshes. Although soft lithography and photolithography techniques have been extensively used to create 3D microfluidic networks, one limitation of these techniques is that the fabricated channels contain a rectangular cross-section and sharp transitions between channels, which do not mimic physiological properties. Furthermore, the presence of rectangular walls and corners, in addition to nonphysiological blood flow conditions, results in asymmetric shear stress around the vessel perimeter, limiting the ability to form confluent endothelial cell layers. To this end, Borenstein et al. used an electroplating process to obtain sheets with semicircular cross-sections and smooth transitions at bifurcations and changes in diameter (Borenstein et al., 2010). To obtain closed circular microchannel networks, the sheets were aligned and bonded by applying an adhesive to the outer edges of the sheets. Cell viability and spreading of human umbilical vein endothelial cells (HUVECs) cultured within the circular channel networks was confirmed after 24 h of culture in channels containing multiple bifurcations of various diameters.

Although the techniques just discussed have been successful in incorporating ECs inside channel lumens, to mimic the more complex architecture of vasculature, various methods have been utilized to fabricate perfusable, cell-laden 3D constructs. One such method, developed by Cuchiara et al., involved a simple and robust multilayer replica molding technique in which PDMS and PEGDA are serially replica molded to develop microfluidic hydrogel networks embedded within independently fabricated PDMS (Cuchiara et al., 2010). Taking rational network design into account, the authors sought to determine the optimal microfluidic vessel spacing to maintain cell viability and maximize construct metabolic density of encapsulated cells to overcome diffusional limitations of nutrients and waste. Indeed, the viability of 3T3 fibroblasts was shown to depend on culture time, distance from perfused channel, and culture conditions. In constructs containing a perfused channel, necrotic cores were shown to be confined to the outermost regions of the scaffold. Needle-based molding techniques have also been utilized to obtain perfusable channels and offer researchers a way to develop experimental models for studying vasculature function with a single channel. Chrobak et al. utilized this approach to generate open tubes of microvascular cells in a collagen gel with vessel diameters ranging from 75 to 150 μm (Chrobak et al., 2006). To obtain microvascular channels, a needle was suspended in a silicone mount while collagen was added into the chamber, surrounding the needle. After the collagen was allowed to gel, the needle was carefully removed to reveal a fluidic channel that matched the dimensions of the needle. By etching the tip of a needle, resulting in reduction of the needle diameter at the tip, abrupt changes in the diameter of vessel can be obtained. Using this technique, the authors demonstrated organization of seeded ECs in the tubes, which allowed the tubes to be perfused without significant leakage (Figure 8.2F). In addition to demonstrating the long-term culture of ECs for up to

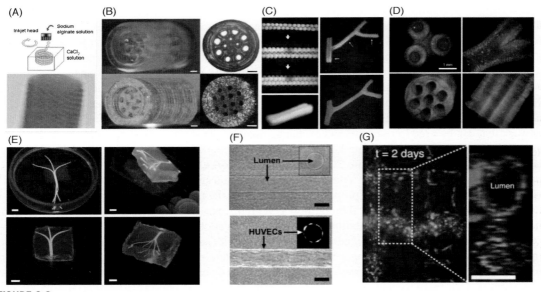

FIGURE 8.2

Microchannel structures with various architecture. (A) Fabrication of tubular structures composed of alginate hydrogel by ink-jet 3D bioprinting. Figure adapted from Nakamura et al. (2008). (B) Multilumen PEG hydrogel conduit obtained by line-scan stereolithography. Scale bars = 1 mm; figure adapted from Arcaute et al. (2006). (C) Assembly of multicellular spheroids into tubular structures with fusion of spheroids after a week in culture can be used to create branched architecture. Figure adapted from Norotte et al. (2009). (D) Fluorescent images of branched and multilumen hyaluronic acid hydrogel fabricated by projection stereolithography. Scale bar = 1 mm; figure adapted from Suri et al. (2011). (E) 3D-printed agarose fibers serve as temporary templates for casting perfusable vasculature. Scale bars = 3 mm; figure adapted from Bertassoni et al. (2014). (F) Endothelial cells can be lined along the lumen of a microchannel, which was fabricated by a needle-based casting approach. Scale bar s = 100 μm; figure adapted from Chrobak et al. (2006). (G) Fluorescent images of vascularized, heterogeneous tissue constructs fabricated by using a 3D bioprinter with multiple independently controlled printheads. Scale bar = 300 μm; figure adapted from Kolesky et al. (2014).

3 weeks, the tubes exhibited normal microvascular functions such as adhesion of leukocytes and response to inflammatory mediators. Additionally, the authors demonstrated feasibility of incorporating cells into the gel by mixing cells with the collagen solution before gelation. Materials such as reconstituted collagen gels often exhibit weak mechanical strength and may not support physiologic flow rates through perfused vasculature. Synthetic materials provide an intriguing alternative, especially when functionalized with photoreactive groups. Photopolymerized hydrogels allow for straightforward tuning of mechanical properties by varying system parameters such as polymer concentration, light intensity, and photoinitiator concentration. To this end, Nichol et al. utilized the needle-based approach to demonstrate GelMA as a potential biomaterial for applications in microfluidic system and fabrication of vascularized engineered tissues (Nichol et al., 2010). Cell-laden constructs with endothelial-lined microchannels were obtained by UV radiation of a solution of GelMA within a rectangular PDMS mold containing a needle. After the needle was withdrawn, resultant gels were perfused and evidence of endothelial cell adhesion and aggregation within surrounding cells, in addition to elongation and reorganization of encapsulated 3T3 cells, was evident. Detailed studies quantifying the toxicity of photoinitiators and light sources have previously been performed (Bryant et al., 2000), allowing the experimentalist to retain cell viability during cell encapsulation.

8.1.4 APPROACHES TO INTEGRATE PATTERNED VASCULATURE *IN VIVO*

Although photolithographic and needle-based approaches to fabricate simple models for studying vasculature function have been successful, more advanced constructs are desirable which closely resemble the heterogeneity and complexity of tissues. Thus, techniques such as ink-jet printing and stereolithography have advanced significantly, providing researchers with high-resolution printing capability and an expanded parameter space of materials from which to choose. Early efforts in ink-jet printing involved fabricating tubular hydrogel structures with a liquid aqueous gelling medium (Nakamura et al., 2008). In this technique, droplets ejected from a print head containing sodium alginate form into alginate microgel beads by contact with calcium ions and fuse to form fibers and sheets as the ink-jet system is moved laterally and vertically (Figure 8.2A). Microgel beads as small as 10 μm in diameter were produced using this approach. To achieve 3D tubular structures on a very small scale, the ink-jet head was moved in a circular pattern as droplets of alginate were ejected onto the substrate in a layer-by-layer fashion to form long tubular structures. The group has also fabricated fibers, 2D sheets, multilayered sheets, and more 3D vessels using this approach (Nishiyama et al., 2009). However, Cui et al. have utilized thermal ink-jet printing technology for microvascular fabrication and have demonstrated alignment and proliferation of cells inside a fibrin substrate (Cui and Boland, 2009). In this study, a fibrin scaffold was printed with cells, resulting in minimal deformation of the scaffold and little, if any, damage to cells. Fabricated microvasculature structures composed of fibrin and ECs resulted in the cells forming contacts with each other and, ultimately, aligning themselves along the fibrin channel to form a confluent lining. In addition, long-term scaffold integrity of the printed microvasculature was demonstrated.

Further techniques to fabricate even more complex heterogeneous structures layer-by-layer involve utilizing CAD-based rapid prototyping methods such as stereolithography. Chan et al. used a custom-modified stereolithography apparatus to fabricate complex 3D structures from photopolymerization of PEG by repetitive deposition and processing of individual layers (Chan et al., 2010). In addition, the authors characterized the penetration depth and critical exposure energy parameters for ideal sequential layer-by-layer hydrogel photopatterning so that layers sufficiently adhere to each other and reduce overcuring of printed layers, as well as swelling behavior and mechanical properties of the laser polymerized PEGDA hydrogels. The authors were also able to demonstrate control over spatial distribution of cells in a multilayered structure with minimal mixing as well as high viability, proliferation, and spreading of entrapped cells, indicating that this methodology can be applied to achieve high spatial control to recapitulate complex 3D tissue microarchitecture. Tissue constructs with open channels can also be obtained by using a light-based projection stereolithography system due to the high fabrication rate and resolution of this technique in printing layer-by-layer (Lin et al., 2013). However, solutions containing cells in the apparatus chamber tend to settle down over time during the printing process due to the inherent higher density of the cells, resulting in constructs with nonhomogenous encapsulation of cells. To address this issue, the authors included optimized concentration of Percoll to increase the density of the precursor solution to match the density of the cells, making them neutrally buoyant. However, addition of Percoll resulted in some undesirable curing since the authors observed that fabrication of porous structures resulted in partial pore occlusion. The authors also fabricated open channel scaffolds with encapsulated cells and observed high cell viability and metabolic activity in the fabricated hydrogels, even though the scaffolds did not contain any adhesive peptides.

Other methods for fabricating vessel-like tissue constructs involve dispensing cell containing macrofilaments or spheroids and then allowing the cells to fuse into whole tissue constructs. One strategy for direct-write bioprinting of a cell-laden ECM hydrogel dispensed prepolymerized cell laden GelMA hydrogel fibers from glass capillary tubes (Bertassoni, Cardoso, et al., 2014). The bioprinting process involves aspirating the hydrogel precursor into a glass capillary by the upward movement of a metallic piston, followed by photopolymerization of the precursor inside the capillary by exposure to UV light, and then dispensing the hydrogel fiber as the metallic piston is pushed down against the cross-linked gel. The authors demonstrated bioprinting of macroscale 3D cell-laden constructs by positioning hydrogel fibers in one plane and stacking a second layer of perpendicular fibers on the plane above. More complex constructs were fabricated that contained GelMA hydrogel blocks with impregnated planar and 3D bifurcating fiber networks to achieve tissue analogs that mimic vasculature. To fabricate constructs with hollow fibers, a sacrificial layer was dispensed during the printing process and layer removed, resulting in microchannels within the fabricated construct. The authors demonstrated relatively high cell viability and proliferation rates of encapsulated cells in the printed gels.

Additionally, other direct-write processes utilize a scaffold-free approach to circumvent some hurdles in tissue engineering approaches that utilize a biomaterial scaffold, such as a plethora of materials to choose from, constructing viable tissues with high cell density, and potential uncontrollable scaffold mechanical effects on cell behavior. Thus, utilization of a scaffold-free approach can avoid some of these issues. Furthermore, a scaffold-free approach is inspired by the self-assembly process of cells during early morphogenetic processes, in which individual cells organize into multicellular subunits (Chang et al., 2013). To fabricate bioartificial vessel-like grafts using a scaffold-free approach, an open-source printer, modified to hold microcapillary tubes for bioprinting of hydrogel macrofilaments, was used to dispense cylindrical macrofilaments layer-by-layer (Skardal et al., 2010). Culture of the filaments resulted in discrete structures, as the filaments fused with each other to form a continuous structure. With this approach, the resolution of the printed tissue construct was determined by the capillary internal diameter, which was 500 μm. To better facilitate the fusion of dispensed cell-containing materials, Jakab et al. utilized a scaffold-free approach by dispensing multicellular spheroids and cylinders to allow cells to self-organize into functional living structures of prescribed shape (Jakab et al., 2008). Multicellular spheroids and cylinders were prepared as bioink and deposited with print heads onto biopaper containing a collagen gel, which serves as building blocks of a molding template. The approach involves centrifuging trypsinized cells and then incubating them in capillary micropipettes, which are transferred into an apparatus that extrudes a cell-laden fiber that is automatically cut into equal-sized cylinders. After culturing for 60 h, the authors observed fusion of cylindrical aggregates into well-defined tubular structures. More advanced constructs, such as branched tubes that better resemble vessels, were also obtained using this method. Further work using cylindrical bioink to obtain multilayered tubular vascular grafts has been developed by the same group, demonstrating versatility of the approach (Norotte et al., 2009). Postprinting fusion of the cellular cylinders was observed to take place in 2–4 days, as opposed to the 5–7 days for spheroids, and resulted in formation of uniform tubes (Figure 8.2C). However, the authors observed sparsely distributed apoptotic cells throughout the vascular wall of their engineered vessels.

On-demand capability of 3D bioprinting techniques to fabricate large, clinically relevant 3D structures at high resolution could permit patient-specific, highly customized tissue constructs. Initial work on fabricating larger multilayered tissue conduits with multiple lumens utilizing stereolithography techniques was achieved by Arcaute et al. (Arcaute et al., 2006) (Figure 8.2B). Large blocks comprising 5 cm in length with embedded channels with a bifurcation were fabricated to demonstrate the feasibility of obtaining geometries representative of the arterial system. The authors demonstrated high cell viability in encapsulated hydrogels photocrosslinked using SL over 24 h, indicating that the cells can survive the stereolithography process. Progress to obtain branched constructs was demonstrated by Suri et al. who used stereolithography to fabricate 3D scaffolds of hyaluronic acid with defined architecture mimicking the native tissue microenvironment (Suri et al., 2011). This approach involves a digital-mirror device (DMD) to create structures in a layer-wise fashion by fabricating an entire layer simultaneously. Hyaluronic acid scaffolds with different geometries with varying pore shapes and sizes were fabricated in a layer-by-layer fashion and were shown to retain functionality after the fabrication process (Figure 8.2E). The authors also demonstrated that the fabricated scaffolds support cell adhesion, spreading, and proliferation after seeding the branched scaffolds. While the scaffolds were designed for nerve tissue engineering purposes, similar clinically relevant large-scale scaffolds can be fabricated for vascular applications. Duan et al. implemented an onboard photocrosslinking system on an extrusion-based open-source printer to enable simultaneous 3D extrusion printing and curing of PEG hydrogels with a mounted UV-LED cross-linking module (Duan et al., 2014). By doing so, a solution was gelled immediately after dispensing and onset of light exposure, without requiring a biopaper to initiate gelation. Trileaflet valve hydrogels were fabricated to test the response of encapsulated human valvular interstitial cells (VICs) within the printed leaflets. The authors observed that encapsulated cells remained spherical in shape after 3 days in culture but began to exhibit more spread morphology on day 7. The same system was used by the group to rapidly engineer complex, heterogeneous aortic valve scaffolds (Hockaday et al., 2012). The authors printed an axisymmetric aortic valve geometry, including the sinus and leaflets, to replicate native valve geometry and regional mechanical heterogeneity. They characterized the mechanical properties of hydrogels and observed high viability of cells seeded onto the valve scaffold for over 3 weeks. The authors also demonstrated the ability to print a complex model obtained from a medical image, suggesting the use of this technology for printing of patient-specific tissue constructs. Additionally, the same group demonstrated the versatility of the approach by using another material system to fabricate mechanically robust living trileaflet valves with multiple cell populations using 3D printing (Duan et al., 2013). In simple alginate/gelatin hydrogels, high cell viability for encapsulated VICs and smooth muscle cells (SMCs) was observed over 7 days in culture with specific markers for VIC expression. The authors fabricated a heterogeneous aortic valve conduit with a root consisting of SMCs, which was deposited first, and then a leaflet region consisting of VICs, to demonstrate fabrication of living valve conduits with anatomical resemblance to the native valve based on alginate/gelatin hydrogel system via 3D bioprinting, obtaining 3D hydrogel constructs that can maintain high cell viability with clinically relevant thickness. This approach allowed the researchers to obtain anatomically complex, mechanically heterogeneous valve scaffolds rapidly, without any processing post-fabrication.

8.1.5 PATTERNING MULTISCALE VASCULATURE WITH ENDOTHELIAL FUNCTION

Although fluid flow through perfusable microvasculature can help mimic physiologic transport on a small scale, the mammalian cardiovascular system is comprised of a wide range of diameters of blood vessels connected through a well-organized fractal hierarchy. Researchers are also investigating ways to generate multiscale vasculature—vessel networks containing an array of channel diameters and their accompanying fluidic junctions.

Work utilizing a direct-writing approach involved fabricating microchannels with barrier function by casting a hydrogel precursor over bioprinted agarose fibers, which serve as template material (Bertassoni, Cecconi, et al. 2014). After photopolymerization of constructs, the template fibers can be removed to result in perfusable networks (Figure 8.2E). The authors demonstrated the ability to fabricate microchannel networks with various architectural features inside a range of photopolymerizable hydrogel systems. Microchannels with diameters ranging from 1000 μm down to 150 μm could be obtained using this approach. By parallel overlapping of multiple template fibers over one another, the authors demonstrated versatility of the approach to fabricate even larger channel diameters that branched out into lateral individual channels of narrower diameters. The authors also demonstrated high efficiency of bioprinted microchannels to form cell-laden tissue constructs with improved functionality due to the channels allowing for improved nutrient transport, resulting in increased cell viability and cell differentiation. Furthermore, formation of endothelial monolayers inside hydrogel constructs was observed, allowing the constructs to remain fully perfusable.

We have also demonstrated an approach to rapidly cast patterned vascular networks in engineered tissue, a technique compatible with a wide range of cell types, synthetic and natural ECMs, and cross-linking strategies (Miller et al., 2012). In this approach, rigid 3D filament networks of carbohydrate glass are printed, cast into an ECM, and then dissolved to obtain a monolithic cellularized tissue construct. Multiscale filament architecture was obtained by varying only the translational velocity of the extrusion nozzle. After sacrificing the carbohydrate glass, smooth elliptical intervessel junctions were left behind by the glass interfilament fusions. As a result, fluidic connections between adjoining vascular channels can be obtained. Encapsulation of cells in the ECM was achieved by mixing a suspension of cells in the ECM prepolymer, while endothelialization was achieved by injecting a suspension of ECs into the vascular architecture. Encapsulated cells were shown to be viable, spread, and migrate normally in the channeled scaffolds. Seeded ECs quickly lined the walls of the network, even in constructs containing vessels of differing diameters (Figure 8.3F). Complete endothelialization of the vascular scaffolds resulted in the ability to perfuse human blood under high pressure by either laminar or pulsatile flow. Additionally, we found that highly sensitive cells, such as primary hepatocytes, can maintain metabolic activity at or near physiologic cell densities in tissue constructs with perfusable vascular channels and junctions.

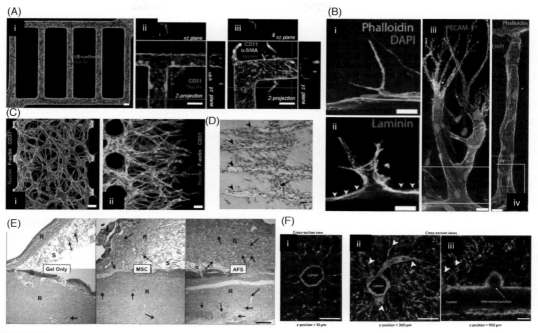

FIGURE 8.3

Demonstrating vasculogenic and angiogenic vessel formation in different systems. (A) Microfluidic vessel network formed by injection molding techniques, which was then stacked and sealed to result in an enclosed fluidic structure: (i) overall network structure, (ii) angiogenesis formation of sprouting endothelial cells, (iii) endothelial sprouting with perivascular interaction. Scale bars = 100 μm; figure adapted from Zheng et al. (2012). (B) Endothelial cells shown to sprout over time (early sprouts: (i and ii); late sprouts: (iii)), forming into matured neovessels that are lumenized end-to-end (iv). Scale bars = 25 μm; figure adapted from Nguyen et al. (2013). (C) Microfluidic device fabricated from soft lithography and replica molding: (i) cells encapsulated inside a channel demonstrate microvascular networks formed through vasculogenesis after 2 days in culture, (ii) cells coated inside a channel demonstrate angiogenic sprouts after 2 days in culture. Scale bars = 20 μm; figure adapted from Kim et al. (2013). (D) Cords constructed from a microtissue molding approach demonstrating formation of new capillaries along the length of the patterned cords after 7 days of implantation (arrows point to sprouts, arrowheads point to cords). Scale bar = 25 μm; figure adapted from Baranski et al. (2013). (E) Histological analysis of bioprinted cells onto skin wounds of a mice indicating thicker regeneration of tissue with more blood vessels in mesenchymal stem cell and amniotic fluid-derived stem cell treated patches compared to gel only (arrows point to vessels formed *in vivo*). Scale bars = 50 μm; figure adapted from Skardal et al. (2012). (F) Perfusable 3D engineered tissues fabricated from casting of patterned networks: (i) Endothelial lining of microchannel surrounded by 10T1/2 cells after 9 days in culture, (ii and iii) endothelial sprouts from patterned vasculature at different positions along the same vessel network (arrowheads point to sprouts). Scale bars = 200 μm; figure adapted from Miller et al. (2012).

8.1.6 ANGIOGENESIS, VASCULOGENESIS, AND *IN VIVO* INTEGRATION

Over the past few years, significant progress has been made to achieve perfusable microchannels containing cells; however, for translational applications, angiogenesis, vasculogenesis, and *in vivo* integration of fabricated scaffolds will need to be demonstrated. Work by Zheng et al. involved the use of lithographic processes to form defined 3D constructs with endothelialized vessels with an emphasis on their biofunctionality *in vitro* (Zheng et al., 2012). Their methodology entailed creation of a micro-structured silicone stamp, which served as a master, to mold collagen, injection and gelation *in situ* of collagen, followed by sealing of collagen layers to form an enclosed fluidic structure. Microfluidic vascular networks seeded with ECs demonstrated expression of endothelial phenotype and localization of cells to maintain cellular junctions between ECs, resulting in a microfluidic vascular network that exhibits a barrier function (Figure 8.3A). Interestingly, the cross-sectional area of the engineered microvessels changed profile from the original square cross-section defined lithographically toward an elliptical cross-section during the first 3 days of culture. Long endothelial sprouting containing lumens was evident in the ECM scaffold by 2 weeks in culture, indicating that the proposed system allows for the initiation of angiogenesis from native-like vessels. When microfluidic vascular networks were cocultured with ECs and perivascular cells, and subsequently perfused with vasculogenic media, the response of the networks to proangiogenic signals was found to be modulated by pericytes. Additionally, the authors demonstrated that perfusion of microfluidic vascular network with blood resulted in majority of platelets flowing past the endothelial surface without adhering. Kim et al. further studied the formation of vasculogenesis- and angiogenesis-derived microvascular networks by fabricating microfluidic devices using soft lithography and replica molding (Kim et al., 2013). The microfluidic chip design involved channels containing cells, partitioned by microposts, flanked with fluidic channels to emulate complex endothelial dynamics to grow native 3D vascular networks that can be readily perfused (Figure 8.3C). The approach yielded the presence of a continuous, hollow lumen along the length of the vessels with continuous cell–cell junctions lining the intersection of the endothelial cells. Quickly after seeding the microvascular networks with endothelial cells, the authors observed that the cells displayed elongated morphology and assembly into tubule-like structures encompassing a nascent lumen. Soon after, perfusable microvascular networks were obtained, followed by the enlargement of the lumen. Additionally, the authors were able to induce angiogenesis sprouting in the matrix as evidenced by tip cell formation and angiogenic sprouting along the openings of channels within 24 h of coculture with ECs and fibroblasts. After 3 days in culture, the leading tips were observed to have formed into lumenized vessels, establishing fluidic connections with flanking channels. To study potential flow-induced endothelial responses, the authors perfused the microvascular networks and observed that the fluid flow induced cytoskeletal reorganization of the ECs within a few hours and that networks perfused for only 1 h resulted in a significant increase in the synthesis of nitric oxide.

To further mimic angiogenesis processes in engineered tissue, Chiu et al. developed a method to engineer vascular network consisting of branching vessels that form a microvascular bed suitable for rapid vascularization of engineered tissues *in vitro* (Chiu et al., 2012). In this approach, biomaterials were used to create a niche that allowed sprouting and anastomosis of the vasculature from the branching vessels *in vitro* by control release of angiogenic and cardioprotective peptide Tβ4. Micropatterned PDMS substrates were fabricated by standard soft lithography to obtain substrates consisting of lanes with varied width and height. A solution of hydrogel was then evenly distributed onto each substrate to coat the surface and allowed to gel. The authors exposed mouse arterial and venous explants to Tβ4 encapsulated collagen-chitosan hydrogels and were interested in guiding and controlling the sprouting from explants through the integrated use of topographical cues involving different width grooves. Indeed, the authors observed that the enabling factor in tube formation and guidance was the presence of topographical cues, which was essential for the formation of branches with open lumens. The grooves not only provide physical guidance for the sprouting capillaries, but also influence the local concentration of autocrine growth factors released from the ECs actively participating in angiogenesis. Organotypic models were also fabricated to investigate angiogenic sprouting and neovessel formation within a 3D ECM by flowing combinations of factors next to an endothelium-lined channel (Nguyen et al., 2013). In this approach, a 3D microfluidic device with two parallel channels was fabricated by casting collagen into a PDMS mold with needles placed across the casting chamber, allowing the collagen to polymerize, and then removing the needles to result in hollow cylindrical channels in the matrix. After ECs were seeded into one of the channels, the second channel was perfused with angiogenic factors to establish a gradient across the collagen matrix to the endothelium. The authors observed that when a complex combination of proangiogenic factors was perfused into the second channel, more substantial multicellular sprout-like structures were formed which mirror major steps of *in vivo* angiogenesis (Figure 8.3B). After 1 week of culture, tip cells breached the source channel and resulted in a connection between the two parallel channels. Perfusion studies showed that beads flowed through the neovessels without leakage into the interstitial space, indicating fully developed, continuously endothelialized lumens. Additionally, the authors observed maturation of secondary branches with their own new tip cells in the fabricated organotypic model.

In addition to observing angiogenesis and vasculogenesis *in vitro*, some efforts demonstrating these processes *in vivo* have emerged. One such technique, laser-induced forward transfer, involves transfer of material from a donor slide, covered with a laser-absorbing gold layer, onto a collector slide by pulsing the laser. Transfer of material to the collector slide is achieved through generation of high gas pressure, which propels material on the donor slide toward the lower collector slide. This technique was used by Gaebel et al. to create specific vascular patterns consisting of different cell types for cardiac regeneration (Gaebel et al., 2011). In addition to the gold layer, the donor slide also contained a layer of cell-containing material to be transferred while the collector slide contained a polyester urethane urea patch immersed in Matrigel®. A cardiac patch composed of polyester urethane urea was spatially patterned with ECs and mesenchymal stem cells (MSCs) *in vitro* and then transplanted to infarcted zone of rat hearts. After 8 weeks of implantation, the authors observed increased vessel formation and functional improvement of infarcted hearts of transplanted implants. Additionally, the authors mentioned that histological analysis revealed integration of the implanted patch with the host myocardium. Tissue constructs fabricated by ink-jet printing technology were also implanted into mice and then analyzed for vasculature formation by using MRI (Xu et al., 2008). In this approach, the print head consisted of calcium chloride and cells, while the chamber consisted of sodium alginate and collagen solution mixture. Upon ejection of the cell-calcium chloride drops, gelation occurred rapidly, resulting in construction of 3D structures. After implanting the printed tissue constructs, the authors used MRI with contrast perfusion to image the vasculature formation *in vivo* and performed histological analysis to determine host cell invasion into the transplanted construct. The authors observed a vascular network in the superficial area of the constructs printed with cells, which express EC-specific marker vWF, while few vascular networks were found in the cell-free implant.

Interestingly, Skardal et al. (2012) also printed patches directly onto the wound site to accelerate the healing process of large skin wounds. In this study, bioprinting technology was used to deposit fibrin/collagen gels, with either MSCs or amniotic-fluid-derived stem (AFS) cells, directly onto the defect site of mice. The printed patch provided full coverage over the wound area and evidence of re-epithelialization was observed by histological analysis (Figure 8.3E). Compared to a gel-only treatment, treatment with MSC and AFS resulted in thicker regenerating tissue with a greater number of blood vessels observed in the regenerating skin. The AFS treatment resulted in larger, mature vessels that strongly stained for SMA and was found to contain RBCs inside the vasculature in the regenerating skin. However, the authors noted that since the cells in the printed patch remained at the wound surface, they did not migrate into the underlying tissue, thus indicating that the printed cells did not permanently integrate with the host tissue. In addition to patches, geometrically defined endothelial cords were developed and transplanted into mice by Baranski et al. (Baranski et al., 2013). Endothelialized cords were micropatterned using a PDMS mold and then transplanted to mice after the cords were polymerized between fibrin layers, encasing the cords. Indeed, formation of blood vessels, which were large and poorly organized on day 3 but remodeled into smaller, lumenized capillaries as early as 7 days, was evident in implanted endothelialized cords (Figure 8.3D). Additionally, the authors demonstrated that the cellular organization of the cords was vital for the vascularization response of the implant due to the presence of blood and vessels throughout the length of the cords. Anastomosis of neovessels was verified by perfusion studies and imaging of colabeled capillaries, which indicated the presence of chimeric composition of host and implanted cells in some vessels. The authors also developed endothelialized cords containing hepatocyte aggregates to demonstrate that the encapsulated cells exhibited enhanced hepatic function due to enhanced vascular supply as a result of neovessels formation along the tissue constructs.

8.1.7 ADVANCED TECHNOLOGIES WHICH MAY ASSIST IN VASCULAR TISSUE FABRICATION

Although much has been accomplished to fabricate 3D constructs with multiscale vasculature, 3D bioprinting technologies will need to advance to become more rapid, and achieve high throughput and resolution for translational purposes. Additionally, further techniques may need to be developed to pattern complex biochemical and biomechanical patterns within scaffolds to better mimic the *in vivo* environment.

To significantly increase the throughput in 3D bioprinting technology, Hansen et al. developed a method involving microvascular multinozzle print heads for printing of planar and multilayered architectures composed of single and multicomponent materials (Hansen et al., 2013). The multinozzle print heads are obtained by patterning a bifurcating network into a clear acrylic substrate using CNC machining, solvent-welding the patterned substrate to a flat acrylic substrate, resulting in a monolithic block that contains the embedded microvascular network. The authors were able to fabricate nozzle heads consisting of six branching generations, where each channel width and depth is 200 μm. The authors also demonstrated that their design is scalable for large-area, rapid fabrication of planar and 3D functional microarchitectures with microscale features. They also validated uniform ink-flow within the fabricated microvascular print heads to achieve high-fidelity, reproducible printed features. Further work for creating rapid vascularized, heterogeneous tissue constructs was also demonstrated by the same group by developing a custom-designed, large-area 3D bioprinter with four independently controlled print heads (Kolesky et al., 2014). These print heads contain inks consisting of cells and an aqueous fugitive ink that was removed from the final tissue construct. The authors showed the ability to print multiple cell types in the fabricated constructs, which are subsequently embedded in a pure GelMA matrix. 1D microchannel arrays were printed with final diameters ranging from 100 μm to 1 mm. Engineered construct channels were seeded with endothelial cells, and after 2 days, high viability of cells as well as assembly into a nearly confluent layer were observed (Figure 8.2G).

In addition to obtaining tissue constructs rapidly and in a high-throughput manner, higher resolution 3D bioprinting systems will be desirable to obtain higher structural resolution, permitting printing of capillaries into constructs. By using two-photon laser scanning (TPLS), Ovsianikov et al. demonstrated the feasibility of obtaining 100 nm structural resolution (Ovsianikov et al., 2008). This technique is demonstrated to precisely reproduce scaffold structure and properties by using CAD-designed 3D scaffolds. Although the authors fabricated a 3D vascular microcapillary structure using this technique, their cellular studies were performed on much simpler structures made out of photosensitive organic–inorganic hybrid polymer ORMOCER and epoxy-based SU-8 materials. The authors showed that cells seeded on a cylindrical structure made out of these materials support cell growth and are biocompatible. Additionally, two-photon photopolymerization can be utilized for biochemical and biomechanical patterning of hydrogels to guide cell behavior (Hahn et al., 2006). Similar techniques were also utilized to fabricate 3D micropatterning of biomolecules immobilized in a preformed network for 3D spatial control of cell migration (Lee et al., 2008). To obtain 3D patterns of covalently immobilized biomolecules within hydrogel networks, cells encapsulated in fibrin clusters were photopolymerized into collagenase-sensitive PEG hydrogels. The hydrogels were then soaked in PEG-RGDS, and, finally, TPLS was used to irradiate hydrogels according to predesigned virtual patterns. By doing so, PEG-RGDS was conjugated in a 3D network of hydrogels in predetermined patterns, allowing the researchers to achieve microscale control of 3D arrangements of biomolecules and cells. Using this method, channels of RGDS were conjugated adjacent to fibroblast clusters to allow guided cell migration out from the cell clusters. Branching of cells and subsequent migration into two separate paths were observed when the migrating cells encountered an intersection of the patterned areas.

Innovations in patterning and 3D printing technologies have advanced the field of tissue engineering to fabricate engineered tissue constructs with complex microstructures. However, debates remain as to the extent of patterning that will be required for reconstructing whole tissue and organ replacements for human patients. Indeed, there are still many technological hurdles involving more precise control of architecture and mass transport for fabrication of clinically relevant tissue constructs. Hence, efforts must be focused on statistically significant explorations of tissue architecture which can be validated to direct future work.

REFERENCES

Albrecht-Buehler, G., 1976. Filopodia of spreading 3T3 cells. Do they have a substrate-exploring function? The Journal of cell biology 69 (2), 275–286, Available at: http://www.pubmedcentral.nih.gov/articlerender.fcgi?artid=2109684&tool=pmcentrez&rendertype=abstract [Accessed July 1, 2014].

Arcaute, K., Mann, B.K., Wicker, R.B., 2006. Stereolithography of three-dimensional bioactive poly(ethylene glycol) constructs with encapsulated cells. Annals of biomedical engineering 34 (9), 1429–1441, Available at: http://www.ncbi.nlm.nih.gov/pubmed/16897421 [Accessed February 28, 2014].

Baranski, J.D., et al., 2013. Geometric control of vascular networks to enhance engineered tissue integration and function. Proceedings of the National Academy of Sciences of the United States of America 110 (19), 7586–7591, Available at: http://www.pubmedcentral.nih.gov/articlerender.fcgi?artid=3651499&tool=pmcentrez&rendertype=abstract [Accessed June 20, 2014].

Bertassoni, L.E., Cardoso, J.C., et al., 2014. Direct-write bioprinting of cell-laden methacrylated gelatin hydrogels. Biofabrication 6 (2), 024105, Available at: http://iopscience.iop.org/1758-5090/6/2/024105 [Accessed June 9, 2014].

Bertassoni, L.E., Cecconi, M., et al., 2014. Hydrogel bioprinted microchannel networks for vascularization of tissue engineering constructs. Lab on a chip, 20–22, Available at: http://www.ncbi.nlm.nih.gov/pubmed/24860845 [Accessed May 27, 2014].

Bettinger, C.J., et al., 2005. Three-Dimensional Microfluidic Tissue-Engineering Scaffolds Using a Flexible Biodegradable Polymer. Advanced materials (Deerfield Beach, Fla.) 18 (2), 165–169, Available at: http://onlinelibrary.wiley.com/doi/10.1002/adma.200500438/full [Accessed July 1, 2014].

Borenstein, J.T., et al., 2010. Functional endothelialized microvascular networks with circular cross-sections in a tissue culture substrate. Biomedical microdevices 12 (1), 71–79, Available at: http://www.ncbi.nlm.nih.gov/pubmed/19787455 [Accessed February 8, 2014].

Bryant, S.J., Nuttelman, C.R., Anseth, K.S., 2000. Cytocompatibility of UV and visible light photoinitiating systems on cultured NIH/3T3 fibroblasts in vitro. Journal of biomaterials science. Polymer edition 11 (5), 439–457, Available at: http://www.tandfonline.com/doi/abs/10.1163/156856200743805 [Accessed July 1, 2014].

Chan, V., et al., 2010. Three-dimensional photopatterning of hydrogels using stereolithography for long-term cell encapsulation. Lab on a chip 10 (16), 2062–2070, Available at: http://www.ncbi.nlm.nih.gov/pubmed/20603661 [Accessed February 28, 2014].

Chang, D.R., et al., 2013. Lung epithelial branching program antagonizes alveolar differentiation. Proceedings of the National Academy of Sciences of the United States of America 110 (45), 18042–18051, Available at: http://www.pubmedcentral.nih.gov/articlerender.fcgi?artid=3831485&tool=pmcentrez&rendertype=abstract [Accessed May 29, 2014].

Chiu, L.L.Y., et al., 2012. Perfusable branching microvessel bed for vascularization of engineered tissues. Proceedings of the National Academy of Sciences of the United States of America 109 (50), E3414–E3423, Available at: http://www.pubmedcentral.nih.gov/articlerender.fcgi?artid=3528595&tool=pmcentrez&rendertype=abstract [Accessed January 28, 2014].

Chiu, Y.-C., et al., 2009. Formation of Microchannels in Poly(ethylene glycol) Hydrogels by Selective Degradation of Patterned Microstructures. Chemistry of Materials 21 (8), 1677–1682, Available at: http://www.pubmedcentral.nih.gov/articlerender.fcgi?artid=2810413&tool=pmcentrez&rendertype=abstract [Accessed June 9, 2014].

Chrobak, K.M., Potter, D.R., Tien, J., 2006. Formation of perfused, functional microvascular tubes in vitro. Microvascular research 71 (3), 185–196, Available at: http://www.ncbi.nlm.nih.gov/pubmed/16600313 [Accessed February 28, 2014].

Cuchiara, M.P., et al., 2010. Multilayer microfluidic PEGDA hydrogels. Biomaterials 31 (21), 5491–5497, Available at: http://www.ncbi.nlm.nih.gov/pubmed/20447685 [Accessed February 24, 2014].

Cui, X., Boland, T., 2009. Human microvasculature fabrication using thermal inkjet printing technology. Biomaterials 30 (31), 6221–6227, Available at: http://www.ncbi.nlm.nih.gov/pubmed/19695697 [Accessed January 20, 2014].

Dagalakis, N., et al., 1980. Design of an artificial skin. Part III. Control of pore structure. Journal of biomedical materials research 14 (4), 511–528, Available at: http://www.ncbi.nlm.nih.gov/pubmed/7400201 [Accessed July 1, 2014].

Duan, B., et al., 2013. 3D bioprinting of heterogeneous aortic valve conduits with alginate/gelatin hydrogels. Journal of biomedical materials research. Part A 101 (5), 1255–1264, Available at: http://www.pubmedcentral.nih.gov/articlerender.fcgi?artid=3694360&tool=pmcentrez&rendertype=abstract [Accessed January 30, 2014].

Duan, B., et al., 2014. Three-dimensional printed trileaflet valve conduits using biological hydrogels and human valve interstitial cells. Acta biomaterialia 10 (5), 1836–1846, Available at: http://www.ncbi.nlm.nih.gov/pubmed/24334142 [Accessed June 16, 2014].

Gaebel, R., et al., 2011. Patterning human stem cells and endothelial cells with laser printing for cardiac regeneration. Biomaterials 32 (35), 9218–9230, Available at: http://www.ncbi.nlm.nih.gov/pubmed/21911255 [Accessed May 14, 2014].

Gauvin, R., et al., 2012. Microfabrication of complex porous tissue engineering scaffolds using 3D projection stereolithography. Biomaterials 33 (15), 3824–3834, Available at: http://www.pubmedcentral.nih.gov/articlerender.fcgi?artid=3766354&tool=pmcentrez&rendertype=abstract [Accessed January 22, 2014].

Golden, A.P., Tien, J., 2007. Fabrication of microfluidic hydrogels using molded gelatin as a sacrificial element. Lab on a chip 7 (6), 720–725, Available at: http://www.ncbi.nlm.nih.gov/pubmed/17538713 [Accessed January 26, 2014].

Gou, M., et al., 2014. Bio-inspired detoxification using 3D-printed hydrogel nanocomposites. Nature communications 5, 3774, Available at: http://www.nature.com/ncomms/2014/140508/ncomms4774/full/ncomms4774.html?message-global=remove [Accessed May 26, 2014].

Hahn, M.S., Miller, J.S., West, J.L., 2006. Three-Dimensional Biochemical and Biomechanical Patterning of Hydrogels for Guiding Cell Behavior. Advanced Materials 18 (20), 2679–2684, Available at: http://dx.doi.org/10.1002/adma.200600647.

Hansen, C.J., et al., 2013. High-throughput printing via microvascular multinozzle arrays. Advanced materials (Deerfield Beach, Fla.) 25 (1), 96–102, Available at: http://www.ncbi.nlm.nih.gov/pubmed/23109104 [Accessed February 21, 2014].

Harley, B.a, et al., 2010. Design of a multiphase osteochondral scaffold III: Fabrication of layered scaffolds with continuous interfaces. Journal of biomedical materials research. Part A 92 (3), 1078–1093, Available at: http://www.ncbi.nlm.nih.gov/pubmed/19301263 [Accessed July 2, 2014].

Hasan, A., et al., 2014. Microfluidic techniques for development of 3D vascularized tissue. Biomaterials 35 (26), 7308–7325, Available at: http://linkinghub.elsevier.com/retrieve/pii/S014296121400489X [Accessed July 1, 2014].

He, J., et al., 2013. Fabrication of nature-inspired microfluidic network for perfusable tissue constructs. Advanced healthcare materials 2 (8), 1108–1113, Available at: http://www.ncbi.nlm.nih.gov/pubmed/23554383 [Accessed June 25, 2014].

Hockaday, L.a, et al., 2012. Rapid 3D printing of anatomically accurate and mechanically heterogeneous aortic valve hydrogel scaffolds. Biofabrication 4 (3), 035005, Available at: http://www.pubmedcentral.nih.gov/articlerender.fcgi?artid=3676672&tool=pmcentrez&rendertype=abstract [Accessed January 30, 2014].

Jakab, K., et al., 2008. Tissue engineering by self-assembly of cells printed into topologically defined structures. Tissue engineering. Part A 14 (3), 413–421, Available at: http://www.ncbi.nlm.nih.gov/pubmed/18333793 [Accessed January 30, 2014].

Jun, H.-W., West, J.L., 2005. Endothelialization of microporous YIGSR/PEG-modified polyurethaneurea. Tissue engineering 11 (7-8), 1133–1140, Available at: http://www.ncbi.nlm.nih.gov/pubmed/16144449.

Kanchanawong, P., et al., 2010. Nanoscale architecture of integrin-based cell adhesions. Nature 468 (7323), 580–584, Available at: http://www.pubmedcentral.nih.gov/articlerender.fcgi?artid=3046339&tool=pmcentrez&rendertype=abstract [Accessed May 26, 2014].

Kane, R.S., et al., 1999. Patterning proteins and cells using soft lithography. Biomaterials 20 (23–24), 2363–2376, Available at: http://www.ncbi.nlm.nih.gov/pubmed/10614942 [Accessed June 20, 2014].

Kim, S., et al., 2013. Engineering of functional, perfusable 3D microvascular networks on a chip. Lab on a chip 13 (8), 1489–1500, Available at: http://pubs.rsc.org/EN/content/articlehtml/2013/lc/c3lc41320a [Accessed June 6, 2014].

King, K.R., et al., 2004. Biodegradable Microfluidics. Advanced Materials 16 (22), 2007–2012, Available at: http://doi.wiley.com/10.1002/adma.200306522 [Accessed February 8, 2014].

Kolesky, D.B., et al., 2014. 3D bioprinting of vascularized, heterogeneous cell-laden tissue constructs. Advanced materials (Deerfield Beach, Fla.) 26 (19), 3124–3130, Available at: http://www.ncbi.nlm.nih.gov/pubmed/24550124 [Accessed June 26, 2014].

Lee, S.-H., Moon, J.J., West, J.L., 2008. Three-dimensional micropatterning of bioactive hydrogels via two-photon laser scanning photolithography for guided 3D cell migration. Biomaterials 29 (20), 2962–2968, Available at: http://www.ncbi.nlm.nih.gov/pubmed/18433863 [Accessed February 28, 2014].

Lin, H., et al., 2013. Application of visible light-based projection stereolithography for live cell-scaffold fabrication with designed architecture. Biomaterials 34 (2), 331–339, Available at: http://www.pubmedcentral.nih.gov/articlerender.fcgi?artid=3612429&tool=pmcentrez&rendertype=abstract [Accessed February 28, 2014].

Liu Tsang, V., et al., 2007. Fabrication of 3D hepatic tissues by additive photopatterning of cellular hydrogels. FASEB journal : official publication of the Federation of American Societies for Experimental Biology 21 (3), 790–801, Available at: http://www.ncbi.nlm.nih.gov/pubmed/17197384 [Accessed May 27, 2014].

Liu, V., Bhatia, S., 2002. Three-dimensional photopatterning of hydrogels containing living cells. Biomedical microdevices 4 (4), 257–266, Available at: http://link.springer.com/article/10.1023/A: 1020932105236 [Accessed June 17, 2014].

McGuigan, A.P., Sefton, M.V., 2007. The influence of biomaterials on endothelial cell thrombogenicity. Biomaterials 28 (16), 2547–2571, Available at: http://www.ncbi.nlm.nih.gov/pubmed/20300848 [Accessed July 1, 2014].

Miller, J.S., et al., 2012. Rapid casting of patterned vascular networks for perfusable engineered three-dimensional tissues. Nature materials 11 (9), 768–774, Available at: http://www.pubmedcentral.nih.gov/articlerender.fcgi?artid=3586565&tool=pmcentrez&rendertype=abstract [Accessed January 22, 2014].

Miller, J.S., 2014. The billion cell construct: will three-dimensional printing get us there? PLoS biology 12 (6), e1001882, Available at: http://www.ncbi.nlm.nih.gov/pubmed/24937565 [Accessed July 1, 2014].

Murray, C.D., 1926. The Physiological Principle of Minimum Work: I. The Vascular System and the Cost of Blood Volume. Proceedings of the National Academy of Sciences of the United States of America 12 (3), 207–214, Available at: http://www.pubmedcentral.nih.gov/articlerender.fcgi?artid=1084489&tool=pmcentrez&rendertype=abstract [Accessed June 25, 2014].

Nakamura, M., et al., 2008. Ink Jet Three-Dimensional Digital Fabrication for Biological Tissue Manufacturing : Analysis of Alginate Microgel Beads Produced by Ink Jet Droplets for Three. Journal of Imaging Science and Technology 52 (6), 1–6.

Nguyen, D.-H.T., et al., 2013. Biomimetic model to reconstitute angiogenic sprouting morphogenesis in vitro. Proceedings of the National Academy of Sciences of the United States of America 110 (17), 6712–6717, Available at: http://www.pubmedcentral.nih.gov/articlerender.fcgi?artid=3637738&tool=pmcentrez&rendertype=abstract [Accessed June 2, 2014].

Nichol, J.W., et al., 2010. Cell-laden microengineered gelatin methacrylate hydrogels. Biomaterials 31 (21), 5536–5544, Available at: http://www.pubmedcentral.nih.gov/articlerender.fcgi?artid=2878615&tool=pmcentrez&rendertype=abstract [Accessed February 20, 2014].

Nishiyama, Y., et al., 2009. Development of a three-dimensional bioprinter: construction of cell supporting structures using hydrogel and state-of-the-art inkjet technology. Journal of biomechanical engineering 131 (3), 035001, Available at: http://www.ncbi.nlm.nih.gov/pubmed/19154078 [Accessed February 28, 2014].

Norotte, C., et al., 2009. Scaffold-free vascular tissue engineering using bioprinting. Biomaterials 30 (30), 5910–5917, Available at: http://www.pubmedcentral.nih.gov/articlerender.fcgi?artid=2748110&tool=pmcentrez&rendertype=abstract [Accessed January 23, 2014].

Ovsianikov, A., et al., 2008. Two-photon polymerization technique for microfabrication of CAD-designed 3D scaffolds from commercially available photosensitive materials, 443–449.

Pham, Q.P., Sharma, U., Mikos, A.G., 2006. Electrospun poly(epsilon-caprolactone) microfiber and multilayer nanofiber/microfiber scaffolds: characterization of scaffolds and measurement of cellular infiltration. Biomacromolecules 7 (10), 2796–2805, Available at: http://www.ncbi.nlm.nih.gov/pubmed/17025355.

Shaner, N.C., Steinbach, P.A., Tsien, R.Y., 2005. A guide to choosing fluorescent proteins. Nature methods 2 (12), 905–909, Available at: http://www.ncbi.nlm.nih.gov/pubmed/16299475 [Accessed May 23, 2014].

Sherman, T.F., 1981. On connecting large vessels to small. The meaning of Murray's law. The Journal of general physiology 78 (4), 431–453, Available at: http://www.jgp.org/cgi/doi/10.1085/jgp.78.4.431 [Accessed July 1, 2014].

Skardal, A., et al., 2012. Bioprinted amniotic fluid-derived stem cells accelerate healing of large skin wounds. Stem cells translational medicine 1 (11), 792–802, Available at: http://europepmc.org/articles/PMC3659666 [Accessed June 16, 2014].

Skardal, A., Zhang, J., Prestwich, G.D., 2010. Bioprinting vessel-like constructs using hyaluronan hydrogels crosslinked with tetrahedral polyethylene glycol tetracrylates. Biomaterials 31 (24), 6173–6181, Available at: http://www.ncbi.nlm.nih.gov/pubmed/20546891 [Accessed February 28, 2014].

Suri, S., et al., 2011. Solid freeform fabrication of designer scaffolds of hyaluronic acid for nerve tissue engineering. Biomedical microdevices 13 (6), 983–993, Available at: http://www.ncbi.nlm.nih.gov/pubmed/21773726 [Accessed May 6, 2014].

Truskett, V.N., Watts, M.P.C., 2006. Trends in imprint lithography for biological applications. Trends in biotechnology 24 (7), 312–317, Available at: http://www.ncbi.nlm.nih.gov/pubmed/16759722 [Accessed June 4, 2014].

Vozzi, G., et al., 2003. Fabrication of PLGA scaffolds using soft lithography and microsyringe deposition. Biomaterials 24 (14), 2533–2540, Available at: http://linkinghub.elsevier.com/retrieve/pii/S0142961203000528 [Accessed June 5, 2014].

Whitesides, G.M., et al., 2001. Soft lithography in biology and biochemistry. Annual review of biomedical engineering 3, 335–373, Available at: http://www.ncbi.nlm.nih.gov/pubmed/11447067 [Accessed June 25, 2014].

Wu, W., et al., 2010. Direct-write assembly of biomimetic microvascular networks for efficient fluid transport. Soft Matter 6 (4), 739, Available at: http://xlink.rsc.org/?DOI=b918436h [Accessed June 25, 2014].

Xu, T., et al., 2008. Characterization of Cell Constructs Generated With Inkjet Printing Technology Using In Vivo Magnetic Resonance Imaging. Journal of Manufacturing Science and Engineering 130 (2), 021013, Available at: http://manufacturingscience.asmedigitalcollection.asme.org/article.aspx?articleid=1452028 [Accessed February 28, 2014].

Yannas, I.V., et al., 1982. Wound tissue can utilize a polymeric template to synthesize a functional extension of skin. Science (New York, N.Y.) 215 (4529), 174–176, Available at: http://www.ncbi.nlm.nih.gov/pubmed/7031899 [Accessed July 2, 2014].

Zheng, Y., et al., 2012. In vitro microvessels for the study of angiogenesis and thrombosis. Proceedings of the National Academy of Sciences of the United States of America 109 (24), 9342–9347, Available at: http://www.pubmedcentral.nih.gov/articlerender.fcgi?artid=3386137&tool=pmcentrez&rendertype=abstract [Accessed February 20, 2014].

CRANIOFACIAL AND DENTAL TISSUE

Michael Larsen, Ruchi Mishra, Michael Miller and David Dean
Department of Plastic Surgery, The Ohio State University, Columbus, OH, USA

9.1 INTRODUCTION

Craniofacial and dental structures are complex tissues that perform vital functions, such as seeing, hearing, breathing, chewing, tasting, speaking, and protecting the brain and eyes (Costello et al., 2010). Standard-of-care craniofacial and dental regenerative medicine consists largely of musculoskeletal reconstructive techniques analogous to orthopedic and general surgical reconstructive techniques, with special attention to the complex 3D structure (Figure 9.1) (Sanchez-Lara and Warburton, 2012). Most of these procedures are aimed at restoring function (e.g., mastication) and structure in an aesthetically appropriate reconstruction. What has changed over the past decade is the expected role that tissue engineering has and will have in regenerative medicine. This is illustrated by the conceptual move away from research in whole organ regeneration to structure- and defect-specific "regenerative medicine," which includes elements of grafting, bone and dental substitute materials, and off-the-shelf or custom devices. Initial research success in all three of these areas is being translated to the repair of particular tissue deficits with a focus on the restoration of function through the augmentation of standard-of-care procedures. Many if not most reconstructive procedures continue to involve transplantation and alloplastic hardware or tissue substitutes.

Tissue engineering remains a scientific activity that has the potential to improve therapies for specific indications. However, tissue engineering has yet to contribute to the craniofacial therapies involving bone, glands, sense organs, joints, muscles, or dental tissues. Careful study of growth factors and progenitor cells in restoring failed craniofacial tissues is now done mostly in parallel to research into surgical therapies, material research, and mass market or patient-specific device research. All of this research is aimed at improving the treatment of craniofacial patients who suffer from traumatic injuries, cancer, or congenital deformity.

Custom devices do not yet involve biological components; thus, there is no true "bioprinting" However, there is initial use of growth factors, such as BMP-2, in some craniofacial therapies. There is much use of nanotechnology in the development of biomaterials for these implants and other surgical devices. A new wave of developments is occurring in developing biomaterials with a wide range of properties that can be rendered in 3D printers. It is likely that those materials will eventually be used as resorbable models, or scaffolds, for tissue engineering applications that utilize craniofacially relevant stem cells and growth factors.

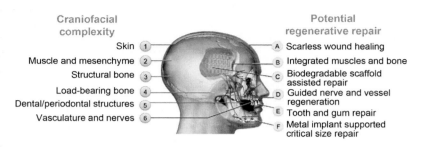

Craniofacial complexity

- Skin (1)
- Muscle and mesenchyme (2)
- Structural bone (3)
- Load-bearing bone (4)
- Dental/periodontal structures (5)
- Vasculature and nerves (6)

Potential regenerative repair

- (A) Scarless wound healing
- (B) Integrated muscles and bone
- (C) Biodegradable scaffold assisted repair
- (D) Guided nerve and vessel regeneration
- (E) Tooth and gum repair
- (F) Metal implant supported critical size repair

FIGURE 9.1

Current craniofacial regenerative procedures.

(Figure 9.1 from Sanchez-Lara et al., 2011).

The next section of this chapter will look at the range of craniofacial therapies, third is a look at research into craniofacial and dental regenerative medicine, fourth is a look at current research into bone tissue engineering, and finally we state our conclusions from this review of craniofacial and dental tissue bioprinting and nanotechnology.

9.2 CLINICAL NEED FOR CRANIOFACIAL AND DENTAL REGENERATIVE MEDICINE

The craniofacial region is composed of a complex interconnection of multiple organ systems. The structure consists of a musculoskeletal framework that is both intricate enough to compliment the function of the other systems as well as adequately robust to provide protection to the delicate components therein. The craniofacial region is the beginning of the respiratory and alimentary tracts, functioning to warm and humidify air and commence the digestion of food in these respective systems. This area is also the sensory headquarters of the body, with all sensory modalities being manifest here; more particularly, the four special sensory systems (visual, auditory, olfactory, and gustatory) exclusively operate here. Stimuli from the environment are received here and are even processed here. The signals received by the sensory organs are then transmitted nearby to the most vital of organs, the brain. Thus, the craniofacial region is the central interface between the individual and their environment; it is also responsible for sending communicative signals by embodying voice into speech, expressing emotions, disclosing identity, displaying aesthetics, and cueing gender, race, and age. With the craniofacial region playing so many key functions, defects and deformities of this area can severely affect an individual's quality of life, leading to both significant physical impairments as well as psychological sequelae. Indeed, it has been shown that patients who have undergone resection of tumors of the head and neck actually more readily adjust to their dysfunction than to their disfigurement (Dropkin, 1999; Gamba et al., 1992). For all these reasons, reconstruction and restoration of craniofacial defects demands state-of-the-art techniques and advanced materials.

9.2.1 MAJOR DIAGNOSES AND CAUSES

Defects and deformities of the craniofacial region arise from dental disease, trauma, aging, cancer, and congenital malformations. Each of these will presently be discussed.

9.2.1.1 Dental Disease

Statistics show that 100% of adults have dental caries, and it is estimated that 15–20% of adults between the ages of 35–44 suffer from severe periodontal disease that will result in tooth loss. Dental disease is cumulative with age, resulting in 30% of people aged 65–74 becoming fully edentulous. The mandible and maxilla of the edentulous naturally atrophy with time, becoming much thinner and more susceptible to fracture.

9.2.1.2 Trauma

The primary cause of death in people under 40 years of age is trauma. Approximately 25,000 people per year require surgery to fix maxillofacial trauma in the United States. Maxillofacial trauma results from assault, motor vehicle collisions, falls, and sporting accidents. The relative incidence of these causes has been shifting. With increasing widespread use of airbags and seatbelts, the number and severity of maxillofacial fractures from motor vehicle collisions has been decreasing. On the other hand, as the population is aging and the elderly are seeking active lifestyles, the number of fractures from falls is trending up (Martinez et al., 2014).

9.2.1.3 Aging

As people age, the collagen framework loosens, the dermis thins, there is a cumulative solar elastosis, and muscles and bones of the face thin and atrophy. These result in an inferomedial descent of the soft tissue and the stigmata of aging including wrinkles, furrows, jowls, decreased dental show, the downward tilt of the corners of the mouth, and sunken cheeks.

9.2.1.4 Cancer

In the United States more than 40,000 people are diagnosed with oral and pharyngeal cancers each year (The Oral Cancer Foundation; http://oralcancerfoundation.org/facts/). An additional 3.5 million people are diagnosed with a new skin cancer (basal cell, squamous cell, or melanoma). A majority of these occur in the highly sun-exposed craniofacial region. The ablation of craniofacial malignancies often requires resection of multiple surrounding tissue types (skin, subcutaneous fat, bone, and mucosa), ensuring adequate margins but leaving the patient with exposed bone, vessels, and nerves and defects that cannot be closed primarily. Radiation treatments further complicate wound healing by compromising the local vasculature.

9.2.1.5 Congenital

In the U.S. approximately 38,000 infants undergo surgery each year for the treatment of congenital defects. One in 1000–2000 infants are born with sporadic or inherited craniosynostosis (fusion of cranial plates), which can lead to a malformed cranium, intracranial hypertension, and restricted brain growth. An additional 1:1000 children are born with cleft lip and palate. Certain syndromes can result in hypoplasia of the maxilla (i.e., Apert's or Crouzon's) or mandible (i.e., Treacher Collins or Pierre Robin Sequence) and airway compromise. Furthermore, many are born with microtia or contour irregularities which they find aesthetically displeasing (CDC, 2008).

9.2.2 STANDARD-OF-CARE PROCEDURES

The standard of care for these craniofacial defects and deformities, with the exception of teeth, has historically centered around manipulating and transplanting the patient's own tissues. Autogenous tissue has the characteristics of an ideal biomaterial: biocompatible, nontoxic, nonallergenic, noninflammatory,

Table 9.1   Autograft and allograft characteristics

Bone graft	Structural strength	Osteoconduction	Osteoinduction	Osteogenesis
Autograft				
Cancellous	No	+++	+++	+++
Cortical	+++	++	++	++
Allograft				
Cancellous				
Frozen	No	++	+	No
Freeze-dried	No	++	+	No
Cortical				
Frozen	+++	+	No	No
Freeze-dried	+	+	No	No

mechanically reliable, resistant to microorganism growth, and provides permanency and consistently reproducible results (Dickinson et al., 2011). Autogenous bone has the additional properties of being osteoconductive (framework conducive to new bone growth), osteoinductive (stimulates osteoprogenitor cells to differentiate and form new bone), and osteogenic (contains osteoblasts) (Tables 9.1 and 9.2). Treatment strategy is largely determined by the size of the defect and the tissues involved. The tissue categories discussed here are teeth, bone and cartilage, and soft tissue.

9.2.2.1 Teeth
Humans have been replacing lost teeth for thousands of years, as evidenced by archeological discoveries of mandibles with implants of carved bamboo, precious metal pegs, and shaped seashells. In the modern era, titanium implants have been used since the 1960s when Dr. Branemark first coined the term "osseointegration." Today, implants are made from surgical grade 5 titanium (Ti-6Al-4V) (de Lavos-Valereto et al., 2002). These implants are crowned with materials such as glass-ceramics, zirconia, and alumina, all of which perform very well (Oilo et al., 2014), demonstrating appropriate wear of opposing natural crown surfaces (Kim et al., 2012b). Unfixed dentures or bridges represent less-involved treatment options. Teeth that are chipped or have defects can be bonded using resins, while misshapen or discolored incisors can be etched on their anterior surface and bonded to veneers. Thus, for the most part, routine dental defects have been solved by biomaterial implants and prostheses; yet, solutions are still needed for reconstruction in compromised conditions, such as when there is insufficient bone or mucosa or poor healing capabilities (Joos, 2009).

9.2.2.2 Bone and Cartilage
Simple craniofacial fractures (whether caused by trauma or therapeutically made by the surgeon), with favorable mechanical forces, merely need to be splinted (e.g., interdental wiring). If the fracture pattern is complicated or is subjected to forces that cause splaying, rotation, or movement, the

Table 9.2   Osteoconductive scaffolds

Type	Graft	Osteoconduction	Osteoinduction	Osteogenesis	Advantages
Bone	Autograft	3	2	2	"Gold standard"
	Allograft	3	1	0	Availability in many forms
Biomaterials	DBM	1	2	0	Supplies osteoinductive BMPs, bone graft extender
	Collagen	2	0	0	Good as delivery vehicle system
Ceramics	TCP, hydroxyapatite	1	0	0	Biocompatible
	Calcium phosphate cement (CPC)	1	0	0	Some initial structural support
Composite grafts	β-TCP/BMA composite	3	2	2	Amply supply
	BMP/synthetic composite	—	3	—	Potentially limitless supply

Score: 0 (none) to 3 (excellent). DBM: demineralised bone matrix, TCP: tricalcium phosphate, BMA: bone marrow aspirate, BMP: bone morphogenetic protein

bones are fixated with screws and plates to avoid malunion, malocclusion, and fibrous (rather than boney) healing. Likewise, when a bone defect is large enough (critical-size or larger), it will heal by fibrous scar formation. In critical-size defects, a nonvascularized bone graft from the iliac crest, tibial plateau, outer table of the calvarium, or olecranon can be employed to fill the gap. Allografts (e.g., demineralized freeze-dried bone), xenografts (coral), metals (titanium), and alloplasts (e.g., hydroxyapatite, methylmethacrylate) are also occasionally used (see **Section 9.4.1**); however, caution is employed because of the risks of nonunion, infection, and extrusion. When a defect >6 cm is encountered or the wound is compromised (i.e., irradiated tissues), then a vascularized bone graft of radius, fibula, rib, or iliac crest is used (Franklin et al., 1980; Song et al., 1982; Taylor, 1983; Taylor et al., 1975, 1979). Large maxillary defects are often filled with an obturator (a removable prosthetic) or a vascularized soft tissue flap; however, soft tissue is not ideal since it can result in asymmetry and contour abnormalities, and will not be able to accept osseointegrated implants for dental reconstruction (Hanasono et al., 2010). Furthermore, autogenous bone or soft tissue grafts have significant drawbacks: limited supply; inaccuracy in the 3D shaping and inset; and their need for increased operative time, larger surgical teams, more equipment, and extra operative site with its complications (Ling and Peng, 2012).

Patients with craniofacial hypoplasia are treated with distraction osteogenesis, where osteotomies (bone fractures) are iatrogenically made of the involved bones, and a device is applied that will distract the bone segments apart 1 mm/day. This rate is slow enough to allow bone to form in the intervening gap. This is continued until the desired bone lengthening is achieved.

The cranial bones of craniosynostosis patients are cut apart, repositioned, and then fixated according to the principles mentioned (usually with plates and screws with or without bone graft).

Ear deformities (e.g., microtia, atresia) can be reconstructed by harvesting autogenous rib cartilage, carving it into the shape of an ear, and implanting it under the skin. However, the reconstructed ear often lacks symmetry with, and the aesthetic definition of, the contralateral ear. Thus, recent data reported by Reinisch et al. showing optimal aesthetic outcomes and minimal complications with a medpor (high-density porous polyethylene) ear implant could lead to medpor becoming the gold standard treatment for this deformity (Reinisch and Li, 2014).

9.2.2.3 Soft Tissue

The treatment paradigm for soft tissue defects has traditionally adhered to the reconstructive ladder, where simple and more local solutions are used if possible; but when these options are nonviable, the surgeon steps up to a feasible but more complex solution (Figure 9.2). Many surgeons have also begun expanding their personal reconstructive ladder to include skin substitutes, such as Integra (Integra NeuroSciences, Plainsboro, NJ), which is a bilayered construct that contains an inner layer consisting of a collagen/glycosaminoglycan matrix that acts as a scaffold for cellular ingrowth and dermal regeneration.

Some early products for facial augmentation were made of expanded polytetrafluoroethlene (ePTFE, W.L. Gore & Associates Inc., Newark, DE; PTFE is best known as Teflon®, Dupont Co., Wilmington, DE). These products were largely abandoned because of extremely high complication

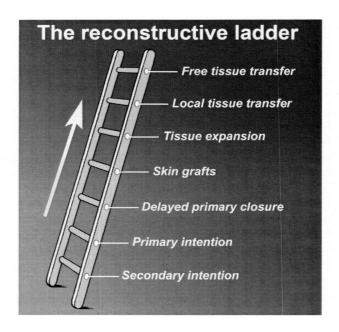

FIGURE 9.2

The reconstructive ladder.

rates, including palpability, migration, shrinkage, and extrusion (Maloney et al., 2012; Truswell, 2007). Injectable fillers have since come to the market and have rapidly become one of the most common cosmetic procedures, with 2.2 million injections in the U.S. last year (13% most recent annual increase) (2014, ASPS, 2014). Various materials are used as fillers: hyaluronic acid, calcium hydroxylapatite, poly-L-lactic acid, silicone, and polymethylmethacrylate (Figure 9.3) (de Vries and Geertsma, 2013). Autologous fat injections have also been used with success (Barton et al., 2007). In addition, newer ePTFE implants have been developed with a softer, more natural quality, and a higher porosity (allowing more cellular ingrowth), that have both led to low complication rates (Niamtu, 2006).

While autografts are currently the gold standard for many craniofacial therapies, it can be seen that the standard of care is shifting. The significant disadvantages of autografts demand the development of new biomaterials, further advances in craniofacial regenerative medicine, and a continued shift in the standard of care.

Filler	What it is made of	What it treats	Avg. length of effectiveness
Juvéderm® Ultra	Hyaluronic Acid	• Folds around the mouth • Temples • Cheeks • Lips	6–18 months
Juvéderm® Ultra Plus	Hyaluronic Acid	• Folds around the mouth • Temples • Cheeks • Lips	6–18 months
Restylane®	Hyaluronic Acid	• Folds around the mouth • Under eyes • Scars • Nose • Lips	6–31 months
Perlane®	Hyaluronic Acid	• Folds around the mouth • Cheeks • Lips	6–18 months
Radiesse®	Calcium Hydroxylapatite	• Moderate to deep facial wrinkles and folds • Under eyes • Cheeks • Nose	Up to 1–2 Years
Sculptra®	Poly-L-lactic Acid	• Moderate to deep facial wrinkles and folds • Cheeks	Up to 1–2 Years
ArteFill®	Collagen and Polymethyl Methacrylate	• Moderate to deep facial wrinkles and folds • Cheeks	Up to 5+ Years

FIGURE 9.3

Injectable fillers.

9.3 CRANIOFACIAL AND DENTAL REGENERATIVE MEDICINE RESEARCH

Despite the great need, there is very little penetration of tissue engineering, especially additive manufacturing or cell printing for tissue engineering, in standard-of-care craniofacial surgical and/or dental therapeutic procedures that repair critical size defects (Schmitz and Hollinger, 1986). However, the significant use in these specialties of bone and tooth substitute materials, especially in regard to the additive manufacture of patient-specific medical implants, may eventually forge a link between tissue engineering and craniofacial and dental care.

9.3.1 NOVEL MATERIALS

Implants and surgical reconstruction hardware for the craniofacial and dental regions have typically used the same materials as elsewhere in the body. However, there are many advanced materials that are used fairly exclusively in the craniofacial region. Resorbable Lactosorb® (Lorenz [Biomet], Jacksonville, FL), which is 82% poly(-L lactic acid) and 18% poly(glycolic acid) is one such example (Biomet microfixation, 2010). These plates and screws resorb by hydrolysis within a year, allowing uninhibited growth of the bones and obviating the need for a secondary procedure to remove the hardware. These plates have particular application in softer bones, such as in the orbit and in infants. Solid-state hyaluronic acid in the form of a thread is another advanced material used in the craniofacial region. This promising new product allows for more controlled wrinkle effacement and soft tissue augmentation (Franklin, 2014). Nitinol-shape memory metals are used in orthodontic wire; these wires are superelastic at body temperature. As mentioned previously, titanium dental implant bone screws and abutments are very successful. However, there is a great deal of innovation with other alloys, zirconia, and coatings designed to improve osseointegration. Single crown or fixed prosthesis implant dental systems are currently more expensive than nonfixed systems (e.g., traditional dentures), but they also provide for better mechanical function, stimulation, and maintenance of oral cavity muscles and skeletal structures. However, the most exciting and cutting-edge application of such advanced biomaterials is in their use in the additive manufacture (3D printing) of patient-specific implants in teeth, craniofacial bone, and the temporomandibular joint.

9.3.2 TEETH

The Sirona CEREC1 (Chairside Economical Restorations of Esthetic Ceramics), the first system to combine digital scanning of the teeth, and computer-aided design (CAD) of prosthetic crowns, appeared in 1985. Most of these crown restorations are currently prepared in the prosthodontist's office or by a nearby service with computerized milling devices. These systems are sufficiently accurate (Figure 9.4), therefore there seems to be no benefit in terms of quality or price at this time in moving to more advanced fabrication strategies (Schaefer et al., 2013).

Ng et al., (2014) noted that it is only recently that restorations prepared in this way have become more economical and perform better than traditional manual methods. Given the success of these post-and-crown systems, the challenge for oral health has reverted back to the issue of sufficient bone bulk to support dental posts. This is especially challenging in edentulous individuals (Cho and Raigrodski, 2014). Dental implants and single crown or fixed prostheses may require autologous and/or alloplastic bone graft materials. This depends on the size of the alveolar cavity left following dental extraction. It is unclear at this time whether these expensive fixed prostheses will become standard of care under the United States' Affordable Care Act in nontraumatic situations (O'Brien, 2014).

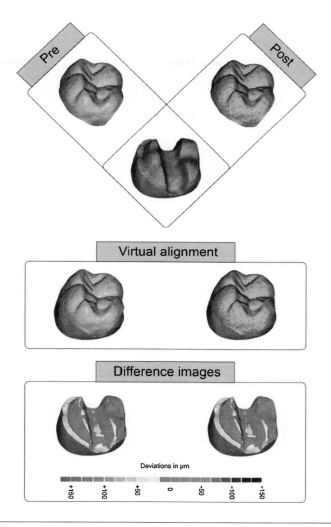

FIGURE 9.4

This comparison of image-based crown design (pre) and rendered crown (post) shows high accuracy of these milled crowns.

(Schaefer et al., 2013).

9.3.3 **BONE**

In cases of craniofacial bone shortage, there are no tissue engineered therapies. Alloplastic materials have been explored. Polymethylmethacrylate (PMMA) is the most common material used followed by grade 5 surgical titanium (Ti-6Al-4V). These materials have been used to make custom cranio-plasties (skull plates) for more than a decade (Dean et al., 2003a). Cranial plates are now commonly 3D printed directly from CAD files in PEEK (Lethaus et al., 2012) with a newer material, PEKK (Baum, 2013), having more appropriate material properties for a bone substitute implant. If alloplastic materials are used, a key consideration is how well they transfer strain to and from the surrounding skull

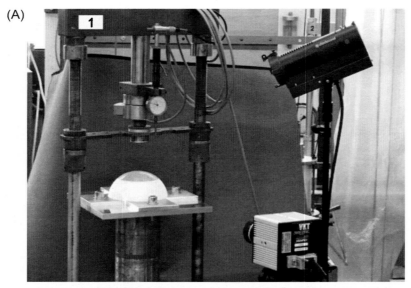

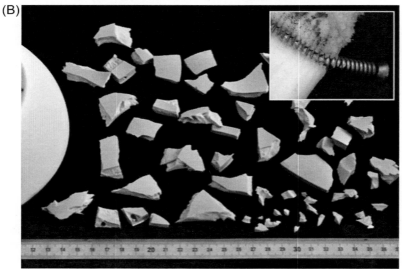

FIGURE 9.5

Custom cranial implant material properties: A. Mechanical testing device. B. Results of mechanical testing.

(Lethaus et al., *2012).*

(Figure 9.5). Creating adaptive, rather than stress-shielding or stress-concentrating, reconstructions is a major consideration in the treatment of mandibular segmental defects if permanent surgical grade 5 titanium fixation is used (Zoumalan et al., 2009). The topological optimization of very stiff implant components in craniofacial implants has been studied by Sutradhar et al., (2010).

Mandibular segmental defects are now commonly being repaired with fibular bone grafts where fibular and mandibular cutting guides are provided and Ti-6Al-4V immobilization plates are prebent. This saves operating room time, and enhances reproducibility and precision, which in turn aids boney union and prevents fibrous ingrowth of the defect space (Wang et al., 2013; Saad et al., 2013). Others have presented their experience with the replacement of most (Zhou et al., 2010) to all (van Kroonenburgh et al., 2012) of the mandible with 3D printed titanium prosthetics. The long-term outcomes of such cases remain to be seen; it is known that the use of titanium or acrylic instead of bone graft or Guided Tissue Regeneration (Park and Wang, 2007) devices directly under the gingivae may risk dehiscence (Chiapasco and Zaniboni, 2009).

9.3.4 TEMPOROMANDIBULAR JOINT

Involvement of the temporomandibular joint (TMJ) in the repair of segmental mandibular defects elevates the complexity. The TMJ includes the head of the mandible, articular disc, or the articular tubercle (of the zygomatic arch). This joint has been cited as one of the most active joints in the body (Guarda-Nardini et al., 2008). Early attempts at total joint replacement involving Teflon joint surfaces were associated with foreign-body giant cell reactions resulting in damage to the surrounding host structures and an unstable joint (Abramowicz et al., 2008). Recent experience has found a wide variety of stable implant configurations, with good performance by patient- specific devices (Mercuri, 2012) and using metal-on-polyethylene TMJ replacements (Figure 9.6) with stable, long-term outcomes (Leiggener et al., 2012).

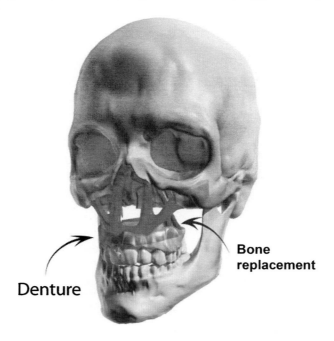

Denture

Bone replacement

FIGURE 9.6

Topological Optimization: Optimization of stress-strain trajectories suggests the need for appropriate strength in the displayed struts.

(Sutradhar et al., *2010).*

9.4 BONE TISSUE ENGINEERING STRATEGIES

Tissue engineering strategies have the potential to augment traditional restorative approaches that utilize transplanted tissues and bone substitute materials. Some key considerations in the development of tissue-engineered strategies for craniofacial/dental tissue are as follows: (1) are there any stem cells that can be used; (2) can aesthetic demands be met; (3) can tissue-engineered constructs be used as part of a composite strategy; (4) can adequate vascularization demands be met; and (5) can the high risk of infection in the oral and nasal cavities be managed (Moioli et al., 2007). These aspects have also been discussed in detail by Mikos et al., (2006) (Table 9.3). Craniofacial tissue-engineered strategies are likely to be required to regenerate more than one tissue phenotype at the same time (Moioli et al., 2007; Rahaman and Mao, 2005).

Bone tissue engineering-based strategies for craniofacial/dental tissue regeneration can be divided broadly into three main categories based on the scaffold, cellular, or growth factor component that plays the major role in the expected regeneration phenomenon. Bioreactor preculture or surface functionalization of these implants may also be used to limit the attachment of cells to a specific type and to help those cells commit to play a role after implantation by having a functioning tissue type (e.g., by producing extracellular matrix) or by prevascularizing the tissue via coculture with endothelial cells (Temple et al., 2013).

Table 9.3 The key aspects to be considered during the development of tissue engineering strategies for oral and craniofacial tissues

Permission pending from (Mikos et al., 2006)

Tissue engineering aspect	Challenge	
Wound healing environment	Regeneration of lost tissues in essentially "contaminated" conditions.	
Scaffold design requirements	• Injectability. • Ability to encapsulate cells. • Ability to harden to a state which mimics the mechanical properties of bone. • Degradation at a rate fast enough to allow the ingrowth of surrounding tissue.	
Cell-surface interactions	Ability to selectively promote the adhesion of specific cell populations while excluding the invasion of others.	
Growth-factor delivery	• Increased efficiency by loading physiological amounts of growth factors into a scaffold allowing for improved cost-effectiveness. • Spatial and temporal control over the release of multiple growth factors using multiple kinetic rates.	
Assessment	Development of clinically relevant animal models for *in vivo* qualitative and quantitative assessment of implanted materials.	
Scale-up	• Ability to use tissue-engineered constructs in large defects where diffusional limitations may limit the viability of encapsulated cells. • Ability to determine the translatability of results seen with *in vivo* animal models in the clinical setting.	
Specific clinical issues which remain unaddressed	• Control over the morphology of regenerated bone. • Periodontal ligament anchorage and effective gingival seal around titanium dental implants. • Regeneration of entire teeth with supporting structures in extraction sockets.	

9.4.1 SCAFFOLDS

Artificial scaffold materials for craniofacial/dental tissue regeneration may act like the extracellular matrix of the desired tissue by supporting cell attachment, proliferation, and differentiation of host or seeded cells in the presence of biological cues. The scaffolds are commonly resorbable and are expected to resorb at a rate similar to that of new bone formation (Hutmacher, 2000). These scaffolds can be used in different forms such as porous 3D solids (Dean et al., 2012; Kim et al., 2011), nanofibers (Gupte and Ma, 2012; Li et al., 2014), cement/putty (Kim et al., 2012a), among others. Various types of scaffolds used for craniofacial/dental tissue regeneration can be divided into three main classes: ceramic/bioactive glasses, natural/synthetic polymers, and composites (Salgado et al., 2004). Illustrations of all three scaffold material classes follow:

9.4.1.1 Ceramic/bioactive Glasses

The ceramic/bioactive glasses are a class of biomaterials made up of inorganic materials having high compressive strength, but low tensile strength (brittle) characteristics. Calcium phosphate-based materials have been extensively used for craniofacial/dental tissue engineering in the form of injectable calcium phosphate cement (CPC) (Chen et al., 2014; Thein-Han et al., 2012) or implanted scaffolds (Chan et al., 2009). Hydroxyapatite in the form of BoneSource® cement (Stryker, Kalamazoo, MI) has also been used for applications in craniofacial tissue engineering and reconstruction (Costantino et al., 1992; Friedman et al., 1998). Hydroxyapatite scaffolds prepared by additive manufacturing (a.k.a. 3D printing) have been studied for their repair capability in minipig mandibular defects for 6 and 18 weeks (Hollister et al., 2005). Bioactive glasses have also been studied as biomaterials for craniofacial reconstruction (Cho and Gosain, 2004; Gosain, 2004). One exciting new product which promises to be quite adaptive is being developed by Filardo et al. Using a novel bioceramization process, they have been able to turn wood into a strong, highly porous product with an elastic modulus that approaches that of bone (Filardo et al., 2014).

9.4.1.2 Natural/synthetic Polymers

Biopolymers can be prepared synthetically or from naturally occurring sources. Polymers tend to be more flexible or ductile in nature than ceramics, although they usually have less compressive strength than ceramics. Biopolymers derived from natural sources used for craniofacial regeneration include collagen (Narotam et al., 2007), fibrin, hyaluronic acid (Kretlow et al., 2009), alginate, silk (Ye et al., 2011), and chitosan (Canter et al., 2010). Biopolymers of synthetic origin studied in bone tissue engineering applications include poly(propylene fumarate) (PPF) (Dean et al., 2012, 2003b), polylactic acid (PLA) (Di Bella et al., 2008), polyglycolic acid (PGA), poly(lactic-co-glycolic acid) (PLGA) (Kaigler et al., 2006), polycaprolactone (PCL) (Schantz et al., 2003), and polyethylene glycol (Terella et al., 2010). Most of these biopolymers have also been used for dental tissue engineering (Horst et al., 2012).

9.4.1.3 Composites

Blending ceramic, polymeric, and possibly metal or graft tissue components may result in an implant with the desired material properties that is at least partially resorbable. Polymers and ceramics are most likely to mimic bone extracellular matrix which in itself is a blend of inorganic (hydroxyapatite) and organic (collagen type I) components. Some of the composites that have been used in various combinations for craniofacial/dental tissue engineering are: beta-tricalcium phosphate, collagen, and autologous bone fragments (Kishimoto et al., 2006); polycaprolactone (PCL)-tricalcium phosphate

(TCP) (Probst et al., 2010); biocomposite cryogels (Mishra et al., 2014; Mishra and Kumar, 2014); and chitosan and inorganic phosphates (Stephan et al., 2010).

9.4.2 GROWTH FACTORS

Bone morphogenetic proteins (BMPs) are the most studied signaling molecules related to bone maturation. Twenty types of BMPs encoded by the human genome have been identified (Nakashima and Reddi, 2003). Out of these, the BMPs that have been most widely studied for craniofacial/dental regeneration are BMP-2 and BMP-7. These BMPs have been studied for cranial regeneration (Guda et al., 2014) and alveolar tissue as well as oral or dental implant osseointegration (Dunn et al., 2005; Wikesjo et al., 2005). GMP (Good Manufacturing Practice) level recombinant human bone morphogenetic protein-2 (rhBMP-2) provided by R&D Systems (Minneapolis, MN) and Akron Biotechnology (Boca Raton, FL) has been approved for bone applications such as interbody spinal fusion, open tibial fractures, and sinus and alveolar ridge augmentations. It is available commercially in the INFUSE® Bone Graft product (McKay et al., 2007). Infuse is composed of rhBMP-2, which is present on an absorbable collagen sponge (ACS) functioning as a carrier. For certain craniofacial/dental applications such as for defects related with extraction sockets, INFUSE® Bone Graft received FDA approval for sinus augmentations and localized alveolar ridge augmentations in 2007 (McKay et al., 2007). Problematic sequellae have been reported for approved and "off-label" use of BMP-2 such as hematoma, seroma, and swelling, *among others*. The safety of BMP-2 has been discussed (Epstein, 2013; Neovius et al., 2013; Smucker et al., 2006). Other growth factors involved in vasculogenesis, such as vascular endothelial growth factors (VEGFs), are also used for craniomaxillofacial regeneration because bone is a highly vascular tissue and vasculogenesis is expected to facilitate bone regeneration (Kaigler et al., 2006). Section 9.4.5 discusses bioreactor administration of growth factors. The use of bioreactors avoids in vivo use and therefore any unintended pleiotropic effects such as those observed with BMP-2. Where growth factor pharmacokinetics and sensitivity are well understood, there are a variety of delivery mechanisms (Vo et al., 2012).

9.4.3 CELL-BASED THERAPIES

The cells used for cell-based therapies for craniofacial/dental tissue engineering are mostly stem cells (Krebsbach and Robey, 2002). Stem cells are attractive candidates for tissue engineering because they are clonogenic and capable of self-renewal, therefore small populations of stem cells can proliferate to provide the desired cell number within a shorter time span. They have the potential to differentiate into different cell types (Rosa et al., 2012). Some of the cells that have been used for craniofacial/dental tissue engineering (Figure 9.7) are mesenchymal stem cells (MSCs) derived from bone marrow or adipocytes (Marra and Rubin, 2012), dental pulp stem cells (DPSCs), differentiated osteoblasts, perivascular cells, stem cells from exfoliated deciduous teeth (SHEDs), stem cells from apical papilla (SCAPs), periodontal ligament stem cells (PDLSCs), and dental follicle precursor cells (DFPCs) (Bhatt and Le Anh, 2009; Costello et al., 2010; Machado et al., 2012; Risbud and Shapiro, 2005). Dental tissue derived mesenchymal stem cells is a relatively new area of research that may find a wider role in craniofacial/dental repair (Yang et al., 2014). However, it is still not determined whether MSCs from bone marrow, adipocytes, or dental sources will be more favorable for stem cell-based repair of craniofacial/dental tissue (Estrela et al., 2011; Mao et al., 2006; Marra and Rubin, 2012). Recently, constructs developed by cells alone (scaffoldless) have been studied as a potential cell source in dental tissue engineering (Syed-Picard et al., 2014).

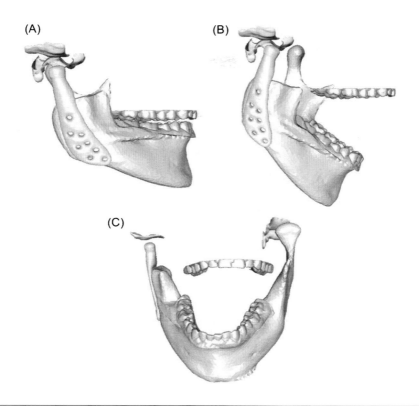

FIGURE 9.7

Mandibular positions during chewing following unilateral total TMJ replacement: **A.** mouth closed (lateral view). **B.** Maximum gape (lateral view). **C.** Maximum gape (frontal view).

(From: Leiggener et al., 2012).

9.4.4 NEW CRANIOFACIAL TISSUES

Two of the more difficult craniofacial tissues are now being explored to see if stem cells can be used in regenerative therapies. The first tissue is retina, where retinal stem cells have received attention given their role in producing sight (Hambright et al., 2012). This work is currently focused on elucidating the mechanisms of retinal tissue repair and treatments for degenerative diseases. The regeneration of sight may be somewhat further off (Yip, 2014). The second tissue is oral mucosa, a highly specialized tissue that is at a high risk of failure and infection following masticatory and dental reconstructive surgery. Good integration between oral mucosa, underlying reconstructed structures, and around percutaneous implants is the goal of this work (Jones and Klein, 2013).

9.4.5 BIOREACTORS

Preculturing of cells inside a bioreactor, with and without shear stress (i.e. from flow or external compression or tension), nutrients, and signaling molecules has been studied as a way of preculturing cells. One of the advantages of preculturing is the production of sufficient cells (starting from low cell

(A)   Pluripotent stem cells

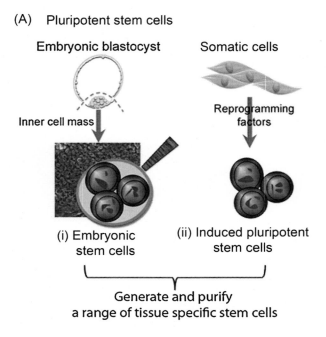

(B)   Adult stem cells

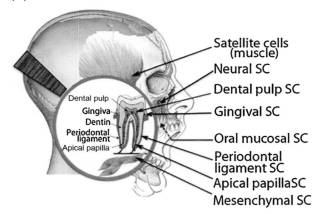

FIGURE 9.8

Specialized tissues where regenerative scaffolds may benefit from preculturing and stem cell differentiation, perhaps in a bioreactor.

(Figure 9.1 from: Sanchez-Lara et al., 2011).

densities) to coat the scaffold and to bring about their maturation to facilitate the host's recognition and integration of the implant as osseous tissue (Mikos et al., 2006). Such a biomaterial represents a tissue-engineered graft (Wallace et al., 2014). This is especially important in clinical settings where it may be difficult to carry out rigorous capture and handling of autologous cells in a distributed, as opposed to a centralized, fashion. It may be especially difficult to characterize the potency (i.e., ability to produce the desired tissue) of all sources of autologous cells.

Bioreactor culturing of well-characterized cells may also offer the advantages of acclimatizing cells to the scaffolds prior to implantation, uniform distribution of oxygen and nutrients throughout the clinical-size bone substitutes, and controlling the amount, nature, and duration of mechanical stimulation applied to seeded cells (de Peppo, 2014). Bioreactor culture of cells has been found to be favorable for craniofacial applications such as the generation of the mandibular condyle (Alhadlaq and Mao, 2003), auricular cartilage (Alhadlaq et al., 2004), the entire TMJ condylar bone (Grayson et al., 2010), and dental applications such as periodontal tissue regeneration (Jin et al., 2003). The protocol for the bioreactor cultivation of cells in tissue engineering of a TMJ by Grayson et al., (2010) is demonstrated in Figures 9.8 and 9.9.

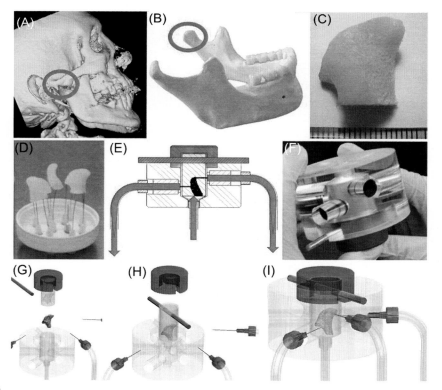

FIGURE 9.9

Bioreactor cultivation for the development of human temporomandibular joint (TMJ) condyles: (a–d) scaffold development; (d) image showing scaffold complexity; (e) The seeding of hMSCs in a bioreactor; (f) photograph of perfusion bioreactor; (g–h) images representing bioreactor assembly.

Permission pending from (Grayson et al., 2010).

9.5 CONCLUSIONS

Bioprinting and nanotechnology research into craniofacial and dental regenerative medicine therapies has utilized growth factors; however, the use of resorbable scaffolds and autologous or banked cells—the remaining elements of tissue engineering—have not yet made their way to the clinic. Craniofacial and dental bioprinting and nanotechnology are concentrated on patient-specific inert devices. This is a large industry that has recently shown dramatic improvements in therapeutic efficacy and safety. There is a revolution occurring in developing 3D printable materials that cover the range of material properties needed for patient-specific implants to perform mechanically like, indeed in tandem with, the tissues to which they are attached. As advances are made in bone, retina, and other areas of craniofacial and dental tissue engineering, they are likely to be incorporated into standard-of-care therapies, unless they are novel enough to shift the current paradigm. Bioprinting and regenerative medicine promise true quality care by facilitating the physician in "doing the right thing, at the right time, in the right way, for the right person—and having the best possible results." However, the current paradigm is relatively new; therefore, it remains to be seen whether the benefits of patient-specific medicine will be made available throughout society through mechanisms like the Affordable Care Act in the United States.

ACKNOWLEDGMENT

Partial support for some of the authors' research reported in this chapter was derived from NIH grants R01-DE013740 and R01-AR061460.

REFERENCES

Abramowicz, S., Dolwick, M.F., Lewis, S.B., Dolce, C., 2008. Temporomandibular joint reconstruction after failed teflon-proplast implant: case report and literature review. Int J Oral Maxillofac Surg 37, 763–767.

Alhadlaq, A., Mao, J.J., 2003. Tissue-engineered neogenesis of human-shaped mandibular condyle from rat mesenchymal stem cells. J Dent Res 82, 951–956.

Alhadlaq, A., Elisseeff, J.H., Hong, L., Williams, C.G., Caplan, A.I., Sharma, B., Kopher, R.A., Tomkoria, S., Lennon, D.P., Lopez, A., Mao, J.J., 2004. Adult stem cell driven genesis of human-shaped articular condyle. Ann Biomed Eng 32, 911–923.

ASPS. 2014. Plastic Surgery Procedures Continue Steady Growth in U.S. l ASPS.

Barton, Jr., F.E., Carruthers, J., Coleman, S., Graivier, M., 2007. The role of toxins and fillers in perioral rejuvenation. Aesthetic surgery journal / the American Society for Aesthetic Plastic surgery 27, 632–640.

Baum, S. 2013. Skull implant produced with 3-D printing technology could shake up orthopedics industry. Medcity News.

Bhatt, A., Le Anh, D., 2009. Craniofacial tissue regeneration: where are we? J Calif Dent Assoc 37, 799–803.

Biomet microfixation. 2010.

Canter, H.I., Vargel, I., Korkusuz, P., Oner, F., Gungorduk, D.B., Cil, B., Karabulut, E., Sargon, M.F., Erk, Y., 2010. Effect of use of slow release of bone morphogenetic protein-2 and transforming growth factor-Beta-2 in a chitosan gel matrix on cranial bone graft survival in experimental cranial critical size defect model. Ann Plast Surg 64, 342–350.

Centers for Disease Control and Prevention (CDC), 2008. Update on overall prevalence of major birth defects – Atlanta, Georgia, 1978–2005. MMWR Morb. Mortal. Wkly. Rep. 57 (1), 1–5.

Chan, W.D., Perinpanayagam, H., Goldberg, H.A., Hunter, G.K., Dixon, S.J., Santos, Jr., G.C., Rizkalla, A.S., 2009. Tissue engineering scaffolds for the regeneration of craniofacial bone. J Can Dent Assoc 75, 373–377.

Chen, W., Thein-Han, W., Weir, M.D., Chen, Q., Xu, H.H., 2014. Prevascularization of biofunctional calcium phosphate cement for dental and craniofacial repairs. Dent Mater 30, 535–544.

Chiapasco, M., Zaniboni, M., 2009. Clinical outcomes of GBR procedures to correct peri-implant dehiscences and fenestrations: a systematic review. Clin Oral Implants Res 20 (Suppl 4), 113–123.

Cho, Y.R., Gosain, A.K., 2004. Biomaterials in craniofacial reconstruction. Clin Plast Surg 31, 377–385, v.

Cho, Y., Raigrodski, A.J., 2014. The rehabilitation of an edentulous mandible with a CAD/CAM zirconia framework and heat-pressed lithium disilicate ceramic crowns: A clinical report. J Prosthet Dent 111(6), 443–447.

Costantino, P.D., Friedman, C.D., Jones, K., Chow, L.C., Sisson, G.A., 1992. Experimental hydroxyapatite cement cranioplasty. Plast Reconstr Surg 90, 174–185, discussion 186–191.

Costello, B.J., Shah, G., Kumta, P., Sfeir, C.S., 2010. Regenerative medicine for craniomaxillofacial surgery. Oral Maxillofac Surg Clin North Am 22, 33–42.

de Lavos-Valereto, I.C., Deboni, M.C., Azambuja, Jr., N., Marques, M.M., 2002. Evaluation of the titanium Ti-6Al-7Nb alloy with and without plasma-sprayed hydroxyapatite coating on growth and viability of cultured osteoblast-like cells. J Periodontol 73, 900–905.

de Vries, C.G.J.C.A., Geertsma, R.E., 2013. Clinical data on injectable tissue fillers: a review. Expert Review of Medical Devices 10, 835–853.

Dean, D., Jonathan, W., Siblani, A., Wang, M.O., Kim, K., Mikos, A.G., Fisher, J.P., 2012. Continuous Digital Light Processing (cDLP): Highly Accurate Additive Manufacturing of Tissue Engineered Bone Scaffolds. Virtual Phys Prototyp 7, 13–24.

Dean, D., Min, K.J., Bond, A., 2003a. Computer aided design of large-format prefabricated cranial plates. J Craniofac Surg 14, 819–832.

Dean, D., Topham, N.S., Meneghetti, S.C., Wolfe, M.S., Jepsen, K., He, S., Chen, J.E., Fisher, J.P., Cooke, M., Rimnac, C., Mikos, A.G., 2003b. Poly(propylene fumarate) and poly(DL-lactic-co-glycolic acid) as scaffold materials for solid and foam-coated composite tissue-engineered constructs for cranial reconstruction. Tissue Eng 9, 495–504.

de Peppo, G.M., Vunjak-Novakovic, G., Marolt, D., 2014. Cultivation of human bone-like tissue from pluripotent stem cell-derived osteogenic progenitors in perfusion bioreactors. Methods Mol Biol 1202, 173–184.

Di Bella, C., Farlie, P., Penington, A.J., 2008. Bone regeneration in a rabbit critical-sized skull defect using autologous adipose-derived cells. Tissue Eng Part A 14, 483–490.

Dickinson, B.P., Roy, I., Lesavoy, M.A., 2011. Temporalis Fascia for Lip Augmentation. Annals of Plastic Surgery 66, 114–117.

Dropkin, M.J., 1999. Body image and quality of life after head and neck cancer surgery. Cancer Practice 7, 309–313.

Dunn, C.A., Jin, Q., Taba, Jr., M., Franceschi, R.T., Bruce Rutherford, R., Giannobile, W.V., 2005. BMP gene delivery for alveolar bone engineering at dental implant defects. Mol Ther 11, 294–299.

Epstein, N.E., 2013. Complications due to the use of BMP/INFUSE in spine surgery: The evidence continues to mount. Surg Neurol Int 4, S343–352.

Estrela, C., Alencar, A.H., Kitten, G.T., Vencio, E.F., Gava, E., 2011. Mesenchymal stem cells in the dental tissues: perspectives for tissue regeneration. Braz Dent J 22, 91–98.

Filardo, G., Kon, E., Tampieri, A., Cabezas-Rodriguez, R., Di Martino, A., Fini, M., Giavaresi, G., Lelli, M., Martinez-Fernandez, J., Martini, L., Ramirez-Rico, J., Salamanna, F., Sandri, M., Sprio, S., Marcacci, M., 2014. New Bio-Ceramization Processes Applied to Vegetable Hierarchical Structures for Bone Regeneration: An Experimental Model in Sheep. Tissue Engineering Part A 20, 763–773.

Franklin, R. 2011. Solid Hyaluronic Acid Tailored to Fix Fine Wrinkles. Medscape. Available at http://www.medscape.com/viewarticle/750397. Accessed Jan 12, 2015

Franklin, J.D., Shack, R.B., Stone, J.D., Madden, J.J., Lynch, J.B., 1980. SINGLE-STAGE RECONSTRUCTION OF MANDIBULAR AND SOFT-TISSUE DEFECTS USING A FREE OSTEOCUTANEOUS GROIN FLAP. American Journal of Surgery 140, 492–498.

Friedman, C.D., Costantino, P.D., Takagi, S., Chow, L.C., 1998. BoneSource hydroxyapatite cement: a novel biomaterial for craniofacial skeletal tissue engineering and reconstruction. J Biomed Mater Res 43, 428–432.

Gamba, A., Romano, M., Grosso, I.M., Tamburini, M., Cantu, G., Molinari, R., Ventafridda, V., 1992. PSYCHOSOCIAL ADJUSTMENT OF PATIENTS SURGICALLY TREATED FOR HEAD AND NECK-CANCER. Head and Neck-Journal for the Sciences and Specialties of the Head and Neck 14, 218–223.

Gosain, A.K., 2004. Bioactive glass for bone replacement in craniomaxillofacial reconstruction. Plast Reconstr Surg 114, 590–593.

Grayson, W.L., Frohlich, M., Yeager, K., Bhumiratana, S., Chan, M.E., Cannizzaro, C., Wan, L.Q., Liu, X.S., Guo, X.E., Vunjak-Novakovic, G., 2010. Engineering anatomically shaped human bone grafts. Proc Natl Acad Sci U S A 107, 3299–3304.

Guarda-Nardini, L., Manfredini, D., Ferronato, G., 2008. Temporomandibular joint total replacement prosthesis: current knowledge and considerations for the future. Int J Oral Maxillofac Surg 37, 103–110.

Gupte, M.J., Ma, P.X., 2012. Nanofibrous scaffolds for dental and craniofacial applications. J Dent Res 91, 227–234.

Hambright, D., Park, K.Y., Brooks, M., McKay, R., Swaroop, A., Nasonkin, I.O., 2012. Long-term survival and differentiation of retinal neurons derived from human embryonic stem cell lines in un-immunosuppressed mouse retina. Mol Vis 18, 920–936.

Hanasono, M.M., Skoracki, R.J., Yu, P.R., 2010. A Prospective Study of Donor-Site Morbidity after Anterolateral Thigh Fasciocutaneous and Myocutaneous Free Flap Harvest in 220 Patients. Plastic and Reconstructive Surgery 125, 209–214.

Hollister, S.J., Lin, C.Y., Saito, E., Schek, R.D., Taboas, J.M., Williams, J.M., Partee, B., Flanagan, C.L., Diggs, A., Wilke, E.N., Van Lenthe, G.H., Muller, R., Wirtz, T., Das, S., Feinberg, S.E., Krebsbach, P.H., 2005. Engineering craniofacial scaffolds. Orthod Craniofac Res 8, 162–173.

Horst, O.V., Chavez, M.G., Jheon, A.H., Desai, T., Klein, O.D., 2012. Stem cell and biomaterials research in dental tissue engineering and regeneration. Dent Clin North Am 56, 495–520.

Hutmacher, D.W., 2000. Scaffolds in tissue engineering bone and cartilage. Biomaterials 21, 2529–2543.

Jin, Q.M., Zhao, M., Webb, S.A., Berry, J.E., Somerman, M.J., Giannobile, W.V., 2003. Cementum engineering with three-dimensional polymer scaffolds. J Biomed Mater Res A 67, 54–60.

Jones, K.B., Klein, O.D., 2013. Oral epithelial stem cells in tissue maintenance and disease: the first steps in a long journey. Int J Oral Sci 5, 121–129.

Joos, U. 2009. Tissue Engineering Strategies in Dental Implantology. In: Meyer U, Meyer Th, Handschel J, and Wiesmann HP (eds): Fundamentals of Tissue Engineering and Regenerative Medicine. http://link.springer.com/chapter/10.1007%2F978-3-540-77755-7_58

Kaigler, D., Wang, Z., Horger, K., Mooney, D.J., Krebsbach, P.H., 2006. VEGF scaffolds enhance angiogenesis and bone regeneration in irradiated osseous defects. J Bone Miner Res 21, 735–744.

Kim, J., McBride, S., Fulmer, M., Harten, R., Garza, Z., Dean, D.D., Sylvia, V.L., Doll, B., Wolfgang, T.L., Gruskin, E., Hollinger, J.O., 2012a. Fiber-reinforced calcium phosphate cement formulations for cranioplasty applications: a 52-week duration preclinical rabbit calvaria study. J Biomed Mater Res B Appl Biomater 100, 1170–1178.

Kim, K., Dean, D., Wallace, J., Breithaupt, R., Mikos, A.G., Fisher, J.P., 2011. The influence of stereolithographic scaffold architecture and composition on osteogenic signal expression with rat bone marrow stromal cells. Biomaterials 32, 3750–3763.

Kim, M.J., Oh, S.H., Kim, J.H., Ju, S.W., Seo, D.G., Jun, S.H., Ahn, J.S., Ryu, J.J., 2012b. Wear evaluation of the human enamel opposing different Y-TZP dental ceramics and other porcelains. J Dent 40, 979–988.

Kishimoto, M., Kanemaru, S., Yamashita, M., Nakamura, T., Tamura, Y., Tamaki, H., Omori, K., Ito, J., 2006. Cranial bone regeneration using a composite scaffold of Beta-tricalcium phosphate, collagen, and autologous bone fragments. Laryngoscope 116, 212–216.

Krebsbach, P.H., Robey, P.G., 2002. Dental and skeletal stem cells: potential cellular therapeutics for craniofacial regeneration. J Dent Educ 66, 766–773.

Kretlow, J.D., Young, S., Klouda, L., Wong, M., Mikos, A.G., 2009. Injectable biomaterials for regenerating complex craniofacial tissues. Adv Mater 21, 3368–3393.

Leiggener, C.S., Erni, S., Gallo, L.M., 2012. Novel approach to the study of jaw kinematics in an alloplastic TMJ reconstruction. Int J Oral Maxillofac Surg 41, 1041–1045.

Lethaus, B., Safi, Y., ter Laak-Poort, M., Kloss-Brandstatter, A., Banki, F., Robbenmenke, C., Steinseifer, U., Kessler, P., 2012. Cranioplasty with customized titanium and PEEK implants in a mechanical stress model. J Neurotrauma 29, 1077–1083.

Li, G., Zhang, T., Li, M., Fu, N., Fu, Y., Ba, K., Deng, S., Jiang, Y., Hu, J., Peng, Q., Lin, Y., 2014. Electrospun fibers for dental and craniofacial applications. Curr Stem Cell Res Ther 9, 187–195.

Ling, X.F., Peng, X., 2012. What Is the Price to Pay for a Free Fibula Flap? A Systematic Review of Donor-Site Morbidity following Free Fibula Flap Surgery. Plastic and Reconstructive Surgery 129, 657–674.

Machado, E., Fernandes, M.H., Gomes Pde, S., 2012. Dental stem cells for craniofacial tissue engineering. Oral Surg Oral Med Oral Pathol Oral Radiol 113, 728–733.

Maloney, B.P., W. Truswell, and S. R. Waldman. 2012. Lip Augmentation Discussion and Debate. *Facial Plastic Surgery Clinics of North America* 20(3), 327–346.

Mao, J.J., Giannobile, W.V., Helms, J.A., Hollister, S.J., Krebsbach, P.H., Longaker, M.T., Shi, S., 2006. Craniofacial tissue engineering by stem cells. J Dent Res 85, 966–979.

Marra, K.G., Rubin, J.P., 2012. The potential of adipose-derived stem cells in craniofacial repair and regeneration. Birth Defects Res C Embryo Today 96, 95–97.

Martinez, A.Y., Como, J.J., Vacca, M., Nowak, M.J., Thomas, C.L., Claridge, J.A., 2014. Trends in Maxillofacial Trauma: A Comparison of Two Cohorts of Patients at a Single Institution 20 Years Apart. Journal of Oral and Maxillofacial Surgery 72, 750–754.

McKay, W.F., Peckham, S.M., Badura, J.M., 2007. A comprehensive clinical review of recombinant human bone morphogenetic protein-2 (INFUSE Bone Graft). Int Orthop 31, 729–734.

Mercuri, L.G., 2012. Alloplastic temporomandibular joint replacement: rationale for the use of custom devices. Int J Oral Maxillofac Surg 41, 1033–1040.

Mikos, A.G., Herring, S.W., Ochareon, P., Elisseeff, J., Lu, H.H., Kandel, R., Schoen, F.J., Toner, M., Mooney, D., Atala, A., Van Dyke, M.E., Kaplan, D., Vunjak-Novakovic, G., 2006. Engineering complex tissues. Tissue Eng 12, 3307–3339.

Mishra, R., Kumar, A., 2014. Osteocompatibility and osteoinductive potential of supermacroporous polyvinyl alcohol-TEOS-Agarose-CaCl2 (PTAgC) biocomposite cryogels. J Mater Sci Mater Med 25, 1327–1337.

Mishra, R., Goel, S.K., Gupta, K.C., Kumar, A., 2014. Biocomposite cryogels as tissue-engineered biomaterials for regeneration of critical-sized cranial bone defects. Tissue Eng Part A 20, 751–762.

Moioli, E.K., Clark, P.A., Xin, X., Lal, S., Mao, J.J., 2007. Matrices and scaffolds for drug delivery in dental, oral and craniofacial tissue engineering. Adv Drug Deliv Rev 59, 308–324.

Nakashima, M., Reddi, A.H., 2003. The application of bone morphogenetic proteins to dental tissue engineering. Nat Biotechnol 21, 1025–1032.

Narotam, P.K., Reddy, K., Fewer, D., Qiao, F., Nathoo, N., 2007. Collagen matrix duraplasty for cranial and spinal surgery: a clinical and imaging study. J Neurosurg 106, 45–51.

Neovius, E., Lemberger, M., Docherty Skogh, A.C., Hilborn, J., Engstrand, T., 2013. Alveolar bone healing accompanied by severe swelling in cleft children treated with bone morphogenetic protein-2 delivered by hydrogel. J Plast Reconstr Aesthet Surg 66, 37–42.

Ng, J., Ruse, D., Wyatt, C., 2014. A comparison of the marginal fit of crowns fabricated with digital and conventional methods. J Prosthet Dent 112(3), 555–560.

Niamtu, J., 2006. Advanta ePTFE facial implants in cosmetic facial surgery. Journal of Oral and Maxillofacial Surgery 64, 543–549.

O'Brien, E. 2014. "Obamacare isn't good for your teeth: Health reform hasn't reduced adult dental costs." How to protect your teeth and wallet. Market Watch: The Wall Street Journal.

Oilo, M., Hardang, A.D., Ulsund, A.H., Gjerdet, N.R., 2014. Fractographic features of glass-ceramic and zirconia-based dental restorations fractured during clinical function. Eur J Oral Sci 122(3), 238–244.

Park, S.H., Wang, H.L., 2007. Clinical significance of incision location on guided bone regeneration: human study. J Periodontol 78, 47–51.

Probst, F.A., Hutmacher, D.W., Muller, D.F., Machens, H.G., Schantz, J.T., 2010. [Calvarial reconstruction by customized bioactive implant]. Handchir Mikrochir Plast Chir 42, 369–373.

Rahaman, M.N., Mao, J.J., 2005. Stem cell-based composite tissue constructs for regenerative medicine. Biotechnol Bioeng 91, 261–284.

Reinisch, J.F., Li, W.-Y., 2014. Abstract 2: medpor ear reconstruction: a twenty-three year experience with 1042 ears. Plastic and reconstructive surgery 133, 974–1974.

Risbud, M.V., Shapiro, I.M., 2005. Stem cells in craniofacial and dental tissue engineering. Orthod Craniofac Res 8, 54–59.

Rosa, V., Della Bona, A., Cavalcanti, B.N., Nor, J.E., 2012. Tissue engineering: from research to dental clinics. Dent Mater 28, 341–348.

Saad, A., Winters, R., Wise, M.W., Dupin, C.L., Hilaire, St., H., 2013. Virtual surgical planning in complex composite maxillofacial reconstruction. Plastic and Reconstructive Surgery 132, 626–633.

Salgado, A.J., Coutinho, O.P., Reis, R.L., 2004. Bone tissue engineering: state of the art and future trends. Macromol Biosci 4, 743–765.

Sanchez-Lara, P.A., Warburton, D., 2012. Impact of stem cells in craniofacial regenerative medicine. Front Physiol 3, 188.

Sanchez-Lara, Pedro A., Zhao S Hu, Bajpai S Ruchi, Abdelhamid, Alaa I., Warburton, David, 2011. Impact of stem cells in craniofacial regenerative medicine. Frontiers in Physiology 3, 188–188.

Schaefer, O., Kuepper, H., Thompson, G.A., Cachovan, G., Hefti, A.F., Guentsch, A., 2013. Effect of CNC-milling on the marginal and internal fit of dental ceramics: a pilot study. Dent Mater 29, 851–858.

Schantz, J.T., Hutmacher, D.W., Lam, C.X., Brinkmann, M., Wong, K.M., Lim, T.C., Chou, N., Guldberg, R.E., Teoh, S.H., 2003. Repair of calvarial defects with customised tissue-engineered bone grafts II. Evaluation of cellular efficiency and efficacy in vivo. Tissue Eng 9 (Suppl 1), S127–139.

Schmitz, J.P., Hollinger, J.O., 1986. The critical size defect as an experimental model for craniomandibulofacial nonunions. Clin Orthop Relat Res, 299–308.

Smucker, J.D., Rhee, J.M., Singh, K., Yoon, S.T., Heller, J.G., 2006. Increased swelling complications associated with off-label usage of rhBMP-2 in the anterior cervical spine. Spine (Phila Pa 1976) 31, 2813–2819.

Song, R.Y., Gao, Y.Z., Song, Y.G., Yu, Y.S., Song, Y.L., 1982. THE FOREARM FLAP. Clinics in Plastic Surgery 9, 21–26.

Stephan, S.J., Tholpady, S.S., Gross, B., Petrie-Aronin, C.E., Botchway, E.A., Nair, L.S., Ogle, R.C., Park, S.S., 2010. Injectable tissue-engineered bone repair of a rat calvarial defect. Laryngoscope 120, 895–901.

Sutradhar, A., Paulino, G.H., Miller, M.J., Nguyen, T.H., 2010. Topological optimization for designing patient-specific large craniofacial segmental bone replacements. Proc Natl Acad Sci U S A 107, 13222–13227.

Syed-Picard, F.N., Ray, Jr., H.L., Kumta, P.N., Sfeir, C., 2014. Scaffoldless tissue-engineered dental pulp cell constructs for endodontic therapy. J Dent Res 93, 250–255.

Taylor, G.I., 1983. THE CURRENT STATUS OF FREE VASCULARIZED BONE-GRAFTS. Clinics in Plastic Surgery 10, 185–209.

Taylor, G.I., Miller, G.D.H., Ham, F.J., 1975. FREE VASCULARIZED BONE GRAFT - CLINICAL EXTENSION OF MICROVASCULAR TECHNIQUES. Plastic and Reconstructive Surgery 55, 533–544.

Taylor, G.I., Townsend, P., Corlett, R., 1979. SUPERIORITY OF THE DEEP CIRCUMFLEX ILIAC VESSELS AS THE SUPPLY FOR FREE GROIN FLAPS - CLINICAL-WORK. Plastic and Reconstructive Surgery 64, 745–759.

Temple, J.P., Yeager, K., Bhumiratana, S., Vunjak-Novakovic, G., Grayson, W.L., 2014. Bioreactor Cultivation of Anatomically Shaped Human Bone Grafts. Methods Mol Biol 1202, 57–78.

Terella, A., Mariner, P., Brown, N., Anseth, K., Streubel, S.O., 2010. Repair of a calvarial defect with biofactor and stem cell-embedded polyethylene glycol scaffold. Arch Facial Plast Surg 12, 166–171.

Thein-Han, W., Liu, J., Xu, H.H., 2012. Calcium phosphate cement with biofunctional agents and stem cell seeding for dental and craniofacial bone repair. Dent Mater 28, 1059–1070.

Truswell, W.H., 2007. Using permanent implant materials for cosmetic enhancement of the perioral region. Facial plastic surgery clinics of North America 15(4), 433–444, vi.

van Kroonenburgh, I., Beerens, M., Engel, C., Mercells, P., Lambrichts, I., Poukens, J., 2012. Doctor and engineer creating the future for 3D printed custom made implants. Digital Dental News, April, pp. 60–65.

Vo, T.N., Kasper, F.K., Mikos, A.G., 2012. Strategies for controlled delivery of growth factors and cells for bone regeneration. Adv Drug Deliv Rev 64, 1292–1309.

Wang, L., Yoon, D.M., Spicer, P.P., Henslee, A.M., Scott, D.W., Wong, M.E., Kasper, F.K., Mikos, A.G., 2013. Characterization of porous polymethylmethacrylate space maintainers for craniofacial reconstruction. J Biomed Mater Res B Appl Biomater 101, 813–825.

Wikesjo, U.M., Polimeni, G., Qahash, M., 2005. Tissue engineering with recombinant human bone morphogenetic protein-2 for alveolar augmentation and oral implant osseointegration: experimental observations and clinical perspectives. Clin Implant Dent Relat Res 7, 112–119.

Yang, M., Zhang, H., Gangolli, R., 2014. Advances of mesenchymal stem cells derived from bone marrow and dental tissue in craniofacial tissue engineering. Curr Stem Cell Res Ther 9, 150–161.

Ye, J.H., Xu, Y.J., Gao, J., Yan, S.G., Zhao, J., Tu, Q., Zhang, J., Duan, X.J., Sommer, C.A., Mostoslavsky, G., Kaplan, D.L., Wu, Y.N., Zhang, C.P., Wang, L., Chen, J., 2011. Critical-size calvarial bone defects healing in a mouse model with silk scaffolds and SATB2-modified iPSCs. Biomaterials 32, 5065–5076.

Yip, H.K., 2014. Retinal stem cells and regeneration of vision system. Anat Rec (Hoboken) 297, 137–160.

Zhou, L.B., Shang, H.T., He, L.S., Bo, B., Liu, G.C., Liu, Y.P., Zhao, J.L., 2010. Accurate reconstruction of discontinuous mandible using a reverse engineering/computer-aided design/rapid prototyping technique: a preliminary clinical study. J Oral Maxillofac Surg 68, 2115–2121.

Zoumalan, R.A., Hirsch, D.L., Levine, J.P., Saadeh, P.B., 2009. Plating in microvascular reconstruction of the mandible: can fixation be too rigid? J Craniofac Surg 20, 1451–1454.

CRANIOFACIAL BONE

Ben P. Hung[1], Pinar Yilgor Huri[1], Joshua P. Temple[1], Amir Dorafshar[2] and Warren L. Grayson[1]

[1]*Department of Biomedical Engineering, Translational Tissue Engineering Center, Johns Hopkins University School of Medicine, Baltimore, MD, USA*
[2]*Department of Plastic and Reconstructive Surgery, Johns Hopkins University School of Medicine, Baltimore, MD, USA*

10.1 INTRODUCTION

According to the Centers for Disease Control and Prevention (CDC), approximately one million cases requiring bone transplantation occur annually in the United States, with around 20% of these in the craniofacial region. This incurs an annual economic burden in excess of $3 billion (Desai, 2007). Such cases may be caused by congenital disease, such as cleft lip (Parker et al., 2010), trauma such as combat injuries (Breeze et al., 2011; Tong and Beirne, 2013), or cancer resection. Regardless of origin, they cause numerous quality-of-life afflictions, such as hindered psychosocial well-being due to facial deformity or the inability to speak or eat due to mandibular or maxillary defects.

Since the available supply of donor bone is far outstripped by this demand, there exists a pressing need for alternative sources of tissue. In tissue engineering (TE), the traditional paradigm is to combine tissue-forming cells, appropriate bioactive factors, and a scaffold to construct tissue *de novo* (Langer and Vacanti, 1993). Bioactive factors can include biochemical, mechanical, or electromagnetic cues while the scaffold provides structural and geometrical guidance to the newly forming tissue at the macro-, micro-, and nanoscales. The macroscale can be defined as geometry on the order of millimeters and above, while microscale considerations are concerned with submillimeter length scales down to the micrometer range. Below this submicrometer length scale is the nanoscale. Micro- and nanoscale considerations, in particular, have to do with *porosity*: the amount of interconnected void space available for cellular population and tissue infiltration; as well as *nanotopography*: the surface features within the walls or struts of the scaffold that are smaller than the cells and can also instruct and guide cellular behavior.

For bone in particular, the scaffold's structural support role is very crucial, as one of the primary functions of bone is mechanical—it supports load, both from the organism's weight as well as from forces exerted by attached muscles. In the craniofacial region, the geometrical role of the scaffold is also of special import: not only does the shape of the scaffold affect the resulting shape of the reconstructed face, but it also affects the pattern in which mechanical forces are distributed throughout the craniofacial skeleton. An in-depth discussion of why precise recapitulation of craniofacial bone geometry is important will follow this section.

To capture these shapes, several modalities have been developed. The first is molding, in which a negative of the desired shape is produced and molten material is poured into the mold. Once the liquid solidifies, the mold is removed and the scaffold is recovered. Such a technique has been used to fabricate scaffolds using polymers with well-defined physical properties, such as poly[lactic acid] (PLA) and poly[ethylene glycol] (PEG) in powdered form (Rahman et al., 2014). With this technique, porosity is generally controlled using porogens, in which a space-filling substance is mixed with the polymer and subsequently evaporated to give rise to void spaces (Mikos et al., 1993). Porogen/molding methods are quite common (see Sadiasa et al., 2014 for a recent example and Xu et al., 2010 for a more in-depth case study); however, the pore structure is not easily controlled and the method has an upper size limitation, as above a certain length scale, the porogens become trapped. Methods used to remove the porogen often require potentially cytotoxic solvents, limiting the usability of this method for craniofacial scaffolds. These drawbacks will be demonstrated in the case study section (Xu et al., 2010).

Another approach employs computer-numeric-controlled (CNC) milling, in which a cutting tool is used to sculpt out desired geometries from a bulk material stock based on programmed paths in three-dimensional space. CNC milling has been used with great success in shaping decellularized trabecular bone to produce scaffolds with appropriate size and shape. A recent study successfully recapitulated the complex shape of the temporomandibular joint (TMJ) condyle using decellularized trabecular bone and a CNC mill (Grayson et al., 2010), a feat that illustrated the possibility of producing complex geometries if a sufficiently large bulk substrate is available. This approach is not suitable for defects greater than 1–2 cm since it is difficult to procure large pieces of trabecular bone that can be milled. Additionally, reproducing certain complex geometries may be limited by the number of degrees of freedom in the movement of the cutting tool and the resolution is governed by the size of the mill.

The recent developments in three-dimensional printing (3DP) techniques provide the capabilities to overcome various limitations associated with molding and milling. Though the term 3DP encompasses a wide variety of methods, it generally refers to the building up of a structure layer-by-layer. 3DP technology is sufficiently advanced such that essentially any 3D geometry can be fabricated; this includes molds for the indirect molding technique discussed earlier and direct fabrication of scaffolds. For direct fabrication, high-resolution printers can achieve finely controlled pore architecture and porosity. Therefore, there is no theoretical size limitation in building porous structures using 3DP techniques as was discussed for the molding methods, and 3DP can be used to build shapes that would be very challenging to produce using CNC milling. Additionally, 3DP provides increased resolution in the control of macro- and microscaled geometries making it a very attractive technique to engineer scaffolds for craniofacial reconstruction.

This chapter will discuss specific features of craniofacial bone that must be recapitulated in a scaffold for successful use in TE. It will be followed by a discussion of how 3DP can achieve those features as well as a section on nanotechnological approaches in the engineering of craniofacial bone.

10.2 CRANIOFACIAL BONE STRUCTURE
10.2.1 CRANIOFACIAL MACROGEOMETRY

Both macrogeometry and microstructure are important to recapitulate in scaffold design. We illustrate the advantages of 3DP approaches that can more readily capture the geometries on both length scales relative to other approaches. On the macroscale, most of the 22 bones of the craniofacial skeleton feature unique geometries that affect two categories of a patient's well-being: (1) facial structure and (2) mechanical transduction.

10.2.1.1 Importance of Average and Symmetric Facial Structure on Psychosocial Health

The craniofacial skeleton provides each person's face with its shape, and consequently, craniofacial defects necessarily result in outward facial deformities. The current gold standard for craniofacial repair, the *"free flap,"* utilizes fragments of autologous bones (most commonly the fibula, but also ribs and pelvis) to achieve craniofacial reconstruction (Figure 10.1). There are obvious drawbacks—autologous bone tissue is in limited supply and donor-site morbidity is unavoidable. It is even clearer that autologous bone segments do not match well with the complex geometries found in the craniofacial skeleton; as such, reconstructed bone using this method still result in a deformed outward facial appearances.

Studies on the societal view of facial structure have established that society values facial geometries that are similar to the *population average* (Langlois and Roggman, 1990; Rhodes et al., 1999) as well as faces with high degrees of symmetry (Mealey et al., 1999), illustrating an established bias on acceptable facial structures. This provides one motivation for reconstructive therapies that mimic the native shape of craniofacial bone as closely as possible. In general, the autologous bone grafts do not result in "normal" or symmetrical faces. Therefore, 3DP approaches can provide major advantages in providing bone grafts that successfully match the facial structure and restore normal appearance following bone resection and trauma.

10.2.1.2 Effect of Craniofacial Bone Shape on Mechanical Transduction of Forces

The major forces acting on the craniofacial skeleton result from mastication and are induced by the action of four muscles—medial and lateral pterygoids, temporalis, and masseter—inserting in specific locations on the posterior mandible. It was once postulated that the shapes of the bones within the craniofacial skeleton evolved to distribute chewing forces evenly. This view was ultimately abandoned, as the anterior root of the zygomatic arch experiences ~ 10 times more strain than does the posterior root during mastication (Hylander and Johnson, 1997). However, it has been demonstrated that the forces resulting from the four muscles of mastication are distributed in a very defined spatiotemporal pattern: through the zygomatic arch, the lateral orbital wall, the supraorbital bone, the infraorbital bone, and finally through the anterior root of the zygomatic arch in that order (Ross et al., 2011). The shape of these bones and the connections between them greatly affect the way they transduce and distribute mechanical stress. Therefore, for proper mechanical function of the craniofacial skeleton, it is crucial that the scaffolds used as craniofacial grafts recapitulate the native geometries. Consequently, fibular flaps do not appropriately transduce forces throughout the skull. Besides affecting mechanics of jaw movement, altered mechanical environments affect cellular behavior, as the osteocytes that comprise the

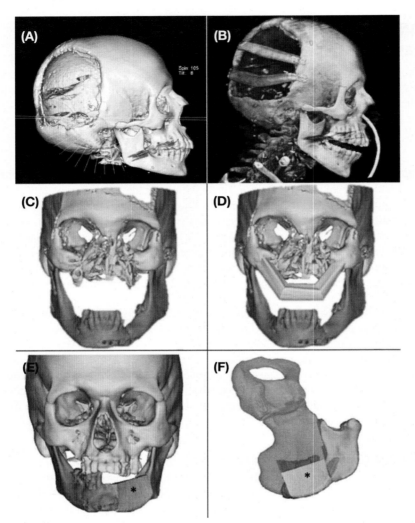

FIGURE 10.1 Autologous bone grafts for craniofacial reconstruction.

Current methods of repairing massive craniofacial bone loss defects require the use of vascularized bone flaps. Several CT images are shown depicting various types of craniofacial defects and treatment approaches. (A, B) A large circular defect in the calvarium defect (A), is bridged with two ribs to provide protection to the brain (B). (C, D) Surgical planning to repair a large region of the maxilla. In this case, the fibula will be cut into multiple pieces and combined as shown in (D) to mimic the structure of the maxilla. (E, F) A mandibular defect reconstructed with a segment of the pelvis (pelvic segment is denoted by the asterisk). In all cases the native craniofacial geometry is not well recapitulated and the approaches result in severe donor-site morbidity.

majority of bone are highly responsive to their mechanical environment (Hung et al., 2013). Scaffolds that recapitulate the appropriate anatomical geometries will also help to provide suitable mechanical environments for cells within the tissues.

10.2.2 CRANIOFACIAL MICROSTRUCTURE

The craniofacial skeleton arises from intramembranous ossification while the long bones in the body arise from endochondral ossification. The bones of the craniofacial skeleton are composed of a hard cortical shell and an inner trabecular region. The cortical shell is composed of concentric lamellar sheets about 5 μm in thickness, while the trabecular network within is made of struts about 150 μm in diameter and pores around 500 μm in diameter (Rho et al., 1998). Trabecular ultrastructure resembles that of osteons; comprised of concentric lamellar sheets but without the central Haversian canals that are found in cortical bone. This microarchitecture is thought to result from a mechanism in which lamellar sheets are deposited sequentially over time (Cohen and Harris, 1958). Generally, TE scaffolds for bone regeneration mimic the inherent porosity of trabecular bone. Modern printers can achieve very finely controlled pore sizes mimicking the 500 μm diameter pores found in trabecular bone as well as the strut networks reminiscent of the trabecular network. These struts may serve as nucleation sites for cell-mediated mineral deposition and for mechanical support. For applications where compressive strength is a priority, it should be noted that increased porosity comes at the expense of mechanical integrity since the large voids within the scaffold structure are not load bearing. As a reference, average maximum bite force which a scaffold within the mandible must withstand is around 1 kN (Hidaka et al., 1999) applied toward the posterior angle. As such, the organization of the strut network is important and the advantages of 3DP techniques in their ability to control porosity and pore structure as functions of spatial location within the scaffolds become clear. One example later in this chapter will explore the use of a perpendicular lattice network to create porosity while generating a "microtruss" for the scaffold's overall mechanical integrity.

10.3 DIFFERENT 3D PRINTING TECHNIQUES AND THEIR APPLICATION TO CRANIOFACIAL SCAFFOLDS

3DP approaches can meet the demands of constructing the complex shapes of the craniofacial skeleton while finely controlling porous microarchitecture. There are several different 3DP techniques, each with specific relative advantages. A more in-depth discussion on 3DP methods can be found in an earlier chapter. Here we focus on a smaller number of common techniques and their advantages and disadvantages when applied to constructing craniofacial scaffolds. A summary table of the techniques discussed here can be found in Table 10.1.

One early method of 3DP was the ink-jet/binder system, in which a nozzle dispenses a binder solution onto a powder bed. Powder that contacts the binding solution is bound, while the other powder remains free and serves as support. For bone, the powder is generally the mineral β-tricalcium phosphate to mimic the mineral phase of bone and the binder is generally some type of acid, such as citric acid (Khalyfa et al., 2007) or phosphoric acid (Inzana et al., 2014). While this method does not require support structures, thereby allowing the printing of overhangs, the main drawback is that the low viscosity of the acid binder solutions result in low resolution and residual acid trapped within large scaffolds that can compromise cell viability. The mechanical integrity of scaffolds printed by this method also depends highly on the interaction between the binder and the powder; in general, scaffolds printed using this technique tend to be brittle. One key advantage of the ink-jet/binder method, however, is the ability to print structures at room temperature, allowing for the incorporation of cells or growth factors. A demonstration of these considerations is included in Case Study 2 (Klammert et al., 2010).

Table 10.1   Summary of 3DP techniques

Method	Description	Advantages	Disadvantages
Ink-jet/binder	Binder solution is dispensed from a nozzle onto a powder bed, binding powder together at spatially defined locations.	Can be printed at room temperatures. Does not require support structures.	Depending on binder solution viscosity, resolution may be low. Binder solution may be cytotoxic.
Selective laser sintering (SLS)	Laser sinters powder together, binding powder at spatially defined locations.	Does not require support structures. Laser provides high resolution.	Laser results in locally elevated temperatures
Thixotropy	Materials flow under shear and are extruded from a nozzle, and solidify once they stop flowing.	Can be printed at room temperature.	Requires support structures. Few materials exhibit thixotropy. Mechanical properties may be ill-suited for load-bearing.
Melt extrusion (ME)	Pressurized nozzle extrudes molten thermoplastic, which solidifies by cooling on the print bed.	Can feature high resolution. Simplest and most common 3DP method. Many materials can be utilized	Requires support structures. Requires elevated temperatures.

A similar technique is selective laser sintering (SLS), which retains the powder bed but uses a laser to sinter the particles together. The resolution is significantly higher than with the binder solution, as there are no problems associated with viscosity and flow. Here, the resolution depends on the diameter of the laser, which is usually on the order of several hundred micrometers (Williams et al., 2005; Eshraghi and Das, 2010), enough to produce struts and pore spaces of comparable sizes to that found in native trabecular bone. SLS, like the ink-jet/binder technique, does not require support structures as the unsintered powder acts as support; however, the high temperatures required to sinter particles preclude the addition of bioactive molecules or living cells.

3DP has also been performed using extrusion of a viscous liquid from a nozzle and allowing that liquid to solidify, thereby building a structure up layer-by-layer. One method to do this uses a special class of materials with the thixotropy property, in which the material has low viscosity when under shear (i.e. while being pressure-extruded out the nozzle) but becomes much more resistant to flow under static conditions. 3DP of thixotropic materials has the benefit of not requiring high temperatures, as the flow properties of the materials are only shear-dependent; however, there are few materials with this property and the method is restricted to certain types of porous glass (Luo et al., 2013; Fu et al., 2011) or mixtures of polymers with organic solvents (Serra et al., 2013). Neither of these classes of materials lends itself well to building bone scaffolds. Furthermore, the thixotropic property may lead to scaffold deformation in response to compressive stresses. As an additional drawback, extrusion-based methods require support structures, limiting the geometries they can produce relative to the powder-based methods.

Another extrusion-based technique utilizes molten thermoplastics. Once extruded, the thermoplastic cools and solidifies. This technique is known as melt extrusion (ME) and is commonly used for poly[ε-caprolactone] (PCL) (Seyednejad et al., 2012; Temple et al., 2014), which has a low melting point of 60 °C (Ang et al., 2007). This property is important because the polymer must cool and solidify quickly once it has been extruded. Unless the polymer rapidly cools to temperatures below its melting point, it will be impossible to stack the printed layers on top of each other. Consequently, polymers with melting points closer to ambient temperatures are most suitable for ME applications. The resolution of ME depends on both the thermoplastic's thermal properties (e.g. viscosity at print temperature) and the diameter of the nozzle. Generally the resolution for ME is good, producing pores and struts on the order of 0.5–1 mm, comparable to that of native trabecular bone. ME requires a support structure for overhangs and operates at high temperatures, but it is perhaps the most intuitive of the 3DP methods and the most widely used. It is also very versatile, as any material that can be melted and extruded can be used for ME-based 3DP. A demonstration of ME applied to the production of craniofacial scaffolds will serve as the final case study (Temple et al., 2014).

In the following sections, several case studies of applying 3DP to the construction of craniofacial scaffolds will be examined. We have selected three key studies that adopt some of the above 3DP processes to create anatomically shaped scaffolds for craniofacial TE applications.

10.3.1 CASE STUDY 1: INDIRECT MOLD METHOD

To demonstrate how 3DP can be used to create a negative mold for scaffold construction, we refer to the example of Xu et al. (Xu et al., 2010), who constructed a poly[glycolic acid] and poly[lactic acid] (PGA/PLA) hybrid scaffold in the shape of a canine mandibular condyle (Figure 10.2A). To obtain the shape, a computed tomography (CT) scan of the animal was performed and scan was reconstructed to a 3D model, which can be programmed into printers to construct the shape. In this case, a commercial Z-Corp printer was used to print the shape from stock material and this was impressed into soft gypsum to create the negative mold. Because the positive model is only used to make the mold, there are less stringent restrictions for the material properties. PGA fibers and a solution of PLA and dichloromethane were added to the mold and later removed after setting. Laser scanning the scaffold and digitally comparing it to the original reconstructed CT demonstrated a mean error in shape no greater than 300 μm. This scaffold casting technique takes advantage of the power of 3DP to create high-fidelity shapes but allows for scaffold materials that may not be amenable to the printing process. As a drawback, scaffold casting diminishes control over the internal structure of the scaffold, limiting the ability to tailor porosity or pore architecture. Also in this case, the use of dichloromethane—a potentially cytotoxic solvent—limits the potential of this technique in applications involving living cells. While the molding technique itself has these drawbacks, it should be noted that the creation of a mold, which requires precise control over geometry, benefits from a 3DP approach.

10.3.2 CASE STUDY 2: INK-JET/BINDER METHOD

An application of printing by dispensing a binder solution onto particles is demonstrated by Klammert et al. (Klammert et al., 2010). Using the same method of obtaining shapes from CT scans and the same commercial Z-Corp printer, the authors printed phosphoric acid onto a tricalcium phosphate powder. These two phases react upon printing to form a hard calcium phosphate cement known as brushite

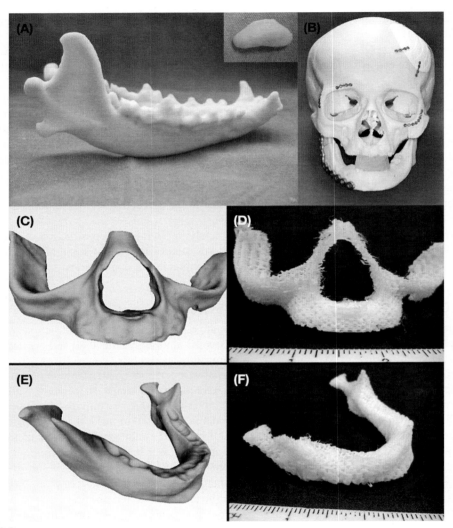

FIGURE 10.2

(A) Ink-jet/binder print of the mandible out-of-stock material used to create the mold. Inset: PGA/PLA composite scaffold made by casting the polymer composite in a mold. Adapted from ref (Xu et al., 2010). (B) 3D powder print of TCP particles with a phosphoric acid binder. Scaffolds are implanted in a human cadaver skull. Adapted from ref (Klammert et al., 2010). (C) Reconstructed model of a human maxilla used to create the print to the right. (D) Final porous PCL maxilla scaffold at full scale. (E) Reconstructed model of a human mandible used to create the print to the right. (F) Final porous PCL mandible scaffold at full scale. Panels C-F adapted from Temple et al. (2014).

($CaHPO_4 \cdot 2H_2O$). Exposing the scaffolds to high temperature (120 °C) in an autoclave for 2 h resulted in the formation of monetite ($CaHPO_4$), also a cement. Both these techniques mimic the mineral composition of native bone. With this 3DP method they were able to fabricate segments of the mandible, calvarium, zygoma, and orbital rim—demonstrating the ability to reproduce highly complex anatomical shapes (Figure 10.2B). Of the two cement scaffolds, brushite was stronger but less porous than monetite; however, the bending strengths of these scaffolds were still at least two orders of magnitude below that of cortical bone, underscoring the limited mechanical properties of scaffolds fabricated using this method.

10.3.3 CASE STUDY 3: MELT EXTRUSION METHOD

In the final case study, Temple et al. (Temple et al., 2014) created porous PCL scaffolds using ME. Because ME-based printers print the entire cross-section of each layer, this method allows complete control over the pore size and porosity. Consequently, the final anatomically shaped scaffolds had consistent rectangular pores throughout. The authors printed defined void volumes that ranged from 20–80%, and selected 60% to match cell seeding studies indicating uniform cell distribution at this void volume. With this setup, the authors were able to seed adipose-derived stem cells (ASCs) throughout the scaffold and induce the formation of bone and vascular tissue, suggesting the potential of this technique for future vascularized bone grafts. The authors were also able to print the human maxilla and mandible geometries with high fidelity and accuracy at full scale despite the requirement for support structures, directly demonstrating the feasibility of using ME to produce craniofacial scaffolds that could be combined with other TE components for bone regeneration (Figure 10.2C).

10.4 NANOTECHNOLOGY IN CRANIOFACIAL GRAFT DESIGN

Inspired by the nanotopography of the native bone extracellular matrix (ECM), nanoscale technologies have impacted the engineering strategies of craniofacial bone. These include the production of nanocomposites to mimic bone ECM composition, and scaffolds composed of nanofibers to better recapitulate the aspects of the native structure. Nanocarriers of bioactive agents were also incorporated in scaffolds to regulate cellular behavior.

10.4.1 NANOCOMPOSITES IN CRANIOFACIAL BONE REGENERATION

A nanocomposite is a material with at least two distinct components and at least one of these components is in the nanometer scale. Natural bone ECM is essentially a nanocomposite material composed of collagen fibers, hydroxyapatite (HAp) nanocrystals, and proteoglycans, having a high degree of structural hierarchy starting from the nanoscale. Therefore, approaches have been developed in order to mimic this nanoscale composite structure of bone ECM.

As collagen type I and HAp are the main components of the bone tissue, their composites have been extensively studied as scaffold materials in the engineering of bone grafts (Wahl and Czernuszka, 2006; He et al., 2010; Kim et al., 2012). In these studies, collagen type I was blended with nano HAp (nHAp) crystals and the resulting nanocomposites were shown to be effective in the regeneration of several craniofacial defect models including immature beagle cranium (diameter: 2.5 cm) (He et al., 2010), rabbit calvaria (diameter: 8 mm) (Kim et al., 2012), and rat infraorbital bone (diameter: 3 mm) (Amaro Martins and Goissis, 2000).

A wide range of other materials has also been employed for the engineering of craniofacial bone. Among these, several natural and synthetic biodegradable polymers were blended with collagen and/or nHAp. Natural polymers such as fibroin (Riccio et al., 2012), chitosan (Costa-Pinto et al., 2012), and calcium alginate (Tan et al., 2011) were used as the continuous phase of the composites due to their osteoconductivity and biochemical similarity to native tissues. In the study of Tan *et.al.*, alginate was used in combination with nHAp/collagen particles (Tan et al., 2011). Two months postimplantation, bone regeneration within the 5 mm diameter rat cranial defects was observed in the presence of mineralized material. The effectiveness of nHAp/collagen/kappa-carrageenan (a linear sulfated polysaccharide) gel as a bone filler was shown in the regeneration of 9 mm wide defects created between premolar teeth and mental foremen of New Zealand white rabbits using distraction osteogenesis (Wang et al., 2010). Natural polymers were also blended with other inorganic components, such as nanocalcium sulfate/alginate (nCS/A) (He et al., 2013) and nanocalcium phosphate/alginate (Lee et al., 2014). When injectable nCS/A was used together with BMP-2 expressing MSCs and endothelial progenitor cells, successful bridging of critical- size (diameter: 8 mm) rat calvarial defects with vascularized bone was observed (He et al., 2013).

Synthetic biocompatible polymers are used to provide the necessary structural integrity for bone reconstruction. These materials are more readily available, processable, and generally have better mechanical strength as compared to natural polymers. Synthetic polymers such as PLA (Li et al., 2011a), its copolymers with PGA (PLGA) (Levi et al., 2010), polycaprolactone (PCL) (Lee et al., 2009), and polyamide (PA) have been prepared as custom matrices. As these synthetic polymers generally lack osteoinductive cues, they were used together with collagen and/or nHAp to enhance osteoconductivity in the regeneration of craniofacial components. For example, PA was used in combination with nHAp. nHAp/PA composite seeded with bone marrow MSCs showed enhanced bone regeneration in the marrow-poor rabbit mandibular angle defects 3 months after implantation (Guo et al., 2012). nHAp/PA has also been used in the preparation of custom-made human mandibular condyle scaffolds, and was successfully used in the clinical treatment of a 27-year-old woman. At 24 months after the surgery, the patient regained proper jaw appearance and joint function (Li et al., 2011b) (Figure 10.3A–C). In another study, nHAp/collagen/PLLA nanocomposite was used to repair rat calvarial defects (diameter: 5 *mm*) (Li et al., 2011a). The material was used in combination with BMP-2 related peptide and the osteoconductivity of the scaffold along with the osteoinductivity through BMP-2 was shown in the cranial bone defect regeneration (Figures 10.3D–F).

Scaffolds intended for use in craniofacial regeneration, especially those in the mandibular region, should have proper mechanical properties as they will be subject to high stresses and strains mainly due to mastication. As such, ceramic components (mainly nHAp and tricalcium phosphates) were blended with biodegradable polymers as a way to reinforce the scaffold material to achieve enhanced mechanical competency as well as enhanced osteoconductivity. For example, it was reported that the compressive strength of PCL matrices could be significantly enhanced by the addition of nHAp (Biqiong Chen, 2005) or tricalcium phosphate (Pilia et al., 2013) nanocrystals.

10.4.2 NANOFIBROUS SCAFFOLDS IN CRANIOFACIAL RECONSTRUCTION

Other than the organization of organic and inorganic components of bone tissue, its fibrous architecture in the nanometer scale has also inspired nanoscale scaffold production methods. Among the conventional scaffold production techniques currently available (e.g. 3DP, solvent casting-particulate leaching, fiber

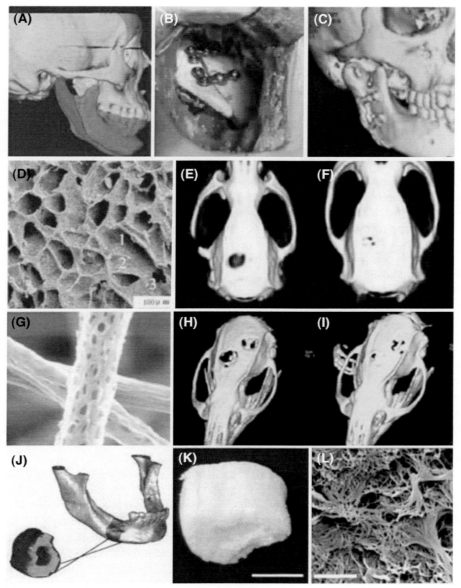

FIGURE 10.3 Nanocomposites are used to mimic the composition of natural bone ECM.

(A) nHAp/PA nanocomposite was used to produce anatomical mandibular condyl graft by 3DP. (B) The scaffold is inserted in the defect area of the patient and fixed with miniplates. (C) CT scan 28 days post-op showing the reconstruction of the condylar defect. Adapted from Li et al. (2011b). nHAp/collagen/PLLA (nHAC/PLLA) composite scaffolds were applied on full thickness 5 mm diameter rat cranial defects. (D) SEM image of the nanocomposite scaffold, (E) unloaded and (F) 1 μg rhBMP-2 loaded nHAC/PLLA scaffolds in the rat cranial defect 12 weeks post surgery. Adapted from ref Li et al. (2011a). **Nanofibers recapitulate the architecture of natural bone**. (G) 3D nonwoven mesh of electrospun PLLA nanofibers were implanted on full thickness 5 mm diameter rat cranial defects. SEM image of nonwoven electrospun nanofiber mesh. (H) unloaded and (I) 175 ng rhBMP-2 loaded nanofibrous scaffolds in the rat cranial defect 12 weeks post surgery. Adapted from Schofer et al. (2011). Fabrication of patient-specific nanofibrous scaffolds. (J) 3D reconstruction of human mandible and preparation of wax mold by rapid prototyping. (K) Anatomically shaped nanofibrous PLLA scaffold prepared by phase-separation technique. (L) SEM image showing nanofibrous-macroporous architecture. Adapted from Chen et al. (2006).

bonding, melt molding, and gas foaming), only a few lead to the formation of 3D structures with nanoscale features to mimic the natural bone ECM structure. These include production of nanofibers by various techniques including electrospinning, self-assembly, and thermally induced phase separation. Electrospinning is based on production of fibers on a grounded collector from a polymer solution by the application of an electric field and can be applied with both natural (e.g. fibrinogen on a thrombin collection bath to form electrospun fibrin fibers) (Zhang et al., 2014) and synthetic such as poly(ethersulfone) (Ren et al., 2013)) polymers to produce scaffolds with nano- and microfibers. Thin fiber layers (mats) obtained through electrospinning are generally used as multilamellar 3D structures (Chen et al., 2011). An alternative approach to producing 3D structures with electrospinning has been to add leachable porogens to the system either in the form of co-electrospun fibers (Baker et al., 2008) or salt particles (Nam et al., 2007) to increase cellular infiltration and attachment throughout the inner regions of the scaffold. Electrospun nanofiber scaffolds have been used in the regeneration of craniofacial bone defects. Nonwoven electrospun PLLA nanofibers (mean diameter 775 ± 294 nm) (Schofer et al., 2009) were collected in molds (diameter: 5 mm, height: 1 mm) and were implanted within same size defects created in the dorsal part of the cranium (Schofer et al., 2011). At 12 weeks post-surgery, PLA nanofiber scaffolds and, more robustly, those applied together with 175 ng rhBMP-2 helped bridge the critical-size cranial bone defects (Figures 10.3G–I). In another study, nHAp/chitosan composite was electrospun to prepare fibrous mats with mean fiber diameter in between 200–300 nm (Liu et al., 2013). Three layers of fiber mats were then held together and implanted within 5 mm × 5 mm rectangular full thickness defects in the rat cranium. Better bone ingrowth and defect bridging was observed in the nHAp/chitosan nanofibers as compared to chitosan nanofibers alone by the end of 20 weeks postimplantation.

Thermally induced phase separation together with porogen leaching has been used as another method to produce macroporous scaffolds with nanofiber features. In this technique, a polymer such as PLA was dissolved in an organic solvent such as tetrahydrofuran and the resulting solution was separated into distinct phases (PLA nanofibers and solvent) at low temperatures (Gupte and Ma, 2012). The solvent was then removed and the resulting nanofibrous scaffold (fibers on the order of 100 nm) was collected. Interconnected porosity within the 3D polymer matrix was obtained by using sacrificial porogen materials such as sugar or paraffin. Various studies have shown that this technique can produce nanofibrous 3D scaffolds having interconnected micropores suitable for bone tissue engineering with a variety of cell types and composite materials (Li et al., 2002; Liu et al., 2009; Wei and Ma, 2006). This technique was also used to produce anatomically shaped nanofibrous mandibular scaffolds. This was achieved by forming a wax mold from the CT images using rapid prototyping technology and then by using this mold to prepare the phase-separated nanofibrous PLLA scaffold as described above (Chen et al., 2006) (Figures 10.3J–L).

10.4.3 NANOCARRIERS IN BIOACTIVE AGENT DELIVERY

Growth factors have important regulatory functions in bone regeneration, from induction of cell recruitment and vascularization toward the defect area to differentiation of osteogenic precursors to mature osteoblasts. Multiple growth factors act in time- and concentration-dependent manners during these processes. Therefore, mimicking the growth factor dose and availability during bone regeneration is central to achieving functional constructs.

Controlled release technology has developed substantially over the last decades and enables the encapsulation of growth factors in carrier structures to protect their bioactivity and to prolong and localize their bioavailability. In tissue engineering strategies, there have been several attempts to incorporate growth factor carriers on scaffolds to increase osteoinductivity and enhance the effectiveness of the construct (Yilgor et al., 2009, 2012; Young et al., 2009; Vo et al., 2012).

Various growth factors have been used within 3D or injectable carrier structures to achieve enhanced bone regeneration in the craniofacial region. For example, in a multicentric clinical study, 160 patients were followed up for 5 years in a total of 21 US centers. Patients received 1.50 mg/ml BMP-2 within collagen sponges for maxillary sinus floor augmentation (Triplett et al., 2009). Patients revealed proper tissue integration and functional recovery starting from 6 months after application. TGF-β3 (Ripamonti et al., 2009) and BMP-7 (Abu-Serriah et al., 2006) were used in preclinical animal models, while TGF-β3 (1.5 μg/application) was especially reported to be a strong mediator of mandibular defect regeneration in primates.

Combined delivery of multiple growth factors was studied recently in order to recapitulate the natural timing of growth factor bioavailability. The coadministration of BMP-2/BMP-7 with a total BMP amount of 50 ng/ml was shown to be effective in the cementoblastic differentiation of dental follicle cells *in vitro* (Kemoun et al., 2007). Other growth factor cocktails have also been shown to be effective: the cocktail of IGF-1/PDGF (total: 5 μg/ml) within methylcellulose gel carriers showed that bone/implant contact and ossification could be enhanced when implanted into extraction sockets in a canine model (Stefani et al., 2000).

10.5 SUMMARY

Osseous defects in the craniofacial skeleton are a large clinical and economic burden. Current repair strategies rely on a limited source of donor tissue and cannot address the special requirements of craniofacial bone geometry, which affects both facial features and mechanical function of mastication. A TE approach—building a bone graft *de novo* from cells seeded within scaffolds and signaled with appropriate bioactive factors—may overcome these drawbacks. For use in craniofacial repair, the scaffold must be porous enough to allow cell and tissue growth on the microscale and capture precise craniofacial geometry on the macroscale. To guide cell fate, scaffolds must also possess appropriate nanoscale features.

One major advance in meeting these demands is the advent of 3DP technology, which allows the construction of any shape in a layer-by-layer fashion. While there are several different methods of 3DP, all methods feature the ability to construct the complex geometries of the craniofacial skeleton while controlling the scaffold's microarchitecture. To functionalize the scaffold, several nanotechnologies including the use of nanoscale components in the scaffolding material, the construction of nanosize features within the scaffold, and the release of bioactive compounds from nanoparticles. Both 3DP and nanoscale technologies have been applied to craniofacial TE and hold promise for the future.

REFERENCES

Abu-Serriah, M., Ayoub, A., Wray, D., et al., 2006. Contour and volume assessment of repairing mandibular osteoperiosteal continuity defects in sheep using recombinant human osteogenic protein 1. J Craniomaxillofac Surg 34, 162–167.

Amaro Martins, V.C., Goissis, G., 2000. Nonstoichiometric hydroxyapatite-anionic collagen composite as support for the double sustained release of gentamicin and norfloxacin/ciprofloxacin. Artif Organs 24, 224–230.

Ang, K.C., Leong, K.F., Chua, C.K., Chandrasekaran, M., 2007. Compressive properties and degradability of poly(epsilon-caprolactone)/hydroxyapatite composites under accelerated hydrolytic degradation. J Biomed Mater Res 80, 655–660.

Baker, B.M., Gee, A.O., Metter, R.B., et al., 2008. The potential to improve cell infiltration in composite fiber-aligned electrospun scaffolds by the selective removal of sacrificial fibers. Biomaterials 29, 2348–2358.

Biqiong Chen, K.S., 2005. Mechanical and dynamic viscoelastic properties of hydroxyapatite reinforced poly(ε-caprolactone). Polymer Testing 24, 978–982.

Breeze, J., Gibbons, A.J., Shieff, C., et al., 2011. Combat-related craniofacial and cervical injuries: a 5-year review from the British military. J Trauma 71, 108–113.

Chen, V.J., Smith, L.A., Ma, P.X., 2006. Bone regeneration on computer-designed nano-fibrous scaffolds. Biomaterials 27, 3973–3979.

Chen, L., Zhu, C., Fan, D., et al., 2011. A human-like collagen/chitosan electrospun nanofibrous scaffold from aqueous solution: electrospun mechanism and biocompatibility. J Biomed Mater Res A 99, 395–409.

Cohen, J., Harris, W.H., 1958. The 3-Dimensional anatomy of Haversian systems. J Bone Joint Surg Am 40, 419–434.

Costa-Pinto, A.R., Correlo, V.M., Sol, P.C., et al., 2012. Chitosan-poly(butylene succinate) scaffolds and human bone marrow stromal cells induce bone repair in a mouse calvaria model. J Tissue Eng Regen Med 6, 21–28.

Desai, B.M., 2007. Osteobiologics. Am J Orthop 36, 8–11.

Eshraghi, S., Das, S., 2010. Mechanical and microstructural properties of polycaprolactone scaffolds with one-dimensional, two-dimensional, and three-dimensional orthogonally oriented porous architectures produced by selective laser sintering. Acta Biomater 6, 2467–2476.

Fu, Q., Saiz, E., Tomsia, A.P., 2011. Direct ink writing of highly porous and strong glass scaffolds for load-bearing bone defects repair and regeneration. Acta Biomater 7, 3547–3554.

Grayson, W.L., Frohlich, M., Yeager, K., et al., 2010. Engineering anatomically shaped human bone grafts. Proc Natl Acad Sci U S A 107, 3299–3304.

Guo, J., Meng, Z., Chen, G., et al., 2012. Restoration of critical-size defects in the rabbit mandible using porous nanohydroxyapatite-polyamide scaffolds. Tissue Eng Part A 18, 1239–1252.

Gupte, M.J., Ma, P.X., 2012. Nanofibrous scaffolds for dental and craniofacial applications. J Dent Res 91, 227–234.

He, D., Genecov, D.G., Herbert, M., et al., 2010. Effect of recombinant human bone morphogenetic protein-2 on bone regeneration in large defects of the growing canine skull after dura mater replacement with a dura mater substitute. J Neurosurg 112, 319–328.

He, X., Dziak, R., Yuan, X., et al., 2013. BMP2 genetically engineered MSCs and EPCs promote vascularized bone regeneration in rat critical-sized calvarial bone defects. PLoS One 8, e60473.

Hidaka, O., Iwasaki, M., Saito, M., Morimoto, T., 1999. Influence of clenching intensity on bite force balance, occlusal contact area, and average bite pressure. J Dent Res 78, 1336–1344.

Hung, B.P., Hutton, D.L., Grayson, W.L., 2013. Mechanical control of tissue-engineered bone. Stem Cell Res Ther 4, 10.

Hylander, W.L., Johnson, K.R., 1997. In vivo bone strain patterns in the zygomatic arch of macaques and the significance of these patterns for functional interpretations of craniofacial form. Am J Phys Anthropol 102, 203–232.

Inzana, J.A., Olvera, D., Fuller, S.M., et al., 2014. 3D printing of composite calcium phosphate and collagen scaffolds for bone regeneration. Biomaterials 35, 4026–4034.

Kemoun, P., Laurencin-Dalicieux, S., Rue, J., et al., 2007. Human dental follicle cells acquire cementoblast features under stimulation by BMP-2/-7 and enamel matrix derivatives (EMD) in vitro. Cell Tissue Res 329, 283–294.

Khalyfa, A., Vogt, S., Weisser, J., et al., 2007. Development of a new calcium phosphate powder-binder system for the 3D printing of patient specific implants. J Mater Sci Mater Med 18, 909–916.

Kim, J.W., Jung, I.H., Lee, K.I., et al., 2012. Volumetric bone regenerative efficacy of biphasic calcium phosphate-collagen composite block loaded with rhBMP-2 in vertical bone augmentation model of a rabbit calvarium. J Biomed Mater Res A 100, 3304–3313.

Klammert, U., Gbureck, U., Vorndran, E., et al., 2010. 3D powder printed calcium phosphate implants for reconstruction of cranial and maxillofacial defects. J Craniomaxillofac Surg 38, 565–570.

Langer, R., Vacanti, J.P., 1993. Tissue engineering. Science 260, 920–926.

Langlois, J.H., Roggman, L.A., 1990. Attractive faces are only average. Psychological Science 1, 115–121.

Lee, C.H., Marion, N.W., Hollister, S., Mao, J.J., 2009. Tissue formation and vascularization in anatomically shaped human joint condyle ectopically in vivo. Tissue Eng Part A 15, 3923–3930.

Lee, K., Weir, M.D., Lippens, E., et al., 2014. Bone regeneration via novel macroporous CPC scaffolds in critical-sized cranial defects in rats. Dent Mater 30, e199–207.

Levi, B., James, A.W., Nelson, E.R., et al., 2010. Human adipose derived stromal cells heal critical size mouse calvarial defects. PLoS One 5, e11177.

Li, W.J., Laurencin, C.T., Caterson, E.J., et al., 2002. Electrospun nanofibrous structure: a novel scaffold for tissue engineering. J Biomed Mater Res 60, 613–621.

Li, J., Hong, J., Zheng, Q., et al., 2011a. Repair of rat cranial bone defects with nHAC/PLLA and BMP-2-related peptide or rhBMP-2. J Orthop Res 29, 1745–1752.

Li, J., Hsu, Y., Luo, E., et al., 2011b. Computer-aided design and manufacturing and rapid prototyped nanoscale hydroxyapatite/polyamide (n-HA/PA) construction for condylar defect caused by mandibular angle ostectomy. Aesthetic Plast Surg 35, 636–640.

Liu, X., Smith, L.A., Hu, J., Ma, P.X., 2009. Biomimetic nanofibrous gelatin/apatite composite scaffolds for bone tissue engineering. Biomaterials 30, 2252–2258.

Liu, H., Peng, H., Wu, Y., et al., 2013. The promotion of bone regeneration by nanofibrous hydroxyapatite/chitosan scaffolds by effects on integrin-BMP/Smad signaling pathway in BMSCs. Biomaterials 34, 4404–4417.

Luo, Y., Wu, C., Lode, A., Gelinsky, M., 2013. Hierarchical mesoporous bioactive glass/alginate composite scaffolds fabricated by three-dimensional plotting for bone tissue engineering. Biofabrication 5, 015005.

Mealey, L., Bridgstock, R., Townsend, G.C., 1999. Symmetry and perceived facial attractiveness: A monozygotic co-twin comparison. J Pers Soc Psychol 76, 151–158.

Mikos, A.G., Sarakinos, G., Leite, S.M., et al., 1993. Laminated 3-dimensional biodegradable foams for use in tissue engineering. Biomaterials 14, 323–330.

Nam, J., Huang, Y., Agarwal, S., Lannutti, J., 2007. Improved cellular infiltration in electrospun fiber via engineered porosity. Tissue Eng 13, 2249–2257.

Parker, S.E., Mai, C.T., Canfield, M.A., et al., 2010. Updated National Birth Prevalence estimates for selected birth defects in the United States, 2004-2006. Birth Defects Res 88, 1008–1016.

Pilia, M., Guda, T., Appleford, M., 2013. Development of composite scaffolds for load-bearing segmental bone defects. Biomed Res Int 2013, 458253.

Rahman, C.V., Ben-David, D., Dhillon, A., et al., 2014. Controlled release of BMP-2 from a sintered polymer scaffold enhances bone repair in a mouse calvarial defect model. J Tissue Eng Regen Med 8, 59–66.

Ren, Y.J., Zhang, S., Mi, R., et al., 2013. Enhanced differentiation of human neural crest stem cells towards the Schwann cell lineage by aligned electrospun fiber matrix. Acta Biomater 9, 7727–7736.

Rho, J.Y., Kuhn-Spearing, L., Zioupos, P., 1998. Mechanical properties and the hierarchical structure of bone. Med Eng Phys 20, 92–102.

Rhodes, G., Sumich, A., Byatt, G., 1999. Are average facial configurations attractive only because of their symmetry? Psychological Science 10, 52–58.

Riccio, M., Maraldi, T., Pisciotta, A., et al., 2012. Fibroin scaffold repairs critical-size bone defects in vivo supported by human amniotic fluid and dental pulp stem cells. Tissue Eng Part A 18, 1006–1013.

Ripamonti, U., Ferretti, C., Teare, J., Blann, L., 2009. Transforming growth factor-beta isoforms and the induction of bone formation: implications for reconstructive craniofacial surgery. J Craniofac Surg 20, 1544–1555.

Ross, C.F., Berthaume, M.A., Dechow, P.C., et al., 2011. In vivo bone strain and finite-element modeling of the craniofacial haft in catarrhine primates. J Anat 218, 112–141.

Sadiasa, A., Nguyen, T.H., Lee, B.T., 2014. In vitro and in vivo evaluation of porous PCL-PLLA 3D polymer scaffolds fabricated via salt leaching method for bone tissue engineering applications. J Biomater Sci Polym Ed 25, 150–167.

Schofer, M.D., Boudriot, U., Wack, C., et al., 2009. Influence of nanofibers on the growth and osteogenic differentiation of stem cells: a comparison of biological collagen nanofibers and synthetic PLLA fibers. J Mater Sci Mater Med 20, 767–774.

Schofer, M.D., Roessler, P.P., Schaefer, J., et al., 2011. Electrospun PLLA nanofiber scaffolds and their use in combination with BMP-2 for reconstruction of bone defects. PLoS One 6, e25462.

Serra, T., Planell, J.A., Navarro, M., 2013. High-resolution PLA-based composite scaffolds via 3-D printing technology. Acta Biomater 9, 5521–5530.

Seyednejad, H., Gawlitta, D., Kuiper, R.V., et al., 2012. In vivo biocompatibility and biodegradation of 3D-printed porous scaffolds based on a hydroxyl-functionalized poly(epsilon-caprolactone). Biomaterials 33, 4309–4318.

Stefani, C.M., Machado, M.A., Sallum, E.A., et al., 2000. Platelet-derived growth factor/insulin-like growth factor-1 combination and bone regeneration around implants placed into extraction sockets: a histometric study in dogs. Implant Dent 9, 126–131.

Tan, R., Feng, Q., Jin, H., et al., 2011. Structure and biocompatibility of an injectable bone regeneration composite. J Biomater Sci Polym Ed 22, 1861–1879.

Temple, J.P., Hutton, D.L., Hung, B.P., et al., 2014. Engineering anatomically shaped vascularized bone grafts with hASCs and 3D-printed PCL scaffolds. J Biomed Mater Res.

Tong, D., Beirne, R., 2013. Combat body armor and injuries to the head, face, and neck region: a systematic review. Mil Med 178, 421–426.

Triplett, R.G., Nevins, M., Marx, R.E., et al., 2009. Pivotal, randomized, parallel evaluation of recombinant human bone morphogenetic protein-2/absorbable collagen sponge and autogenous bone graft for maxillary sinus floor augmentation. J Oral Maxillofac Surg 67, 1947–1960.

Vo, T.N., Kasper, F.K., Mikos, A.G., 2012. Strategies for controlled delivery of growth factors and cells for bone regeneration. Adv Drug Deliv Rev 64, 1292–1309.

Wahl, D.A., Czernuszka, J.T., 2006. Collagen-hydroxyapatite composites for hard tissue repair. Eur Cell Mater 11, 43–56.

Wang, L., Cao, J., Lei, D.L., et al., 2010. Application of nerve growth factor by gel increases formation of bone in mandibular distraction osteogenesis in rabbits. Br J Oral Maxillofac Surg 48, 515–519.

Wei, G., Ma, P.X., 2006. Macroporous and nanofibrous polymer scaffolds and polymer/bone-like apatite composite scaffolds generated by sugar spheres. J Biomed Mater Res A 78, 306–315.

Williams, J.M., Adewunmi, A., Schek, R.M., et al., 2005. Bone tissue engineering using polycaprolactone scaffolds fabricated via selective laser sintering. Biomaterials 26, 4817–4827.

Xu, H., Han, D., Dong, J.S., et al., 2010. Rapid prototyped PGA/PLA scaffolds in the reconstruction of mandibular condyle bone defects. Int J Med Robot 6, 66–72.

Yilgor, P., Tuzlakoglu, K., Reis, R.L., et al., 2009. Incorporation of a sequential BMP-2/BMP-7 delivery system into chitosan-based scaffolds for bone tissue engineering. Biomaterials 30, 3551–3559.

Yilgor, P., Yilmaz, G., Onal, M.B., et al., 2012. An in vivo study on the effect of scaffold geometry and growth factor release on the healing of bone defects. J Tissue Eng Regen Med.

Young, S., Patel, Z.S., Kretlow, J.D., et al., 2009. Dose effect of dual delivery of vascular endothelial growth factor and bone morphogenetic protein-2 on bone regeneration in a rat critical-size defect model. Tissue Eng Part A 15, 2347–2362.

Zhang, S., Liu, X., Barreto-Ortiz, S.F., et al., 2014. Creating polymer hydrogel microfibres with internal alignment via electrical and mechanical stretching. Biomaterials 35, 3243–3251.

ADDITIVE MANUFACTURING FOR BONE LOAD BEARING APPLICATIONS

11

Mihaela Vlasea[1], Ahmad Basalah[1], Amir Azhari[1], Rita Kandel[2] and Ehsan Toyserkani[1]

[1]Department of Mechanical and Mechatronics Engineering, University of Waterloo, Waterloo, ON, Canada
[2]Mount Sinai Hospital, University of Toronto, Toronto, ON, Canada

11.1 NEED FOR BONE SUBSTITUTES

The incentive behind fabricating constructs with a direct application in bone and joint reconstruction surgeries lies in understanding the demand for such devices. Bone and cartilage conditions, such as arthritis, osteoporosis, traumatic musculoskeletal injuries, spinal injuries, and spinal deformities (Hutchinson, 2009), although mostly nonlife-threatening, can become very incapacitating, diminishing the quality of life of the affected individuals by causing ongoing pain, discomfort, inflammation, and restrictions in range of motion (Hutchinson, 2009). Furthermore, these conditions represent a major financial burden on the healthcare sector (Hutchinson, 2009; Cheng et al., 2013). The current conventional treatment for advanced joint and bone trauma is to fully or partially replace the affected area with tissue grafts (Gikas et al., 2009) or with artificial prosthetics (Bartel et al., 2006) to restore near-normal functions. Current state-of-the-art prosthetic implants fail to meet structural and functional requirements that would render them as permanent remediation solutions (Bartel et al., 2006). As a result, thousands of patients undergo painful and costly subsequent surgeries for implant replacements or readjustments. Cell-based or tissue graft solutions have been proven to ameliorate the quality of life of patients, but are limited in terms of size and anatomical shape of defect that can be addressed, as well as the availability of healthy donor tissue and morbidity of the donor site (Koh, 2004; Gikas et al., 2009; Vasiliadis et al., 2010). There is a pressing need for more successful bone and osteochondral reconstruction approaches that take into account biochemical, morphological, and anatomical factors. One such approach focuses on manufacturing biocompatible and/or bioresorbable bone substitutes with complex internal and external architecture and appropriate biochemical cues that can enhance or replace the defect area, gradually mature, and seamlessly integrate with the native tissue (Hutmacher, 2000). The bone substitute would serve as a biocompatible template that would encourage cell migration, proliferation, and differentiation, ideally acting as a temporary bioresorbable porous support until the bone matrix is regenerated (Bohner et al., 2011). A vast amount of work has been done in materials research and manufacturing methodologies in this field, specifically in constructing bone substitutes for load bearing applications; however, there is still a gap in understanding the ideal relationship between the scaffold morphology (pore size, shape, and interconnectivity), transient biochemical interactions, and mechanical properties (Bohner et al., 2011; Butscher et al., 2011).

11.2 COMPOSITIONAL, STRUCTURAL AND MECHANICAL PROPERTIES OF BONE

11.2.1 COMPOSITIONAL PROPERTIES OF BONE AND REQUIREMENTS FOR BONE SUBSTITUTES

Bone is a dynamic and complex organ, encompassing a variety of tissues such as mineralized osseous tissue, cartilage, endosteum, periosteum, marrow, nerves, and blood vessels (Porter et al., 2009). The main role of the bone network is in providing the necessary mechanical support, movement, and protection, with other roles ranging from blood production, to storage of mineral materials, pH regulation, and housing multi progenitor cells (Porter et al., 2009; Szpalski et al., 2012). Due to the complex nature of the bone as biological system, in the context of fabricating bone substitute implants, the focus is generally on understanding the biochemical and structural makeup of the bone extracellular matrix (ECM), as well as the interaction of the ECM with cells and the environment in which they reside (Szpalski et al., 2012). The bone ECM is in essence a composite material comprised of carbonated apatite ($\sim$69% of the ECM), mainly hydroxyapatite ($Ca_{10}(PO_4)_6(OH)_2$) crystals, entrapped in an organic matrix ($\sim$22% of the ECM) of mostly type I collagen, and water ($\sim$9 % of the ECM) (Szpalski et al., 2012; Bose and Tarafder, 2012). Lipids, proteins, and osteogenic factors also reside in the ECM organic matrix (Bose and Tarafder, 2012). From a compositional point of view, the bone substitute matrix should be at least biocompatible with the ECM, cellular, and chemical environment, osteoconductive to encourage fast bone ingrowth from surrounding healthy tissue (Pilliar et al., 2001), as well as nontoxic, nonmutagenic, noncarcinogenic, and nonteratogenic (Leong et al., 2003). Ideally, the material should be osteoinductive to promote formation of new bone at the site (Porter et al., 2009; Yang et al., 2001).

11.2.2 STRUCTURAL PROPERTIES OF BONE AND REQUIREMENTS FOR BONE SUBSTITUTES

Structurally, the bone ECM is comprised of two main zones with very different morphological properties. Trabecular bone, also known as cancellous bone, is a highly porous bone matrix, with interconnected porosities between 50–90% and visible macropores in the range of 500–1000 μm (Karageorgiou and Kaplan, 2005). Trabecular bone has a complex and organized porous architecture, with trabeculae following the direction of mechanical stress (Porter et al., 2009) as a direct result of adaptations to mechanical loading, as postulated by Wolff's law (Bartel et al., 2006) and the mechanostat theory (Frost 2003). Trabecular bone encloses bone marrow and is enclosed by cortical bone. Cortical bone ECM has a compact solid-like structure, with enclosed vascular Haversian canals, having a low porosity between 3–12%, and pores <500 μm (Karageorgiou and Kaplan, 2005). Cortical bone has a solid structure with a series of voids, for example Haversian canals, with a 3–12% porosity (typical apparent density values for proximal tibial trabecular bone 0.30 ± 0.10 g/cm3). The bone ECM is constantly remodeled by the cells that reside in it, where osteoblasts are responsible for producing and mineralizing new bone matrix, osteocytes work on maintaining the matrix, and osteoclasts are responsible for

Table 11.1   Range of mechanical properties for human cancellous and cortical bone (Porter et al. 2009; Yang et al. 2001)

Bone type	Tensile strength (MPa)	Compressive strength (MPa)	Young's modulus (GPa)
Cancellous bone	N/A	4–12	0.01–0.5
Cortical bone	60–160	130–225	3–30

resorbing the matrix (Karageorgiou and Kaplan, 2005). From a structural standpoint, in designing a bone substitute, it is necessary to consider a gradient in porosity and mechanical properties, from a dense external configuration matching the characteristics of cortical bone to the highly porous region with interconnected porosity matching the characteristics of cancellous bone (Mehrali et al., 2013). This means that an ideal bone substitute must have a heterogeneous porous structure, with varying physical and mechanical characteristics. In addition, the implant must also be designed to have an anatomically accurate three-dimensional shape in order to maintain a natural contact load distribution post implantation (Koh, 2004).

11.2.3 MECHANICAL PROPERTIES OF BONE AND REQUIREMENTS FOR BONE SUBSTITUTES

The high level of porosity and pore interconnectivity that is ideal for a bone substitute may be limited by the mechanical strength requirements for that specific implant, especially in the case of load bearing applications. The bone substitute should provide physical support, starting from the seeding process *in vitro* until the tissue is remodeled *in vivo*. Furthermore, the implant must provide sufficient mechanical support to endure *in vivo* stresses and load bearing cycles (Hutmacher, 2000). Table 11.1 summarizes the range in mechanical properties of human cancellous and cortical bone.

11.3 DIFFICULTIES IN ACHIEVING AN IDEAL BONE SUBSTITUTE

Manufacturing of optimal porous bone substitutes from a biochemical, structural, and mechanical properties point of view is highly complex due to a collection of factors. From an architectural standpoint, the bone substitute supports biological and mechanical functions (Bohner et al., 2011), which may be in conflict. For example, for increasing the load-bearing property of the material, a denser material is needed, which conflicts with the requirement of having a highly porous matrix to encourage bone ingrowth and fluid permeability (Karageorgiou and Kaplan, 2005). What is generally defined as an optimization of scaffold properties is likely a tuning of a single parameter with little regard to how other scaffold properties are modified. Furthermore, characterizing, digitizing, and manufacturing the scaffold architecture are difficult tasks. Using characterization methods to reveal pore surface, pore volume, pore shape, interconnectivity, and volume porosity in bone tissues (Bohner et al., 2011), and furthermore translating such data into a digital format that can be interpreted into fabrication methodologies in a continuous or discrete fashion can be a challenge. Typically, the interconnected macroporosity should be >50 μm (Bohner et al., 2011; Yang et al., 2001; Chang et al., 2000;

Kujala et al., 2003; Karageorgiou and Kaplan, 2005; Bose and Tarafder, 2012), with a specific orientation to match the stress loading conditions and fluid and nutrient transport mechanics (Bohner et al., 2011). Also, what is defined as microporosity, with a diameter ranging between 0.1–10 μm (Bohner et al., 2011; Karageorgiou and Kaplan, 2005; Bose and Tarafder, 2012) has shown an effect on the biological response of scaffolds, thus the pores at this scale should be characterized and integrated in the final design. From a structural standpoint, it is also necessary to implement a gradient in porosity and mechanical properties, from a dense external configuration matching the characteristics of cortical bone to the highly porous region with interconnected porosity matching the characteristics of cancellous bone (Mehrali et al., 2013; Butscher et al., 2011; Porter et al., 2009). Manufacturing methodologies that can incorporate the interpretation and implementation of digital data at a macro- and micropore scale are of concern.

Another issue related to manufacturing the appropriate bone substitute architecture lies in the development of appropriate software interpreter design strategies (Lin et al., 2004; Cai and Xi, 2008; Sanz-Herrera et al., 2009) that can convert the desired structural porous morphology and mechanical properties of the bone to be replaced, into appropriate voxel units that can be fabricated using various manufacturing platforms. Such voxels may be computed mathematically using topology optimization algorithms (Lin et al., 2004; Hollister, 2005) or numerical simulation (Sanz-Herrera et al., 2009).

From a material standpoint, difficulties arise in designing structures that can bioresorb *in vivo* at an appropriate rate matching bone remodeling. In the context of regenerative medicine, the terminology of materials with biodegradable, bioresorbable, bioerodible, and bioabsorbable (Hutmacher, 2000) properties are often used. The biodegradation pathway will have an effect on the mechanical, structural, and biochemical properties of the scaffold, and needs to be fully understood (Bohner et al., 2011). Some of the parameters that affect the degradation rate are pore size, pore interconnectivity, permeability, scaffold shape, and volume, as well as implantation location within the musculoskeletal system. Furthermore, the long-term native tissue response to the degradation products should also be considered (Bohner et al., 2011). To add to the difficulty of producing an ideal implant, the overall biochemical, structural, and mechanical properties of the bone substitute should match patient-specific needs such as age, gender, health, metabolism, implant location, and loading conditions (Bohner et al., 2011).

11.4 METALLIC BONE SUBSTITUTES
11.4.1 METALLIC MATERIALS, LIMITATIONS AND OPPORTUNITIES

For a long time, metals were the main material utilized for orthopedic implants. This interest in metals resulted from the excellent physical and mechanical properties that are intrinsic to metals. At present, the interest in nonmetallic materials has prompted the fabrication of tissue scaffolds. These materials are mostly polymer or ceramic, and are used to produce biodegradable scaffolds. Biodegradable scaffolds can be useful for young patients, because they have high growth rates of tissue to restore the functionality of the damaged area. However, the case is completely different for senior citizens, who have very low tissue growth rates. When faced with a certain degradation rate of the materials, this may cause a mismatch in terms of the mechanical properties (Yarlagadda et al., 2005). Therefore, permanent metallic bone substitutes are more appropriate in the case of older patients. Several metals have been used for implants, such as stainless steels (316L), Co-Cr-Mo, pure titanium, titanium alloys, and tantalum. Each metal has advantages and disadvantages that can either expand or limit its usage.

Table 11.2   Comparison of the mechanical properties of different metals (Krishna et al. 2007)

Material	Bone	Magnesium	Co–Cr–Mo and alloys	Ti and alloys	Stainless steels
Density (g cc^{-1})	1.8–2.1	3.1	8.3–9.2	4.4–4.5	7.9–81
Compressive strength (MPa)	130–180	65–100	450–1896	590–1117	170–310
Elastic modulus (GPa)	3–20	41–45	200–253	55–117	189–205
Toughness (MPam$^{1/2}$)	3–6	15–40	100	55–115	50–200

Stainless steel is considered one of the first metals used in the orthopedic field as plates and screws for bone fixation in the early twentieth century (Stevens et al., 2008). The most popular stainless steel alloy used in prosthesis fixation is (316L), with moderate strength and toughness in comparison to other metals as shown in Table 11.2. This alloy is distinguished by good corrosion resistance in comparison to other steel alloys, since the 12% Cr in its content forms a corrosion protective layer Cr_2O_3 on the surface (Navarro et al., 2008). This metal is both widely available and economically effective in terms of processing and manufacturing (Long and Rack, 1998; Dabrowski et al., 2010). However, as the wear resistance of stainless steel is very low, its usage in hip replacement was stopped (Navarro et al., 2008). Nowadays, stainless steel is rarely used for orthopedic implants. Instead, stainless steel is used for temporary fixation devices such as nails, screws, and plates due to the superiority of other metals such as Ti, Ti alloys, and Co–Cr alloy in terms of mechanical properties and corrosion resistance (Navarro et al., 2008).

Another widely used metal in the orthopedic industry is Co-Cr-Mo alloy that is characterized by a high level of mechanical strength, fatigue strength, wear resistance, and low cost of production (Reclaru et al., 2005; Navarro et al., 2008; Dabrowski et al., 2010). While these advantages are considered beneficial for some applications, this is not generally true for orthopedic implants, since this metal's high strength and elastic modulus has caused a stress shielding between the implant and the bone due to the mismatch of mechanical properties. Moreover, the high level of metal ions that are released and the nanoparticle debris which is caused by the wear, negatively affect the biocompatibility of this metal (Billi and Campbell, 2010; Dabrowski et al., 2010). Despite these drawbacks, this metal is the most preferable in hip replacement due to its high level of wear resistance compared to the other metals. In particular, this material is used to fabricate the femoral head and acetabular cup which is an area of high friction (Navarro et al., 2008). In addition, dentists have some interest in this metal as a coating for some dental devices (Long and Rack, 1998).

NiTi (nitinol) is a shape memory alloy, which means it can restore its original shape after a plastic deformation by using a heat treatment. This alloy was introduced in 1960 and is characterized by high strength, superelasticity, and good corrosion resistance (Shishkovsky et al., 2008; Michiardi et al., 2006; Navarro et al., 2008). However, the release of the Ni ions limits its usage as an orthopedic prosthesis due to the possibility of toxicity and the inflammation effect of the surrounding tissue. This has led researchers to treat the surface through exposure to the oxidization process in an attempt to create a protective layer free of Ni (Wataha et al., 2001; Navarro et al., 2008; Michiardi et al., 2006).

Alternatively, degradable metals have captured the interest of some researchers. One of these metals is magnesium (Mg), which is characterized by having low elastic modulus. In addition, Mg is an osteoconductive material and it does not show any inflammatory effect following fixation (Staiger et al., 2006). However, as Mg is a degradable material, the dissolved particles may cause toxicity, which could limit its usage in orthopedic applications (Alvarez and Nakajima, 2009).

Titanium (Ti) was introduced in the orthopedic field in 1965 in the form of screws and plates. This long period of usage illustrates the preference of using Ti, as supported by long-term clinical data, which indicates this material's good biocompatibility. Moreover, Ti is characterized by a nano-oxide layer covering its surface that increases *in vivo* osteointegration (Frosch and Stürmer, 2006). Indeed, from a biological aspect, Ti has proved its compatibility more so than stainless steel or CoCr, as proven by the cell culturing of these metals (Frosch & Stürmer, 2006). Overall, Ti is considered one of the strongest and highest corrosion-resistant metals. Moreover, Ti is characterized by an excellent strength-to-weight ratio and good toughness (Ryan et al., 2006; Navarro et al., 2008).

Recently, it has been noted that the science community is concentrating on Ti alloys, in particular, Ti-6Al-4V, to fabricate bone substitute scaffolds. This interest originates from the good mechanical properties and biocompatibility of this alloy. The presence of aluminum (Al) and vanadium (V) in this alloy improves the mechanical properties compared to commercially pure Ti (CP Ti) (Navarro et al., 2008). One of the most important benefits of Ti and its alloy (Ti-6Al-4V) is its relatively low elastic modulus which is the half of the Co–Cr modulus, which could assist in reducing the effect of the stress shielding (Ryan et al., 2006). Despite the fact that Ti has a low elastic modulus in comparison to other metals, the mismatch between bone and implant still exists (Ryan et al., 2006). Arguably, the presence of V in this alloys is a source of concern, since V is toxic (Okazaki, 2001; Alvarez and Nakajima, 2009). Further drawbacks of Ti and its alloys include the expensive machining cost and the sophisticated heat treatment process (Navarro et al., 2008). Nowadays, most dental implants are fabricated from CP Ti, and Ti-6Al-4V are used in orthopedic applications (Navarro et al., 2008).

In order to eliminate the effect of V in Ti6Al4V alloy, new Ti alloys have been developed to overcome the toxicity issue, such as Ti-6Al-7Nb, Ti-5Al-2.5Fe, Ti-35Nb-5Ta-7Zr, Ti-35Nb-5Ta-7Zr-0.4O, and Ti-15Zr-4Nb-4Ta. For instance, Ti-15Zr-4Nb-4Ta has shown excellent mechanical properties, biocompatibility, and good corrosion resistance. In addition, it is likely that more bone was formed than that which formed around the Ti 6Al 4V when implanted on the bone marrow of the rat tibia (Okazaki, 2001). Most of these alloys are still in the development stage in order to characterize the mechanical properties and biocompatibility for application in bone regenerative medicine (Navarro et al., 2008).

Another material, tantalum (Ta) is considered one of the best metals in the orthopedic field due to its fairly low Young's modulus, high corrosion resistance, and excellent biocompatibility (Matassi et al., 2013). A recent study comparing Ta and Ti showed that Ta is superior compared to Ti in regards to cell integration with a controllable Young's modulus in the range of 1.5–20 GPa by changing the porosity of the structure (Balla et al., 2010). However, the high cost of production and fabrication of this metal is the main obstacle limiting its usage in the orthopedic industry along with short-term clinical data (Dabrowski et al., 2010; Matassi et al., 2013).

Generally, the major constraints which can govern the success of a load bearing implant are biocompatibility, porosity, and proper mechanical properties. The highly important mechanical properties in the designing of a load bearing implant are Young's modulus and compressive strength. Other mechanical properties such as bending strength are less relevant in load bearing implant. However, certain properties such as fatigue strength are necessary for a long lasting implant, but are not as crucial as Young's modulus and compressive strength, since these properties may cause short-term failure of implant (Oh et al., 2003).

11.4.2 AM OF METALS FOR BONE SUBSTITUTES

Historically, conventional techniques were used to fabricate porous metals for several industrial applications. Once porous metal was introduced in early 1972 in the orthopedic field (Weber and White, 1972), scientists looked for ways to control and improve the properties of this metal foam. Consequently, several conventional techniques emerged. However, these techniques are not capable of precisely controlling the porosity and producing a scaffold with a predefined microstructure. Also, these techniques cannot produce a scaffold with a contoured external shape. In order to overcome these obstacles, additive manufacturing (AM) techniques are considered the most promising alternative for precise fabrication of internal and external features of scaffold architecture.

There are many AM techniques that have been used in the fabrication of metal scaffolds, such as three-dimensional printing (3DP), electron beam melting (EBM), selective laser melting (SLM), direct metal deposition (DMD), and selective laser sintering (SLS) (Ryan et al., 2008; Murr et al., 2009; Mullen et al., 2009; Dinda et al., 2008; Shishkovsky et al., 2008). Each technique has different opportunities and limitations. Some of these techniques rely on laser technology, and others rely on the injection of the slurry. The costs of production through these techniques are varied according to complexity. For example, the laser sintering methods are highly expensive due to their use of laser technology and the capability of producing a precise microstructure. Other techniques appropriate for fabricating a scaffold have acceptable microstructure details, such as fiber deposition and 3DP.

SLS is an AM technique (Figure 11.1) that works by sintering very fine layers of metal powders layer over layer using a CO_2 laser beam, either directly or indirectly. In direct SLS, the powder mixture is a compound of two metals: a low sintering temperature metal and the main metal (Campanelli

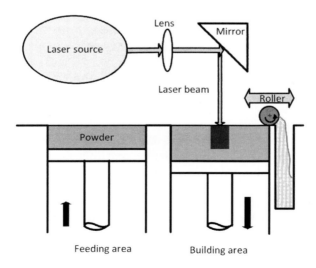

FIGURE 11.1

Schematic presentation of working principle of SLS technique.

et al., 1994). In this technique, the laser beam melts the low sintering temperature metal and uses it to bind the main metal particles to each other. For instance, in a study where NiTi dental implant was fabricated using DMLS, the base metal was Ti and the binding agent was Ni (Shishkovsky et al., 2008). The second technique is called indirect SLS, where a preprocessing of the powder is required to coat the metal powder particles with some polymer to work as a binding agent of the green sample. This technique requires a postprocessing of the green sample by debinding the polymer from the green sample followed by the sintering of the metal particles in a shielded environment at a very high temperature. The diversity of materials produced by this manufacturing method is one of the most important advantages of this technique (Wong and Hernandez, 2012a). The final product resolution is moderate since heat transfers to the adjacent area and fuses extra particles to the targeted area, which might lead to a limitation in the accuracy according to the size of particles (Woesz, 2008a; Wong and Hernandez, 2012b). Economically speaking, fabrication using SLS is costly (Dabrowski et al., 2010). Technically, SLS is considered a lengthy manufacturing method, since it requires preprocessing of the powder and the laser sintering process is time-consuming. In addition, the resulting product may require a postprocessing heat treatment that usually takes hours to complete (Campanelli et al., 1994; Woesz, 2008b).

Another complex technique, EBM, works based on the layering process similar to SLS. EBM differs from SLS in its use of an electron beam rather than a laser to sinter the powders, which is generated in a tungsten filament and is accelerated and controlled with a magnetic field (van Noort 2012). This manufacturing technique is relatively fast and the building of samples takes place under vacuum (Alvarez and Nakajima, 2009). It offers an attractive opportunity in the fabrication of fully dense or porous parts due to several controllable parameters such as beam current, scan rate, and sequence variations (Murr et al., 2009). In this technique, an additive or fluxing agent is not required to fulfill the melting process because the electron beam is powerful enough to raise the temperature of the particles to the melting point (van Noort 2012). However, there are still some limitations in this method such as the low dimensional accuracy and surface quality due to shrinkage. Additionally, this technique is costly, and removing the excess material inside the structure can prove difficult (Wen et al., 2002). Different metals have been fabricated using EBM, such as Cp-Ti, Ti-6Al-4V, and Co/Cr for orthopedic implant and maxillofacial surgery (Alvarez & Nakajima 2009; van Noort, 2012).

Similarly, SLM utilizes the same mechanism of the EBM, and a wide range of materials in powder form can be used (Alvarez and Nakajima, 2009). Moreover, SLM is able to produce solid and porous parts based on the laser energy density (Mullen et al., 2009). Also, this technique is free of binders and fluxing agents, so there is no need for a postprocessing step (Pupo et al., 2013; Alvarez and Nakajima, 2009; Campanelli et al., 1994). Unlike EBM, SLM uses an ytterbium fiber laser 200 W power and uses Ar or N in the building chamber, which may increase the thermal conductivity and maintain a consistent rapid cooling of the printed zone more than EBM (Murr et al., 2012; Mullen et al., 2009). However, high production cost, lengthy fabrication time, and difficulty in removing the trapped powder are the major limitations of this technique (Dabrowski et al., 2010; Alvarez and Nakajima, 2009; Campanelli et al., 1994). Furthermore, the vaporization phenomenon is one of the drawbacks of this technique. This phenomenon is generated due to the high temperature of the molten pool caused by the laser beam evaporating the particles. This leads to an overpressure in the molten pool, which results in spewing of some molten metals out of the pool (Campanelli et al., 1994). Today, this technique is utilized in the aerospace industry, orthopedics prostheses, and dental implants (Pupo et al. 2013).

In the same context, laser engineering net shaping (LENS) and direct metal deposition (DMD) are both AM techniques used in manufacturing bone substitutes. These techniques mainly depend on laser technology to fuse the metal powder. LENS systems use a neodymium–yttrium–aluminum–garnet (Nd:YAG) laser, while DMD uses the CO_2 laser beam. The powder feeding systems are completely different, since the powder feeding in the DMD comes through a concentric ring at the tip of the laser nozzle, while in the LENS the powder comes through different powder feeders to the melting zone. Both techniques use an inert gas during the manufacturing process to avoid oxidation (Dinda et al., 2008). With these techniques, it is possible to use a wide range of metal powders such as stainless steel, nickel-based alloys, Ti and its alloys, tooling steel, copper alloys, alumina, or a combination of these (Lin et al., 2009; Wong and Hernandez, 2012b; Woesz, 2008b). Thus, LENS can be useful for the repairing or re-manufacturing process that cannot be implemented by other AM techniques (Wong & Hernandez, 2012b). LENS and DMD work through the process of depositing the metal layer-over-layer. Both are similar in terms of processing; however, the high cost of DMD has boosted LENS's popularity. The need for support structures for overhanging features is one of the limitations of the use of these techniques in the fabrication of orthopedic prostheses (Woesz, 2008b). Another drawback is the residual stress caused by the rapid heating and solidifying process (Wong and Hernandez, 2012b).

Many studies have been conducted using two techniques: fiber deposition (FD) and 3DP. The two have been chosen mainly due to their manufacturing simplicity compared with SLS or EBM and cost efficiency. The base material of FD is metal powder. In the FD method, the powders are mixed with a solution to form the slurry which can be deposited from the machine onto a substrate. The scaffold is built by the process of layering from bottom to top. The FD technique requires postprocessing of the product by the sintering of the produced scaffold in a very high temperature furnace. When using FD in the fabrication of a scaffold, several parameters can influence the strength and the porosity of the structure, such as the gap between fibers, the fiber's lay down angle, and the nozzle's diameter (P. J. Li et al., 2007; Li et al., 2005). The dimensional accuracy of the final product is poor compared to other techniques such as 3DP due to the wet nature of the fiber slurry, which leaves a high level of shrinkage after drying and sintering (Basalah et al., 2012). In fact, this technique is still in the experimental stage.

One of the simplest rapid prototyping techniques is the ink-jet-based 3DP, as shown in Figure 11.2. In this technique, the printer functionality is similar to an ink-jet printer, where instead of ink, a binder

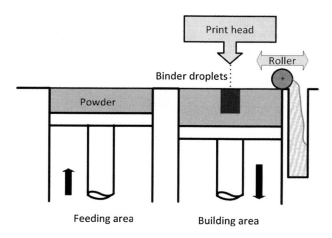

FIGURE 11.2

Simple sketch showing the working principle of the 3DP machine.

is dispersed through a print head on each layer of powder. Similar to previous methods, this printer works on a layer-by-layer basis, starting from the bottom and going up. The printer has two different powder zones: one for building the structure and the other for feeding the powder, so that when the feeding area goes up, the building area lowers down, moving a distance equivalent to the required layer thickness. A counter-rotating roller spreads the powder from the feeding bed over to the building bed, followed by the injection of the binder based on an image corresponding with the slice layer data of the 3D structure. This technique is very fast in the building process and is significantly cheaper than other techniques (Woesz, 2008b). This machine was used indirectly to produce a porous metal scaffold by printing a sacrificial mold from alumina powder to cast the Co-Cr alloy which is then removed through several thermal and chemical steps (Curodeau et al., 2000). Similarly, the same principle has been used to create a Ti scaffold and wax template as an alternative sacrificial mold (Ryan et al., 2008). This method of manufacturing provides a precise surface texture. However, the time- consuming process through the multiple stages of manufacturing is considered to be the main disadvantage of the indirect 3DP technique. This attempt has fascinated and motivated many researchers to replicate this success through direct printing of the desired metal scaffold, followed by sintering of the green part under a high temperature vacuum furnace (Hutmacher et al., 2004; Basalah et al., 2012). This technique has fewer stages than indirect printing. With 3DP, several parameters are able to form the microstructure of the desired part such as powder size, sintering temperature, and duration (Basalah et al., 2012). The layer thickness is primarily chosen based on the particle size and cannot be thinner than the largest particle in the powder (Hutmacher et al., 2004). However, the resolution is lower than the laser-based techniques because the binder penetrates the powders adjacent to the targeted area (Woesz, 2008b). Difficulties in removing the trapped powder from the green part are the main disadvantage of this technique (Hutmacher et al., 2004; Woesz, 2008b; Basalah et al., 2012).

Using the 3DP method and CP Ti as a building material, the authors' group has been able to develop a set of optimized processing conditions to assure control over a microstructure and the associated mechanical and physical properties of the structure (Basalah et al., 2012). Material powder size is one of the most influential parameters on changing the strength and density of the structure, as shown in Figure 11.3. Also, the physical appearance of the final product is crucial in the fabrication of the orthopedic implant, since a high level of shrinkage in the implant is undesirable, and the particle size influences the shrinkage, as shown in Figure 11.4.

As a whole, AM techniques offer a good control over the manufacturing of the external and internal details of metallic bone substitute scaffolds, with limitations in the creation of complex internal interconnected macro pores (Yarlagadda et al., 2005).

11.5 BIOCERAMIC BONE SUBSTITUTES

The appropriate materials for constructing bone substitutes for tissue regeneration or augmentation should adhere to a basic set of criteria stating that: (i) the material must be biocompatible to avoid the expression of unwanted immune responses, and must be osteoconductive to encourage rapid integration with surrounding bone, and ideally osteoinductive to promote formation of new bone at the site (Porter et al., 2009; Yang et al., 2001); (ii) the material should be bioresorbable such that it is capable of bulk degradation and resorption through natural pathways (Hutmacher, 2000; Porter et al., 2009; Bohner et al., 2012); and (iii) the material should provide the physical support, starting from the seeding process

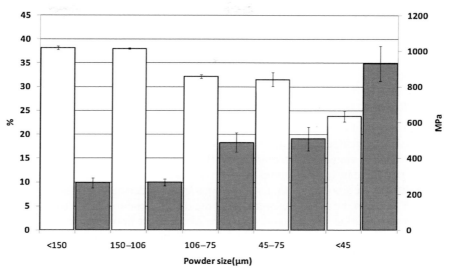

FIGURE 11.3

Compressive strength and porosity of the porous structures fabricated by varied sizes of Ti powder.

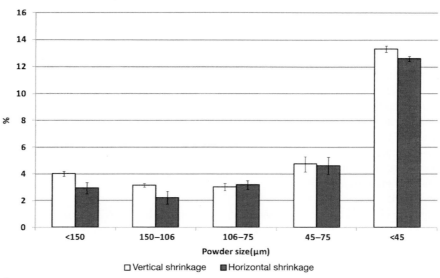

FIGURE 11.4

Vertical and horizontal shrinkage of Ti samples fabricated by varied sizes of Ti powder.

in vitro until the tissue is remodeled *in vivo,* to support stresses and load bearing cycles (Hutmacher, 2000; Bohner et al., 2012; Bose and Tarafder, 2012). Bioceramics, such as crystalline ceramics, amorphous glasses, and ceramic composites, have been studied in the context of regenerative medicine as bone substitute materials as they show promise in meeting the outlined basic material criteria.

11.5.1 BIOCERAMIC, BIOACTIVE GLASSES AND COMPOSITE MATERIALS

Depending on the chemical composition and porous architecture of the bioceramic material, the main resorption mechanisms reported in literature can be classified as chemical dissolution, cell-mediated dissolution, hydrolysis, or enzymatic decomposition (Bohner et al., 2012), where resorption is understood here as the gradual removal of material from the bone substitute construct over time. To select an appropriate bioceramic for a bone substitute, it is important to understand the resorption mechanism, as well as the link between the rate of resorption and the rate of bone formation to ensure mechanical stability of the implant during bone remodeling. One of the most important resorption pathways is cell-mediated dissolution, since it is a biologically controlled mechanism, enabling the rate of degradation of the underlying bone substitute to be controlled by the host cells (Bohner et al., 2012; Detsch et al., 2010). Cell-mediated dissolution of materials is ideally validated *in vivo,* however *in vitro* tests can be done to observe so-called "dissolution" pits produced by osteoclast or osteoclast-like cells on the material (Bohner et al., 2012).

Calcium phosphates bioceramics are very promising materials in producing bone substitute components, as it was found that the degradation of these ceramics produces calcium and phosphate ions, which regulate bone metabolism, as they promote new bone formation through osteoinduction (Bohner et al., 2012; Bohner, 2010). For calcium phosphates, the main *in vivo* resorption is done through cell-mediated dissolution, also called osteoclastic resorption, where the osteoclast cells in contact with the material release hydrochloric acid in small amounts, enough to lower the local pH at the site, triggering calcium phosphate dissolution (Detsch et al., 2010; Bohner, 2010). Generally, calcium ions influence osteoblast proliferation and osteoclast regulation, while phosphate ions regulate osteoblast apoptosis and mineralization rate (Shanjani, 2011; Bohner et al., 2012). *In vitro*, the degradation of calcium phosphates in aqueous solutions generally occurs through chemical dissolution (Pilliar et al., 2001) and the rate of degradation is influenced mainly by the crystallinity of the material, solubility, porosity, surface area, and pH of the solution (Shanjani, 2011). There is a range of calcium phosphates ceramics, depending on the molar ratio of Ca:P, with different resorption properties, as summarized in Table 11.3. Calcium phosphates are popular materials in bone regenerative medicine as they have great versatility in terms of biodegradability and bioactivity, and they have chemical similarity to bone minerals (Bose and Tarafder, 2012).

Hydroxyapatite (HA) is a calcium phosphate compound with a strong compositional similarity to bone (Bohner et al., 2012; Bose and Tarafder, 2012), and with a proven osteoinductive and osteoconductive properties (Bose and Tarafder, 2012), forming strong bonds with the surrounding bone tissue. As a biomaterial in regenerative medicine, HA has been found to have the lowest resorption rate amongst calcium phosphates (Gbureck, et al., 2007), in the order of years, causing mechanical incompatibility at the implantation site (Bohner et al., 2012). To mitigate the slow degradation rate, HA is typically used in conjunction with other faster-resorbing calcium phosphates to produce so-called biphasic materials, or with polymers to produce composites. For example, mixtures of HA with tricalcium phosphate (β-TCP) are commonly used commercially as bone fillers (Bone Save, Stryker,

Table 11.3   Bioceramics and bioactive glasses for bone substitutes

Class	Material name	Chemical formula	Main resorption	References
Calcium Phosphates	Calcium polyphosphate (CPP)	$[Ca(PO_3)_2]_n$	Cell mediated dissolution (moderate)	(a)
	Dicalcium phosphate (DCP)	$CaHPO_4$	Chemical dissolution (fast)	(b)
	Dicalcium phosphate dihydrate (DCPD)	$CaHPO_4*2H_2O$	Chemical dissolution (fast)	(c)
	Octacalcium phosphate (OCP)	$Ca_8(HPO_4)_2(PO_4)_4*5H_2O$	Cell mediated dissolution (slow)	(d)
	Tricalcium phosphate (TCP)	$Ca_3(PO_4)_2$	Cell mediated dissolution (moderate)	(e)
	Hydroxyapatite (HA)	$Ca_{10}(PO4)_6(OH)_2$	Cell mediated dissolution (slow)	(f)
	Tetracalcium phosphate (TTCP)	$Ca_4P_2O_9$	Cell mediated dissolution (moderate)	(g)
Calcium Carbonate	Calcium carbonate	$CaCO_3$	Chemical dissolution or cell mediated	(h)
Calcium Sulfates	Hemihydrate (Plaster of Paris)	$CaSO_4*0.5H_2O$	Chemical dissolution	(i)
	Dihydrate (Gypsum)	$CaSO_4*2H_2O$	Chemical dissolution	(i)
Bioactive Glasses	Bioglass® 45S5	$Na_2O(24.5\%) \, CaO \,(24.5\%) \, SiO2 \,(45\%) \, P_2O_5 \,(6\%)$	Partial dissolution (very low)	(j)
	Bioglass 58S	$CaO \,(36\%) \, SiO_2 \,(60\%) \, P_2O_5 \,(4\%)$	Partial dissolution (low)	(k)
	Borate glass 13-93B3	$Na_2O \,(5.8\%) \, K_2O \,(11.7\%) \, MgO \,(4.9\%) \, CaO \,(19.5\%) \, SiO_2 \,(34.4\%) \, B_2O_3 \,(19.9\%)$	Partial dissolution (low)	(l)
	Phosphate glass $P_{50}C_{35}N_{15}$	$Na_2O \,(9.3\%) \, CaO \,(19.7\%) \, P_2O_5 \,(71.0\%)$	Partial dissolution (moderate)	(m)

(a) (Pilliar et al., 2013; Kandel et al., 2006; Grynpas et al., 2002; Pilliar et al., 2007; Shanjani et al., 2013) (b) (Gbureck, Hölzel, Klammert, et al., 2007); (c) (Gbureck, Hölzel, Klammert, et al. 2007; Alge et al., 2012); (d) (Anada et al., 2008); (e) (Kim et al., 2014; Friesenbichler et al., 2014); (f) (Bohner et al., 2012; Bose and Tarafder, 2012); (g) (Moseke and Gbureck, 2010); (h) (Bohner, 2010; Bohner et al., 2012); (i) (Bohner, 2010); (j) (Rahaman et al., 2011; Gerhardt and Boccaccini, 2010); (k) (Pereira et al., 2014); (l) (Rahaman et al., 2011; Gerhardt and Boccaccini, 2010); (m) (Rahaman et al., 2011; Gerhardt and Boccaccini, 2010)

80%TCP, 20% HA) and experimentally in literature, where ceramic mixtures of HA and β-TCP (Bose and Tarafder, 2012; Sánchez-Salcedo et al., 2008; Maté-Sánchez de Val et al., 2014; Kobayashi and Murakoshi, 2014) or β-TCP and nanoscale HA (Shuai et al., 2013) were used to take advantage of the intrinsic strength provided by HA, and the resorbable properties of β-TCP. Octacalcium phosphate (OCP) has been known to transition to HA (Anada et al., 2008) and is further resorbed as HA.

Dicalcium phosphate (DCP) and dicalcium phosphate dihydrate (DCPD), commonly referred to as monetite and brushite respectively, are generally resorbed by direct chemical dissolution, and have some of the higher resorption rates among the calcium phosphates (Bohner et al., 2012). It has been found that DCP has a higher resorption rate *in vivo* when compared to DCPD, as hydroxyapatite is formed and precipitated during the degradation process for DCPD (Gbureck et al., 2007). These materials by themselves are brittle, with mechanical properties below the required limits for bone augmentation (Alge et al., 2012). To mitigate the fast resorption rate of this class of calcium phosphates, research groups have used DCPD as a cement, reinforced with a polymer, for example poly(propylene fumarate) (Alge et al., 2012), to increase mechanical properties, while maintaining biocompatibility and osteointegration capabilities.

Tricalcium phosphate (TCP) is generally produced in three allotropic forms, β-TCP, α-TCP, and α'-TCP, differing in crystalline structure and mechanical properties (D. Liu et al., 2013). The β-TCP formulation is appropriate for bone scaffolding due to the biological response and mechanical properties of this allotropic phase, with care given to avoid the α-phase during sintering or material preparation (D. Liu et al., 2013). β-TCP has been used as granules, as an injectable paste (Matsuno et al., 2008), or mixed with HA as it has proven osteoconductive and resorbable properties (Vorndran et al., 2008). Products such as the Integra MozaikStrip™ porous scaffolds (Integra, 80% TCP, 20% Type I collagen) and Integra MozaikPutty™ (Integra, 80% TCP, 20% Type I collagen) are commercially available for repairing small bone defects and voids. Ongoing research focuses on manufacturing methodologies that would enable TCP to be used in load bearing applications (Vorndran et al., 2008) as well as on biological augmentation using embedded biomolecules or cells to improve osteointegration (Kim et al., 2014). TCP is a very popular bioceramic that promotes long-term osteointegration at a rate similar to allografts (Kim et al., 2014). Concerns have been raised over products such as GeneX® paste composed of calcium sulfate and β-TCP due to inflammation response (Friesenbichler et al., 2014).

For calcium polyphosphate (CPP), the decreased Ca:P ratio allows the molecules to organize in a chain-like configuration, similar to a polymer, where the oxygen-bridged links can form linear structures (polyphosphates), ring structures (metaphosphates), or cage structures (ultraphosphates) with a random distribution in an amorphous state or in an organized distribution in a crystalline state (Pilliar et al., 2001). CPP can be sintered at different temperatures to produce a range of crystalline phases, with decreasing rates of degradation in order of CPP > α-CPP > β-CPP > γ-CPP (Qiu et al., 2006). CPP has been proven to be osteoconductive and osteoinductive *in vivo* (Kandel et al., 2006; Pilliar et al., 2007; Shanjani et al., 2013) via release of calcium and phosphate ions, with the main resorption mechanism being cell-mediated.

Aside from bioceramics, amorphous glasses, such as the bioactive glass, form a family of glass composites used in biomedical applications for bone augmentation or replacement (Rahaman et al., 2011). In spite of their inherent brittleness, bioactive amorphous glass materials have been extensively investigated, specifically the commercially available Bioglass® (45S5 glass) (Rahaman et al., 2011; Gerhardt and Boccaccini, 2010) due to their proven osteoconductivity and bonding to bone, as the material forms a HA-like layer once implanted at the site, which encourages bonds with the surrounding bone (Rahaman et al., 2011). Bioglass® is biocompatible; however, it has a limited resorption rate, making it difficult to tune the resorption rate with the rate of new bone formation (Bohner, 2010; Rahaman et al., 2011). Borate bioactive glasses are also biocompatible and show faster degradation rates; however, there are some concerns over the toxicity of the borate ions *in vitro* and *in vivo* (Rahaman et al., 2011). Phosphate bioactive glasses offer the possibility of tuning the resorption rate by changing the composition (Rahaman et al., 2011). Bioactive glasses generally have an inherent brittleness (Rahaman et al., 2011) and it is also difficult to form them into porous 3D bone substitute structures or scaffolds (Rahaman et al., 2011; Gerhardt and Boccaccini, 2010).

There is a general trend toward producing composites of bioceramic or bioactive glass mixed with natural or synthetic polymers in order to meet the complex mechanical requirements in terms of compressive, tensile, shear, and fatigue properties for load bearing applications, as well as maintaining the biocompatible and bioresorbable characteristic of the bone substitute. Natural polymers such as chitosan, agarose, or collagen as well as synthetic polymers such as poly(ε-caprolactone) (PCL), poly(lactide-co-glycolide) (PLGA), poly(L-lactic acid) (PLLA), poly(L-lactide) (PLL), and poly (propylene fumarate) (PPF) are commonly used. The diversity of research done on composites, such as HA/PCL (Dorj et al., 2013), HA/PLGA (Kim et al., 2006), HA/PLLA (Wei and Ma, 2004), HA/β-TCP/collagen (Maté-Sánchez de Val et al., 2014), HA/β-TCP/agarose (Sánchez-Salcedo et al., 2008), HA/chitosan (Zhang et al., 2003), β-TCP/PLLA (D. Liu et al., 2013), β-TCP/chitosan (Dessì et al., 2013), bioglass/PLL (Zhang et al., 2004), and many more, are popular and can be used to control the resorption rate and mechanical properties of the material system. The prolific work done in this field may suggest that the optimal bone substitute scaffold material, structure, and properties have not yet been achieved. There are still limitations in achieving an acceptable range in mechanical and biochemical properties for resorbable ceramic and ceramic composites (Bohner et al., 2012). The key to solving this issue lies not only in considering the material itself, but also in implementing the appropriate complex 3D internal architecture of the construct from a structural and morphological point of view (Bohner et al., 2012; Butscher et al., 2011).

11.5.2 AM OF BIOCERAMIC MATERIALS: SEVERAL TECHNIQUES, LIMITATIONS, AND OPPORTUNITIES

Materials with controlled porous internal properties and complex 3D external characteristics are referred to as designer structures (Hollister, 2005). AM approaches are being refined toward achieving the goal of fabricating such designer structures in the context of producing bone substitutes. AM methodologies allow for parts to be built incrementally, layer-by-layer, based on information provided from a computer-aided design (CAD) program (Leong et al., 2003; Hutmacher et al., 2004). AM technologies strive to have each layer built to have the specific morphological configuration that would result in the desired micro- and macrostructure of the final part. In essence, AM offers the possibility of automated manufacturing of highly reproducible custom-shaped parts with controlled internal structure and custom external 3D architecture (Hutmacher et al., 2004). There are various AM techniques that can be used to construct bioceramic bone substitutes, depending on the type of raw materials used. Based on the ASTM F2792-12a standard terminology for AM technologies (ASTM International 2012), the general categories of AM methodologies are: (i) liquid-based methods such as direct light processing (DLP) and stereolithography (SL); (ii) solid or slurry extrusion-based methods such as fused deposition modeling (FDM) and low temperature deposition modeling (LDM); and (iii) powder-based methods such as SLS, DMLS, EBM, 3DP. These methods are summarized in Figure 11.5.

11.5.2.1 Liquid-based AM Approaches

One class of AM systems relies on photopolymerization of liquid-based materials. In this AM fabrication approach, radiant energy is used to excite and initiate the polymerization process in low molecular

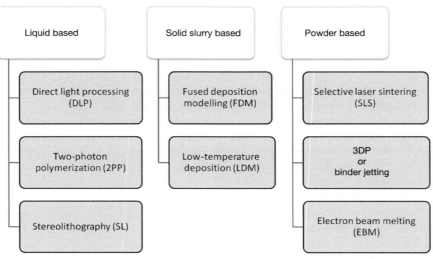

FIGURE 11.5

Classification of additive manufacturing methods for fabricating porous bioceramic bone substitutes.

weight monomers that are capable of forming long chain polymers (Yang et al., 2002). The family of polymers is called photopolymers. Biophotopolymers are a type of photopolymers that can biologically degrade by hydrolytic cleavage (Liska et al., 2007). A photopolymerization approach is DLP, where an array of micromirrors is used to direct light onto a photosensitive resin only at specific locations within each layer (Woesz et al., 2005; Skoog et al., 2014). SL is another photopolymerization method, where a UV laser moves in the x–y plane to irradiate and solidify a layer of UV curable polymer at specific locations (Lee et al., 2008; Skoog et al., 2014). The most common polymers used in SL for load bearing applications are poly(propylene fumarate) (PPF) (typically used with a diethylfumarate (DEF) cross-linking agent) and poly(D, L-lactide) (PDLLA), as they are biocompatible and bioresorbable, with attractive mechanical properties after cross-linking, specifically in producing bone substitutes for load bearing applications (Skoog et al., 2014). Skoog et al. (Skoog et al., 2014) have provided an excellent overview of the current state of the art on SL in tissue engineering, specifically for bone substitutes. They have outlined the benefits of SL in having a high degree of control over the internal and external architecture of the scaffolds, as well as the possibility of integrating bioactive ceramics fillers into the polymer matrix to enhance the mechanical and biological properties of the composite, at the cost of increasing the difficulty of SL processing, as the resulting resins are more viscous and more difficult to work with. For instance, HA powder was used with PPF/DEF (Lee et al., 2009) to produce intricate scaffolds with increased bioactivity. The most common use of SL in fabricating bone substitutes via indirect fabrication, where a mold is produced via SL and then infiltrated with bioceramic slurries such as β-TCP (X. Li et al., 2007) or HA/β-TCP (Sánchez-Salcedo et al., 2008) followed by a postprocessing protocol intended to remove the intermediate mold and to cure the bioceramics (Skoog et al., 2014). In general, photopolymerization is one of the most accurate AM techniques, with a high resolution, but it is limited to using photopolymeric materials (Stevens et al., 2008) and is generally used to manufacture bioceramic bone substitutes via indirect manufacturing by producing a mold.

11.5.2.2 Solid- or Slurry-based AM Approaches

Solid- or slurry-based AM methodologies rely on extrusion of molten materials or solids suspended in slurries. The most common method is FDM. A typical FDM machine deposits molten materials extruded through a heated nozzle displaced in the x–y plane to produce one layer following a specific geometric path based on CAD data. The layer is then displaced in the z-direction and a new layer can be built sequentially (Hutmacher et al., 2004). LDM and robocasting are methods that have a non heating liquefying deposition head to deposit a slurry that solidifies due to a low-temperature manufacturing environment (Xiong et al., 2002) or due to chemical setting (Dorj et al., 2013), respectively. Dorj et al. (Dorj et al., 2013) used a robocasting technique to prepare PCL/HA composites with increased mechanical properties by incorporating positively charged carbon nanotubes into the slurry-producing samples with a pore size of ~230 μm, with a compressive strength of up to 50 MPa, showing a considerable increase in mechanical properties with the addition of the carbon nanotubes. Liu et al. (Liu et al., 2009) used a PLGA/TCP slurry, while Xiong et al. (Xiong et al., 2002) employed PLLA/TCP slurry in the LDM process to produce bone substitutes with promising biological and mechanical properties. The most important benefit of these methods is the fact that there is a low risk of residual build materials being trapped inside the part. There is, however, a limited selection of materials that can be used for fabrication using these approaches (Hutmacher et al., 2004). These techniques also require the integration of extra support structures for building complex shapes with overhanging features. Using these methods, the nozzle dimension is one of the key limiting parameters that influence the feature size, as well as in dictating the speed of fabrication of the components.

11.5.2.3 Powder-based AM Approaches

Another class of AM techniques applies to powder-based materials. One such method is SLS, where a laser beam, typically a CO_2 laser, is used to locally raise the temperature of composite powders up to the glass transition temperature, resulting in fusion of neighboring particles and previous deposition layer (Yang et al., 2002). The powder can be a polymer, ceramic, metal, or composites (Leong et al., 2003). The laser is displaced in the x–y plane to create a layer based on CAD slice data. A roller mechanism spreads a new layer of powder and the process is repeated. The benefit of SLS is that there is generally no need for postprocessing steps. The process also allows for a wide variety of materials to be used.

For load bearing applications, researchers have looked at ways of using SLS to produce bioceramic or bioactive glass bone substitutes with appropriate mechanical and structural properties. For bioceramics, the task at hand is to enable the laser beam to bring the exposed powder ceramic material up to the sintering temperature to fuse particles together, without creating undesirable crystalline phases or inducing cracks in the part. Liu et al. (D. Liu et al., 2013) have investigated the appropriate material composition of a β-TCP powder blended with small amounts of PLLA (0.5–3 wt%) that would induce a liquid phase during sintering to decrease the sintering temperature of β-TCP, in order to avoid α-TCP phase transitions, achieving an increase of 18.8% in fracture toughness (1.43 ± 0.02 MPa m$^{1/2}$) and 4.5% in compressive strength (17.67 ± 0.04 MPa) when compared to the pure sintered β-TCP. Shuai et al. have also looked at improving the mechanical properties of pure β-TCP constructs by changing the SLS fabrication parameters, resulting in 3.59 GPa hardness and 1.16 MPa m$^{1/2}$ fracture toughness (Shuai et al., 2013). SLS fabrication using bioactive glasses typically results in weaker constructs, intended for maxillofacial applications, such as Bioglass 58S and poly (D, L) lactide (PDLLA) scaffolds with a maximum reported compressive strength of 1.68 MPa (Pereira et al., 2014). Similarly, Bioglass® 45S5 bone substitutes have also been investigated using SLS, with special consideration given to phase transitions in the glass material during sintering, targeting the $Na_2Ca_2Si_3O_9$ final phase (J. Liu et al., 2013) and resulting in a maximum fracture toughness of 0.57 MPa m$^{1/2}$.

The major disadvantage of SLS techniques is that it is difficult to remove the powder trapped within the parts, limiting the overall feature size that can be achieved using this method (Leong et al., 2003). Also, because of the high temperature involved during the manufacturing process, the materials may thermally degrade or produce undesirable phases, reducing the mechanical strength or biological compatibility of the part (Leong et al., 2003; D. Liu et al., 2013). High temperatures will also lead to unwanted bonding among adjoining powder particles, decreasing resolution, accuracy, and repeatability, and making it difficult to build features less than 400 μm (Yang et al., 2002). The feature size is limited by the laser spot size, and generally the SLS methodology suffers from slow fabrication rates. In addition, SLS systems are inherently expensive as they employ a laser system.

3DP, also known as adhesion bonding or binder jetting, is another category of powder-based AM methodologies (Yang et al., 2002). 3DP allows parts to be manufactured in a layer-by-layer fashion based on information provided by a software program. A 3D model is first designed in CAD software and then sectioned into bottom-up images corresponding to parallel slices or layers. 3DP employs powders as the substrate material for constructing the designed parts in a layer-by-layer fashion (Butscher et al., 2011). Within each layer, the powder particles are first spread onto a building area using a counter-rotating roller and then glued together at locations corresponding to the slice layer image by injecting a binder using an ink-jet printhead or similar printing technology (Cima et al., 1993, 1994. 1995). The building area moves down, allowing for new layers of powder to be spread sequentially until the part is completed. The resulting product is referred to as the green part. Further postprocessing is typically required to fully cure or anneal the green parts (Hutmacher, 2000; Yang et al., 2002).

3DP is a very popular methodology in manufacturing bone substitutes using bioceramic and bioglass powders. Typically, in this methodology, the bioceramic material is either coated or mixed with another polymer that can dissolve when exposed to the printed liquid binder and act as a glue to enhance the green part strength. Calcium phosphates have been widely used in 3DP by different groups as the powder material, where researchers have used CPP blended with a small wt% of PVA powder (Shanjani et al., 2011), where in a subsequent study Hu et al. (Hu et al., 2014) reported an increase in compressive strength from 21 ± 4.5 MPa for conventionally fabricated CPP/PVA samples to 50.2 ± 4.74 MPa for 3DP CPP/PVA samples, with good osteointegration *in vivo* (Shanjani et al., 2013). Instead of bioceramic/polymer powder blending, other research groups have used polymer coatings of ceramic powders such as HA with a maltodextrin coating (Chumnanklang et al., 2007), resulting in porosity ranges between 59–65% and an upper limit on flexural strength of approximately 1.5 MPa. Another approach is to set the bioceramic material into a cement by jetting a reagent, for example phosphoric acid to cement TCP (Gbureck et al., 2008, 2007, 2007), with reported compressive strength of up to 21.7 ± 1.1 MPa, porosity of 38–44%, and good osteointegration (Habibovic et al., 2008).

A significant benefit of the 3DP methodology is that a wide variety of materials in powder form such as polymers, ceramics, metals, and composites can be used to manufacture products (Butscher et al., 2011; Stevens et al., 2008; Yang et al., 2002; Castilho et al., 2011). Other key advantages include: ease of customizing, scaling, and reproducing complex-shaped designs with no need for support structures (Stevens et al., 2008; Hutmacher et al., 2004; Leong et al., 2003; Yang et al., 2002); green part manufacturing at room temperature (Hutmacher et al., 2004; Leong et al., 2003; Yang et al., 2002; Castilho et al., 2013, 2011); and digital control of 3D shape design and macro features such as embedded interconnected channels (Hutmacher, 2000; Stevens et al., 2008; Hollister, 2005; Hutmacher et al., 2004; Castilho et al., 2013). For some applications, such as tissue scaffold design, the inherent microporosity resulting from the arrangement of powder particles in the AM process is also a highly beneficial feature (Hollister, 2005; Hutmacher et al., 2004; Leong et al., 2003).

One of the most pressing drawbacks is that the smallest feature size of conventional powder-based 3DP techniques is limited by the binder droplet volume (Butscher et al., 2011), powder particle size with appropriate flowability (Butscher et al., 2011; Leong et al., 2003; Yang et al., 2001), powder compaction force (Shanjani, 2011), and the high potential for having trapped particles inside cavities formed within the part (Butscher et al., 2011; Hutmacher et al., 2004; Leong et al., 2003; Yang et al., 2002). Overall, it is difficult to achieve features below 500 μm in size using this fabrication method (Butscher et al., 2011; Hutmacher et al., 2004; Leong et al., 2003; Yang et al., 2002; Castilho et al., 2011). This issue becomes even more pressing in manufacturing constructs with complex conformal channels, as it becomes increasingly difficult to remove trapped support materials from features within the parts. Other emerging limiting aspects of conventional powder-based 3DP techniques are the use of only one powder size or powder type during manufacturing, the application of a constant compaction force during powder layer spreading, the utilization of a single layer thickness setting throughout the part, and the lack of control over the grayscale gradient of binder volume dispersed within each layer. These aspects impose limitations in manufacturing of porous scaffolds with heterogeneous or functionally graded properties as required in various industrial and biomedical applications.

To address the current limitations in 3DP technology, a mechatronic system was designed by the authors to control the fabrication of functionally graded internal features, porosities, and material properties of parts. To this end, a variable porosity AM system via 3DP was developed as seen in Figure 11.6, where a collection of control modules were incorporated to achieve the required performance. Such controlled devices include a counter-rotating roller module, multiple supply bed selection and alignment module, sacrificial porogen particle insertion module, sacrificial polymer deposition module, UV curing module, and an environment control module. The sacrificial porogen particle insertion module and sacrificial polymer deposition module assemblies are used to produce a controlled porous feature size in the range of 100–500 μm, which can be achieved by selectively depositing either porogens or sacrificial polymeric structures on specific layers that can thermally disintegrate during postprocessing, leaving behind controlled porosities and/or interconnected channels. This approach allows for control of internal features by preventing loose support powder material from being trapped inside complex cavities of parts, with results available in literature (Vlasea et al., 2013; Vlasea and Toyserkani, 2013). The system has been tested in the context of fabricating porous bone substitutes using CPP bioceramic powder.

The newly developed multiscale 3DP system allows for the capability of selecting between three powder feed compartments during runtime to produce layers with different material composition at selected locations throughout the part. To the author's knowledge, such a system is not commercially available, nor is it discussed in the literature thus far. As a preliminary work presented as a case study here, to evaluate the performance of the multipowder system configuration, two powder particle sizes were selected, CPP bioceramic powder with 75–150 μm particle size (hereafter referred to as large), and <75 μm particle size (hereafter referred to as small), with a morphology and particle distribution shown in Figure 11.7. Three categories of parts were manufactured, with small, large, and dual (50/50) powder composition, respectively. The large CPP and small CPP powders were blended, respectively, with 10% polyvinyl alcohol (PVA) powder (Alfa Aesar, Ward Hill, MA) and with particle size <63 μm. The fabricated test parts were 4 mm in diameter and 6 mm tall, printed with a layer

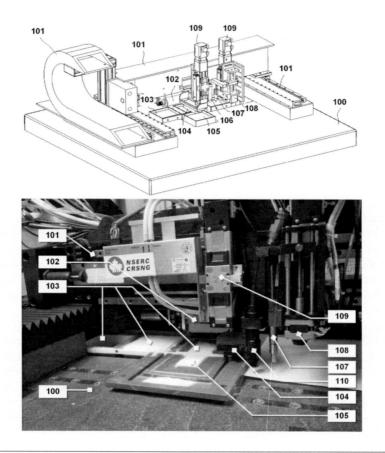

FIGURE 11.6

Novel multiscale 3DP additive manufacturing system for fabrication of functionally graded porous structures. (100) Granite support, (101) Precision *xy* gantry assembly, (102) Counter rotating roller module, (103) Multiple supply bed selection and alignment module, (104) Binder dispensing print head module, (105) Build bed module, (106) Sacrificial porogen particle insertion module, (107) Microsyringe sacrificial polymer deposition module, (108) Microsyringe deposition control, (109) Precision control *z*-axis, (110) UV curing module.

thickness of 150 μm. The green parts were then sintered in a high-temperature furnace (Lindberg/Blue M, ThermoScientific) with an established heat treatment protocol (Pilliar et al., 2009).

Figure 11.8a illustrates the compressive strength measurements plotted against porosity values, with statistical significance derived from one-way ANOVA. The mechanical compression test results were surprising. The category of samples with the highest compressive strength corresponded to parts manufactured using the large powder size, with a compressive strength of 15.5 ± 1.9 MPa, the result being corroborated by bulk density and bulk porosity measurements.

The literature suggests that samples with a lower particle size composition manufactured using 3DP approaches would result in parts with a higher mechanical compressive strength (Will et al., 2008). The discrepancy between expected results and the actual results obtained in this chapter can be explained

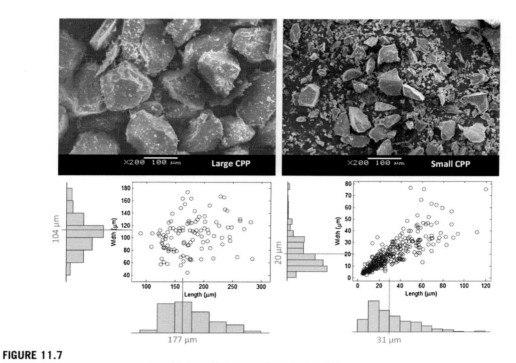

FIGURE 11.7

The particle morphology and corresponding size distribution for large particle size, sieved between 75–150 μm ($n = 105$) and small particle size, sieved at < 75 μm, ($n = 300$).

by looking at powder flowability, layer thickness, binder imbibition, and most noteworthy, the sintering protocol as a function of powder particle size. Small particles tend to have poor flowability (Butscher et al., 2011) and are inclined to agglomerate. This effect causes limited powder compaction during layer spreading (Shanjani et al., 2011) and may explain the poor mechanical strength of the samples manufactured using powder with small particle size in this chapter. The parts with the lowest performance were the dual powder composition samples. This may have been caused by stresses occurring at the interface between the two powder types during sintering, as indicated by the fracture patterns occurring at that location. Another aspect that influences part strength is the sintering process. In this work, all samples were subjected to the same sintering protocol, optimized for the 75–150 μm powder particle range. Figure 11.8b illustrates examples of sintered CPP samples corresponding to the three categories of manufactured parts. It can be seen that the parts have a significantly different color, indicating that sintering had a different effect on samples, depending on powder size composition. A more in-depth investigation into an appropriate sintering protocol for samples with different powder size layer composition should be investigated.

The important achievement of this work is in demonstrating the feasibility of using multiple powders during the multiscale 3DP manufacturing process, which advances the current state of powder-based AM toward new manufacturing opportunities. Further investigations into increasing the performance of this approach are currently underway.

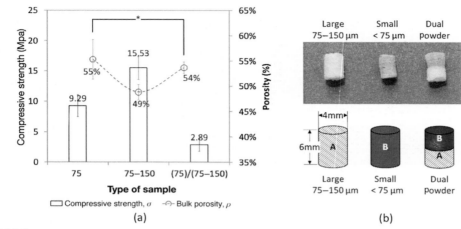

FIGURE 11.8

(a) Bulk porosity characteristics and compressive strength measurements of cylindrical samples with small, large, and dual particle size composition, with ($p < 0.05$). For each trial, ($n = 10$). (*) Illustrates Tukey HSD statistical pairwise similarity. (b) Sintered samples with large, small, and dual powder composition.

11.6 NANOCOMPOSITE BONE SUBSTITUTES
11.6.1 NANOMATERIALS, LIMITATIONS, AND OPPORTUNITIES

Bone is inherently a nanocomposite, which serves as a mechanical support to the body and also a reservoir for essential minerals (James et al., 2011; Murugan and Ramakrishna, 2005).There have been many studies on the development of synthetic bone substitutes such as ceramic-based and polymer-based materials. In addition to certain requirements of these materials like high mechanical strength, biodegradability, and biocompatibility, they should be shaped into 3D porous structures to fit into the defected part of the bone. Osteoconductivity of the scaffolds is of great importance since it determines the rate of osteoblast cell infiltration and proliferation. The scaffolds should also be degraded with a similar rate to the bone healing process and the yield of degradation must not provoke immune responses in the body. Commonly, all these requirements are not obtained through a single-phase material and there is a significant need for the development of multiphase materials (i.e. composites with similar physical and structural properties to bone). Bone is composed of a nanocomposite known as ECM which consists of an organic matrix, osteoid, and an inorganic phase as reinforcement. Fibrous protein and collagen are the main components of the organic matrix, whereas the inorganic phase is mainly made of carbonate-substituted hydroxyapatite (HA) crystals. The mineral is not pure HA and some traces of magnesium, carbonate, and zinc elements are mostly detected in the structure. In addition, calcium-to-phosphorous ratio is not constant and might undergo significant variations at the same bone site, but in different populations or age groups (Rogel et al., 2008; Crystals et al. 2006).

For bone grafting, ceramic materials, such as calcium phosphates and hydroxyapatite, are conventionally employed as fillers and coatings. Although they provide good osteoconductivity, their scaffolds are brittle and undergo detrimental failure under load. They also suffer from slow biodegradability, which impedes an efficient bone healing process. Polymeric materials have been investigated widely as bone grafts due to their high biocompatibility and proper mechanical strength. Polylactic acid (PLA), polyglycolic acid (PGA), poly(ε-caprolactone) (PCL), and their copolymers are the most common polymers used for tissue engineering. It has been reported that ceramic/polymer composites represent improved biological and mechanical properties. Incorporation of synthetic nanocomposites along with a proper micro- and macrostructural design can further enhance the performance of bone grafts. Higher surface area and higher grain boundaries can improve the toughness and mechanical properties of the structures significantly. Additionally, nanocomposites provide enhanced osteoblast cell adhesion and proliferation. There are many reports on the development of ceramic, polymeric, and metallic nano-composites for load-bearing applications. However, manufacturing of porous structures with proper interconnectivity and strength is still challenging and needs more in-depth investigation.

11.6.2 AM OF NANOCOMPOSITES: SEVERAL TECHNIQUES, LIMITATIONS AND OPPORTUNITIES

Zhou et al. (Zhou et al. 2008) reported the synthesis of PLLA/HA nanocomposite with the HA particle size of around 20 nm. The scaffolds were built using SLS at the layer thickness of 100 μm. It was shown that using inappropriate laser power, part bed temperature (PBT) and scan spacing (SS) can lower the resolution and quality of 3D printed structures. *In vitro* cell culture experiments indicated promising results in regards to osteoconductivity and proliferation of osteoblast cells.

Heo et al. (Heo et al., 2009) studied layer-by-layer manufacturing of HA/PCL composite fabricated with nano (n-HPC) and micro HA (m-HPC) particles, respectively, prepared through a solvent casting method. 3D-printed scaffolds were freeze-dried after immersion in distilled water. The SEM images of the n-HCP and m-HCP scaffolds reveal that the n-HCP scaffolds have a smooth surface while that of m-HCP is rough with less consistency. It was also shown that the attachment of cells, alkaline phosphatase activity, calcium content, and mechanical strength were higher in n-HPC compared to that of m-HPC scaffolds.

In another study, calcium phosphate /poly(hydroxybutyrateco-hydroxyvalerate) (PHBV) and HA/PLLA nanocomposites were synthesized and 3D-printed into scaffolds using SLS (Duan and Wang 2010a, 2010b; Duan et al., 2010, 2011). Osteoconductivity and biodegradability are provided by calcium phosphate and polymer matrix, respectively. Proximal femoral condyle sintered scaffolds show high quality prints compared to the model. The mechanical properties of the nanocomposite scaffolds were shown to be higher than their polymeric counterparts in dry state. It was also reported that SaOS-2 osteoblast cell adhesion, proliferation, and ALP activity of the structures were significantly improved using the nanocomposite.

Poly-lactide-co-glycolide acid (PLGA) and nanotitania particles are also known as biocompatible materials. Liu et al. (Liu and Webster, 2011) fabricated titania/PLGA nanocomposite dispersion via a solvent casting technique and used aerosol-based M3D™ 3D printing system (M3D™ developed by OPTOMEC) to produce 3D scaffolds of titania/PLGA nanocomposite. The results indicated better mechanical properties and biological performance than PLGA control samples.

Graphene has been reported to improve mechanical properties and osteoblast cell adhesion and proliferation significantly (Zhang et al., 2013; Zanin et al., 2013; Marques et al., 2012; Biris et al., 2011; Ma et al., 2012; Liu et al., 2012; Rodríguez-Lorenzo, 2009). Azhari et al. (Azhari et al., 2014) investigated the AM of graphene/hydroxyapatite nanocomposite using layer-by-layer 3D-printing technique. As illustrated in Figure 11.9, the rotating roller spread a layer of powder onto the building bed where an aqueous binder is injected to adhere particles based on the sliced CAD model. The process is iterated until the object is built up. Graphene oxide was mixed with HA particles through a wet mixing process

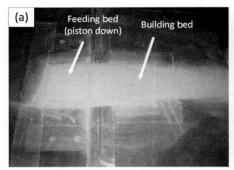
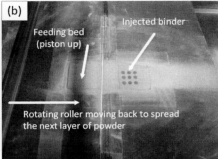

FIGURE 11.9

Fabrication process of 3D printed structures by powder-bed AM, (a) feeding bed and building bed are indicated, (b) the compartment moves forward to inject the binder on the building bed, one layer of binder is injected on the powder bed, compartment moves backward to spread another layer of powder onto the building bed.

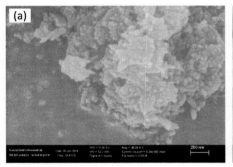

FIGURE 11.10

SEM images of HG4 nanocomposite, (a) graphene-coated Hap particles; (b) showing higher magnifications. The nanoparticles seen in (b) could be attributed to agglomerated graphene oxide sheets.

to fabricate HA/graphene nanocomposite. SEM images of the nanocomposite powder show that HA flakes are fully decorated by graphene particles (see Figure 11.10). This texture can lower the van der Waals attractive forces between the particles. It was shown that the flowability of the nanocomposite powder, which is an essence in the manufacturing process, was improved significantly by the addition of graphene. The samples printed at the layer thickness of 125 μm with the shell binder saturation (SBS) to core binder saturation level (CBS) of 100/400% showed the highest mechanical strength. As seen in Figure 11.11, the compressive strength of the 3D-printed cylinders was increased from ~0.1 to ~7.0 MPa in the samples with the graphene content of only 0.4 wt%.

In summary, it was realized that the flowability of HA particles decorated with graphene was improved significantly. This phenomenon will enhance the resolution and quality of the structures 3D-printed through AM. It was also shown that the mechanical strength of HG4 green specimens was almost 70 times more than HG0 specimens. In future efforts to employ the samples for bioimplantation, the mechanical properties and biocompatibility of the sintered samples should be investigated thoroughly.

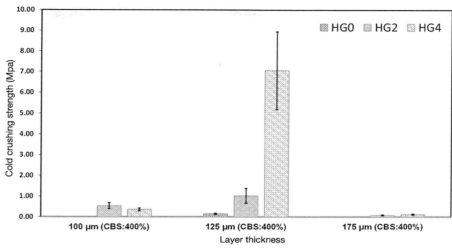

FIGURE 11.11

Cold crushing strength of HG0, HG2, and HG4 specimens 3D-printed at layer thicknesses of 100, 125, and 175 μm with SBS/CBS ratio of 100/400%.

11.7 CONCLUSIONS

Metals still form the bulk of primary bone substitute materials in the orthopedic industry. Interest in metal implants will remain strong in the future, especially for senior patients. The fabrication of metal implants is costly, but with new developments in the AM field, manufacturability may become easier. Attention has been focused on the use of bioceramics in the fabrication of bone substitutes due to their biochemical properties. To improve on the mechanical properties of bioceramics, polymer matrix composites are typically used in various AM approaches. Owing to the high degree of similarity between bone microstructure as a nanocomposite, biocompatible nanocomposites are also arousing interest in bone regenerative medicine and have been employed using AM approaches. Overall, AM methodologies offer the possibility of producing complex bone substitute implants, specifically for load bearing applications, with a complex 3D external shape, and intricate porous internal architecture. The extensive ongoing research in the field of materials and manufacturing methodologies suggests that there are still barriers that need to be overcome to achieve consistently reliable and safe substitutes in regenerating or augmenting bone function.

REFERENCES

Alge, D.L. et al., 2012. Poly(propylene fumarate) reinforced dicalcium phosphate dihydrate cement composites for bone tissue engineering. *Journal of biomedical materials research. Part A*, 100(7), pp. 1792–802. Available at: http://www.ncbi.nlm.nih.gov/pubmed/22489012. [Accessed June 15, 2012].

Alvarez, K. & Nakajima, H., 2009. Metallic Scaffolds for Bone Regeneration. *Materials*, 2(3), pp. 790–832. Available at: http://www.mdpi.com/1996-1944/2/3/790/. [Accessed May 27, 2014].

Anada, T. et al., 2008. Dose-Dependent Osteogenic Effect of Octacalcium Phosphate on Mouse Bone Marrow Stromal Cells. *Tissue Engineering Part A*, 14(6), pp. 965–978. Available at: http://www.liebertonline.com/doi/abs/10.1089/ten.tea.2007.0339. [Accessed June 15, 2014].

ASTM International, 2012. *ASTM F2792-12a Standard Terminology for Additive Manufacturing Technologies*, Available at: http://www.astm.org/Standards/F2792.htm.

Azhari, A. et al., 2014. Additive Manufacturing of Graphene–Hydroxyapatite Nanocomposite Structures. *International Journal of Applied Ceramic Technology*, pp. 1–10. Available at: http://onlinelibrary.wiley.com/doi/10.1111/ijac.12309/full.

Balla, V.K. et al., 2010. Porous tantalum structures for bone implants: fabrication, mechanical and in vitro biological properties. *Acta biomaterialia*, 6(8), pp. 3349–59. Available at: http://www.pubmedcentral.nih.gov/articlerender.fcgi?artid = 2883027&tool = pmcentrez&rendertype = abstract. [Accessed June 20, 2014].

Bartel, D.L., Davy, D.T., Keaveny, T.M., 2006. Orthopaedic biomechanics: mechanics and design in musculoskeletal systems First. Pearson Prentice Hall, Upper Saddle River.

Basalah, A. et al., 2012. Characterizations of additive manufactured porous titanium implants. *Journal of biomedical materials research. Part B, Applied biomaterials*, 100(7), pp. 1970–9. Available at: http://www.ncbi.nlm.nih.gov/pubmed/22865677. [Accessed June 20, 2014].

Billi, F. & Campbell, P., 2010. Nanotoxicology of metal wear particles in total joint arthroplasty: a review of current concepts. *Journal of applied biomaterials & biomechanics: JABB*, 8(1), pp. 1–6. Available at: http://www.ncbi.nlm.nih.gov/pubmed/20740415.

Biris, A.R. et al., 2011. Novel Multicomponent and Biocompatible Nanocomposite Materials Based on Few-Layer Graphenes Synthesized on a Gold/Hydroxyapatite Catalytic System with Applications in Bone Regeneration. *The Journal of Physical Chemistry C*, 115(39), pp. 18967–18976. Available at: http://pubs.acs.org/doi/abs/10.1021/jp203474y.

Bohner, M. et al., 2011. Commentary: Deciphering the link between architecture and biological response of a bone graft substitute. *Acta biomaterialia*, 7(2), pp. 478–84. Available at: http://www.ncbi.nlm.nih.gov/pubmed/20709195. [Accessed November 15, 2012].

Bohner, M., 2010. Resorbable biomaterials as bone graft substitutes. *Materials Today*, 13(1–2), pp. 24–30. Available at: http://linkinghub.elsevier.com/retrieve/pii/S1369702110700146. [Accessed May 1, 2014].

Bohner, M., Galea, L. & Doebelin, N., 2012. Calcium phosphate bone graft substitutes: Failures and hopes. *Journal of the European Ceramic Society*, 32(11), pp. 2663–2671. Available at: http://linkinghub.elsevier.com/retrieve/pii/S0955221912001021. [Accessed October 26, 2012].

Bose, S. & Tarafder, S., 2012. Calcium phosphate ceramic systems in growth factor and drug delivery for bone tissue engineering: A review. *Acta biomaterialia*, 8(4), pp. 1401–21. Available at: http://www.ncbi.nlm.nih.gov/pubmed/22127225. [Accessed March 16, 2012].

Butscher, A. et al., 2011. Structural and material approaches to bone tissue engineering in powder-based three-dimensional printing. *Acta Biomaterialia*, 7(3), pp. 907–20. Available at: http://www.ncbi.nlm.nih.gov/pubmed/20920616. [Accessed March 8, 2013].

Cai, S. & Xi, J., 2008. A control approach for pore size distribution in the bone scaffold based on the hexahedral mesh refinement. Computer-Aided Design, 40(10–11), pp. 1040–1050. Available at: http://linkinghub.elsevier.com/retrieve/pii/S0010448508001838. [Accessed May 21, 2014].

Campanelli, S., et al., 1994. Capabilities and Performances of the Selective Laser Melting Process. In: Meng, J. (Ed.), New Trends in Technologies: Devices, Computer. Communication and Industrial Systems, pp. 233–254.

Castilho, M. et al., 2013. Fabrication of computationally designed scaffolds by low temperature 3D printing. *Biofabrication*, 5(3), p.035012. Available at: http://www.ncbi.nlm.nih.gov/pubmed/23887064. [Accessed September 28, 2013].

Castilho, M. et al., 2011. Structural evaluation of scaffolds prototypes produced by three-dimensional printing. *The International Journal of Advanced Manufacturing Technology*, 56(5–8), pp. 561–569. Available at: http://link.springer.com/10.1007/s00170-011-3219-4. [Accessed March 26, 2013].

Chang, B.S. et al., 2000. Osteoconduction at porous hydroxyapatite with various pore configurations. *Biomaterials*, 21(12), pp. 1291–8. Available at: http://www.ncbi.nlm.nih.gov/pubmed/19845154.

Cheng, S. et al., 2013. *Hip and knee replacements in Canada: Canadian joint replacement registry 2013 annual report*, Ottawa. Available at: https://secure.cihi.ca/free_products/CJRR_2013_Annual_Report_EN.pdf.

Chumnanklang, R. et al., 2007. 3D printing of hydroxyapatite: Effect of binder concentration in pre-coated particle on part strength. *Materials Science and Engineering: C*, 27(4), pp. 914–921. Available at: http://linkinghub.elsevier.com/retrieve/pii/S092849310600378X. [Accessed August 22, 2013].

Cima, M. et al., 1995. Three Dimensional Printing, Patent Number 5,387,380.

Cima, M.J. et al., 1994. Three Dimensional Printing Techniques, Patent Number 5,340,656.

Cima, M.J. et al., 1993. Three-dimensional printing techniques, Patent Number 5,204,055.

Crystals, N.C.H. et al., 2006. Collagen Scaffolds Reinforced with Biomimetic Composite., 12.(9).

Curodeau, a, Sachs, E. & Caldarise, S., 2000. Design and fabrication of cast orthopedic implants with freeform surface textures from 3-D printed ceramic shell. *Journal of biomedical materials research*, 53(5), pp. 525–35. Available at: http://www.ncbi.nlm.nih.gov/pubmed/10984701.

Dabrowski, B. et al., 2010. Highly porous titanium scaffolds for orthopaedic applications. *Journal of biomedical materials research. Part B, Applied biomaterials*, 95(1), pp. 53–61. Available at: http://www.ncbi.nlm.nih.gov/pubmed/20690174. [Accessed June 20, 2014].

Dessì, M. et al., 2013. Novel biomimetic thermosensitive β-tricalcium phosphate/chitosan-based hydrogels for bone tissue engineering. *Journal of biomedical materials research. Part A*, 101(10), pp. 2984–93. Available at: http://www.ncbi.nlm.nih.gov/pubmed/23873836. [Accessed June 15, 2014].

Detsch, R. et al., 2010. The resorption of nanocrystalline calcium phosphates by osteoclast-like cells. *Acta biomaterialia*, 6(8), pp. 3223–33. Available at: http://www.ncbi.nlm.nih.gov/pubmed/20206720. [Accessed May 22, 2014].

Dinda, G.P., Song, L. & Mazumder, J., 2008. Fabrication of Ti-6Al-4V Scaffolds by Direct Metal Deposition. Metallurgical and Materials Transactions A, 39(12), pp. 2914–2922. Available at: http://link.springer.com/10.1007/s11661-008-9634-y. [Accessed June 20, 2014].

Dorj, B. et al., 2013. Robocasting nanocomposite scaffolds of poly(caprolactone)/hydroxyapatite incorporating modified carbon nanotubes for hard tissue reconstruction. *Journal of biomedical materials research. Part A*, 101(6), pp. 1670–81. Available at: http://www.ncbi.nlm.nih.gov/pubmed/23184729. [Accessed June 4, 2014].

Duan, B., Wang, M., Li, Z.Y., et al., 2010. Surface modification of three-dimensional Ca-P/PHBV nanocomposite scaffolds by physical entrapment of gelatin and its in vitro biological evaluation. *Frontiers of Materials Science*, 5(1), pp. 57–68. Available at: http://link.springer.com/10.1007/s11706-011-0101-0. [Accessed June 12, 2014].

Duan, B., Wang, M., Zhou, W.Y., et al., 2010. Three-dimensional nanocomposite scaffolds fabricated via selective laser sintering for bone tissue engineering. *Acta biomaterialia*, 6(12), pp. 4495–505. Available at: http://www.ncbi.nlm.nih.gov/pubmed/20601244. [Accessed June 12, 2014].

Duan, B., Cheung, W.L. & Wang, M., 2011. Optimized fabrication of Ca-P/PHBV nanocomposite scaffolds via selective laser sintering for bone tissue engineering. *Biofabrication*, 3(1), p.015001. Available at: http://www.ncbi.nlm.nih.gov/pubmed/21245522. [Accessed June 12, 2014].

Duan, B. & Wang, M., 2010a. Customized Ca-P/PHBV nanocomposite scaffolds for bone tissue engineering: design, fabrication, surface modification and sustained release of growth factor. *Journal of the Royal Society, Interface / the Royal Society*, 7 Suppl 5, pp.S615–29. Available at: http://www.pubmedcentral.nih.gov/articlerender.fcgi?artid = 3024573&tool = pmcentrez&rendertype = abstract [Accessed June 12, 2014].

Duan, B. & Wang, M., 2010b. Encapsulation and release of biomolecules from Ca–P/PHBV nanocomposite microspheres and three-dimensional scaffolds fabricated by selective laser sintering. *Polymer Degradation and Stability*, 95(9), pp. 1655–1664. Available at: http://linkinghub.elsevier.com/retrieve/pii/S0141391010002247. [Accessed June 12, 2014].

Friesenbichler, J. et al., 2014. Adverse reactions of artificial bone graft substitutes: lessons learned from using tricalcium phosphate geneX®. *Clinical orthopaedics and related research*, 472(3), pp. 976–82. Available at: http://www.ncbi.nlm.nih.gov/pubmed/24078171. [Accessed June 15, 2014].

Frosch, K.-H. & Stürmer, K.M., 2006. Metallic Biomaterials in Skeletal Repair. *European Journal of Trauma*, 32(2), pp. 149–159.

Frost, H.M., 2003. Bone's mechanostat: a 2003 update. *The anatomical record. Part A, Discoveries in molecular, cellular, and evolutionary biology*, 275(2), pp. 1081–101. Available at: http://www.ncbi.nlm.nih.gov/pubmed/14613308. [Accessed August 7, 2013].

Gbureck, U., Hölzel, T., Doillon, C.J., et al., 2007. Direct printing of bioceramic implants with spatially localized angiogenic factors. *Advanced Materials*, 19(6), pp. 795–800. Available at: http://doi.wiley.com/10.1002/adma.200601370. [Accessed August 20, 2013].

Gbureck, U. et al., 2008. Preparation of tricalcium phosphate/calcium pyrophosphate structures via rapid prototyping. *Journal of Materials Science: Materials in Medicine*, 19(4), pp. 1559–63. Available at: http://www.ncbi.nlm.nih.gov/pubmed/18236137. [Accessed August 20, 2013].

Gbureck, U., Hölzel, T., Klammert, U., et al., 2007. Resorbable Dicalcium Phosphate Bone Substitutes Prepared by 3D Powder Printing. *Advanced Functional Materials*, 17(18), pp. 3940–3945. Available at: http://doi.wiley.com/10.1002/adfm.200700019. [Accessed May 26, 2014].

Gerhardt, L.-C. & Boccaccini, A.R., 2010. Bioactive Glass and Glass-Ceramic Scaffolds for Bone Tissue Engineering. *Materials*, 3(7), pp. 3867–3910. Available at: http://www.mdpi.com/1996-1944/3/7/3867/. [Accessed May 30, 2014].

Gikas, P.D. et al., 2009. An overview of autologous chondrocyte implantation. Bone and Joint Surgery - British Volume, 91(8), pp. 997–1006. Available at: http://www.ncbi.nlm.nih.gov/pubmed/19651824. [Accessed July 29, 2010].

Grynpas, M.D. et al., 2002. Porous calcium polyphosphate scaffolds for bone substitute applications in vivo studies. *Biomaterials*, 23(9), pp. 2063–70. Available at: http://www.ncbi.nlm.nih.gov/pubmed/11996048.

Habibovic, P. et al., 2008. Osteoconduction and osteoinduction of low-temperature 3D printed bioceramic implants. *Biomaterials*, 29(7), pp. 944–53. Available at: http://www.ncbi.nlm.nih.gov/pubmed/18055009. [Accessed August 20, 2013].

Heo, S. et al., 2009. In Vitro and Animal Study of Novel Nano- Scaffolds Fabricated by Layer Manufacturing Process., 15(5).

Hollister, S.J., 2005. Porous scaffold design for tissue engineering. *Nature materials*, 4(7), pp. 518–24. Available at: http://www.ncbi.nlm.nih.gov/pubmed/16003400.

Hu, Y. et al., 2014. Porous calcium polyphosphate bone substitutes: additive manufacturing versus conventional gravity sinter processing-effect on structure and mechanical properties. *Journal of biomedical materials research. Part B, Applied biomaterials*, 102(2), pp. 274–83. Available at: http://www.ncbi.nlm.nih.gov/pubmed/23997039. [Accessed April 14, 2014].

Hutchinson, M., 2009. The burden of musculoskeletal diseases in the united states: prevalance, societal and economic cost. *Journal of the American College of Surgeons*, 208(1), pp.e5-e6. Available at: http://linkinghub.elsevier.com/retrieve/pii/S1072751508007357.

Hutmacher, D.W., 2000. Scaffolds in tissue engineering bone and cartilage. *Biomaterials*, 21(24), pp. 2529–43. Available at: http://www.ncbi.nlm.nih.gov/pubmed/11071603.

Hutmacher, D.W., Sittinger, M. & Risbud, M.V., 2004. Scaffold-based tissue engineering: rationale for computer-aided design and solid free-form fabrication systems. *Trends in Biotechnology*, 22(7), pp. 354–62. Available at: http://www.ncbi.nlm.nih.gov/pubmed/15245908. [Accessed July 29, 2010].

James, R. et al., 2011. Nanocomposites and bone regeneration. *Frontiers of Materials Science*, 5(4), pp. 342–357. Available at: http://link.springer.com/10.1007/s11706-011-0151-3. [Accessed June 12, 2014].

Kandel, R.A. et al., 2006. Repair of osteochondral defects with biphasic cartilage-calcium polyphosphate constructs in a sheep model. *Biomaterials*, 27(22), pp. 4120–31. Available at: http://www.ncbi.nlm.nih.gov/pubmed/16564568. [Accessed July 29, 2010].

Karageorgiou, V. & Kaplan, D., 2005. Porosity of 3D biomaterial scaffolds and osteogenesis. *Biomaterials*, 26(27), pp. 5474–91. Available at: http://www.ncbi.nlm.nih.gov/pubmed/15860204. [Accessed August 14, 2013].

Kim, J. et al., 2014. In vivo performance of combinations of autograft, demineralized bone matrix, and tricalcium phosphate in a rabbit femoral defect model. *Biomedical materials (Bristol, England)*, 9(3), p.035010. Available at: http://www.ncbi.nlm.nih.gov/pubmed/24784998. [Accessed June 15, 2014].

Kim, S.-S. et al., 2006. Poly(lactide-co-glycolide)/hydroxyapatite composite scaffolds for bone tissue engineering. *Biomaterials*, 27(8), pp. 1399–409. Available at: http://www.ncbi.nlm.nih.gov/pubmed/16169074.

Kobayashi, S. & Murakoshi, T., 2014. Characterization of mechanical properties and bioactivity of hydroxyapatite / β -tricalcium phosphate composites. *Advanced Composite Materials*, 23(2), pp. 163–177.

Koh, J.L., 2004. The effect of graft height mismatch on contact pressure following osteochondral grafting: a biomechanical study. *American Journal of Sports Medicine*, 32(2), pp. 317–320. Available at: http://journal.ajsm.org/cgi/doi/10.1177/0363546503261730. [Accessed August 7, 2013].

Krishna, B.V., Bose, S. & Bandyopadhyay, A., 2007. Low stiffness porous Ti structures for load-bearing implants. *Acta biomaterialia*, 3(6), pp. 997–1006. Available at: http://www.ncbi.nlm.nih.gov/pubmed/17532277. [Accessed May 26, 2014].

Kujala, S. et al., 2003. Effect of porosity on the osteointegration and bone ingrowth of a weight-bearing nickel-titanium bone graft substitute. *Biomaterials*, 24(25), pp. 4691–4697. Available at: http://linkinghub.elsevier.com/retrieve/pii/S0142961203003594. [Accessed July 29, 2010].

Lee, J.W. et al., 2009. Development of nano- and microscale composite 3D scaffolds using PPF/DEF-HA and micro-stereolithography. *Microelectronic Engineering*, 86(4–6), pp. 1465–1467. Available at: http://linkinghub.elsevier.com/retrieve/pii/S0167931708006473. [Accessed June 18, 2014].

Lee, J.W. et al., 2008. Fabrication and characteristic analysis of a poly(propylene fumarate) scaffold using micro-stereolithography technology. *Journal of Biomedical Materials Research. Part B, Applied Biomaterials*, 87(1), pp. 1–9. Available at: http://www.ncbi.nlm.nih.gov/pubmed/18335437. [Accessed July 29, 2010].

Leong, K., Cheah, C.M. & Chua, C.K., 2003. Solid freeform fabrication of three-dimensional scaffolds for engineering replacement tissues and organs. *Biomaterials*, 24(13), pp. 2363–78. Available at: http://linkinghub.elsevier.com/retrieve/pii/S0142961203000309.

Li, J.P. et al., 2005. Porous Ti6Al4V scaffolds directly fabricated by 3D fibre deposition technique: effect of nozzle diameter. *Journal of materials science. Materials in medicine*, 16(12), pp. 1159–63. Available at: http://www.ncbi.nlm.nih.gov/pubmed/16362216.

Li, P.J. et al., 2007. Bone ingrowth in porous titanium implants produced by 3D fiber deposition. Biomaterials, 28(18), pp. 2810–2820.

Li, X. et al., 2007. Fabrication of bioceramic scaffolds with pre-designed internal architecture by gel casting and indirect stereolithography techniques. *Journal of Porous Materials*, 15(6), pp. 667–671. Available at: http://link.springer.com/10.1007/s10934-007-9148-9. [Accessed June 18, 2014].

Lin, C.Y., Kikuchi, N. & Hollister, S.J., 2004. A novel method for biomaterial scaffold internal architecture design to match bone elastic properties with desired porosity. *Journal of biomechanics*, 37(5), pp. 623–36. Available at: http://www.ncbi.nlm.nih.gov/pubmed/15046991. [Accessed April 28, 2014].

Lin, J.G. et al., 2009. Degradation of the strength of porous titanium after alkali and heat treatment. *Journal of Alloys and Compounds*, 485(1–2), pp. 316–319. Available at: http://linkinghub.elsevier.com/retrieve/pii/S0925838809009724. [Accessed May 30, 2014].

Liska, R. et al., 2007. Photopolymers for rapid prototyping. *Journal of Coatings Technology and Research*, 4(4), pp. 505–510. Available at: http://www.springerlink.com/index/10.1007/s11998-007-9059-3. [Accessed July 29, 2010].

Liu, D., et al., 2013a. Mechanical properties' improvement of a tricalcium phosphate scaffold with poly- L -lactic acid in selective laser sintering. Biofabrication 5, 1–10.

Liu, H. et al., 2012. Simultaneous Reduction and Surface Functionalization of Graphene Oxide for Hydroxyapatite Mineralization. *The Journal of Physical Chemistry C*, 116(5), pp. 3334–3341. Available at: http://pubs.acs.org/doi/abs/10.1021/jp2102226.

Liu, H. & Webster, T.J., 2011. Enhanced biological and mechanical properties of well-dispersed nanophase ceramics in polymer composites: From 2D to 3D printed structures. *Materials Science and Engineering: C*, 31(2), pp. 77–89. Available at: http://linkinghub.elsevier.com/retrieve/pii/S0928493110001761. [Accessed May 27, 2014].

Liu, J. et al., 2013. Fabrication and Characterization of Porous 45S5 Glass Scaffolds via Direct Selective Laser Sintering. *Materials and Manufacturing Processes*, 28(6), pp. 610–615. Available at: http://www.tandfonline.com/doi/abs/10.1080/10426914.2012.736656. [Accessed June 17, 2014].

Liu, L. et al., 2009. A Novel Osteochondral Scaffold Fabricated via Multi-nozzle Low-temperature Deposition Manufacturing. *Journal of Bioactive and Compatible Polymers*, 24(1 Suppl), pp. 18–30. Available at: http://jbc.sagepub.com/cgi/doi/10.1177/0883911509102347. [Accessed July 6, 2010].

Long, M., Rack, H.J., 1998. Titanium alloys in total joint replacement—a materials science perspective. Biomaterials 19, 1621–1639.

Ma, H. et al., 2012. Preparation and cytocompatibility of polylactic acid/hydroxyapatite/graphene oxide nanocomposite fibrous membrane. *Chinese Science Bulletin*, 57(23), pp. 3051–3058. Available at: http://link.springer.com/10.1007/s11434-012-5336-3. [Accessed September 5, 2013].

Marques, P. a. a. P. et al., 2012. Graphene Oxide and Hydroxyapatite as Fillers of Polylactic Acid Nanocomposites: Preparation and Characterization. *Journal of Nanoscience and Nanotechnology*, 12(8), pp. 6686–6692. Available at: http://openurl.ingenta.com/content/xref?genre=article&issn=1533-4880&volume=12&issue=8&spage=6686. [Accessed September 5, 2013].

Matassi, F. et al., 2013. Porous metal for orthopedics implants. Clinical cases in mineral and bone metabolism: the official journal of the Italian Society of Osteoporosis, Mineral Metabolism, and Skeletal Diseases, 10(2), pp. 111–115. Available at: http://www.pubmedcentral.nih.gov/articlerender.fcgi?artid=3796997&tool=pmcentrez&rendertype=abstract.

Maté-Sánchez de Val, J.E. et al., 2014. Comparison of three hydroxyapatite/β-tricalcium phosphate/collagen ceramic scaffolds: an in vivo study. Journal of biomedical materials research. Part A, 102(4), pp. 1037–46. Available at: http://www.ncbi.nlm.nih.gov/pubmed/23649980. [Accessed May 27, 2014].

Matsuno, T. et al., 2008. Preparation of injectable 3 D-formed beta-tricalcium phosphate bead/alginate composite for bone tissue engineering. *Dental materials journal*, 27(6), pp. 827–34. Available at: http://www.ncbi.nlm.nih.gov/pubmed/19241692.

Mehrali, M. et al., 2013. Dental implants from functionally graded materials. *Journal of biomedical materials research. Part A*, 101(10), pp. 3046–57. Available at: http://www.ncbi.nlm.nih.gov/pubmed/23754641. [Accessed November 4, 2013].

Michiardi, a et al., 2006. New oxidation treatment of NiTi shape memory alloys to obtain Ni-free surfaces and to improve biocompatibility. *Journal of biomedical materials research. Part B, Applied biomaterials*, 77(2), pp. 249–56. Available at: http://www.ncbi.nlm.nih.gov/pubmed/16245290. [Accessed May 23, 2014].

Moseke, C. & Gbureck, U., 2010. Tetracalcium phosphate: Synthesis, properties and biomedical applications. *Acta biomaterialia*, 6(10), pp. 3815–23. Available at: http://www.ncbi.nlm.nih.gov/pubmed/20438869. [Accessed June 15, 2014].

Mullen, L. et al., 2009. Selective Laser Melting: a regular unit cell approach for the manufacture of porous, titanium, bone in-growth constructs, suitable for orthopedic applications. *Journal of biomedical materials research. Part B, Applied biomaterials*, 89(2), pp. 325–34. Available at: http://www.ncbi.nlm.nih.gov/pubmed/18837456. [Accessed June 20, 2014].

Murr, L.E. et al., 2012. Metal Fabrication by Additive Manufacturing Using Laser and Electron Beam Melting Technologies. *Journal of Materials Science & Technology*, 28(1), pp. 1–14. Available at: http://linkinghub.elsevier.com/retrieve/pii/S1005030212600164. [Accessed June 2, 2014].

Murr, L.E. et al., 2009. Microstructures and mechanical properties of electron beam-rapid manufactured Ti–6Al–4V biomedical prototypes compared to wrought Ti–6Al–4V. *Materials Characterization*, 60(2), pp. 96–105. Available at: http://linkinghub.elsevier.com/retrieve/pii/S1044580308002076. [Accessed June 7, 2014].

Murugan, R. & Ramakrishna, S., 2005. Development of nanocomposites for bone grafting. *Composites Science and Technology*, 65(15–16), pp. 2385–2406. Available at: http://linkinghub.elsevier.com/retrieve/pii/S026635380500285X. [Accessed May 30, 2014].

Navarro, M. et al., 2008. Biomaterials in orthopaedics. *Journal of the Royal Society, Interface / the Royal Society*, 5(27), pp. 1137–58. Available at: http://www.pubmedcentral.nih.gov/articlerender.fcgi?artid=2706047&tool=pmcentrez&rendertype=abstract. [Accessed May 23, 2014].

Van Noort, R., 2012. The future of dental devices is digital. *Dental materials: official publication of the Academy of Dental Materials*, 28(1), pp. 3–12. Available at: http://www.ncbi.nlm.nih.gov/pubmed/22119539. [Accessed June 12, 2014].

Oh, I.-H. et al., 2003. Mechanical properties of porous titanium compacts prepared by powder sintering. *Scripta Materialia*, 49(12), pp. 1197–1202. Available at: http://linkinghub.elsevier.com/retrieve/pii/S1359646203005244. [Accessed June 14, 2014].

Okazaki, Y., 2001. A New Ti–15Zr–4Nb–4Ta alloy for medical applications. *Current Opinion in Solid State and Materials Science*, 5(1), pp. 45–53. Available at: http://linkinghub.elsevier.com/retrieve/pii/S1359028600000255.

Pereira, R. do V. et al., 2014. Scaffolds of PDLLA / Bioglass 58S Produced via Selective Laser Sintering. *Materials Research*, ahead.(ahead).

Pilliar, R.M. et al., 2007. Osteochondral defect repair using a novel tissue engineering approach: sheep model study. *Technology and Health Care: Official Journal of the European Society for Engineering and Medicine*, 15(1), pp. 47–56. Available at: http://www.ncbi.nlm.nih.gov/pubmed/17264412.

Pilliar, R.M. et al., 2013. Porous calcium polyphosphate as load-bearing bone substitutes: in vivo study. *Journal of biomedical materials research. Part B, Applied biomaterials*, 101(1), pp. 1–8. Available at: http://www.ncbi.nlm.nih.gov/pubmed/23143776. [Accessed February 13, 2014].

Pilliar, R.M. et al., 2001. Porous calcium polyphosphate scaffolds for bone substitute applications - in vitro characterization. *Biomaterials*, 22(9), pp. 963–72. Available at: http://www.ncbi.nlm.nih.gov/pubmed/11311015.

Pilliar, R.M., Hong, J. & Santerre, P.J., 2009. Method of manufacture of porous inorganic structures, Patent Number 7494614.

Porter, J.R., Ruckh, T.T. & Popat, K.C., 2009. Bone tissue engineering: a review in bone biomimetics and drug delivery strategies. *Biotechnology progress*, 25(6), pp. 1539–60. Available at: http://www.ncbi.nlm.nih.gov/pubmed/19824042. [Accessed March 1, 2012].

Pupo, Y. et al., 2013. Scanning Space Analysis in Selective Laser Melting for CoCrMo Powder. *Procedia Engineering*, 63, pp. 370–378. Available at: http://linkinghub.elsevier.com/retrieve/pii/S1877705813014410. [Accessed June 20, 2014].

Qiu, K. et al., 2006. Fabrication and characterization of porous calcium polyphosphate scaffolds. *Journal of Materials Science*, 41(8), pp. 2429–2434. Available at: http://link.springer.com/10.1007/s10853-006-5182-2. [Accessed May 26, 2014].

Rahaman, M.N. et al., 2011. Bioactive glass in tissue engineering. *Acta biomaterialia*, 7(6), pp. 2355–73. Available at: http://www.pubmedcentral.nih.gov/articlerender.fcgi?artid=3085647&tool=pmcentrez&rendertype=abstract. [Accessed May 27, 2014].

Reclaru, L. et al., 2005. Electrochemical corrosion and metal ion release from Co-Cr-Mo prosthesis with titanium plasma spray coating. *Biomaterials*, 26(23), pp. 4747–56. Available at: http://www.ncbi.nlm.nih.gov/pubmed/15763254. [Accessed May 31, 2014].

Rodríguez-Lorenzo, L., 2009. Synthesis and Biocompatibility of Hydroxyapatite in a Graphite Oxide Matrix. *Key Engineering* .... Available at: http://www.scientific.net/KEM.396-398.477. [Accessed September 3, 2013].

Rogel, M.R., Qiu, H. & Ameer, G. a., 2008. The role of nanocomposites in bone regeneration. *Journal of Materials Chemistry*, 18(36), p. 4233. Available at: http://xlink.rsc.org/?DOI=b804692a. [Accessed June 12, 2014].

Ryan, G., Pandit, A. & Apatsidis, D.P., 2006. Fabrication methods of porous metals for use in orthopaedic applications. *Biomaterials*, 27(13), pp. 2651–70. Available at: http://www.ncbi.nlm.nih.gov/pubmed/16423390. [Accessed May 27, 2014].

Ryan, G.E., Pandit, A.S. & Apatsidis, D.P., 2008. Porous titanium scaffolds fabricated using a rapid prototyping and powder metallurgy technique. *Biomaterials*, 29(27), pp. 3625–35. Available at: http://www.ncbi.nlm.nih.gov/pubmed/18556060. [Accessed May 27, 2014].

Sánchez-Salcedo, S., Nieto, A. & Vallet-Regí, M., 2008. Hydroxyapatite/β-tricalcium phosphate/agarose macroporous scaffolds for bone tissue engineering. *Chemical Engineering Journal*, 137(1), pp. 62–71. Available at: http://linkinghub.elsevier.com/retrieve/pii/S1385894707006183. [Accessed July 29, 2010].

Sanz-Herrera, J. a, García-Aznar, J.M. & Doblaré, M., 2009. A mathematical approach to bone tissue engineering. Philosophical transactions. Series A, Mathematical, physical, and engineering sciences, 367(1895), pp. 2055–78. Available at: http://www.ncbi.nlm.nih.gov/pubmed/19380325. [Accessed May 18, 2014].

Shanjani, Y. et al., 2011. Mechanical characteristics of solid-freeform-fabricated porous calcium polyphosphate structures with oriented stacked layers. *Acta Biomaterialia*, 7(4), pp. 1788–96. Available at: http://www.ncbi.nlm.nih.gov/pubmed/21185409. [Accessed February 11, 2013].

Shanjani, Y. et al., 2013. Solid freeform fabrication of porous calcium polyphosphate structures for bone substitute applications: in vivo studies. *Journal of biomedical materials research. Part B, Applied biomaterials*, 101(6), pp. 972–80. Available at: http://www.ncbi.nlm.nih.gov/pubmed/23529933. [Accessed March 12, 2014].

Shanjani, Y., 2011. Solid freeform fabrication of porous calcium polyphosphate structures for use in orthopaedics.

Shishkovsky, I.V. et al., 2008. Porous biocompatible implants and tissue scaffolds synthesized by selective laser sintering from Ti and NiTi. *Journal of Materials Chemistry*, 18(12), p.1309. Available at: http://xlink.rsc.org/?DOI=b715313a. [Accessed June 20, 2014].

Shuai, C., Feng, P., Zhang, L., et al., 2013. Correlation between properties and microstructure of laser sintered porous β -tricalcium phosphate bone scaffolds. *Science and Technology of Advanced Materials*, 14(5), p.055002. Available at: http://stacks.iop.org/1468-6996/14/i=5/a=055002?key=crossref.7011781bc8806989cf2f7e8be93edb63. [Accessed June 10, 2014].

Shuai, C., Feng, P., Nie, Y., et al., 2013. Nano-Hydroxyapatite Improves the Properties of (-tricalcium Phosphate Bone Scaffolds. International Journal of Applied Ceramic Technology, 10(6), pp. 1003–1013. Available at: http://doi.wiley.com/10.1111/j.1744-7402.2012.02840.x. [Accessed May 27, 2014].

Skoog, S. a, Goering, P.L. & Narayan, R.J., 2014. Stereolithography in tissue engineering. *Journal of materials science. Materials in medicine*, 25(3), pp. 845–56. Available at: http://www.ncbi.nlm.nih.gov/pubmed/24306145. [Accessed May 29, 2014].

Staiger, M.P. et al., 2006. Magnesium and its alloys as orthopedic biomaterials: a review. *Biomaterials*, 27(9), pp. 1728–34. Available at: http://www.ncbi.nlm.nih.gov/pubmed/16246414. [Accessed May 24, 2014].

Stevens, B. et al., 2008. A review of materials, fabrication methods, and strategies used to enhance bone regeneration in engineered bone tissues. *Journal of biomedical materials research. Part B, Applied biomaterials*, 85(2), pp. 573–82. Available at: http://www.ncbi.nlm.nih.gov/pubmed/17937408. [Accessed July 29, 2010].

Szpalski, C., et al., 2012. Bone tissue engineering: current strategies and techniques—Part I: Scaffolds. Tissue Engineering Part B-Reviews 18 (4), 246–257.

Vasiliadis, H.S., Wasiak, J. & Salanti, G., 2010. Autologous chondrocyte implantation for the treatment of cartilage lesions of the knee: a systematic review of randomized studies. *Knee Surgery, Sports Traumatology, Arthroscopy*, 18(12), pp. 1645–1655. Available at: http://www.ncbi.nlm.nih.gov/pubmed/20127071. [Accessed July 29, 2010].

Vlasea, M. et al., 2013. A combined additive manufacturing and micro-syringe deposition technique for realization of bio-ceramic structures with micro-scale channels. *The International Journal of Advanced Manufacturing Technology*, 68(9–12), pp. 2261–2269. Available at: http://link.springer.com/10.1007/s00170-013-4839-7. [Accessed November 3, 2013].

Vlasea, M., Toyserkani, E., 2013. Experimental characterization and numerical modeling of a micro-syringe deposition system for dispensing sacrificial photopolymers on particulate ceramic substrates. Journal of Materials Processing Technology 213 (11), 1970–1977.

Vorndran, E. et al., 2008. 3D Powder Printing of β-Tricalcium Phosphate Ceramics Using Different Strategies. *Advanced Engineering Materials*, 10(12), pp.B67–B71. Available at: http://doi.wiley.com/10.1002/adem.200800179. [Accessed August 20, 2013].

Wataha, J.C., et al., 2001. Relating Nickel-Induced Tissue Inflammation to Nickel Release in vivo. Journal of Biomedical Materials Research 58 (5), 537–544.

Weber, J.N., White, E.W., 1972. Carbon-metal graded composites for permanent osseous attachment of non-porous metals. Materials Research Bulletin 7 (9), 1005–1016.

Wei, G. & Ma, P.X., 2004. Structure and properties of nano-hydroxyapatite/polymer composite scaffolds for bone tissue engineering. *Biomaterials*, 25(19), pp. 4749–57. Available at: http://www.ncbi.nlm.nih.gov/pubmed/15120521.

Wen, C.E. et al., 2002. Novel titanium foam for bone tissue engineering. *JOURNAL OF MATERIALS RESEARCH*, 17 10 2633-2639.

Will, J. et al., 2008. Porous ceramic bone scaffolds for vascularized bone tissue regeneration. *Journal of Materials Science: Materials in Medicine*, 19(8), pp. 2781–90. Available at: http://www.ncbi.nlm.nih.gov/pubmed/18305907. [Accessed August 22, 2013].

Woesz, A., 2008a. Scaffolds with controlled architecture for possible use in bone tissue engineering.

Woesz, A., 2008b. Scaffolds with controlled architecture for possible use in bone tissue engineering.

Woesz, A., et al., 2005. Towards bone replacement materials from calcium phosphates via rapid prototyping and ceramic gelcasting. Materials Science and Engineering C 25, 181–186.

Wong, K. V & Hernandez, A., 2012a. *A Review of Additive Manufacturing,.*

Wong, K. V & Hernandez, A., 2012b. A Review of Additive Manufacturing. International Scholarly Research Network Mechanical Engineering, 2012, pp. 1–10.

Xiong, Z. et al., 2002. Fabrication of porous scaffolds for bone tissue engineering via low-temperature deposition. *Scripta Materialia*, 46(11), pp. 771–776. Available at: http://linkinghub.elsevier.com/retrieve/pii/S1359646202000714.

Yang, S. et al., 2001. The design of scaffolds for use in tissue engineering. Part I. Traditional factors. *Tissue Engineering*, 7(6), pp. 679–89. Available at: http://www.ncbi.nlm.nih.gov/pubmed/11749726.

Yang, S., Du, Z., Chua, C.K., 2002. Review The Design of Scaffolds for Use in Tissue Engineering. Part II. Rapid Prototyping Techniques. Tissue Engineering 8 (1), 1–11.

Yarlagadda, P.K.D. V, Chandrasekharan, M. & Shyan, J.Y.M., 2005. Recent advances and current developments in tissue scaffolding. *Bio-medical materials and engineering*, 15(3), pp. 159–77. Available at: http://www.ncbi.nlm.nih.gov/pubmed/15911997.

Zanin, H. et al., 2013. Fast preparation of nano-hydroxyapatite/superhydrophilic reduced graphene oxide composites for bioactive applications. *Journal of Materials Chemistry B*. Available at: http://xlink.rsc.org/?DOI=c3tb20550a. [Accessed August 29, 2013].

Zhang, K. et al., 2004. Processing and properties of porous poly(l-lactide)/bioactive glass composites. *Biomaterials*, 25(13), pp. 2489–2500. Available at: http://linkinghub.elsevier.com/retrieve/pii/S0142961203007695. [Accessed May 28, 2014].

Zhang, L. et al., 2013. A tough graphene nanosheet/hydroxyapatite composite with improved in vitro biocompatibility. *Carbon*, 61, pp. 105–115. Available at: http://linkinghub.elsevier.com/retrieve/pii/S0008622313003837. [Accessed June 2, 2014].

Zhang, Y., et al., 2003. Calcium Phosphate – Chitosan Composite Scaffolds for Bone. Tissue Engineering 9 (2.).

Zhou, W.Y. et al., 2008. Selective laser sintering of porous tissue engineering scaffolds from poly(L: -lactide)/carbonated hydroxyapatite nanocomposite microspheres. *Journal of materials science. Materials in medicine*, 19(7), pp. 2535–40. Available at: http://www.ncbi.nlm.nih.gov/pubmed/17619975. [Accessed June 6, 2014].

CARTILAGE 3D PRINTING

Shawn P. Grogan, Erik W. Dorthé and Darryl D. D'Lima

Shiley Center for Orthopaedic Research and Education at Scripps Clinic, La Jolla, CA, USA

12.1 GENERAL INTRODUCTION

12.1.1 ARTICULAR CARTILAGE: FUNCTION AND ORGANIZATION

Healthy adult knee human articular cartilage is a remarkable living tissue that permits almost frictionless articulation in synovial joints and protects the subchondral bone from mechanical stress (Cucchiarini and Madry, 2005; O'Driscoll, 1998). This load-bearing tissue is devoid of neural innervation, vasculature, or a lymphatic system, yet is comprised of chondrocytes that produce and maintain an extensive extracellular matrix (ECM) network.

The major components of the cartilaginous matrix include collagen fibrils and proteoglycans (glycosaminoglycans; GAGs), which bind much of the water in cartilage (70–80%). Collagen type II is the predominant collagen, but other collagens such as type VI, IX, XI, and XIV are also present. Other noteworthy macromolecules that make up the cartilage ECM include link protein, cartilage oligomeric matrix protein (COMP), decorin, tenascin, fibromodulin, and fibronectin (Hunziker et al., 1997). Articular cartilage is not a simple mixture of these components; rather it is exquisitely structured and organized into "zones" of cells and ECM extending from the articular surface transitioning down to the underlying subchondral bone (Figure 12.1). Zonal variations in the collagen network orientation and depth-dependent variation in the concentration of GAGs contribute to the complex compressive, shear, and tensile properties (Benninghoff, 1925; Schinagl et al., 1997).

The upper, most specialized zone (upper 10–50um or 10% of the full thickness) is the superficial zone (SZ) that experiences both compressive and shear forces (Buckley et al., 2008; Wong et al., 2010), harbors the highest cell density, and contains cells that are small, flattened, and without an extensive pericellular matrix (PCM) (Hunziker et al., 2002; Siczkowski and Watt, 1990). SZ cells secrete a number of unique proteins including the synovial fluid "lubricant" molecule known as superficial zone protein or lubricin (Flannery et al., 1999; Jay et al., 1998; Schumacher et al., 1999), clusterin (Khan et al., 2001; Malda et al., 2010), and developmental endothelial locus-1 (Del1) (Pfister et al., 2001). In the SZ, chondrocytes secrete less collagen type II (Darling et al., 2004; Hidaka et al., 2006; Khan et al., 2001) and the ECM contains the least amount of GAGs (Buckwalter and Mankin, 1998). The SZ cells, however, produce higher levels of keratan sulfate and other specialized proteoglycans in comparison to the deep zone (DZ) cells (Aydelotte et al., 1988; Aydelotte and Kuettner, 1988; Siczkowski and Watt, 1990; Waldman et al., 2003; Zanetti et al., 1985). The specialized proteoglycans known as small leucine-rich proteoglycans (SLRPs) (Miosge et al., 1994; Poole et al., 1996) are found in higher concentrations in the SZ, including decorin, biglycan, asporin, lumican, and fibromodulin, which modulate many processes including metabolism, growth factor signaling, and maintaining tissue integrity (Kizawa et al., 2005; Pelletier et al., 2000; Poole et al., 1996; Roughley, 2006). Recently a fibulin-like ECM protein (EFEMP1) was found predominately expressed in the SZ (Grogan et al., 2013a) and was postulated to maintain the immature status of chondroprogenitor cells (Wakabayashi et al., 2010), which contributes to the mounting evidence of the presence of progenitor cell populations residing in this region (Alsalameh et al., 2004; Dowthwaite et al., 2004; Grogan et al., 2009), which may be critical for replenishing cells and maintaining tissue homeostasis.

The middle zone (MZ) region (40–60% of cartilage thickness; Figure 12.1) contains chondrocytes with more PCM and collagen fibrils that are more randomly dispersed. In comparison to the SZ, the MZ ECM contains higher levels of aggrecan (Bayliss et al., 1983) and is specifically high in hyaluronic acid (HA), dermatan sulfate, and collagen type II (Asari et al., 1994; Buschmann et al., 2000; Franzen et al., 1981; Maroudas et al., 1969; Ratcliffe et al.; 1984; Schinagl et al., 1997; Wong et al., 1996). The main role of aggrecan is to maintain tissue hydration, which is important for endowing cartilage its mechanical properties (Dudhia, 2005). Cartilage intermediate layer protein (CILP) is uniquely located in the MZ (Lorenzo et al., 1998). The ECM protein cartilage oligomeric matrix protein (COMP) is also mainly seen in MZ and DZ (DiCesare et al., 1995; Murray et al., 2001).

The deepest 30–40% of cartilage thickness is classified as the radial or deep zone (DZ), which contains the lowest cell density and collagen content compared to the SZ and MZ (Venn, 1978). On the other hand, the proteoglycan content, including aggrecan, keratan sulfate, and chondroitin sulfate, is much higher than the upper zones (Asari et al., 1994; Jacoby and Jayson, 1975; Mitrovic and Darmon, 1994;

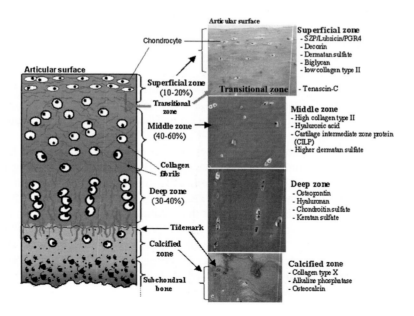

FIGURE 12.1

Cartilage zonal organization and extracellular matrix (ECM) composition.

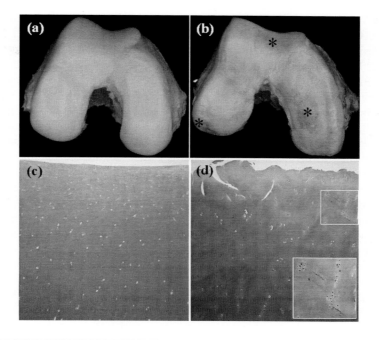

FIGURE 12.2

Macro and histology images of human cartilage: (a) photograph of a normal articular femoral knee joint; (b) osteoarthritic femoral knee joint with several surface erosions (*); (c) histological Safranin O-fast green stained section of normal cartilage with a smooth intact surface; and (d) osteoarthritic cartilage section with fibrillated surface, lower cell density, reduced Safranin O staining, and cell cluster formation (inset). (Magnification C and D = 10×).

Muir et al., 1970). Although the total collagen content in the DZ is lower than that in the other zones, the collagen fibrils are of larger diameter and, similar to the cells that are commonly arranged into stacks or columns, are oriented perpendicular to the articular surface (Siczkowski and Watt, 1990) (Figure 12.1). Osteopontin is exclusively expressed in the DZ (Pullig et al., 2000; Schnapper and Meyer, 2004).

Integration between articular cartilage and the much stiffer subchondral bone begins at the base of the deep zone where a 5 μm thick undulating "tidemark" is observed, which marks the transition into a layer of calcified cartilage (20–250 μm thick) containing perpendicular chondrocyte-derived collagen type II fibers that become structurally cemented in the collagen type I osteoid produced by osteoblasts (Hoemann et al., 2012; Orth et al., 2013) (Figure 12.1). The calcified cartilage contains small cells in a chondroid matrix that is characterized by increased mineral density speckled with apatitic salts (Burr, 2004; Mow et al., 2012).

12.1.2 CARTILAGE INJURY, DISEASE, AND TREATMENT OPTIONS

Cartilage function is often compromised following acute or chronic injury or as part of the aging process and eventually leading to tissue degeneration manifested as osteoarthritis (OA). Photographs and histology images of normal and degenerated OA human articular cartilage are shown in Figure 12.2.

For patients more than 60 years of age with end-stage arthritis, artificial joint replacement is the recommended treatment (Grayson and Decker, 2012; Noble et al., 2005; Wylde et al., 2012). For younger patients with focal lesions, biological or cell-based procedures are employed such as microfracture (Goyal et al., 2014, 2013; Sherman et al., 2014), autologous chondrocyte implantation (Kon et al., 2012), or osteochondral grafting (Sherman et al., 2014). However, these approaches do not restore long-lasting healthy cartilage with the major issues being lack of regeneration of the zonal structure of hyaline articular cartilage and poor integration with host tissue (Clar et al., 2005; Rasanen et al., 2007; Zeifang et al., 2010).

Over the past 10–14 years, much effort has been devoted to prefabrication of neotissues in several laboratories using various scaffolds systems and bioreactors (Tuan et al., 2013). These tissue engineering approaches typically involve cell harvesting and isolation, cell expansion, and an *in vitro* phase of 3D culture to produce a neotissue graft (Roelofs et al., 2013). The general concept of maturing a cultured 3D graft tissue before implantation is to ensure development of the desired cartilage mechanical properties (Mohanraj et al., 2013; Pabbruwe et al., 2009; Theodoropoulos et al., 2011). However, integration of mature cartilage with the host tissue is limited or not consistently demonstrated due to a number of parameters including lack of vascularity, mismatch between the properties of the ECM structure of the native tissue and implanted graft, cell death, inadequate differentiation of the cells, and the type of biomaterial or scaffold used (Khan et al., 2008). Only recently have efforts toward translating engineered cartilage tissue into human clinical trials been initiated, such as Denovo ET (Engineered Tissue Graft) and NeoCart (see clinicaltrials.gov). Outcomes of the NeoCart phase-II prospective, randomized clinical trial were similar to that of microfracture surgery and were associated with greater clinical efficacy at two years after treatment (Crawford et al., 2012; Fedorovich et al., 2012). However, longer-term clinical outcomes are pending.

Overall, current strategies to engineer cartilage have failed to fabricate new repair tissue *in vivo* that is indistinguishable from native cartilage in terms of mechanical properties, zonal organization, and ECM (Klein et al., 2009; Schuurman et al., 2013a). The detailed characterization of cartilage ECM composition throughout cartilage (as detailed earlier); knowledge of cartilage biology in terms of cartilage structure and organization (Grogan et al., 2009; Miosge et al., 1994; Poole et al., 1996; Waldman et al., 2003), and cartilage zonal phenotype (Grogan et al., 2013a); and the availability of various promising cells sources (e.g. MSC and ESC) as alternative sources for cartilage tissue regeneration (Bulman et al., 2013; Filardo et al., 2013; Olee et al., 2014); coupled with advances in biomaterials (scaffolds, hydrogels) that support cartilage formation and possess mechanical properties approaching cartilage (Spiller et al., 2011; Tuan et al., 2013) are collectively very valuable in resolving the present challenges facing cartilage repair. 3D bioprinting is an emerging technology with the ability to deliver cells and ECM proteins to organized repair tissues with the desired cell densities, zones, and variation in ECM properties to mimic normal articular cartilage. If successful, 3D bioprinting can overcome the major hurdles preventing successful repair such as inadequate reproduction of the tissue structure and organization, and poor integration with host tissue.

12.1.3 3D BIOPRINTING AND CELL PRINTING APPROACHES

A number of techniques have been developed for 3D bioprinting, most of which are based on traditional 3D printing or rapid prototyping technologies. The primary methods in development today are extrusion and ink-jet; however, stereolithography and laser-assisted direct-writing techniques are also active areas of research. The modes of action for each technology are outlined in Figure 12.3.

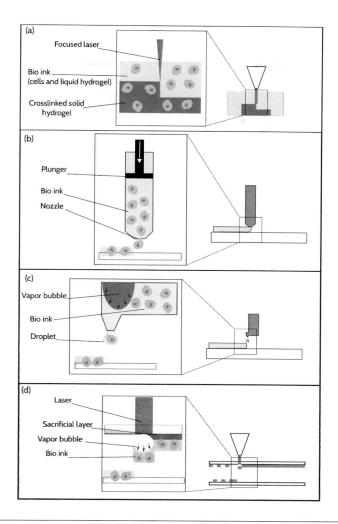

FIGURE 12.3

Overview of the mechanisms of action for each 3D printing method: (a) stereolithography traps cells that are suspended in a liquid photosensitive monomer hydrogel, dark gray represents the focused laser-induced cross-linked solid hydrogel; (b) extrusion pushes suspended cells from a column through a contacting nozzle; (c) thermal ink-jet uses a pressure increase formed by a vapor bubble to force the mix of cells and ink from a nozzle; (d) laser induced forward transfer (LIFT) vaporizes a sacrificial layer of cells suspended on a polymer to push cells toward the substrate.

All techniques share a key characteristic: the reliance on some form of bioink. This can take many forms; in the simplest case, it is a biocompatible, printable fluid. In most cases, the bioink is intended as a carrier for cells, not just a tissue engineering scaffold material. In these instances, the fluid must provide both structure for the 3D object and a favorable environment for cell viability. A major component of many bioinks is a hydrogel, often agarose or HA (Billiet et al., 2012; Jabbari, 2011; Schuurman et al., 2013a).

In some bio-inks, cells are not simply suspended, they are clustered into spheroids or other micro-structures (Mironov et al., 2009; Norotte et al., 2009). The goal in these cases is to print with small, preassembled tissue-like constructs and decrease the time necessary for cells to self-organize after printing.

Of the various 3D printing technologies, the earliest was stereolithography (Billiet et al., 2012). While it revolutionized the prototyping process for some industries, it has not been as effective in the biomedical sciences. In stereolithography, a laser is rastered over a reservoir of photocrosslinkable polymer, solidifying the material (Figure 12.3A). After each layer, the printing substrate is lowered or the reservoir depth is increased, causing the solidified material to be covered in uncrosslinked polymer. The process is repeated to form each layer in the object. This can be easily adapted to photocrosslinkable biomaterials, and has been used for simple implant and scaffold creation (Billiet et al., 2012; Grogan et al., 2013b; Soman et al., 2013). Bio-inks have been made possible by the development of photocrosslinkable hydrogels. If cells are suspended in the bioink, they are trapped within the hydrogel matrix at the time of cross-linking. It has been demonstrated that cells can survive the relatively brief UV laser exposure during printing (Lin et al., 2013). Incorporating 2D projection techniques, for example with digital micromirrors, has dramatically increased the print speed while preserving resolution and limiting cell exposure to UV light (Beke et al., 2012). Despite these advantages, stereolithography requires a large amount of bioink in the reservoir in which layers can be constructed, and the feasibility of multi-ink printing is limited (Melchels et al., 2012).

Extrusion printing is the most familiar 3D printing technology, bearing a strong resemblance to consumer-grade plastic 3D printers. In the consumer systems, a thermoplastic is heated above the glass transition temperature and is extruded from a nozzle, building an object one layer at a time (Figure 12.3B). Early uses of 3D printing for biomedical applications used this process directly; parts were made from biocompatible polymers, either for implantation or for use as a tissue-engineering scaffold (Fedorovich et al., 2012; Seyednejad et al., 2011).

For bioprinting applications, the cells suspended in a bioink are loaded into a delivery system and extruded to build layers. The delivery system can be as simple as a syringe or a capillary tube with a fine plunger (Skardal et al., 2010). The relatively large nozzle sizes and high extrusion pressure allow extrusion of viscous inks with high cell density. Bioinks can be developed with some complex structures and properties, such as tissue spheroids. This advantage can also be seen as a disadvantage: large nozzle sizes and bulk extrusion volumes limit the minimum resolution. Extrusion printing requires contact with the print substrate, limiting potential *in situ* applications and increasing the possibility of contamination. Organovo (San Diego, CA) has commercialized extrusion printing for the pharmaceutical industry. Organovo's bioprinter is capable of placing cell aggregates, such as spheroids, with high accuracy, although resolution of the actual printed construct is still limited by the size of the aggregates (Khatiwala et al., 2012). Organovo's approach generates analogs of human tissue, such as liver, that can be used for drug discovery.

Ink-jet bioprinting draws on the same technology as ink-jet document printing. A low viscosity fluid is ejected from a nozzle toward the printing substrate. Generally, the necessary pressure is achieved by a thermal bubble or a piezoelectric effect (Figure 12.3C). Like the other 3D printing techniques, this method has been used to create scaffolds for tissue engineering purposes (Butscher et al., 2011; Lee et al., 2005). Both thermal and piezoelectric ink-jets are capable of printing viable cells (Cui et al., 2012a, 2012b; Saunders et al., 2008; Tirella et al., 2011). Ink-jet does not require contact with the print substrate or previous printed layers, which can be an advantage in certain medical applications. A single system can print with multiple bioinks making it feasible to construct complex tissues containing multiple cell types (Xu et al., 2013). The small size of individual droplets increases the resolution of the printed constructs and reduces the amount of bioink needed. However, the higher resolution comes at the cost of increased printing duration and limits the application to low viscosity inks. Cell settling and clumping can clog the microchannels and nozzles in an ink-jet system (Chahal et al., 2012).

Laser-induced forward transfer (LIFT), biological laser printing (BioLP), matrix-assisted pulsed laser evaporation-direct writing (MAPLE-DW), and laser-assisted bioprinting (LAB) use a laser to print bioink, which is spread across an optically transparent "ribbon" (Figure 12.3D). The ribbon is suspended with the bioink orientated downward, over the print substrate or "bio-paper." A laser then superheats and vaporizes a thin layer of material on the underside of the ribbon, inducing a jet of bioink on the target substrate (Barron et al., 2004; Guillemot et al., 2010; Guillotin et al., 2010; Ringeisen et al., 2006). The need for carefully prepared ribbons and the small ejection size limits the throughput of the process and printing large multilayered 3D constructs using LIFT techniques has not yet been reported.

While the vast majority of bioprinting research involves one of the four major approaches just mentioned, other unique techniques are being explored. For instance, patterned chips have been used for cell placement (Jing et al., 2011; Rosenthal et al., 2007). Cells are preferentially trapped in wells on the chip, which is then overturned to drop the cells in the well pattern on another substrate. While this technique is not immediately obvious as a 3D manufacturing method, it has shown potential for investigating cell–cell interactions and other spatially varying phenomenon. Another uncommon method for cell placement uses microfluidic channels to deposit cells one layer at a time (Tan and Desai, 2004). Having control of cell and matrix types at the resolution of microfluidic chips has advantages for both laboratory investigation and construct development, although the technique is limited to shapes that can be molded in channels.

12.1.4 APPLYING 3D PRINTING FOR CARTILAGE AND BONE

Initial 3D printing-based approaches have involved forming cell-free 3D scaffolds *in vitro* that were subsequently seeded with cells. Sherwood et al. (Sherwood et al., 2002) used the TheriForm™ process to bind powder particles with a liquid binder to form an osteochondral construct that was subsequently seeded with cells. A mixture of PLGA and PLA polymer powder particles was bound using chloroform to generate a porous scaffold. Adding tricalcium phosphate to the subchondral region generated scaffold mechanical properties that were comparable to bone. Woodfield et al. (Woodfield et al., 2004) utilized a pressure-driven syringe to deposit molten copolymer fibers onto a computer-controlled *x-y-z* table to produce scaffolds comprised of poly(ethylene glycol)-terephthalate-poly(butylene terephthalate) (PEGT/PBT), into which they seeded bovine articular chondrocytes. The mechanical properties of these scaffolds could be altered by changing PEGT/PBT composition and by altering the geometry and porosity. These scaffolds supported a homogeneous cell distribution and subsequent cartilage-like tissue formation. Normal articular cartilage possesses an equilibrium modulus of 0.27 MPa and a dynamic stiffness of 4.10 MPa, and these engineered scaffolds approximated these qualities with an equilibrium modulus ranging from 0.05–2.5 MPa and a dynamic stiffness range spanning 0.16–4.33 MPa (Woodfield et al., 2004). To reconstruct the zonal arrangements of articular cartilage, high-density pellet cultures of cells derived from different zones were seeded into PEGT/PBT scaffolds, although conclusive evidence of zonal structure was not provided (Schuurman et al., 2013b).

The Bioplotter, a device previously described by Moroni et al. (Moroni et al., 2006) was used to extrude Poly[(ethylene oxide) terephthalate-co-poly(butylene) terephthalate] (PEOT/PBT) 3D fibers for cartilage regeneration (Moroni et al., 2008). The Bioplotter extruded the highly viscous polymeric fibers layer by layer to construct a rectangular porous block. Each fiber was approximately 300 μm thick, fiber spacing was 800 μm, and the layer height was 225 μm. Bovine chondrocytes were seeded into 4 mm high scaffolds and a cartilaginous matrix with a Safranin O positive stain was deposited over 21 days. Scanning electron micrograph studies also revealed an uneven distribution of cells throughout the scaffold, which is one of the major issues that confront prefabricated scaffold systems (Carletti et al., 2011; Wendt et al., 2006).

Simultaneous printing of cells and scaffold, achieved by bioprinting, has several advantages. 3D bioprinting of cartilage and bone tissue engineering was described by Fedorovich et al. (Fedorovich et al., 2012), who used a pneumatic dispensing system (the BioScaffolder) to deposit human chondrocytes and bone marrow-derived mesenchymal stem cells (MSCs) in alginate into various patterns up to 10 layers thick (0.8 mm high). The alginate encapsulating MSCs were also supplemented with osteoinductive materials to promote osteogenic differentiation. *In vitro* culture generated evidence of cartilage formation (deposition of collagens type II and VI) and a bone-like phenotype (collagen type I, alkaline phosphatase, osteonectin, and Alizarin red positive). *In vivo* implantation over 6 weeks in nude mice resulted in limited bone formation perhaps due to the confining nature of alginate.

12.1.5 DIRECT *IN SITU* PRINTING

The vast majority of 3D printing techniques are directed at engineering tissue *in vitro* with the objective of subsequent implantation into the joint. One attractive alternative is printing engineered tissue directly into the human body. Cohen et al. demonstrated proof of concept by delivering alginate and demineralized bone matrix via syringe extrusion into surgically created articular defects (Cohen et al., 2010). CT scans of a cadaver bovine femur were obtained to extract the geometry of the defect. For repair of a chondral defect, precrosslinked alginate hydrogel was extruded through a syringe. For repair of an osteochondral defect, a paste made of demineralized bone matrix mixed in gelatin was first extruded into the bony defect followed by alginate extrusion layered over the demineralized bone matrix. Although no cells were involved and the femur had been devitalized, this report demonstrated proof of concept of the potential for additive manufacturing for direct *in situ* printing.

Proof of concept for *in situ* 3D printing of live human articular chondrocytes for cartilage repair was demonstrated by Cui et al. (Cui et al., 2012b). A modified thermal ink-jet printer deposited chondrocytes suspended in photocrosslinkable poly(ethylene) glycol dimethacrylate (PEGDMA) into a cartilage defect created within a bovine osteochondral (cartilage and bone) tissue explant for cartilage repair. This study demonstrated the utility of direct cartilage repair and bioprinting by successfully controlling the placement of individual cells with high viability, by providing a supportive environment for secretion of cartilage ECM protein, by printing material with adequate mechanical properties, and by generating tissue that integrated directly with the host tissue.

Poly(ethylene glycol) (PEG) macromers are biocompatible and water soluble with low viscosity making them ideal for bioprinting. Cross-linking PEG generates a compressive modulus that matches that of human cartilage (Bryant et al., 2004). Most importantly, PEG has also been shown to maintain chondrocyte viability and support the generation of cartilaginous ECMs including the essential collagen type II and GAGs (Bryant and Anseth, 2002; Elisseeff et al., 2000). In the

Cui et al. (Cui et al., 2012b) study, a cylindrical cartilage defect size, 4 mm in diameter and 2 mm deep (extending to the subchondral bone), was filled in less than 2 min. Following simultaneous polymerization with UV exposure during printing of each layer (approx. 18 μm each) the cell viability was high (~90%). In contrast, if the defect was first completely filled with PEGDMA, crosslinking, took a minimum of 11 min, which also significantly reduced cell viability (~63%) (Kim et al., 2003). Cross-linking during ink-jet printing also prevented cell settling at the interface between layers due to gravity, as observed in other studies (Kim et al., 2003; Sharma et al., 2007). Feasibility of controlling the 3D location of each printed cell can be very useful in reproducing the native organization of cells and ECM. Examples of potential applications are: placing specific cell types (chondroprogenitors and osteoprogenitors) in appropriate locations; placing zonally derived chondrocytes in their proper zone (Schuurman et al., 2013b); and encapsulating cells in spatially varying ECM for better defined zonal differentiation states (Grogan et al., 2014; Grogan et al., 2013a; Klein et al., 2009; Schuurman et al., 2013b). After 6 weeks of culture, Cui et al. (Cui et al., 2012b) observed that the printed PEG gel and chondrocytes produced cartilaginous neotissue with decreased collagen type I and increased collagen type II (gene expression level), substantial GAG deposition (Safranin O stain), and evidence of histologic integration with host tissue. By 6 weeks the compressive modulus of the printed tissue approached 400 kPa, within the range of that reported for articular cartilage (Lai and Levenston, 2010). Such mechanical properties may be suitable for surviving in a loaded environment.

12.1.6 MAJOR CHALLENGES AND PITFALLS

The scientific and technical challenges facing successful bioprinting can be broadly classified into cellular, bioink, and printing technology.

12.1.6.1 Cell source

One has to address the scarcity of obtaining normal autologous human chondrocytes in sufficient quantities. One approach is to use autologous adult stem cells or progenitors that can be expanded into sufficient quantities before differentiation into matrix-producing chondrocytes. Another approach is to use an allogeneic cell source. Advantages of the former are lack of allogeneic response with lower risk for regulatory approval. Advantages of the latter are a well-defined and characterized source of cells with potential for off-the-shelf treatment. In addition to maintaining cell viability that survives the printing process, one has to ensure an appropriate and enduring phenotype.

12.1.6.2 Scaffolds

Many challenges have to be overcome before printable scaffolds can be clinically useful. For successful tissue regeneration, the scaffold must be biocompatible with minimal biological side effects upon degradation or release of degradation products. The bioink has to rapidly undergo liquid-to-solid transition to support efficient printing. Traditional methods used in commercial 3D printers, such as extrusion of molten polymer or powder-based printing, are not likely to be compatible for obvious reasons. Novel approaches to chemical or photocrosslinking biomaterials are presently being explored. Other challenges which the appropriate bioink must overcome include the properties to survive implantation and postoperative use (such as mechanical strength), and the appropriate biochemical and biophysical cues to maintain the phenotype of the printed cells.

12.1.6.3 Delivery

The present state-of-the-art includes a variety of delivery systems. At a minimum, these delivery methods need to be optimized to deal with the conflicting requirements of spatial resolution, viscosity of bioink, efficiency of printing, and cell density. Maintaining cell suspensions in the print-reservoir to avoid cell clumping during printing, or worse, clogging of the print nozzles, is important. Multiple nozzles may be required for chemical cross-linking of the bioink, for delivery of different cell types, or different ingredients in the bioink (multi-ink printing). Finally, a robust platform is needed that can accurately control the spatial location and orientation of the print head, coordinate the delivery of the ink, and maintain the biological integrity of the bioink.

12.2 FUTURE DIRECTIONS

It is important to demonstrate that 3D printing can recapitulate the crucial architectural features of articular cartilage. Even more important is generating evidence that printed artificial tissue with these appropriate features can resolve the major obstacles facing cartilage tissue engineering: integration with host tissue and long-term function.

Discoveries in cellular biology especially in the identification and characterization of the various stages between stem cells and progenitor status are likely to be of great significance in advancing the field of bioprinting. Factors influencing cell differentiation including biochemical, biomechanical, and other biophysical cues are critical in determining the fate of printed tissue.

Research in biomaterials is required to enhance printing efficiency and material properties. Other areas worthy of attention are responsive biomaterials that release growth factors or agents that modulate proliferation and other cellular functions such as differentiation and matrix production.

Finally, the ultimate goal is not merely to engineer biomimetic features, but to generate tissue that survives physiologic loading, remodels to adapt to changing conditions, and self-heals when exposed to injury.

ACKNOWLEDGMENTS

We thank Dr. Shantanu Patil for providing knee images, and Judy Blake for manuscript formatting and copyediting. We are grateful for the continued funding support by Donald and Darlene Shiley, and the Shaffer Family Foundation.

REFERENCES

Alsalameh, S., Amin, R., Gemba, T., Lotz, M., 2004. Identification of mesenchymal progenitor cells in normal and osteoarthritic human articular cartilage. Arthritis Rheum 50 (5), 1522–1532.

Asari, A., Miyauchi, S., Kuriyama, S., Machida, A., Kohno, K., Uchiyama, Y., 1994. Localization of hyaluronic acid in human articular cartilage. J Histochem Cytochem 42 (4), 513–522.

Aydelotte, M.B., Greenhill, R.R., Kuettner, K.E., 1988. Differences between sub-populations of cultured bovine articular chondrocytes. II. Proteoglycan metabolism. Connect Tissue Res 18 (3), 223–234.

Aydelotte, M.B., Kuettner, K.E., 1988. Differences between sub-populations of cultured bovine articular chondrocytes. I. Morphology and cartilage matrix production. Connect Tissue Res 18 (3), 205–222.

Barron, J.A., Wu, P., Ladouceur, H.D., Ringeisen, B.R., 2004. Biological laser printing: a novel technique for creating heterogeneous 3-dimensional cell patterns. Biomed Microdevices 6 (2), 139–147.

Bayliss, M.T., Venn, M., Maroudas, A., Ali, S.Y., 1983. Structure of proteoglycans from different layers of human articular cartilage. Biochem J 209 (2), 387–400.

Beke, S., Anjum, F., Tsushima, H., Ceseracciu, L., Chieregatti, E., Diaspro, A., Athanassiou, A., Brandi, F., 2012. Towards excimer-laser-based stereolithography: a rapid process to fabricate rigid biodegradable photopolymer scaffolds. J R Soc Interface 9 (76), 3017–3026.

Benninghoff, A., 1925. Form und bau der Geleknorpel in ihren Bezeihungen zur Funktion. Z Zellforsch Mikrosk Anat 2, 783–825.

Billiet, T., Vandenhaute, M., Schelfhout, J., Van Vlierberghe, S., Dubruel, P., 2012. A review of trends and limitations in hydrogel-rapid prototyping for tissue engineering. Biomaterials 33 (26), 6020–6041.

Bryant, S.J., Anseth, K.S., 2002. Hydrogel properties influence ECM production by chondrocytes photoencapsulated in poly(ethylene glycol) hydrogels. J Biomed Mater Res 59 (1), 63–72.

Bryant, S.J., Chowdhury, T.T., Lee, D.A., Bader, D.L., Anseth, K.S., 2004. Crosslinking density influences chondrocyte metabolism in dynamically loaded photocrosslinked poly(ethylene glycol) hydrogels. Ann Biomed Eng 32 (3), 407–417.

Buckley, M.R., Gleghorn, J.P., Bonassar, L.J., Cohen, I., 2008. Mapping the depth dependence of shear properties in articular cartilage. J Biomech 41 (11), 2430–2437.

Buckwalter, J.A., Mankin, H.J., 1998. Articular cartilage: tissue design and chondrocyte-matrix interactions. Instr Course Lect 47, 477–486.

Bulman, S.E., Barron, V., Coleman, C.M., Barry, F., 2013. Enhancing the mesenchymal stem cell therapeutic response: cell localization and support for cartilage repair. Tissue Eng Part B Rev 19 (1), 58–68.

Burr, D.B., 2004. Anatomy and physiology of the mineralized tissues: role in the pathogenesis of osteoarthrosis. Osteoarthritis Cartilage 12 (Suppl A), S20–S30.

Buschmann, M.D., Maurer, A.M., Berger, E., Perumbuli, P., Hunziker, E.B., 2000. Ruthenium hexaammine trichloride chemography for aggrecan mapping in cartilage is a sensitive indicator of matrix degradation. J Histochem Cytochem 48 (1), 81–88.

Butscher, A., Bohner, M., Hofmann, S., Gauckler, L., Muller, R., 2011. Structural and material approaches to bone tissue engineering in powder-based three-dimensional printing. Acta Biomater 7 (3), 907–920.

Carletti, E., Motta, A., Migliaresi, C., 2011. Scaffolds for tissue engineering and 3D cell culture. Methods Mol Biol 695, 17–39.

Chahal, D., Ahmadi, A., Cheung, K.C., 2012. Improving piezoelectric cell printing accuracy and reliability through neutral buoyancy of suspensions. Biotechnol Bioeng 109 (11), 2932–2940.

Clar, C., Cummins, E., Mcintyre, L., Thomas, S., Lamb, J., Bain, L., Jobanputra, P., Waugh, N., 2005. Clinical and cost-effectiveness of autologous chondrocyte implantation for cartilage defects in knee joints: systematic review and economic evaluation. Health Technol Assess 9 (47), 1–82, iii-iv, ix-x.

Cohen, D.L., Lipton, J.I., Bonassar, L.J., Lipson, H., 2010. Additive manufacturing for in situ repair of osteochondral defects. Biofabrication 2 (3), 035004.

Crawford, D.C., Deberardino, T.M., Williams, 3rd., R.J., 2012. NeoCart, an autologous cartilage tissue implant, compared with microfracture for treatment of distal femoral cartilage lesions: an FDA phase-II prospective, randomized clinical trial after two years. J Bone Joint Surg Am 94 (11), 979–989.

Cucchiarini, M., Madry, H., 2005. Gene therapy for cartilage defects. J Gene Med 7 (12), 1495–1509.

Cui, X., Boland, T., D'Lima, D.D., Lotz, M.K., 2012a. Thermal inkjet printing in tissue engineering and regenerative medicine. Recent Pat Drug Deliv Formul 6 (2), 149–155.

Cui, X., Breitenkamp, K., Finn, M.G., Lotz, M., D'Lima, D.D., 2012b. Direct human cartilage repair using three-dimensional bioprinting technology. Tissue Eng Part A 18 (11-12), 1304–1312.

Darling, E.M., Hu, J.C., Athanasiou, K.A., 2004. Zonal and topographical differences in articular cartilage gene expression. J Orthop Res 22 (6), 1182–1187.

Dicesare, P.E., Morgelin, M., Carlson, C.S., Pasumarti, S., Paulsson, M., 1995. Cartilage oligomeric matrix prote: isolation and characterization from human articular cartilage. J. Orthop Res 13 (3), 422–428.

Dowthwaite, G.P., Bishop, J.C., Redman, S.N., Khan, I.M., Rooney, P., Evans, D.J., Haughton, L., Bayram, Z., Boyer, S., Thomson, B., Wolfe, M.S., Archer, C.W., 2004. The surface of articular cartilage contains a progenitor cell population. J Cell Sci 117 (Pt 6), 889–897.

Dudhia, J., 2005. Aggrecan, aging and assembly in articular cartilage. Cell Mol Life Sci 62 (19-20), 2241–2256.

Elisseeff, J., Mcintosh, W., Anseth, K., Riley, S., Ragan, P., Langer, R., 2000. Photoencapsulation of chondrocytes in poly(ethylene oxide)-based semi-interpenetrating networks. J Biomed Mater Res 51 (2), 164–171.

Fedorovich, N.E., Schuurman, W., Wijnberg, H.M., Prins, H.J., Van Weeren, P.R., Malda, J., Alblas, J., Dhert, W.J., 2012. Biofabrication of osteochondral tissue equivalents by printing topologically defined, cell-laden hydrogel scaffolds. Tissue Eng Part C Methods 18 (1), 33–44.

Filardo, G., Madry, H., Jelic, M., Roffi, A., Cucchiarini, M., Kon, E., 2013. Mesenchymal stem cells for the treatment of cartilage lesions: from preclinical findings to clinical application in orthopaedics. Knee Surg Sports Traumatol Arthrosc 21 (8), 1717–1729.

Flannery, C.R., Hughes, C.E., Schumacher, B.L., Tudor, D., Aydelotte, M.B., Kuettner, K.E., Caterson, B., 1999. Articular cartilage superficial zone protein (SZP) is homologous to megakaryocyte stimulating factor precursor and Is a multifunctional proteoglycan with potential growth-promoting, cytoprotective, and lubricating properties in cartilage metabolism. Biochem Biophys Res Commun 254 (3), 535–541.

Franzen, A., Inerot, S., Hejderup, S.O., Heinegard, D., 1981. Variations in the composition of bovine hip articular cartilage with distance from the articular surface. Biochem J 195 (3), 535–543.

Goyal, D., Keyhani, S., Goyal, A., Lee, E.H., Hui, J.H., Vaziri, A.S., 2014. Evidence-based status of osteochondral cylinder transfer techniques: a systematic review of level I and II studies. Arthroscopy 30 (4), 497–505.

Goyal, D., Keyhani, S., Lee, E.H., Hui, J.H., 2013. Evidence-based status of microfracture technique: a systematic review of level I and II studies. Arthroscopy 29 (9), 1579–1588.

Grayson, C.W., Decker, R.C., 2012. Total joint arthroplasty for persons with osteoarthritis. PM R 4 (5 Suppl), S97–103.

Grogan, S.P., Chen, X., Sovani, S., Taniguchi, N., Colwell, Jr., C.W., Lotz, M.K., D'Lima, D.D., 2014. Influence of cartilage extracellular matrix molecules on cell phenotype and neocartilage formation. Tissue Eng Part A 20 (1-2), 264–274.

Grogan, S.P., Chung, P.H., Soman, P., Chen, P., Lotz, M.K., Chen, S., D'Lima, D.D., 2013b. Digital micromirror device projection printing system for meniscus tissue engineering. Acta Biomater 9 (7), 7218–7226.

Grogan, S.P., Duffy, S.F., Pauli, C., Koziol, J.A., Su, A.I., D'Lima, D.D., Lotz, M.K., 2013a. Zone-specific gene expression patterns in articular cartilage. Arthritis Rheum 65 (2), 418–428.

Grogan, S.P., Miyaki, S., Asahara, H., D'Lima, D.D., Lotz, M.K., 2009. Mesenchymal progenitor cell markers in human articular cartilage: normal distribution and changes in osteoarthritis. Arthritis Res Ther 11 (3), R85.

Guillemot, F., Souquet, A., Catros, S., Guillotin, B., Lopez, J., Faucon, M., Pippenger, B., Bareille, R., Remy, M., Bellance, S., Chabassier, P., Fricain, J.C., Amedee, J., 2010. High-throughput laser printing of cells and biomaterials for tissue engineering. Acta Biomater 6 (7), 2494–2500.

Guillotin, B., Souquet, A., Catros, S., Duocastella, M., Pippenger, B., Bellance, S., Bareille, R., Remy, M., Bordenave, L., Amedee, J., Guillemot, F., 2010. Laser assisted bioprinting of engineered tissue with high cell density and microscale organization. Biomaterials 31 (28), 7250–7256.

Hidaka, C., Cheng, C., Alexandre, D., Bhargava, M., Torzilli, P.A., 2006. Maturational differences in superficial and deep zone articular chondrocytes. Cell Tissue Res 323 (1), 127–135.

Hoemann, C.D., Lafantaisie-Favreau, C.H., Lascau-Coman, V., Chen, G., Guzman-Morales, J., 2012. The cartilage-bone interface. J Knee Surg 25 (2), 85–97.

Hunziker, E.B., Michel, M., Studer, D., 1997. Ultrastructure of adult human articular cartilage matrix after cryotechnical processing. Microsc Res Tech 37 (4), 271–284.

Hunziker, E.B., Quinn, T.M., Hauselmann, H.J., 2002. Quantitative structural organization of normal adult human articular cartilage. Osteoarthritis Cartilage 10 (7), 564–572.

Jabbari, E., 2011. Bioconjugation of hydrogels for tissue engineering. Curr Opin Biotechnol 22 (5), 655–660.

Jacoby, R.K., Jayson, M.I., 1975. Proceedings: Adult human articular cartilage in organ culture. Synthesis of glycosaminoglycan, effect of hyperoxia, and zonal variation of matrix synthesis. Ann Rheum Dis 34 (5), 468.

Jay, G.D., Haberstroh, K., Cha, C.J., 1998. Comparison of the boundary-lubricating ability of bovine synovial fluid, lubricin, and Healon. J Biomed Mater Res 40 (3), 414–418.

Jing, G., Wang, Y., Zhou, T., Perry, S.F., Grimes, M.T., Tatic-Lucic, S., 2011. Cell patterning using molecular vapor deposition of self-assembled monolayers and lift-off technique. Acta Biomater 7 (3), 1094–1103.

Khan, I.M., Gilbert, S.J., Singhrao, S.K., Duance, V.C., Archer, C.W., 2008. Cartilage integration: evaluation of the reasons for failure of integration during cartilage repair. A review. Eur Cell Mater 16, 26–39.

Khan, I.M., Salter, D.M., Bayliss, M.T., Thomson, B.M., Archer, C.W., 2001. Expression of clusterin in the superficial zone of bovine articular cartilage. Arthritis Rheum 44 (8), 1795–1799.

Khatiwala, C., Law, R., Dorfman, B.S.S., Csete, M., 2012. 3D cell bioprinting for regenerative medicine research and therapies. Gene Ther Regul 7 (1), 1–19.

Kim, T.K., Sharma, B., Williams, C.G., Ruffner, M.A., Malik, A., Mcfarland, E.G., Elisseeff, J.H., 2003. Experimental model for cartilage tissue engineering to regenerate the zonal organization of articular cartilage. Osteoarthritis Cartilage 11 (9), 653–664.

Kizawa, H., Kou, I., Iida, A., Sudo, A., Miyamoto, Y., Fukuda, A., Mabuchi, A., Kotani, A., Kawakami, A., Yamamoto, S., Uchida, A., Nakamura, K., Notoya, K., Nakamura, Y., Ikegawa, S., 2005. An aspartic acid repeat polymorphism in asporin inhibits chondrogenesis and increases susceptibility to osteoarthritis. Nat Genet 37 (2), 138–144.

Klein, T.J., Rizzi, S.C., Reichert, J.C., Georgi, N., Malda, J., Schuurman, W., Crawford, R.W., Hutmacher, D.W., 2009. Strategies for zonal cartilage repair using hydrogels. Macromol Biosci 9 (11), 1049–1058.

Kon, E., Filardo, G., Di Martino, A., Marcacci, M., 2012. ACI and MACI. J Knee Surg 25 (1), 17–22.

Lai, J.H., Levenston, M.E., 2010. Meniscus and cartilage exhibit distinct intra-tissue strain distributions under unconfined compression. Osteoarthritis Cartilage 18 (10), 1291–1299.

Lee, M., Dunn, J.C., Wu, B.M., 2005. Scaffold fabrication by indirect three-dimensional printing. Biomaterials 26 (20), 4281–4289.

Lin, H., Zhang, D., Alexander, P.G., Yang, G., Tan, J., Cheng, A.W., Tuan, R.S., 2013. Application of visible light-based projection stereolithography for live cell-scaffold fabrication with designed architecture. Biomaterials 34 (2), 331–339.

Lorenzo, P., Bayliss, M.T., Heinegard, D., 1998. A novel cartilage protein (CILP) present in the mid-zone of human articular cartilage increases with age. J Biol Chem 273 (36), 23463–23468.

Malda, J., Ten Hoope, W., Schuurman, W., Van Osch, G.J., Van Weeren, P.R., Dhert, W.J., 2010. Localization of the potential zonal marker clusterin in native cartilage and in tissue-engineered constructs. Tissue Eng Part A 16 (3), 897–904.

Maroudas, A., Muir, H., Wingham, J., 1969. The correlation of fixed negative charge with glycosaminoglycan content of human articular cartilage. Biochim Biophys Acta 177 (3), 492–500.

Melchels, F.P, M.D., Klein, T.J., Malda, J., Bartolo, P.J., Hutmacher, D.W., 2012. Additive manufacturing of tissue and organs. Prog Polym Sci 37, 1079–1104.

Miosge, N., Flachsbart, K., Goetz, W., Schultz, W., Kresse, H., Herken, R., 1994. Light and electron microscopical immunohistochemical localization of the small proteoglycan core proteins decorin and biglycan in human knee joint cartilage. Histochem J 26 (12), 939–945.

Mironov, V., Visconti, R.P., Kasyanov, V., Forgacs, G., Drake, C.J., Markwald, R.R., 2009. Organ printing: tissue spheroids as building blocks. Biomaterials 30 (12), 2164–2174.

Mitrovic, D.R., Darmon, N., 1994. Characterization of proteoglycans synthesized by different layers of adult human femoral head cartilage. Osteoarthritis Cartilage 2 (2), 119–131.

Mohanraj, B., Farran, A.J., Mauck, R.L., Dodge, G.R., 2013. Time-dependent functional maturation of scaffold-free cartilage tissue analogs. J Biomech.

Moroni, L., Hamann, D., Paoluzzi, L., Pieper, J., De Wijn, J.R., Van Blitterswijk, C.A., 2008. Regenerating articular tissue by converging technologies. PLoS One 3 (8), e3032.

Moroni, L., Schotel, R., Sohier, J., De Wijn, J.R., Van Blitterswijk, C.A., 2006. Polymer hollow fiber three-dimensional matrices with controllable cavity and shell thickness. Biomaterials 27 (35), 5918–5926.

Mow, V.C., Proctor, C.S., Kelly, M.A., 2012. Biomechanics of articular cartilage. In: Basic biomechanics of the musculoskeletal system, Nordin, M., Frankel, V.H., (eds.). 4th ed. Philadelphia: Lippincott Williams & Wilkins.

Muir, H., Bullough, P., Maroudas, A., 1970. The distribution of collagen in human articular cartilage with some of its physiological implications. J Bone Joint Surg Br 52 (3), 554–563.

Murray, R.C., Smith, R.K., Henson, F.M., Goodship, A., 2001. The distribution of cartilage oligomeric matrix protein (COMP) in equine carpal articular cartilage and its variation with exercise and cartilage deterioration. Vet J 162 (2), 121–128.

Noble, P.C., Gordon, M.J., Weiss, J.M., Reddix, R.N., Conditt, M.A., Mathis, K.B., 2005. Does total knee replacement restore normal knee function? Clin Orthop Relat Res 431, 157–165.

Norotte, C., Marga, F.S., Niklason, L.E., Forgacs, G., 2009. Scaffold-free vascular tissue engineering using bioprinting. Biomaterials 30 (30), 5910–5917.

O'driscoll, S.W., 1998. The healing and regeneration of articular cartilage. J Bone Joint Surg Am 80 (12), 1795–1812.

Olee, T., Grogan, S.P., Lotz, M.K., Colwell, Jr., C.W., D'Lima, D.D., Snyder, E.Y., 2014. Repair of cartilage defects in arthritic tissue with differentiated human embryonic stem cells. Tissue Eng Part A 20 (3-4), 683–692.

Orth, P., Cucchiarini, M., Kohn, D., Madry, H., 2013. Alterations of the subchondral bone in osteochondral repair—translational data and clinical evidence. Eur Cell Mater 25, 296–314, 299-316; discussion.

Pabbruwe, M.B., Esfandiari, E., Kafienah, W., Tarlton, J.F., Hollander, A.P., 2009. Induction of cartilage integration by a chondrocyte/collagen-scaffold implant. Biomaterials 30 (26), 4277–4286.

Pelletier, J.P., Jovanovic, D.V., Lascau-Coman, V., Fernandes, J.C., Manning, P.T., Connor, J.R., Currie, M.G., Martel-Pelletier, J., 2000. Selective inhibition of inducible nitric oxide synthase reduces progression of experimental osteoarthritis in vivo: possible link with the reduction in chondrocyte apoptosis and caspase 3 level. Arthritis Rheum 43 (6), 1290–1299.

Pfister, B.E., Aydelotte, M.B., Burkhart, W., Kuettner, K.E., Schmid, T.M., 2001. Del1: a new protein in the superficial layer of articular cartilage. Biochem Biophys Res Commun 286 (2), 268–273.

Poole, A.R., Rosenberg, L.C., Reiner, A., Ionescu, M., Bogoch, E., Roughley, P.J., 1996. Contents and distributions of the proteoglycans decorin and biglycan in normal and osteoarthritic human articular cartilage. J Orthop Res 14 (5), 681–689.

Pullig, O., Weseloh, G., Ronneberger, D., Kakonen, S., Swoboda, B., 2000. Chondrocyte differentiation in human osteoarthritis: expression of osteocalcin in normal and osteoarthritic cartilage and bone. Calcif Tissue Int 67 (3), 230–240.

Rasanen, P., Paavolainen, P., Sintonen, H., Koivisto, A.M., Blom, M., Ryynanen, O.P., Roine, R.P., 2007. Effectiveness of hip or knee replacement surgery in terms of quality-adjusted life years and costs. Acta Orthop 78 (1), 108–115.

Ratcliffe, A., Fryer, P.R., Hardingham, T.E., 1984. The distribution of aggregating proteoglycans in articular cartilage: comparison of quantitative immunoelectron microscopy with radioimmunoassay and biochemical analysis. J Histochem Cytochem 32 (2), 193–201.

Ringeisen, B.R., Othon, C.M., Barron, J.A., Young, D., Spargo, B.J., 2006. Jet-based methods to print living cells. Biotechnol J 1 (9), 930–948.

Roelofs, A.J., Rocke, J.P., De Bari, C., 2013. Cell-based approaches to joint surface repair: a research perspective. Osteoarthritis Cartilage 21 (7), 892–900.

Rosenthal, A., Macdonald, A., Voldman, J., 2007. Cell patterning chip for controlling the stem cell microenvironment. Biomaterials 28 (21), 3208–3216.

Roughley, P.J., 2006. The structure and function of cartilage proteoglycans. Eur Cell Mater 12, 92–101.

Saunders, R.E., Gough, J.E., Derby, B., 2008. Delivery of human fibroblast cells by piezoelectric drop-on-demand inkjet printing. Biomaterials 29 (2), 193–203.

Schinagl, R.M., Gurskis, D., Chen, A.C., Sah, R.L., 1997. Depth-dependent confined compression modulus of full-thickness bovine articular cartilage. J Orthop Res 15 (4), 499–506.

Schnapper, A., Meyer, W., 2004. Osteopontin distribution in the canine skeleton during growth and structural maturation. Cells Tissues Organs 178 (3), 158–167.

Schumacher, B.L., Hughes, C.E., Kuettner, K.E., Caterson, B., Aydelotte, M.B., 1999. Immunodetection and partial cDNA sequence of the proteoglycan, superficial zone protein, synthesized by cells lining synovial joints. J Orthop Res 17 (1), 110–120.

Schuurman, W., Harimulyo, E.B., Gawlitta, D., Woodfield, T.B., Dhert, W.J., Van Weeren, P.R., Malda, J., 2013b. Three-dimensional assembly of tissue-engineered cartilage constructs results in cartilaginous tissue formation without retainment of zonal characteristics. J Tissue Eng Regen Med.

Schuurman, W., Levett, P.A., Pot, M.W., Van Weeren, P.R., Dhert, W.J., Hutmacher, D.W., Melchels, F.P., Klein, T.J., Malda, J., 2013a. Gelatin-methacrylamide hydrogels as potential biomaterials for fabrication of tissue-engineered cartilage constructs. Macromol Biosci 13 (5), 551–561.

Seyednejad, H., Gawlitta, D., Dhert, W.J., Van Nostrum, C.F., Vermonden, T., Hennink, W.E., 2011. Preparation and characterization of a three-dimensional printed scaffold based on a functionalized polyester for bone tissue engineering applications. Acta Biomater 7 (5), 1999–2006.

Sharma, B., Williams, C.G., Kim, T.K., Sun, D., Malik, A., Khan, M., Leong, K., Elisseeff, J.H., 2007. Designing zonal organization into tissue-engineered cartilage. Tissue Eng 13 (2), 405–414.

Sherman, S.L., Garrity, J., Bauer, K., Cook, J., Stannard, J., Bugbee, W., 2014. Fresh osteochondral allograft transplantation for the knee: current concepts. J Am Acad Orthop Surg 22 (2), 121–133.

Sherwood, J.K., Riley, S.L., Palazzolo, R., Brown, S.C., Monkhouse, D.C., Coates, M., Griffith, L.G., Landeen, L.K., Ratcliffe, A., 2002. A three-dimensional osteochondral composite scaffold for articular cartilage repair. Biomaterials 23 (24), 4739–4751.

Siczkowski, M., Watt, F.M., 1990. Subpopulations of chondrocytes from different zones of pig articular cartilage. Isolation, growth and proteoglycan synthesis in culture. J Cell Sci 97 (Pt 2), 349–360.

Skardal, A., Zhang, J., Prestwich, G.D., 2010. Bioprinting vessel-like constructs using hyaluronan hydrogels crosslinked with tetrahedral polyethylene glycol tetracrylates. Biomaterials 31 (24), 6173–6181.

Soman, P., Chung, P.H., Zhang, A.P., Chen, S., 2013. Digital microfabrication of user-defined 3D microstructures in cell-laden hydrogels. Biotechnol Bioeng 110 (11), 3038–3047.

Spiller, K.L., Maher, S.A., Lowman, A.M., 2011. Hydrogels for the repair of articular cartilage defects. Tissue Eng Part B Rev 17 (4), 281–299.

Tan, W., Desai, T.A., 2004. Layer-by-layer microfluidics for biomimetic three-dimensional structures. Biomaterials 25 (7-8), 1355–1364.

Theodoropoulos, J.S., De Croos, J.N., Park, S.S., Pilliar, R., Kandel, R.A., 2011. Integration of tissue-engineered cartilage with host cartilage: an in vitro model. Clin Orthop Relat Res 469 (10), 2785–2795.

Tirella, A., Vozzi, F., De Maria, C., Vozzi, G., Sandri, T., Sassano, D., Cognolato, L., Ahluwalia, A., 2011. Substrate stiffness influences high resolution printing of living cells with an ink-jet system. J Biosci Bioeng 112 (1), 79–85.

Tuan, R.S., Chen, A.F., Klatt, B.A., 2013. Cartilage regeneration. J Am Acad Orthop Surg 21 (5), 303–311.

Venn, M.F., 1978. Variation of chemical composition with age in human femoral head cartilage. Ann Rheum Dis 37 (2), 168–174.

Wakabayashi, T., Matsumine, A., Nakazora, S., Hasegawa, M., Iino, T., Ota, H., Sonoda, H., Sudo, A., Uchida, A., 2010. Fibulin-3 negatively regulates chondrocyte differentiation. Biochem Biophys Res Commun 391 (1), 1116–1121.

Waldman, S.D., Grynpas, M.D., Pilliar, R.M., Kandel, R.A., 2003. The use of specific chondrocyte populations to modulate the properties of tissue-engineered cartilage. J Orthop Res 21 (1), 132–138.

Wendt, D., Stroebel, S., Jakob, M., John, G.T., Martin, I., 2006. Uniform tissues engineered by seeding and culturing cells in 3D scaffolds under perfusion at defined oxygen tensions. Biorheology 43 (3-4), 481–488.

Wong, B.L., Kim, S.H., Antonacci, J.M., McIlraith, C.W., Sah, R.L., 2010. Cartilage shear dynamics during tibio-femoral articulation: effect of acute joint injury and tribosupplementation on synovial fluid lubrication. Osteoarthritis Cartilage 18 (3), 464–471.

Wong, M., Wuethrich, P., Eggli, P., Hunziker, E., 1996. Zone-specific cell biosynthetic activity in mature bovine articular cartilage: a new method using confocal microscopic stereology and quantitative autoradiography. J Orthop Res 14 (3), 424–432.

Woodfield, T.B., Malda, J., De Wijn, J., Peters, F., Riesle, J., Van Blitterswijk, C.A., 2004. Design of porous scaffolds for cartilage tissue engineering using a three-dimensional fiber-deposition technique. Biomaterials 25 (18), 4149–4161.

Wylde, V., Livesey, C., Blom, A.W., 2012. Restriction in participation in leisure activities after joint replacement: an exploratory study. Age Ageing 41 (2), 246–249.

Xu, T., Zhao, W., Zhu, J.M., Albanna, M.Z., Yoo, J.J., Atala, A., 2013. Complex heterogeneous tissue constructs containing multiple cell types prepared by inkjet printing technology. Biomaterials 34 (1), 130–139.

Zanetti, M., Ratcliffe, A., Watt, F.M., 1985. Two subpopulations of differentiated chondrocytes identified with a monoclonal antibody to keratan sulfate. J Cell Biol 101 (1), 53–59.

Zeifang, F., Oberle, D., Nierhoff, C., Richter, W., Moradi, B., Schmitt, H., 2010. Autologous chondrocyte implantation using the original periosteum-cover technique versus matrix-associated autologous chondrocyte implantation: a randomized clinical trial. Am J Sports Med 38 (5), 924–933.

BIOPRINTING FOR SKIN

13

Lothar Koch[1],*, Stefanie Michael[2],*, Kerstin Reimers[2], Peter M. Vogt[2] and Boris Chichkov[1]

[1]*Nanotechnology Department, Laser Zentrum Hannover e.V., Hollerithallee 8, Hannover, Germany*
[2]*Department of Plastic, Hand- and Reconstructive Surgery, Hannover Medical School, Hannover, Germany*
*These authors contributed equally

13.1 SKIN, SKIN SUBSTITUTES, POSSIBLE APPLICATIONS FOR PRINTED SKIN

13.1.1 SKIN: FUNCTION AND STRUCTURE

13.1.1.1 Function

The skin is the largest organ, covering the whole body and having a surface area of 1.5–2 m^2 and a weight of 3–10 kg. Due to its specific structure, it represents a barrier between the external environment and the inside of the person, thereby serving as a highly effective means of protection against environmental harm, such as chemical, mechanical, and thermal influences, irradiation, or pathogens.

The first protection against pathogens is achieved by the acid mantle of the skin, with a pH of 5.5–5.7. It is complemented by antimicrobial peptides (defensins) of the innate immune system and several cell types of the adaptive immune system (e.g. Langerhans cells, macrophages, leucocytes, plasma cells, and mast cells). Furthermore, the apical surface of the skin is inhabited by various bacteria and fungi, which constitute the normal skin flora and contribute to its protective function. As commensals, they protect their habitat—and thereby us—against pathogenic microorganisms (Michael, 2013c).

Apart from its protective function against external influences, the skin is crucial for the maintenance of temperature and water balance homeostasis, avoiding conditions such as dehydration or hypothermia. In case of large burn injuries, the barrier function of the skin is destroyed and liquid and protein loss may result in a life- threatening condition.

Moreover, the skin acts as a sensory organ that can detect pain, contact, pressure, vibration, itching, and temperature. This contributes to our protection, but also to our general perception of the environment by complementing visual or auditory perceptions.

On the social level, the skin is essential in representation, communication, and manifestation of current mood. It is therefore important for daily living, especially for communication and interaction among individuals. In this context, people's skin color may indicate their habits and preferences (e.g. having deeply tanned skin due to regular engagement in outdoor activities or frequent visits to a

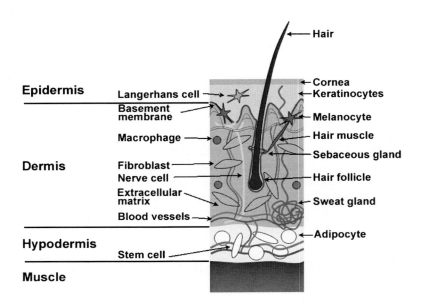

FIGURE 13.1

The image shows the skin with its specific and complex 3D structure. It consists of several layers: the epidermis, which is tightly connected to the dermis via the basement membrane, the dermis, and the hypodermis. Specific cell types and organs are present in the different layers, exhibiting specific functions.

solarium). Also, the hair can send optical signals with social, cultural, psychological, or political content, depending on its form, color, presence, or absence (Michael, 2013c).

13.1.1.2 Structure

The skin is composed of four layers, namely, the epidermis, the basement membrane, the dermis, and the hypodermis. All of these structures, except the basement membrane, contain various specialized cell types and serve different purposes (Figure 13.1).

13.2 EPIDERMIS

The epidermis is the uppermost skin layer and consists of a multilayered stratified squamous epithelium formed by keratinocytes. Depending on its location, the epidermis is normally about 0.05 mm thick, but may reach several millimeters at the palms and the sole of the feet (Michael, 2013c). Throughout life the epidermis regenerates continuously, having a turnover rate of about 30–56 days. It is composed of four layers: the basal layer, the spinous layer, the granular layer, and the cornified layer (from inside to outside). In the basal layer, epidermal stem cells are responsible for the regenerative capacity of the skin. These cells are highly proliferative, are characterized by cytokeratin 5 and 14 expression, and comprise interfollicular stem cells, hair follicle stem cells, and sebaceous gland stem cells (Koster, 2009). They are the source of all keratinocytes that form the epidermis.

During their life span, the keratinocytes undergo terminal differentiation, which is accompanied by many biochemical and morphological changes (Koster, 2009; Houben et al., 2007; Proksch et al., 2008). After a few more replications, the cells lose their replicative capability and start to produce different proteins (e.g. cytokeratins 1 and 10). Filaggrin helps the cytokeratins form tight bundles of intermediate filaments, which are important for the structural integrity of the cell and promote the collapse of the cell into a flattened shape (Houben et al., 2007). A so-called cornified envelope (CE) develops, including proteins like involucrin, loricrin, and filaggrin as well as different lipids (Houben et al., 2007). At that stage, both proteins and lipids are covalently linked (Eckert et al., 2005). During this transition to so-called corneocytes, the cells undergo programmed cell death and are finally embedded in a self-produced intercellular lipid matrix, comprising cholesterol, free fatty acids, and other lipids. These tightly packed lipids are responsible for the low permeability of skin to water and the formation of a permeability barrier. Nevertheless, the lower layers also contribute to the barrier function via the formation of a tight junction. Several cells together form horny flakes, the uppermost layer of the skin, which are shed continuously in humans (Houben et al., 2007).

If the balance between proliferation and differentiation is impaired, a disturbance of the skin barrier may occur, resulting in the formation of inflammation, dermatitis, ichthyosis, and psoriasis, as well as the entry of pathogens (Proksch et al., 2008). Therefore, in healthy persons, this balance is tightly controlled.

To form a stable epithelium, the cells are connected via cell-cell and cell-matrix junctions. Adherens junctions and desmosomes provide a mechanical connection and thereby mechanical strength between neighboring keratinocytes, while hemidesmosomes and actin-linked cell-matrix adhesions are responsible for the coupling of cells to the underlying basement membrane. In adherens junctions, so-called classical cadherins are found, especially e-cadherin in skin. The connection to the basement membrane is realized by integrins. Tight junctions also contribute to the barrier function of the skin by closing the intercellular space between the apical and the basal part of the epithelium. The barrier, however, has a selective permeability that can be altered depending on the circumstances. Moreover, gap junctions form channels between the cytoplasms of neighboring cells via transmembrane proteins (connexins). They allow for the exchange of low molecular substances (e.g. hormones, glucose, or vitamins) as well as ions (e.g. electrical coupling in the heart muscle), and are also essential for cell–cell communication.

Other cell types in the epidermis include melanocytes, which reside in the basal layer. Due to their production of melanin and their distribution to the surrounding keratinocytes, melanocytes are responsible for skin color as well as protection against UV irradiation and corresponding mutagenesis. Langerhans cells are also found in the epidermis. They belong to the adaptive immune system and represent the antigen-presenting cells of the epidermis (Michael, 2013c).

Since the epidermis does not contain any blood vessels, it needs to be supplied with nutrition and oxygen via diffusion from the dermis. At the dermal–epidermal junction both layers interdigitate, forming the rete ridges (epidermal part) and the papillae (dermal part). The resulting high increase of the available surface for diffusion enables the adequate supply of the epidermis. Furthermore, the rete ridges/papillae are essential for the mechanical stability of the skin, as they greatly increase the resistance of skin against shear forces (Michael, 2013c).

13.2.1 BASEMENT MEMBRANE

The basement membrane is essential for the attachment of the epidermis to the dermis and consequently the mechanical strength of skin. Moreover, it stabilizes the connection between the epithelial cells of the basal layer and prevents them from sliding apart. Furthermore, it promotes survival, proliferation, and differentiation of the cells. The basement membrane consists of special extracellular matrix (ECM) molecules, and is partly secreted by the epidermal cells and partly by the dermal cells. It contains collagen type IV and VII, glycoproteins like laminin and fibronectin, nidogen, proteoglycans, and integrins, among others. Collagen type IV forms the basis of the membrane, developed as a felt-like network. It gives the basement membrane its high tensile strength. The other components are mostly integrated in this network. To attach the membrane to the underlying dermis, collagen type VII forms anchoring fibrils (Michael, 2013c).

The basement membrane acts as a barrier between the epidermis and the dermis, keeping the keratinocytes and fibroblasts in their spatial surroundings. In contrast, macrophages, lymphocytes, and nerve cells can cross the membrane by secreting specialized enzymes, which are able to degrade components of the membrane. Of course, nutrients, oxygen, and metabolites must be able to pass the barrier (Michael, 2013c).

13.2.2 DERMIS

The dermis is located underneath the basement membrane. Its thickness varies between 0.6 mm at the eyelids and more than 3 mm at the back and the sole of the feet. The dermis is divided into two parts, with the upper papillary region containing most of the cells and the lower reticular region comprising hair follicles and glands.

The main characteristic of the dermis is the abundant presence of ECM. It is secreted by the resident cells, the fibroblasts. The dermis has several different tasks. Via the proteoglycans and glycoproteins of the ECM, liquid in the form of hydrated water is bound. This results in a gel-like substance that can well resist compressive forces, thereby protecting the underlying tissue. Diffusion of nutrients, metabolites, and hormones becomes possible, ensuring the supply of the skin. The tensile strength and mechanical stability of the dermis and therefore the skin is due to collagen type I, which forms long fibers. Equally essential is elasticity, which is enabled by elastic fibers. Altogether, the dermis can absorb mechanical compression or shear forces, while remaining very pliable (Michael, 2013c).

In contrast to the epidermis, the dermis contains many blood vessels. They are organized in two plexuses, one located directly above the hypodermis, the other directly underneath the basement membrane. The latter is responsible for the supply of the epidermis via diffusion, and reaches into the papillae. The blood vessel system of the dermis also takes part in temperature regulation of the whole body.

Furthermore, the dermis contains sebaceous and perspiratory glands, hair follicles, and nerves. The nerves are responsible for the perception of pain, touch, itch, and temperature. Finally, cells of the innate and the adapted immune system can also be found in the dermis: mast cells, macrophages, leucocytes, and plasma cells (Michael, 2013c).

13.2.3 HYPODERMIS

The hypodermis consists of connective tissue and fat, and has several functions. First, it ensures heat insulation, energy storage, and protection of internal organs against external mechanical forces.

Furthermore, the hypodermis connects the upper skin layers to the lower tissues (e.g. muscles, fascia, and bone). It is also where hormones are generated. The blood vessels in the hypodermis are even larger than those in the dermis, and supply the glands and hair bulges with nutrients and oxygen (Michael, 2013c). Cell types, mainly adipocytes and multipotent stem cells (Zuk et al., 2001), can be found in the hypodermis.

13.2.4 CUTANEOUS APPENDAGES

As earlier mentioned, several cutaneous appendages can be found in the skin, which are formed by specific cells: hair bulges/hair, nails, sebaceous glands, perspiratory glands, scent glands, and mammary glands. Mammary glands enables feeding of the offspring. Perspiratory glands produce sweat and are therefore crucial for temperature homeostasis and regulation. Specialized, hormone-dependent types of hair can be found in different parts of the body. These include eyelashes, eyebrows, beard hairs, pubic and axillary hair, and hair in the nose or the outer ear canal. Sebum secreted by the sebaceous glands keeps the skin and hair pliable. Apart from playing a role in temperature control, hair is also involved in tactile sensation. Nails are a further specialized form of hair. They protect us against mechanical injuries and also support the tactile ability of skin (Michael, 2013c).

13.2.5 SKIN SUBSTITUTES, APPLICATIONS FOR PRINTED SKIN

Skin substitutes/equivalents may have several applications. These comprise wound healing after skin injuries as well as research of skin diseases or corresponding drugs.

13.2.6 INJURIES OF THE SKIN

If the skin is harmed superficially, it will normally heal by itself without complications. But under certain circumstances, even the skin cannot regenerate anymore. This is true for large and/or deep burn injuries. Burns result from various situations ranging from severe sun exposure to scalding with hot water/coffee to electrical or chemical accidents. Direct heat, irradiation, or friction can also cause burns (Pallua and von Bülow, 2006; Herndon, 2002, Chapter 2). Unfortunately, treatment of large and deep burns is only of limited success. Autologous split-thickness skin grafts are normally used to cover the wounds, but donor sites are rare in large burns and extensive scarring is often unavoidable. This may lead to aesthetic and functional impairment. In case of deep wounds, dermal substitutes are used, but insufficient vascularization may lead to rejection of the graft. Moreover, available skin substitutes do not contain hair follicles, melanocytes, and sebaceous and sweat glands, which deprives them of functions like temperature regulation as well as aesthetic appeal. This may in turn result in a low quality of life for patients as they struggle with fear of defacement, rejection, and feelings of failure and worthlessness. In severe cases, scar formation may also lead to impairment of mobility and restrictions in daily living.

Another problematic situation results from chronic wounds that are characterized by their pathophysiology. The normally tightly controlled wound healing is impaired, resulting in badly or non-healing wounds. Chronic wounds comprise three categories: an impaired venous drain due to a faulty microcirculation (venous ulcers of the lower extremities); an increased mechanical strain (decubitus); and a vascular, nerval, and/or metabolic tissue injury (diabetes). Often, chronic wounds are the result of several causes (Riedel et al., 2008). Pain and exudates may cause constrictions of mobility, leading to restrictions at work or in everyday life, and eliciting the same fears and negative feelings among burn patients. Nonchronic but deep and large wounds can also result from tumor resections or accidents.

In all described cases, a skin substitute as similar to physiological skin as possible is needed to be able to adequately replace the lost skin.

13.3 RESEARCH

In research, skin equivalents are used for different purposes. First, they can be applied to replace animal experiments in the development of cosmetics, cleaning agents, and drugs. In the European Union, animal testing for cosmetics is now considered illegal. Another reliable test system is thus needed to assess newly developed components. Observed parameters include inflammations, irritancies, or toxic effects (Gibbs, 2009; Mertsching et al., 2008). The disadvantage of skin equivalents in this context is that no complete organism is present and thus no systemic reaction can be determined. On the other hand, human cells can be used for *in vitro* test systems, thereby avoiding erroneous results due to species specificities (Kuhn et al., 2011).

Furthermore, skin equivalents can be used as a model for (skin) diseases, providing insights into the mechanisms of the disease and/or testing corresponding drugs. In this context, they can comprise different kinds of cells and materials. For example, photo carcinogenesis due to UV irradiation can be assessed in skin equivalents including cells from normal persons compared to cells from *xeroderma pigmentosum* patients. These people suffer from an impaired DNA repair system and are especially prone to skin cancer (Bernerd et al., 2001). Invasion and metastatic potential of skin cancer types may be assessed by including these cell types in the skin equivalents. In this context, the 3D structure including the basement membrane is crucial since cancer cells from the epidermis must be able to pass the basement membrane to develop their invasive potential (Kataoka et al., 2010). Less dangerous but very unpleasant are pigmentation disorders, which can also be studied using skin equivalents (Okazaki et al., 2005). Finally, skin equivalents can be used as an *in vitro* model of psoriasis (Fransson, 2000), to assess antifungal drugs for the treatment of candidiasis (Okeke et al., 2001), or to test new treatments for wound healing (Xie et al., 2010).

13.3.1 SKIN SUBSTITUTES GENERATED BY BIOPRINTING

Very important in any case is the organotypic 3D structure of the skin equivalent, mimicking natural skin and ensuring the potential for natural interaction of the different skin cell types. This specific spatial distribution of the several cell types can be achieved with bioprinting.

For printing skin cells, all three most common bioprinting techniques—microdispensing (extrusion printing), ink-jet printing, and laser-assisted bioprinting—have been applied. For ink-jet bioprinting, mostly commercial office ink-jet printers are modified for cell printing. Microdispensing uses extrusion mechanisms (Khalil et al., 2005; Lee et al., 2009). The material is extruded under pressure out of a nozzle with about 50–1000 mm diameter continuously or in droplets by switching a microvalve with a suitable frequency. Laser-assisted bioprinting is explained later (section 2.3).

13.3.2 **HYDROGEL**

For printing living cells using one of these techniques, the cells are embedded in a medium, mostly hydrogel or a hydrogel precursor, also called sol, such as collagen, fibrinogen, hyaluronic acid, alginate, and the like. They are printed as a hydrogel droplet containing between one and several hundred cells or extruded as a hydrogel strand with embedded cells.

If 3D cell patterns or tissue are to be printed, the hydrogel (precursor) has to fulfill several requirements. First, it needs to be suitable for the printing process. It must have an appropriate viscosity, which depends on the concentration, and material-specific properties. Shear thinning gels or gel precursors are advantageous in reducing shear forces during the printing process, while shear thickening gels are problematic for the abovementioned printing technologies, since shear forces increase disproportionally with the velocity of the gel, which is accelerated during the printing process. To avoid cell damage, lower gel viscosities are required. Alginate and hyaluronic acid are shear thinning gels that are often used for cell printing. Just as important are the wetting characteristics of the gel (precursor) on the printing setup's surface materials. Second, the gel should support cell survival through a convenient environment. A neutral pH-value is therefore mandatory. Third, after printing, the gel functions as the ECM, which in 3D patterns requires a certain degree of stiffness and mechanical strength. Fourth, the gel must not adversely affect the cells, since in general, cells respond to environmental cues. There can be manifold stimuli having physical, mechanical, chemical, and biological backgrounds.

To fulfill these requirements, the hydrogel consists of four components. A primary component, offering a suitable environment with nutrients for the cells, is mixed with a secondary component for optimizing viscosity. Depending on the application, a third component stimuli like growth factors or agents are added. In this mixture, called hydrogel precursor or sol, the cells are suspended. This hydrogel precursor with the embedded cells is then printed in the predefined pattern. Afterwards, a cross-linker is printed for gelation to attain the desired stiffness as the ECM.

Examples are (i) fibrinogen mixed with hyaluronic acid (to have a suitable viscosity for printing) and cross-linked with thrombin or (ii) blood plasma mixed with alginate and cross-linked with calcium chloride. By this means, hydrogels and hydrogel precursors with a wide range of rheological properties (Gruene, et al., 2011a; Lin et al., 2009) have been printed. The cross-linker can be printed or sprayed onto the printed hydrogel precursor in a second step; alternatively, the sol can be printed into a cross-linker reservoir (Yan *Huang, and Chrisey*, 2013).

However, some gels of interest like collagen, the most abundant protein in mammals, do not allow postprint gelling. Collagen is gelled by a change of the pH value. Therefore, a complete merging of all gel components before printing is required, since the cells will not survive in the acidic collagen precursor. Thus, the ability to print high-viscosity gels is needed.

However, the printing technologies have different viscosity limitations for printing living cells, since shear forces, especially in printing nozzles may destroy or affect the cells. Thus, ink-jet printing is limited to low viscosities (< 0.1 Pa s) and low cell densities (up to a few million cells per milliliter) to avoid shear stress in the nozzle and cell clogging (Born et al., 1992). For microdispensing of cells, nozzles with about 50–1000 µm diameter have been applied. With smaller nozzle diameters, similar limitations in viscosity and cell density as with ink-jet printers exist, but these limitations can be varied by the extrusion velocity as a further parameter. With larger extrusion nozzle diameters, higher viscosity and cell density can be printed at the expense of resolution (Chang et al., 2008).

13.3.3 EXTRUSION-BASED BIOPRINTED SKIN CELLS

Lee et al. (2009) successfully printed vital human skin fibroblasts and keratinocytes with an extrusion printer. They printed 10 layers of collagen precursor (e.g. rat tail collagen type I, BD Biosciences, and MA). Since this precursor is acidic, they printed it without cells. Each layer was cross-linked with nebulized aqueous sodium bicarbonate (NaHCO$_3$). In between these layers, they printed one layer of fibroblasts and (after six further collagen layers) one layer of keratinocytes, each suspended in cell culture medium. Thus, they circumvented the general problem with collagen printing. As the most abundant protein in the human skin (and in the human body), collagen has become a major area of interest in tissue engineering. However, the collagen precursor is acidic; after neutralization with a base, which is indispensable for cell survival, the viscosity of the collagen gel increases, hence, it can no longer be printed with small nozzles.

On the other hand, in printing the cells, suspended in a cell medium, separately from the collagen, it might prove difficult to achieve a tissue-like (especially epithelium-like) cell density since the intermittent collagen layers serve as a scaffold but do not contain cells. Additionally, the high volume of the cell media layers might cause displacement between the collagen layers and thus prevent the formation of a stiff 3D hydrogel block.

A peculiarity of the work of Lee et al. is the printing in a 3D free-form mold of poly(dimethylsiloxane) (PDMS). The intention was to replicate the surface contour of a skin wound with the mold. The single layers of fibroblasts and keratinocytes were meant for the demonstration of the "ability to print spatially distinctive cell layers." However, they did not show tissue generation or the establishment of intercellular junctions. Furthermore, the gap of about 75 μm between fibroblasts and keratinocytes did not allow the development of an interface (basement membrane), as can be found between the dermis and epidermis in natural skin.

13.3.4 LASER-ASSISTED BIOPRINTED SKIN

13.3.4.1 Schematic of the Laser-assisted Bioprinting Setup

A further technique used by several groups is laser-assisted bioprinting. In principle, a laser-bioprinting setup consists of a pulsed laser and two coplanar glass slides. The upper one (here called donor) is coated with a thin layer of a laser-absorbing material, and subsequently a layer of the biomaterial that will be printed (i.e. the cells embedded in an appropriate hydrogel). Then, this glass slide is mounted upside-down above the second glass slide (here called collector). The distance is set from a few hundred micrometers up to a few millimeters.

Absorption layers such as gold, titanium, and polymers like triazene (Schiele, et al., 2010) or gelatin (Schiele, Chrisey, and Corr, 2011) have been used. Systems with two layers (Lin et al., 2011) or without absorbing layer (Barron et al., 2004) have also been described. In the latter case, the cell-containing hydrogel itself is used as the laser-absorbing material. A detailed description of the different realizations and denominations of laser-bioprinting are given by Schiele et al. (2010).

The laser pulses are focused through the donor glass slide into the absorbing layer, which is evaporated locally in the focal spot. The high pressure vapor expands as a bubble into the gel layer and propels the subjacent gel toward the collector glass slide. This bubble reaches its maximum size and recollapses after a few microseconds, due to the vapor pressure decreasing below the outer pressure. However, the gel underneath the bubble moves forward further on by inertia. A gel jet then forms, which lasts for a few hundred microseconds (Figure 13.2). Via this jet, between a few picoliters and a few nanoliters of gel (with embedded cells) are transferred to the collector glass slide and remain

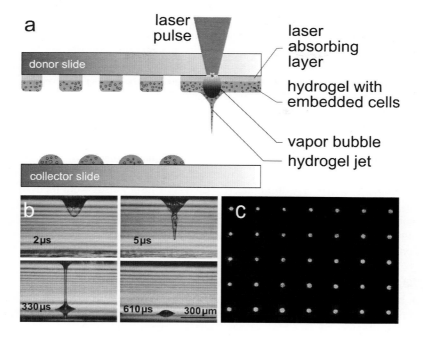

FIGURE 13.2

Laser-assisted bioprinting: (a) schematic sketch of the laser-assisted bioprinting setup. (b) The laser pulse is focused through the donor glass slide into the absorbing layer, which is evaporated immediately. Thereby, a vapor bubble is generated. The subjacent hydrogel is accelerated by the expanding bubble. Although the bubble recollapses within about 6 μs, the hydrogel moves on due to inertia and flows within a few hundred μs as a "jet" onto the donor slide, where it remains as a droplet (Unger et al., 2011). (c) The droplets can be precisely positioned. The diameter of the droplets, each containing several cells, is about 80 μm, the distance between two droplets is 600 μm. The horizontal lines in (b) are interference patterns resulting from the illumination of the process with coherent laser light.

there as a droplet (Duocastella *et al.*, 2009; Unger, et al., 2011). By moving the glass slides and the laser beam, any desired 2D pattern and layer-by-layer 3D pattern can be generated. Figure 13.2 shows a printed pattern of droplets with several fibroblast cells. Fibroblasts labeled with Green Fluorescent Protein (GFP) were embedded in a sol mixture of fibrinogen and hyaluronic acid and printed onto a layer of fibrin with a spacing of 600 μm between droplets.

Often, the collector glass slide is coated with a hydrogel layer before printing to achieve a humid environment for the cells and to cushion the impact. However, the cells survive the printing process also without this hydrogel layer. Instead of the collector glass slide, other items can be put under the upper glass slide to print onto or into (e.g. a scaffold) (Ovsianikov et al., 2010).

The printed droplet volume can be controlled by the thickness and viscosity of the biomaterial layer on the donor glass slide, the thickness of the absorption layer, the laser pulse energy (Gruene et al., 2011a; Guillotin et al., 2010), and the focal spot size. The ratio between printed droplet volume and laser pulse energy is nearly proportional in the relevant energy range at a constant viscosity and layer thickness. In contrast, there is no simple dependency of the printed droplet volume on viscosity or layer thickness at unvaried laser pulse energies. A specific viscosity for every layer thickness exists, at which the printed droplet volume reaches its maximum. This specific viscosity increases with hydrogel layer thickness. Usually, the quantity of cells per droplet is proportional to the droplet volume and the initial cell density in the hydrogel layer on the donor slide, and is subject to statistical variations. The alternative of printing single cell droplets requires a low cell density and is time-consuming, since each cell needs to be targeted separately. The number of printed droplets per second depends on the laser pulse repetition rate and the velocity of the mechanical setup, and might be in the kilohertz range. For printing of tissue as well as high-throughput assembly of cell arrays, studies on multiple cell responses in parallel high printing speeds are needed.

Koch et al. (2010) used a laser-assisted bioprinting technique for the printing of skin. For the printing process, a Nd:YAG-laser (DIVA II; Thales Laser, Orsay, France) with 1064 nm wavelength, about 10 ns pulse duration, and 20 Hz repetition rate was used. The laser pulses were focused with a 60 mm achromatic lens into an ablation spot size with 45 μm diameter (FWHM). The laser pulse energy was set to 40 μJ, corresponding to laser fluency of averaged 1.26 J/cm^2 in the focal spot. As an absorption layer, a 60 nm thin gold layer was deposited on the donor glass slide by sputter coating and a 60 μm thick layer of collagen with embedded skin cells was coated onto the absorption layer by blade coating.

13.3.4.2 The Printing Process Does Not Affect the Cells

Several groups have demonstrated successful cell printing with various cell types. However, for applications it is essential that the printing process does not affect the cells in their vitality, behavior, genotype, and phenotype. The impact of the printing process on skin cells and other cell types was therefore extensively studied. For the skin tissue printing described later, NIH3T3 fibroblasts and HaCaT keratinocytes were used. Investigations into the effects of the printing process on these cell lines are discussed here.

The cell survival rate was examined directly after printing (Koch et al., 2010). A survival rate of 98.4% ± 0.8% for NIH3T3 and 98.6% ± 0.3% for HaCaT was calculated, which is in accordance with studies on other cell types (Hopp et al., 2005). Furthermore, genotoxicity was investigated via a single-cell gel electrophoresis (Comet Assay). It was demonstrated that the printing does not induce DNA strand breaks (Koch et al., 2010). The genotypes of the cells remained unaffected. These findings are consistent with similar investigations by Ringeisen et al. (2004) with P19 pluripotent embryonal carcinoma cells. Apoptosis as a parameter of possible cell death caused by LaBP was assessed by measurement of the activity of caspases 3/7. Up to 48 h after printing, no increase in apoptosis was detected, neither compared to nonprinted control cells nor compared at different test intervals (Koch et al., 2010). Also, the influence of the printing process on cell proliferation was studied by cell counting up to 6 days after printing. In accordance with experiments of other groups on other cell types (Barron et al., 2005; Hopp et al., 2005), no difference in the proliferation behavior of NIH3T3 fibroblasts and HaCaT keratinocytes compared to nonprinted control cells could be found (Koch et al., 2010). Since high temperature is induced in the absorption layer by the laser pulse energy, potential cell damage by heat was investigated via immunocytochemical studies (not shown). No increased expression of heat shock proteins by printed cells was demonstrated (Gruene et al. 2011b). This is in line with experiments of other groups (Barron et al., 2004, 2005). Furthermore, in studies with stem cells, it was shown that the printing process does not affect the immunophenotype (Koch et al., 2010) or the differentiation potential (Gruene, et al. 2011b; Gruene et al., 2011c) of the printed cells. In summary, all studies so far consistently show that the laser printing procedure has no significant effect on the cell; printed cells are viable and fully functional.

13.3.4.3 *Laser-assisted Printing of Skin Tissue – in vitro Culture*

For printing 3D skin tissue, mouse NIH3T3 Swiss albino fibroblast (DSMZ Braunschweig, Germany) and human immortalized HaCaT (DKFZ, Heidelberg, Germany) keratinocyte cell lines were used (Koch et al., 2012; Michael et al., 2013b). These well-established cell lines have been combined in other studies (Bigelow et al., 2005; Delehedde et al., 2001). 3T3 fibroblast cells are often used in keratinocyte cultivation because of secreting growth factors favorable for keratinocytes (Boehnke et al., 2007; Linge, 2004; Schoop et al., 1999).

To approximate native skin as far as possible, collagen type I from rat tail was applied as hydrogel for embedding the cells for the printing process and as ECM afterwards, since this is the main ECM protein in skin (Koch et al., 2012).

As the basis for the printed skin tissue, a 1 mm thick sheet of a collagen-elastin matrix (Matriderm™, MedSkin Solutions Dr. Suwelack AG Billerbeck, Germany) was used for providing mechanical strength directly after printing. Matriderm™ is porous and permeable for cell culture media (Golinski et al., 2009). Successively, 20 layers of NIH3T3 fibroblasts and 20 layers of HaCaT keratinocytes, both embedded in collagen, were printed onto the Matriderm™ substrate. After printing, the skin constructs were cultivated for 10 days submerged in cell media. Histologic sections, depicted in Figure 13.3, showed a tissue-like cell pattern. The printed cells remained in the bilayered 3D structure. Figure 13.3 also depicts the formation of a basal lamina as part of a basement membrane at the interface between keratinocytes and fibroblasts, as it exists between epidermis and dermis in natural skin. Basal laminae are established by epithelial cells like keratinocytes and consist mainly of the proteins collagen type IV and laminin, which is connected to the epithelial cell membranes.

Directly after printing, the 40 printed layers had a total thickness (without the Matriderm™) of about 500 μm. This thickness decreased to 250 μm or 50% within 24 h due to the fibroblasts contracting the surrounding collagen. This shrinkage is well known from literature (Bellows et al., 1982). The remaining thickness of 250 μm is small enough to supply the cells by diffusion.

13.3.4.4 *Tissue Formation In Vitro (Submerged Culture)*

Tissue formation is determined by intercellular junctions (Ko et al., 2000), which can be found as cell–cell and cell–matrix connections in all kinds of tissue, abundantly in epithelium like the epidermis.

Adherens junctions occur in epithelium often as bands that encircle the cell (zonula adherens). In other tissues, only punctate junctions and spots of adhesion can be seen. Adherens junctions are composed mainly of cadherins (calcium-dependent adherent proteins), which are connected to the actin cytoskeleton of the cells via other proteins. At the extracellular side of the cell membrane, cadherins connect with cadherins of adjacent cells by forming homodimers. Adherens junctions are essential for tissue morphogenesis and cohesion (Gumbiner, 1996; Niessen, 2007).

Gap junctions are clusters of cell-cell channels crossing the cell membranes of two adjacent cells and connecting their cytoplasm directly. They enable intercellular communication between neighboring cells by exchanging small molecules ($\leq$ 1000 Dalton) by diffusion and synchronizing electrical as well as physiological activities (Mese et al., 2007; Simon and Goodenough, 1998). Gap junctions are established by two cells, each forming one semichannel (connexon). Connexons consist of normally six connexins (Richard, 2000) in a hexagonal pattern with a free channel in the middle. The connexons of two adjacent cells connect, forming one intercellular channel. Gap junctions play a fundamental role in differentiation, cell cycle progression, and cell survival (Fitzgerald et al., 1994; Schlie et al., 2010).

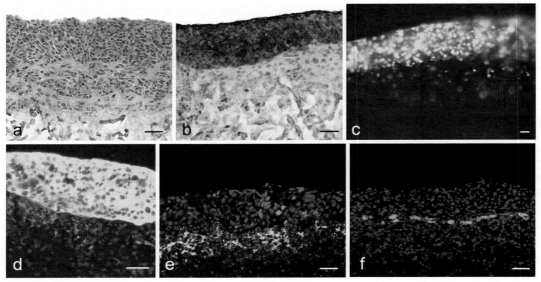

FIGURE 13.3

Laser-printed skin constructs: Skin mimicking bilayered construct: 20 layers of murine fibroblasts (NIH3T3) and 20 layers of human keratinocytes (HaCaT) embedded in collagen type I were printed subsequently on a sheet of a Matriderm™ collagen-elastin matrix. (c) A section through the laser-printed structure, prepared directly after printing, with transduced fibroblasts (red) and keratinocytes (green). (a), (b), (d), (e), and (f) show cryostat sections, prepared 10 days after printing. All scale bars are 50 μm. A hematoxylin and eosin staining (a) shows all printed cells in a tissue-like pattern. Immunoperoxidase staining of cytokeratin 14 in reddish-brown (b) depicts keratinocytes in the bilayered structure while all cell nuclei (fibroblasts and keratinocytes) are counterstained in light blue with haematoxylin. In picture (d) the fibroblasts are stained in red (pan-reticular fibroblast), keratinocytes are stained in green (cytokeratin 14), and cell nuclei are stained in blue (Hoechst 33342). Especially the keratinocytes formed a compact cell organization.

In picture (e) all cell nuclei are stained with Hoechst 33342 (blue), the proliferating cell nuclei are stained with Ki-67 (red), and fibroblasts are stained in green. The fibroblasts and the keratinocytes above are still vital and proliferating. Image (f) depicts an antilaminin staining in green and all cell nuclei in blue (Hoechst 33342). Laminin is a major constituent of the basement membrane in skin. Reprinted from Koch et al. (2012) (Copyright © 2012 Wiley Periodicals, Inc.) A color version of this figure can be viewed online.

Cadherin and connexin localization as well as gap junction coupling are good parameters in the study of tissue properties. Tissue formation of laser-printed 3D skin structures was therefore studied with a focus on cadherins and connexin-43 (Cx43), which is the main connexin in human skin (Fitzgerald et al., 1994; Wiszniewski et al., 2000). The functionality of cell–cell communication via gap junction coupling was assessed with a dye-transfer, a so-called scrape-loading method (Begandt et al., 2010).

The extensive formation of intercellular adherens junctions between printed keratinocytes can be seen in Figure 13.4 by pan-cadherin staining (Koch et al., 2012). Keratinocytes form the dermal epithelium (epidermis) where a higher level of junctions can be found encircling the whole cell (Niessen, 2007). Between fibroblasts there is a minor formation of adherens junctions (not shown).

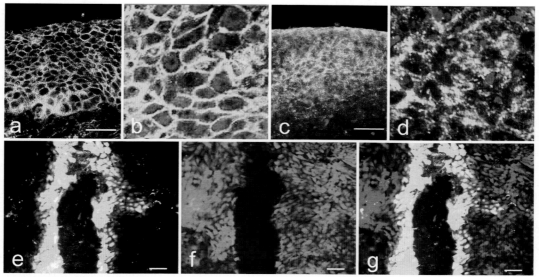

FIGURE 13.4

Laser-printed skin constructs: 10 days after printing, histological sections were prepared to investigate the formation of intercellular junctions, which are essential for tissue function. Fluorescence microscopic images of 3D laser-printed fibroblasts and keratinocytes are depicted as a qualitative analysis of adherens and gap junction formation. (a) pancadherin-staining shows the adherens junctions consisting of cadherins, which are located between the membranes of neighboring cells; (b) pancadherin-staining (green) and cell nuclei staining with Hoechst 33342 (blue); (c,d) connexin-43 (Cx43) staining (green) and Hoechst 33342 nuclei staining (blue), connexins are the constituents of the gap junctions; Picture (d) shows Cx43 distributed in a scattered, punctate fashion, which is a sign for the formation of gap junctions; (e–g) gap junction coupling visualized in green with Lucifer yellow (e,g) after scrape-loading procedure, the keratinocytes, transfected with mCherry, are depicted in red (f,g). All scale bars are 50 μm. Reprint from: Koch et al. (2012) (Copyright © 2012 Wiley Periodicals, Inc.) A color version of this figure can be viewed online.

Immunostaining (Figure 13.4) of connexin 43 (Cx43), the main connexin in human skin, shows its expression in all cells and its localization within the cell membrane 10 days after the cell printing procedure (Koch et al., 2012). As can be seen, Cx43 is distributed in a scattered, punctate fashion, which is a sign for the formation of gap junctions (Morritt, et al., 2007).

The gap junctions' functional capability for cell–cell communication was tested by a dye transfer with a scrape loading method in vital 3D cell constructs. For this purpose, confluently grown 3D-printed keratinocyte cells were scratched with a razor blade and the dye Lucifer yellow (LY) was added, which then could penetrate into the destroyed cells. Since LY molecules are small enough to pass the gap junction channels, the penetration into channel-connected neighboring cells and the diffusion distance through further cells can be analyzed as a measure for gap junction coupling. Twenty layers of keratinocytes embedded in collagen were printed on a glass slide and cultivated for 10 days at 37°C (5% CO_2). After washing with PBS, the slides were submerged in a DPBS solution including 0.25% LY. With a razor blade, one scratch along the whole sample was set. After 5 min of incubation, the slides were washed four times with DPBS for 5 min each. The diffusion distance of LY was documented via fluorescence microscopy (Figure 13.4). The LY dye penetrated from the scratched keratinocyte cells into neighboring cells and further cells, which is only possible through functional gap junctions. The LY went through up to ten cells. This demonstrated that the printed skin grafts possessed epithelium-specific functions with respect to adherens and gap junctions (Koch et al., 2012).

13.3.4.5 Tissue Formation In Vitro (Air–Liquid Interface Culture)

The production of skin equivalents via laser-assisted bioprinting and their culture under submerged conditions has been described. As a first step, this served to assess the printed skin constructs as well as the printing process. As a next step, keratinocyte differentiation, which is crucial for the formation of a multikeratinized epidermis, was induced by different stimuli. First, the skin constructs were raised to the air-liquid-interface (Michael et al., 2013b). This means that the skin equivalents were supplied with nutrients and liquid from below, but had contact to the air from above—like real skin in our bodies. Second, differentiation supporting substances were added (10^{-7} mM hydrocortisone, 10^{-7} mM isoprenaline hydrochloride, and 10^{-7} mM insulin) and the calcium concentration was raised.

Skin constructs were cultured for 11 days and examined on day 0 (day after printing, when raised to the interface), day 5, and day 11. The used cell lines were stably transduced with genes for fluorescent molecules so that the cells could easily be detected via fluorescent microscopy at the end of the experiments.

On day 0, the cells were mostly rounded and still embedded in the collagen, which had served as the printing matrix (Figure 13.5). On days 5 and 10, however, the keratinocytes were connected, forming an epidermis-like tissue. Analogously to the previous experiments, e-cadherin as a marker for adherens junctions could be detected in the epidermis-like layer on days 5 and 10, thereby confirming the formation of an epithelium.

In contrast to the keratinocytes, the printed fibroblasts partly migrated into the underlying Matriderm™, following its fibers closely. Part of the cells stayed at the interface between the Matriderm™ and the keratinocytes, though. This is very promising for future *in vivo* applications.

In the Masson's trichrome staining, inclusions of collagen could be seen in the epidermis, especially directly above the Matriderm™. Probably, the printed cells—especially the fibroblasts—did not completely digest the collagen with which they were printed, but instead migrated out of it, leaving the collagen behind.

In summary, a bilayered skin construct was successfully created by laser-assisted bioprinting and subsequent *in vitro* culture.

13.3.4.6 Laser-assisted Printing of Skin Tissue – In Vivo Culture

On their way to clinical applications, new therapies need to be tested in animal experiments to ensure maximum safety. In the first step, a rodent model is often used since these animals are small, simple to keep and breed, and inexpensive. Different genetically altered strains are also available. As Michael et al. used xenogeneic (keratinocytes) and allogeneic (fibroblasts) cells, they needed to use T-cell-deficient mice to avoid a rejection of the skin transplants (Michael et al., 2013b). Analogously to the *in vitro* experiments, the used cell lines were stably transduced with genes for fluorescent molecules to enable direct detection of the printed cells.

For the *in vivo* evaluation the dorsal skin fold chamber in mice was used. During the preparation of the chamber, the dorsal skin of the anaesthetized mouse was lifted and fixed via two titanium frames (chamber), resulting in a sandwich-like construction. In one side a full-thickness wound was cut, while the opposite skin remained unharmed. The skin construct was then inserted into the wound. The chamber was closed by insertion of a glass cover slip that was fixed with a snap ring. In contrast to humans

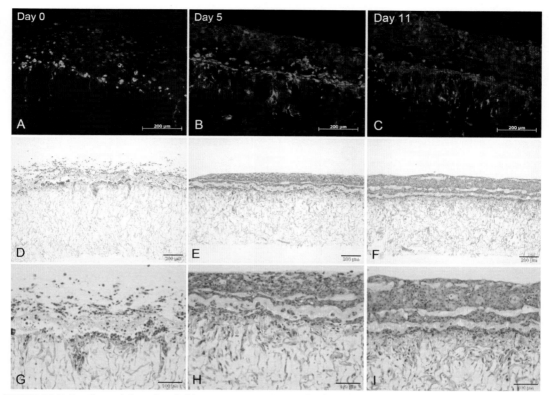

FIGURE 13.5 Sections of the printed skin constructs, cultured *in vitro* at the air-liquid interface.

Printed skin constructs were cultured at the air–liquid interface with differentiation medium for 11 days. Sections (A) to (C) show the printed cells, which directly exhibit fluorescence due to transduction (keratinocytes – red, fibroblasts – green), Masson's trichrome staining shows connective tissue in green and the cells in red. The fibroblasts start to migrate into the Matriderm™ already on day one (A, D, G) and continue to do so later on. The keratinocytes are rounded and not connected on day 1, but already form a dense tissue on day 5. The epidermis even increases in height until day 11. Scale bars represent 200 µm (A–F) and 100 µm (G–I). Reprint from Michael et al. (2013b). A color version of this figure can be viewed online.

whose wounds heal by granulation tissue and re-epithelialization, wounds in rodents are mainly closed by contraction. In the chambers, the skin is fixed and no contraction of the wound area is possible. Furthermore, due to the cover slip, a continuous observation of the wound is possible and no change of dressings is needed. This considerably reduces the stress for the animals.

Skin constructs were printed analogously to the previously described ones and inserted into the chamber the day after printing (day 0). The *in vitro* culture overnight gave the printed cells the opportunity to adhere to their surroundings. The animals were sacrificed after 11 days and analyzed by histology and immunohistochemistry.

At the end of the experiments, the skin constructs were tightly connected to the surrounding host tissue (Figure 13.6). This macroscopic observation could be confirmed on the histological level. The printed cells were alive and could directly be tracked via fluorescence microscopy.

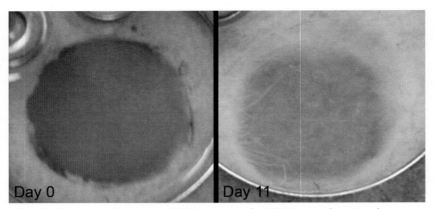

FIGURE 13.6 Printed skin constructs implanted in the dorsal skin fold chamber in nude mice.

Printed skin constructs were inserted into full-thickness skin wounds in the dorsal skin of nude mice and fill the wounds completely after implantation (left). After 11 days, the constructs are connected to the surrounding host tissue (right). Reprint from Michael et al. (2013b).

The printed keratinocytes formed a dense stratified tissue, similar to normal epidermis. In some samples, even a corneal layer could be observed (Figure 13.7). The printed keratinocytes did not exhibit a complete differentiation, though, but a beginning one could be found (Figure 13.8). Since the experiments lasted only 11 days and complete differentiation require about 3 weeks, these results are not unexpected.

Formation of adherens junctions and later tissue/epithelium was confirmed by detection of e-cadherin throughout the whole epithelium (Figure 13.9). Despite the similarity to normal mouse skin, the epidermis developed by the printed keratinocytes was less thick and had no rete ridges. While not optimal for the mechanical stability of the skin, the results looks promising, though, since they represents the first and successful step toward tissue engineering of skin by laser-assisted bioprinting. Thicker epithelium may be printed in the future, including rete ridges.

In contrast to the keratinocytes, only part of the printed fibroblasts remained on top of the Matriderm™, while a large fraction migrated into it. The fibroblasts and keratinocytes together formed a multilayered tissue (Figure 13.7). This is very satisfactory, since an organotypic structure of the skin construct is essential for a future application.

Small blood vessels growing into the Matriderm™ from the depth of the wound bed and the wound edges could be found (Figures 13.7 and 13.10). They were directed toward the printed cells, which may be explained by cytokine expression (e.g. vascular endothelium growth factor or VEGF) of the cells (Ballaun et al., 1995). In control experiments without cells, no vascularization of the constructs at all could be found (Michael et al., 2013a). This strengthens the conclusion that the printed cells induced blood vessel formation.

13.3.4.7 Ink-jet-based In Situ Bioprinted Skin

Instead of printing skin and implanting it afterwards, Binder (2011) used a commercial ink-jet printer to print skin cells directly into wounds *in situ*. He combined the printer with a laser scanning system to measure the wound size and calculate a wound surface model. Then the ink-jet printer, mounted on

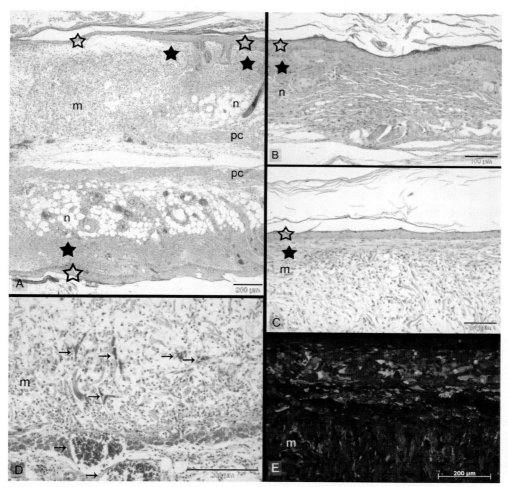

FIGURE 13.7 Histological sections of the printed skin constructs after 11 days in the dorsal skin fold chambers.

Printed skin constructs were implanted in the dorsal skin fold chamber in nude mice. Masson's trichrome staining (A–D) depicts connective tissue in green and cells in red. In an overview, the junction between the skin construct (m – Matriderm™) and the surrounding native mouse skin (n) at the wound edge can be seen (A). The skin construct and the unharmed skin opposite from it are separated by the panniculus carnosus (pc). Both in the native mouse skin (B) and the printed skin (C), a dense epidermis (empty asterisks) and a corneal layer can be seen. The printed keratinocytes form the epidermis in the case of the printed skin constructs, as can be directly detected via their fluorescence (E, keratinocytes – green). The fibroblasts partly migrate into the Matriderm™ (E, fibroblasts – red). Collagen deposition is marked with filled asterisks. Blood vessels (arrows) are present in the printed skin constructs. Scale bars represent 100 μm (A,D,E) and 100 μm (B,C). Reprint from Michael et al. (2013b). A color version of this figure can be viewed online.

a portable XYZ plotting system, was applied to fill the wound with hydrogel and cells. For the printing, they loaded each cell type into an individual cartridge like color inks are contained in different cartridges. Thereby, fibroblasts and keratinocytes could be printed simultaneously.

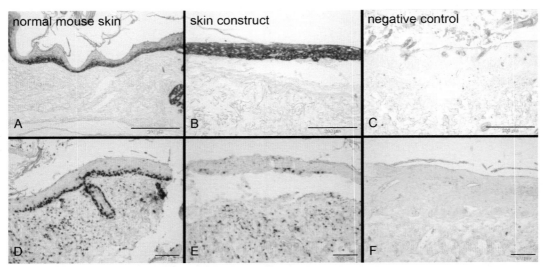

FIGURE 13.8 Detection of cytokeratin 14 and proliferation in printed skin constructs.

Printed skin constructs were cultivated in the dorsal skin fold chamber in mice for 11 days. Cytokeratin 14 expression is only found in the suprabasal layers of the epidermis in native mouse skin (A). In contrast, it is present in the whole epidermis in the printed skin constructs. Proliferation (Ki67) is limited to the suprabasal layers in both normal mouse skin and the printed skin constructs (D,E). Scale bars represent 200 μm (A–C) and 100 μm (D–F). Reprint from Michael et al. (2013b).

With this printer, human skin cells were printed directly into a dorsal full-thickness skin wound on nude immunodeficient mice. For this, the cells were embedded in a hydrogel mixture of fibrinogen and collagen type I. After printing of each layer, the fibrinogen was cross-linked with printed or spray-coated thrombin.

It was demonstrated that the printed human cells remained in the mouse wound area. The printed cells accelerated the closure of the wound compared to printed hydrogel without cells by a factor of up to two. Within 3 weeks the printed skin cells developed a structure with organized dermal collagen and a formed epidermis. The printed human skin cells were still present in the dermis and epidermis of the regenerated skin (Binder, 2011).

As a next step, the printer was used in a porcine wound model. Again, a full-thickness excisional wound was made on the dorsa. The wound was imaged with the laser scanning system and filled with fibroblasts and keratinocytes by printing. Here, the pigs were not immunodeficient and autologous or allogeneic cells were used.

Again, the wounds treated with printed skin cells closed more quickly than the wounds treated with hydrogel alone. Two weeks after printing autologous skin cells, the keratinocytes established areas of epithelialization, which grew and covered the entire wound area within 8 weeks. An epidermis with the same thickness than the surrounding normal skin developed. A dermis-epidermis interface could be observed and the general composition of the printed skin appeared to be similar to native skin. Prelabeled printed skin cells were still visible in the center of the wound of one pig after 8 weeks.

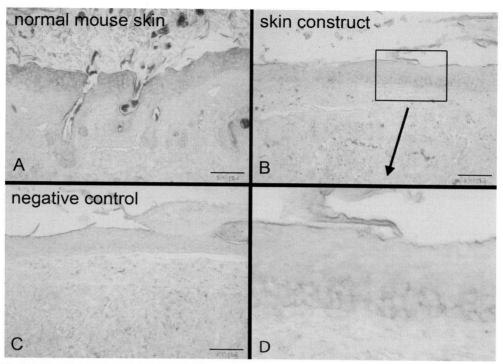

FIGURE 13.9 E-cadherin expression detected by immunohistochemistry.

Printed skin constructs were cultivated in the dorsal skin fold chamber in mice for 11 days. E-cadherin expression can be detected in normal mouse skin (A) as well as in the skin constructs (B, D). Normal mouse skin without first antibody constitutes the negative control (C). All scale bars represent 100 μm. Reprint from Michael et al. (2013b).

13.3.5 DISCUSSION OF THE DIFFERENT BIOPRINTING TECHNIQUES AND CLINICAL APPLICABILITY

13.3.5.1 Optimization of the Skin Equivalents

So far, different bioprinting techniques have been used to print fibroblast and keratinocyte skin cells in a bilayered 3D structure. The formation of tissue with intercellular junctions could be observed *in vitro* and *in vivo* (Koch et al., 2012; Michael et al., 2013b). Implantation of *ex vivo* printed skin equivalents (Michael et al., 2013b) and printing skin cells in wounds *in situ* (Binder, 2011), both result in an ingrowth of the printed tissue into the surrounding natural skin. Thus, the printed skin is capable of covering wounds for preventing liquid or protein loss or infections.

However, these skin equivalents lack the important functions of natural skin, such as effective barrier function, regulation of body temperature by sweat glands, and immune competence. Also, printed skin's function as a sensory organ is very constricted. The appearance of printed skin differs fundamentally from natural skin, particularly in the absence of hair. Moreover, high mechanical stability for printed skin equivalents is needed. Perhaps, some of these functions might be regenerated by immigrating cells from the patient's organism.

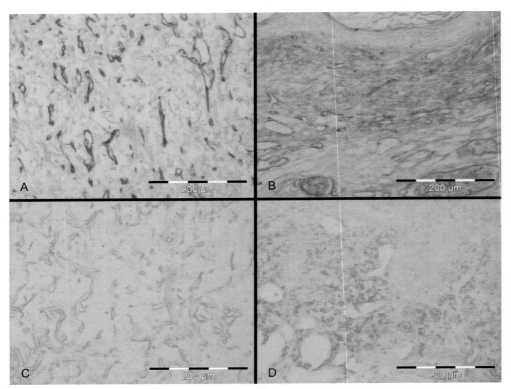

FIGURE 13.10 Presence of blood vessels in the printed skin constructs.

Printed skin constructs were cultivated in the dorsal skin fold chamber in mice for 11 days. Collagen VI expression (brown) indicates blood vessels. In the skin constructs, small vessels can be found in the Matriderm™, reaching from the wound bed in the direction of the cells (A). Small and large vessels can be found in normal mouse skin (B). The respective negative controls are shown in C (Matriderm™) and D (normal mouse tissue). Scale bars represent 200 μm each. Reprint from Michael et al. (2013b).

Nevertheless, skin constructs can be improved with the addition of further cell types such as endothelial cells for quick blood vessel generation, hair follicle cells, melanocytes for pigmentation, immune cells (e.g. Langerhans cells, macrophages, leucocytes, plasma cells, and mast cells), nerve and Schwann cells for perception, or cells present in perspiratory glands to enable sweating and temperature homeostasis. However, simply adding these cell types may not be enough to create all of the necessary cutaneous appendages. Instead, the correct microenvironments need to be created to enable the different cell types to fulfill their respective purposes.

Bioprinting offers the possibility of exactly depositing each cell type at its correct 3D location, helping to create the necessary microenvironment for the cells to fulfill their functions. This makes bioprinting such a promising tool for tissue engineering. While several bioprinting setups, commercially available or individually constructed, are used for tissue engineering research, they have yet to fulfill the requirements for clinical applications in skin tissue generation. For use in clinical skin reconstruction, the printing needs to be safe, rapid, and cost-effective. Thus, several adaptations are needed on the way from bench to bedside and different challenges need to be met.

13.3.6 TECHNICAL AND BIOMEDICAL CHALLENGES

A scale-up of the printed skin substitutes is necessary, especially with regard to large burn injuries. This is partly a technical challenge. The printing area of most printers is too small and the printing velocity of several printers is too low, since often, these printers are not yet throughput-optimized. While scaling up the printing area is just an engineering task, there will be fundamental limits for the printing velocity. Parallel printing with multiple print heads may increase the printing velocity further on toward a high-throughput system. However, a scale-up is also a biological challenge, since a great many of skin cells are needed for large burn wounds. Thus, appropriate cell sources and culture conditions are necessary.

For the production, storage, and delivery of the printed skin substitutes, high standards of quality and safety need to be applied, usually good manufacturing practice (GMP) conditions and the like. The appropriate guidelines include adequate quality management and quality control (including self-inspection), appropriately trained personnel, suitable premises and equipment, and an extensive documentation of all steps of the manufacture, storage, and distribution of products (ISPE, 2014). In terms of printing setups, numerous technical adaptations must therefore be implemented to meet GMP standards. These include the encapsulation of the printing area into a laminar flow hood to ensure sterility as well as the use of disposable printing process and cell culture components to minimize contamination (Chang et al., 2011).

To avoid infection of the cells and the skin tissues, an integration of the different process steps (e.g. isolation of cells from a biopsy, if required; cell culture; bioprinting; and the tissue culture) in one sterile environment would be advantageous. Walles et al. (2014) established an integrated system without a bioprinting system. This "tissue factory" is capable of producing up to 5000 tissue pieces per month fully automated.

13.3.7 STEM CELLS AS POSSIBLE CELL SOURCES FOR BIOPRINTING OF SKIN

On top of the technical challenges, the question on the source of printed cells remains. Ideally, autologous cells should be used to avoid rejection and transmission of (viral) diseases. Although it is quite feasible to propagate enough cells from a skin biopsy prior to a planned surgical intervention of, for example, a small chronic wound, in the case of extensive burn injuries, donor skin for biopsies to gain autologous cells is scarce. Therefore, other cell sources are needed. In this context, the use of stem cells has been proposed. They are abundantly present in the different organs, since most tissue is capable of self-renewal (Metcalfe and Ferguson, 2008). The use of adult stem cells is not restricted by ethical considerations, and adult stem cells can be found in bone marrow, adipose tissue, blood, and skin. In particular, stem cells derived from adipose tissue (ASCs) are very advantageous as they are easy to isolate and propagate, and are abundant in humans (Zuk et al., 2001; Kuhbier et al., 2010). Apart from their differentiation to bone, cartilage, and fat, they can also be differentiated to endothelial cells to promote quick vascularization of skin substitutes (Auxenfans et al., 2011). Moreover, stem cells from sweat glands are easy to gain and propagate; they also enhance vascularization (Danner et al., 2012).

As an alternative to print skin tissue, Skardal et al. (2012) printed stem cells (amniotic fluid-derived stem cells or AFS, and bone marrow-derived mesenchymal stem cells or MSCs embedded in fibrin-collagen gel) in dorsal full-thickness excisional wounds in mice. They demonstrated that these cells considerably stimulate the regeneration of the skin by the mouse organism. Wound closure and re-epithelialization were significantly greater than those in wounds treated by fibrin-collagen gel only. Higher microvessel density, bigger capillary diameters, and thicker epithelial tissue were observed by applying these stem cells. However, the stem cells did not integrate into the skin tissue, but promoted the skin tissue regeneration by secreted trophic factors.

The use of stem cells might be dangerous, though, since stem cells might differentiate into harmful cells or may secrete factors that enhance formation and growth of cancer cells. In this context, the interaction between the tumor cells and their microenvironment is crucial for tumor development, and stem cells may have a large influence on it. On the other hand, mesenchymal stem cells are recruited to and accumulate in tumors. There, they contribute to the angiogenesis of the tumors and suppress anti-tumor immune responses (Cuiffo and Karnoub, 2012). Furthermore, it could be shown that cancer cells are able to induce malignant transformation of MSCs *in vitro* via paracrine effects (Liu et al., 2012).

Generally, the risk of tumor formation depends on the origin of the stem cells, the extent of their *ex vivo* expansion, the induced differentiation, and the site/route of administration. Differences can also be found between embryonic stem cells, which are pluripotent, and adult stem cells, which are multipotent. Therefore, utmost care has to be exercised concerning the choice of the used cells in tissue engineering.

Even if stem cells were safe enough to use, the question remains whether or not the cells should be used in their stem cell phenotype—as a 'growth factor factory'—or whether or not they should be differentiated to skin cells before implantation. And when keratinocytes are used, can they be used in their undifferentiated form, or do they need to be differentiated *in vitro* to form an epithelium before implantation? Thus far, this cannot be answered completely. Often undifferentiated stem cells are employed to utilize their regenerative potential; but these might be involved in tumor formation, as described earlier. Concerning the keratinocytes, Koch et al. (2012) and Michael et al. (2013b) used nondifferentiated cells for bioprinting and implanted them also in a nondifferentiated state. In the animals, they could detect a beginning differentiation. This means that the body with its growth factors could be enough to provide a suitable environment for the differentiation of the cells and the formation of a stable epithelium.

13.4 CONCLUSION

Bioprinting bears the promise of "building" skin in its full complexity, despite the fact that printed skin tissue is far from its natural archetype and lacks most of its functions. The possibility of transferring biomaterial, including cells, macromolecules, and growth factors, to exactly the desired 3D position in a (skin) tissue is the unique feature and major strength of bioprinting. The different printing technologies might meet the requirements for clinical or commercial applications (e.g. high throughput, high resolution, epithelium-like high cell density, and a mechanically stable ECM) to a different degree. However, all of these techniques have further potential for improvement. Skin is particularly suitable for improving bioprinting technologies. Starting with just the two major cell types in skin—fibroblasts and keratinocytes—tissue was already printed. The complexity can be increased gradually by integrating further cell types. Thus, the missing functions may be added bit by bit.

The functional and aesthetical enhancement of printed tissue should be the focus of future research since the printed skin equivalents are still inchoate. The integration of a vascular network is one of the major challenges, but would accelerate wound healing and could prove to be essential since insufficient vascularization often leads to the rejection of transplants, especially very large ones. Moreover, printing skin with a connectable vascular network would enable further applications in drug testing (e.g. in a body-on-a-chip system). First steps toward a printed vascular network, applied *in vivo*, were already done (Gaebel et al., 2011).

ACKNOWLEDGMENTS

The studies described here have been supported by Deutsche Forschungsgemeinschaft, SFB TransRegio 37, REBIRTH Cluster of Excellence (Exc62/1), and by Land Niedersachsen and Volkswagenstiftung in the Biofabrication for NIFE project.

REFERENCES

Auxenfans, C., Lequeux, C., Perrusel, E., Mojallal, A., Kinikoglu, B., Damour, O., 2011. Adipose-derived stem cells (ASCs) as a source of endothelial cells in the reconstruction of endothelialized skin equivalents. Journal of Tissue Engineering and Regenerative Medicine 6 (7), 512–518.

Ballaun, C., Weninger, W., Uthman, A., Weich, H., Tschachler, E., 1995. Human keratinocytes express the three major splice forms of vascular endothelial growth factor. The Journal of Investigative Dermatology 104 (1.), 7–10.

Barron, J.A., Krizman, D.B., Ringeisen, B.R., 2005. Laser printing of single cells: statistical analysis, cell viability, and stress. Annals of Biomedical Engineering 33 (2.), 121–130.

Barron, J.A., Ringeisen, B.R., Kim, H., Spargo, B.J., Chrisey, D.B., 2004. Application of laser printing to mammalian cells. Thin Solid Films. 453-454, 383–387.

Bellows, C.G., Melcher, A.H., Aubin, J.E., 1982. Association between tension and orientation of periodontal ligament fibroblasts and exogenous collagen fibres in collagen gels in vitro. J Cell Sci 58, 125–138.

Begandt, D., Bintig, W., Oberheide, K., Schlie, S., Ngezahayo, A., 2010. Dipyriamole increases gap junction coupling in bovine GM-7373 aortic endothelial cells by a cAMP-protein kinase A dependent pathway. Journal of Bioenergetics and Biomembranes 42, 79.

Bernerd, R., Asselineau, D., Vioux, C., Chevallier-Lagente, O., Bouadjar, B., Sarasin, A., Magnaldo, T., 2001. Clues to epidermal cancer proneness revealed by reconstruction of DNA repair-deficient *xeroderma pigmentosum* skin *in vitro*. Proceedings of the National Academy of Science U S A 98 (14), 7817–7822.

Bigelow, R.L.H., Jen, E.Y., Delehedde, M., Chari, N.S., McDonnell, T.J., 2005. Sonic Hedgehog Induces Epidermal Growth Factor Dependent Matrix Infiltration in HaCaT Keratinocytes. Journal of Investigative Dermatology 124, 457.

Binder, K.W., 2011. In Situ Bioprinting of the Skin. A Thesis submitted in partial fulfillment of the Requirements of Wake Forest University for the Degree of Doctor of Philosophy. Wake Forest University, Winston-Salem, NC, USA.

Boehnke, K., Mirancea, N., Pavesio, A., Fusenig, N.E., Boukamp, P., Stark, H.J., 2007. Effects of fibroblasts and microenvironment on epidermal regeneration and tissue function in long-term skin equivalents. Eur J Cell Biol 86, 731.

Born, C., Zhang, Z., Al-Rubeai, M., Thomas, C.R., 1992. Estimation of Disruption of Animal Cells by Laminar Shear Stress. Biotechnol Bioeng 40, 1004.

Chang, C.C., Boland, E.D., Williams, S.K., Hoying, J.B., 2011. Direct-write bioprinting three-dimensional biohybrid systems for future regenerative therapies. Journal of Biomedical Materials Research Part B: Applied Biomaterials 98 (1), 160–170.

Chang, R., Nam, J., Sun, W., 2008. Effects of Dispensing Pressure and Nozzle Diameter on Cell Survival from Solid Freeform Fabrication-Based Direct Cell Writing. Tissue Eng Part A 14, 41.

Cuiffo, B.G., Karnoub, A.E., 2012. Mesenchymal stem cells in tumor development: emerging roles and concepts. Cell Adhesion and Migration 6 (3), 220–230.

Danner, S., Kremer, M., Petschnik, A.E., Nagel, S., Zhang, Z., Hopfner, U., Reckhenrich, A.K., Weber, C., Schenck, T.L., Becker, T., Kruse, C., Machens, H.G., Egaña, J.T., 2012. The use of human sweat gland-derived stem cells for enhancing vascularization during dermal regeneration. Journal of Investigative Dermatology 132 (6.), 1707–1716.

Delehedde, M., Cho, S.H., Hamm, R., Brisbay, S., Ananthaswamy, H.N., Kripke, M., McDonnell, T.J., 2001. Impact of Bcl-2 and Ha-ras on Keratinocytes in Organotypic Culture. J Invest Dermatol 116, 366.

Duocastella, M., Fernández-Pradas, J.M., Morenza, J.L., Serra, P., 2009. Time-resolved imaging of the laser forward transfer of liquids. J. Appl. Phys 106, 084907.

Eckert, R.L., Sturniolo, M.T., Broome, A.-M., Ruse, M., Rorke, E.A., 2005. Transglutaminase function in epidermis. Journal of Investigative Dermatology 124 (3), 481–492.

Fitzgerald, D.J., Fusenig, N.E., Boukamp, P., Piccoli, C., Mesnil, M., Yamasaki, H., 1994. Expression and function of connexin in normal and transformed human keratinocytes in culture. Carcinogenesis 15, 1859.

Fransson, J., 2000. Tumour necrosis factor-alpha does not influence proliferation and differentiation of healthy and psoriatic keratinocytes in a skin-equivalent model. Acta Dermato-Venereologica 80 (6), 416–420.

Gaebel, R., Ma, N., Liu, J., Guan, J., Koch, L., Klopsch, C., Gruene, M., Toelk, A., Wang, W., Mark, P., Wang, F., Chichkov, B., Li, W., Steinhoff, G., 2011. Patterning human stem cells and endothelial cells with laser printing for cardiac regeneration. Biomaterials 32, 9218–9230.

Gibbs, S., 2009. *In vitro* irritation models and immune reactions. Skin Pharmacology and Physiology 22 (2), 103–113.

Golinski, P.A., Zöller, N., Kippenberger, S., Menke, H., Bereiter-Hahn, J., Bernd, A., 2009. Development of an engraftable skin equivalent based on matriderm with human keratinocytes and fibroblasts. Handchir Mikrochir Plast Chir 41 (6), 327.

Gruene, M., Unger, C., Koch, L., Deiwick, A., Chichkov, B., 2011a. Dispensing pico to nanolitre of a natural hydrogel by laser-assisted bioprinting. Biomedical Engineering Online 10, 19.

Gruene, M., Deiwick, A., Koch, L., Schlie, S., Unger, C., Hofmann, N., Bernemann, I., Glasmacher, B., Chichkov, B., 2011b. Laser printing of stem cells for biofabrication of scaffold-free autologous grafts. Tissue Engineering, Part C Methods 17, 79–87.

Gruene, M., Pflaum, M., Deiwick, A., Koch, L., Schlie, S., Unger, C., Wilhelmi, M., Haverich, A., Chichkov, B., 2011c. Adipogenic differentiation of laser-printed 3D tissue grafts consisting of human adipose-derived stem cells. Biofabrication 3, 015005.

Guillotin B., Souquet A., Catros S., Duocastella M., Pippenger B., Bellance S., Bareille R., Rémy M., Bordenave L., Amédée J., Guillemot F. (2010) Laser assisted bioprinting of engineered tissue with high cell density and microscale organization *Biomaterials*. 31. P.7250-7256.

Gumbiner, B.M., 1996. Cell adhesion: the molecular basis of tissue architecture and morphogenesis. Cell 84, 345.

Herndon, D.N., 2002. Total Burn Care, 2nd Ed. Saunders W. B.

Hopp, B., Smausz, T., Kresz, N., Barna, N., Bor, Z., Kolozsvari, L., Chrisey, D.B., Szabo, A., Nogradi, A., 2005. Survival and proliferative ability of various living cell types after laser-induced forward transfer. Tissue Engineering 11 (11-12), 1817–1823.

Houben, E., De Paepe, K., Rogiers, V., 2007. A keratinocyte's course of life. Skin Pharmacology and Physiology 20 (3), 122–132.

ISPE (2014), International Society for Pharmaceutical Engineering, What is GMP [Online] Available from: http://www.ispe.org/gmp-resources/what-is-gmp.[Accessed: 26th April 2014].

Kataoka, T., Umeda, M., Shigeta, T., Takahashi, H., Komori, T., 2010. A new *in vitro* model of cancer invasion using AlloDerm® a human cadaveric dermal equivalent: a preliminary report. Kobe Journal of Medical Sciences 55 (5.), E106–115.

Khalil, S., Nam, J., Sun, W., 2005. Multi-nozzle deposition for construction of 3D biopolymer tissue scaffolds. Rapid Prototyping J 11, 9.

Ko, K., Arora, P., Lee, W., McCulloch, C., 2000. Biochemical and functional characterization of intercellular adhesion and gap junctions in fibroblasts. American Journal of Physiology - Cell Physiology 279, C147–C157.

Koch, L., Kuhn, S., Sorg, H., Gruene, M., Schlie, S., Gaebel, R., Polchow, B., Reimers, K., Stoelting, S., Ma, N., Vogt, P.M., Steinhoff, G., Chichkov, B., 2010. Laser printing of skin cells and human stem cells, *Tissue Engineering*. Part C Methods 16, 847–854.

Koch, L., Deiwick, A., Schlie, S., Michael, S., Gruene, M., Coger, V., Zychlinski, D., Schambach, A., Reimers, K., Vogt, P.M., Chichkov, B., 2012. Skin tissue generation by laser cell printing. Biotechnology and Bioengineering 109, 1855–1863.

Koster, M.I., (Jul) 2009. Making an epidermis. Annals of the New York Academy of Science 1170, 7–10.

Kuhn, S., Radtke, C., Allmeling, C., Vogt, P.M., Reimers, K., 2011. Keratinocyte culture techniques in medical and scientific applications. In: Khopkar, U. (Ed.), Skin biopsy- perspectives. InTech.

Lee, W., Debasitis, J.C., Lee, V.K., Lee, J.-H., Fischer, K., Edminster, K., Park, J.-K., Yoo, S.-S., 2009. Multi-layered culture of human skin fibroblasts and keratinocytes through three-dimensional freeform fabrication. Biomaterials 30, 1587.

Lin, Y., Huang, Y., Chrisey, D.B., 2009. Droplet formation in matrix-assisted pulsed-laser evaporation direct writing of glycerol-water solution. Journal of Applied Physics 105 (9), 093111.

Lin, Y., Huang, Y., Chrisey, D.B., 2011. Metallic Foil-Assisted Laser Cell Printing. Journal of Biomechanical Engineering 133, 025001.

Linge, C., 2004. Establishment and Maintenance of Normal Human Keratinocyte Cultures. In: Picot, J. (Ed.), Methods in Molecular Medicine, vol. 107: Human Cell Culture Protocols, Second Edition. Humana Press Inc, Totowa, NJ, p. 1.

Liu, J., Zhang, Y., Bai, L., Cui, X., Zhu, J., 2012. Rat bone marrow mesenchymal stem cells undergo malignant transformation via indirect co-cultured with tumour cells. Cell Biochemistry and Function 30 (8), 650–656.

Mertsching, H., Weimer, M., Kersen, S., Brunner, H., 2008. Human skin equivalent as an alternative to animal testing. GMS Krankenhaushygiene interdisziplinär 3 (1), 1863–5245.

Mese, G., Richard, G., White, T.W., 2007. Gap junctions: Basic structure and function. Journal of Investigative Dermatology 127, 2516–2524.

Metcalfe, A.D., Ferguson, M.W., 2008. Skin stem and progenitor cells: using regeneration as a tissue-engineering strategy. Cellular and Molecular Life Sciences 65 (1), 24–32.

Michael, S., Sorg, H., Peck, C.-T., Reimers, K., Vogt, P.M., 2013a. The mouse dorsal skin fold chamber as a means for the analysis of tissue engineered skin. Burns 39 (1), 82–88.

Michael, S., Sorg, H., Peck, C.-T., Koch, L., Deiwick, A., Chichkov, B., Vogt, P.M., Reimers, K., 2013b. Tissue Engineered Skin Substitutes Created by Laser-Assisted Bioprinting Form Skin-Like Structures in the Dorsal Skin Fold Chamber in Mice. PLoS One 8 (3), e57741.

Michael, S., 2013c. Laser induced forward transfer of biomaterials to create a skin substitute for burn patients. Ph.D. Thesis. Technische Informationsbibliothek und Universitätsbibliothek Hannover (TIB), Hannover.

Morritt, A.N., Bortolotto, S.K., Dilley, R.J., Han, X.L., Kompa, A.R., McCombe, D., Wright, C.E., Itescu, S., Angus, J.A., Morrison, W.A., 2007. Cardiac Tissue Engineering in an *in vivo* vascularized chamber. Circulation 115, 353–360.

Niessen, C.M., 2007. Tight junctions/adherens junctions: Basic structure and function. Journal of Investigative Dermatology 127, 2525–2532.

Okazaki, M., Suzuki, Y., Yoshimura, K., Harii, K., 2005. Construction of pigmented skin equivalent and its application to the study of congenital disorders of pigmentation. Scandinavian Journal of Plastic and Reconstructive Surgery and Hand Surgery 39 (6), 339–343.

Okeke, D.N., Tsuboi, R., Ogawa, H., 2001. Quantification of Candida albicans actin mRNA by the LightCycler system as a means of assessing viability in a model of cutaneous candidiasis. Journal of Clinical Microbiology 39 (10), 3491–3494.

Ovsianikov, A., Gruene, M., Pflaum, M., Koch, L., Maiorana, F., Wilhelmi, M., Haverich, A., Chichkov, B., 2010. Laser printing of cells into 3D scaffolds. Biofabrication 2, 014104.

Pallua, N., von Bülow, S., 2006. Methods of burn treatment. Part II: Technical aspects – Behandlungskonzepte bei Verbrenungen. Chirurg 77 (2), 179–186.

Proksch, E., Brandner, J.M., Jensen, J.M., 2008. The skin: and indispensable barrier. Experimental Dermatology 17 (12), 1063–1072.

Richard, G., 2000. Connexins: a connection with the skin. Experimental Dermatology 9, 77.

Riedel, K., Ryssel, H., Koellensperger, E., Germann, G., Kremer, T., 2008. Pathogenesis of chronic wounds – Pathophysiologie der chronischen Wunde. Der Chirurg 79 (6), 526–534.

Ringeisen, B.R., Kim, H., Barron, J.A., Krizman, D.B., Chrisey, D.B., Jackman, S., Auyeung, R.Y.C., Spargo, B.J., 2004. Laser printing of pluripotent embryonal carcinoma cells. Tissue Engineering 10 (3-4), 483–491.

Schiele, N.R., Corr, D.T., Huang, Y., Raof, N.A., Xie, Y., Chrisey, D.B., 2010. Laser-based direct-write techniques for cell printing. Biofabrication 2, 032001.

Schiele, N.R., Chrisey, D.B., Corr, D.T., 2011. Gelatin-Based Laser Direct-Write Technique for the Precise Spatial Patterning of Cells. Tissue Eng. Part C 17 (3), 289–298.

Schlie, S., Mazur, K., Bintig, W., Ngezahayo, A., 2010. Cell cycle dependent regulation of gap junction coupling and apoptosis in GFSHR-17 granulosa cells. Journal of Biomedical Science and Engineering 3, 884–891.

Schoop, V.M., Mirancea, N., Fusenig, N.E., 1999. Epidermal organization and differentiation of HaCaT keratinocytes in organotypic coculture with human dermal fibroblasts. Journal of Investigative Dermatology 112 (3), 343.

Skardal, A., Mack, D., Kapetanovic, E., Atala, A., Jackson, J.D., Yoo, J., Soker, S., 2012. Bioprinted Amniotic Fluid-Derived Stem Cells Accelerate Healing of Large Skin Wounds. Stem Cells Translational Medicine 1, 792–802.

Simon, A.M., Goodenough, D.A., 1998. Diverse functions of vertebrate gap junctions. Trends in Cell Biology 8, 477.

Unger, C., Gruene, M., Koch, L., Koch, J., Chichkov, B., 2011. Time-resolved imaging of hydrogel printing via laser-induced forward transfer. Applied Physics A 103, 271–277.

Walles H. et al. 2014 Information from conversation and online www.tissue-factory.com.[Accessed: 25th April 2014].

Wiszniewski, L., Limat, A., Saurat, J.-H., Meda, P., Salomon, D., 2000. Differential Expression of Connexins During Stratification of Human Keratinocytes. J Invest Dermatol 115, 278.

Xie, Y., Rizzi, S.C., Dawson, R., Lynam, El., Richards, S., Leavesley, D.I., Upton, Z., 2010. Development of a three-dimensional human skin equivalent wound model for investigating novel wound healing therapies. Tissue Engineering Part C Methods 16 (5), 1111–1123.

Yan, J., Huang, Y., Chrisey, D.B., 2013. Laser-assisted printing of alginate long tubes and annular constructs. Biofabrication 5, 015002.

Zuk, P.A., Zhu, M., Mizuno, H., Huang, J., Futrell, J.W., Katz, A.J., Benhaim, P., Lorenz, H.P., Hedrick, M.H., 2001. Multilineage cells from human adipose tissue: implications for cell-based therapies. Tissue Engineering 7 (2), 211–228.

NANOTECHNOLOGY AND 3D BIOPRINTING FOR NEURAL TISSUE REGENERATION

Wei Zhu[1], Nathan J. Castro[1] and Lijie Grace Zhang[1,2]

[1]*Department of Mechanical and Aerospace Engineering, The George Washington University, Washington DC, USA*
[2]*Department of Medicine, The George Washington University, Washington DC, USA*

14.1 INTRODUCTION

Acute traumatic injuries to, and debilitating chronic degenerative diseases of, the nervous system typically lead to the loss of central nervous system (CNS) and peripheral nervous system (PNS) function. Traumatic injury of the CNS may include traumatic brain injuries (TBI) and spinal cord injuries (SCI) (Shoichet et al., 2008), which have had a severe impact on the national healthcare system and the overall quality of life of patients around the world. Annually, 1.7 million Americans sustain TBIs constituting a third (30.5%) of all injury-related deaths in the United States (Faul et al., 2010). In addition to TBI-related nerve damage, the annual incidence of SCIs is estimated to be approximately 40 cases per million population (not including mortalities at the time of accident) or approximately 12,000 new cases each year leading to a total of ~273,000 persons suffering from SCI (Center, 2013). In spite of the development of numerous clinical treatments, therapeutics for successfully bridging injured nerve gaps and full recovering of neural function are still in their infancy thus leading scientists to draw inspiration from neural tissue engineering strategies in the repair of CNS and PNS injuries (Shoichet et al., 2001; Constans, 2004).

Neural tissue engineering is focused on the development of biological substitutes that integrate biomimetic scaffolding with cells for the restoration, maintenance, and improvement of neural tissue function. As critical components in successful nerve regeneration, 3D scaffolds and scaffolding materials provide a necessary physical support to facilitate cell function resulting in better host tissue engraftment and subsequent new tissue development. Ideally, a tissue-engineered nerve scaffold should meet the following criteria: (i) biocompatibility, the scaffold should promote cell adhesion, proliferation, and differentiation in the absence of immune and cytotoxic response; (ii) biodegradability, the scaffold should degrade at a rate closely matching the formation of new tissue and eventually expelled from the body; (iii) conductivity, neural communication depends on the action potential produced at the synapse. Electrical conductivity of scaffolds can thereby promote neurite outgrowth and nerve regeneration; (iv) suitable mechanical properties, the scaffolds must not increase tension in the lesion site or collapse during regular movement; and (v) porous interconnectivity, a porous structure which mimics the extracellular matrix (ECM) of native tissue allowing for good spatial distribution of cells and exchange of nutrients and waste (Cunha et al., 2011; Subramanian et al., 2009). Additionally, it is encouraged to incorporate morphological and chemical features for improved axonal guidance of the tissue in an effort to bridge discontinuous injuries and facilitate nerve regeneration. Figure 14.1 illustrates the key features that an ideal neural scaffold should have and several typical tissue-engineered neural scaffolds for CNS and PNS nerve regeneration.

In order to achieve maximum axonal regeneration and functional recovery in the CNS, it is also necessary to inhibit scar tissue formation after injury, prevent discontinuity during phagocytosis of dying cells, and guarantee adult neuron viability for initial axonal extension (Ellis-Behnke et al., 2006). Researchers have begun to utilize nanotechnology (such as nanomaterials and 3D nanofabrication techniques) to address these obstacles and improve cell proliferation, migration, and differentiation in various biomimetic scaffolds (Kim et al., 2014; Malarkey et al., 2009; Matson and Stupp, 2011; Saito et al., 2009; Wei et al., 2013; Zhang et al., 2005; Ellis-Behnke et al., 2006). In particular, biologically inspired nanomaterials exhibiting similar dimensions with tissue ECM can facilitate neural cell growth and guided neural tissue regeneration (Zhang and Webster, 2009). For instance, carbon-based nanomaterials are extensively investigated in neural regeneration due to their outstanding electrical conductivity, excellent mechanical strength, and modifiable surface chemistry (Hu et al., 2004). In addition, nanotechnology in tissue engineering readily involves the fabrication of 3D nanostructured scaffolds for improved tissue regeneration (Saracino et al., 2013). Currently, two of the widely used nanofabrication techniques typically include self-assembling and electrospinning (Cunha et al., 2011). The bottom–up self-assembly process commonly refers to the spontaneous formation of a scaffold from specific nanoscale amphipathic molecules while electrospinning utilizes biocompatible polymeric materials to fabricate nanofibrous scaffolds. All of these nanostructured scaffolds have illustrated promising results in bridging injured gaps and recovering nerve function (Cao et al., 2009; Iwasaki et al., 2014; Panseri et al., 2008; Xie et al., 2010).

Although conventional 3D nano/microscaffold fabrication approaches have yielded favorable effects in repairing nerve injures, intrinsic limitations exist with regards to adequate control of scaffold outer shape and internal architecture. To circumvent this problem, 3D bioprinting has garnered greater attention in the production of 3D scaffolds with exact spatial distribution and microstructural cues (Schmidt and Leach, 2003). It allows for the printing of viable cells, bioactive factors, and biomaterials individually or in tandem, layer-by-layer, resulting in complex tissue substitutes with controlled shape and structure based on predesigned models (Ozbolat and Yu, 2013). Furthermore, the capacity of 3D bioprinting to manufacture patient-specific scaffolds renders it superior to other traditional 3D tissue manufacturing techniques with rapidly growing interest in the neural tissue engineering field. This chapter provides an overview of the cutting-edge 3D bioprinting and nanotechnology strategies for nerve regeneration. We will focus on (i) recent development of novel nanomaterials, including self-assembling nanomaterials, carbon-based nanomaterials, and conventional biomaterials for nanoneural scaffold design; and (ii) 3D bioprinting techniques for the fabrication of customized neural scaffolds.

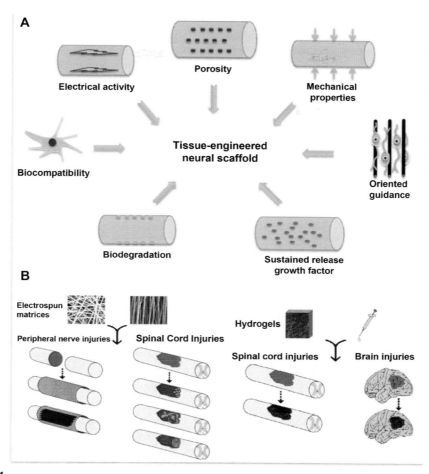

FIGURE 14.1

(A) Schematic illustration of an ideal tissue-engineered neural scaffold. (B) Schematic illustration of tissue-engineered neural scaffolds for nerve repair. In particular, the left image in (B) represents electrospun random or highly aligned scaffolds for repairing peripheral nerve transection and spinal cord injuries. The right image in (B) represents hydrogel scaffolds for healing spinal cord and traumatic brain injuries. Images (B) are adopted from Saracino et al. (2013).

14.2 NANOTECHNOLOGY FOR NEURAL TISSUE REGENERATION
14.2.1 SELF-ASSEMBLING NANOBIOMATERIALS

Self-assembly is a process wherein individual biological components (including viruses, peptides, proteins, and RNA and DNA complexes) organize spontaneously into nanostructures, including nanofibers, nanotubes, vesicles, helical ribbons, and β-sheets (Hosseinkhani et al., 2013; Ko et al., 2010; Smith et al., 2011; Yang et al., 2013b; Zhang et al., 2008). Self-assembling dynamics can be readily influenced by environmental alterations, such as pH, temperature, ionic strength, presence of specific solutes, and mechanical and electrical stimulation (Xu and Kopeček, 2007). These alterations trigger multiple noncovalent interactions, including hydrogen bonding, electrostatic association, and van der Waals forces which drive the formation of self-assembling structures (Stephanopoulos et al., 2013). Natural tissue ECM is composed of various nanostructured components that can spontaneously self-assemble. Therefore, the use of biomimetic self-assembling nanobiomaterials holds high potential in stimulating cell functions and new tissue formation (Zhang and Webster, 2009; Hauser and Zhang, 2010). In the following, we will discuss several self-assembling peptides used for neural tissue engineering. In addition, other self-assembling nanomaterials such as self-assembly bacteriophage and emerging self-assembling nanotubes will be introduced as well.

Peptide amphiphiles are one of the most popular biological molecules for the fabrication of self-assembling nanobiomaterials. Peptides containing both a hydrophilic and hydrophobic moiety have been shown to readily self-assemble into nanofibers wherein the hydrophilic component forms the outer layer and the hydrophobic component is directed toward the core (Yang et al., 2009). They are much more tractable and scalable when compared to complex biomolecules such as proteins. Particularly, β-sheet-forming peptides exhibit the unique capability of assembling into one-dimensional nanostructures via the formation of intermolecular hydrogen bonding (Lock et al., 2013). These one-dimensional nanostructures could be used as building blocks to construct complex 3D networks. In addition, these peptide amphiphiles can be readily modified for enhanced biocompatibility and biodegradability through selecting specific amino acid sequences or controlling over their self-assembling structures (Maude et al., 2013; Cui et al., 2010).

Self-assembling peptide nanofibrous scaffolds (SAPNS) have been used to regenerate PNS and CNS (brain and spinal cord) both *in vitro* and *in vivo*. Zhan et al. reported a novel nanofibrous conduit comprised of a blood vessel and filled with RADA16-I (Ac-RADARADARADARADA-CONH$_2$) SAPNS to repair a 10 mm transection in a rat sciatic nerve (Zhan et al., 2013) (**Figure 14.2**). Compared to a blood vessel only control, the SAPNS-incorporated graft enabled peripheral axons to bridge the 10 mm gap as well as significantly enhance motor neuron protection, axonal regeneration, and remylination. The functional recovery illustrates the great potential of SAPNS-based conduits for the regeneration of peripheral nerve defects. In another work, Liang et al. successfully accelerated brain tissue regeneration via a novel SAPNS assembled by RADA4 (arginine-alanine-aspartic acid-alanine) peptide placed upon a cortical gray matter lesion (Liang et al., 2011). Combined with noninvasive manganese-enhanced magnetic resonance imaging, the chronic optic tract lesion displayed regeneration of axons leading to near-complete repair and return to function. In addition, Ellis-Behnke et al. used RADA16-I SAPNS to create a permissive environment for brain injury regeneration (Ellis-Behnke et al., 2006). Their experiments showed peptide scaffolds have the capacity to direct axons to reconnect to target tissues and return the function of vision using a severed optic tract model in hamsters. Guo et al. investigated the capacity of RADA16-I SAPNS to repair spinal cord injuries (Guo et al., 2007). In their study, neural progenitor cells and Schwann cells were isolated from rats and implanted into the transected dorsal column of the spinal cord after culturing with SAPNS. Results showed host cells migrated well in SAPNS, blood vessels grew, and the spinal cord lesion was further bridged.

In addition to preassembled SAPNS scaffolds, peptides could be directly injected into injured sites and self-assembled *in vivo*. This specific feature makes them an appealing material for the treatment of SCIs. Once liquid self-assembling peptides are injected into the lesions, they will fill the void, regardless of the shape and size, and assemble into a hydrogel. The direct contact between peptides and ECM

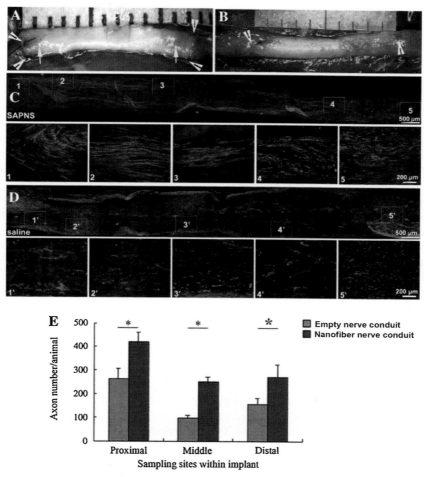

FIGURE 14.2

A sample of blood vessel sheathed self-assembling peptide neural conduit bridging the injured sciatic nerve at the time of (A) immediate after surgery and (B) 16 weeks postimplantation. Arrows indicate the sutured connections. NF200-labeling axons regenerated images in (C) the self-assembling nanofiber filled conduit and (D) empty conduit groups. (E) Axon regeneration comparisons at proximal, middle, and distal part in two neural conduits. Images are adopted from Zhan et al. (2013).

requires that self-assembly occur in a relatively tender environment. In a recent study, Liu et al. showed $K_2(QL)_6K_2$ (QL6) self-assembling scaffolds attenuated post-traumatic inflammation and glial scar formation *in vitro* and *in vivo* (Liu et al., 2013). Unlike most peptides, QL6 is able to self-assemble into β-sheets at neutral pH, which significantly reduced post-traumatic apoptosis, inflammation, and astrogliosis after injection within a clip compression spinal cord injury site. *In vitro* experiments revealed that a QL6 scaffold promoted neuronal differentiation, suppressed astrocytic development, increased conduction velocity of axons, reduced refractoriness, and improved high-frequency conduction.

In designing neural scaffolds, controlling the cellular microenvironment is critical for successful tissue regeneration. Self-assembling scaffolds can also be functionalized with various bioactive molecules and binding peptides such as the laminin-derived IKVAV (isoluceine–lysine–valine–alanine–valine) and cell-adherent RGD (arginine–glycine–aspartic acid) moieties for enhanced cellular attachment and neurite outgrowth (Abdul Kafi et al., 2012; Cheng et al., 2013). In work developed by Tysseling et al., a peptide amphiphile was incorporated on neuroactive epitope IKVAV and injected into a mouse spinal cord injury (Tysseling et al., 2008; Tysseling et al., 2010). After 11 weeks, IKVAV PA nanofibers regenerated both descending motor fibers and ascending sensory fibers. They concluded that 3D self-assembling nanofibers with neuroactive epitopes are able to accelerate and improve regeneration of spinal cord injuries. Moreover, SAPNS can be used as a slow and sustained bioactive factor release system to further enhance tissue regeneration. For instance, Gelain et al. investigated the diffusive mechanisms of active cytokines within RADA16-I (neutral charge), RADA16-DGE (Ac-RADARADARADARADAGGDGEA-CONH2) (negative charge), and RADA16-PFS (Ac-RADARADARADARADAGGPFSSTKT-CONH2) (positive charge) (Gelain et al., 2010). Results demonstrated that protein mobility is directly related to both, physical hindrances and the charge between proteins and peptide nanofibers. Cell studies showed the capability of sustained active cytokine (βFGF) up to 3 weeks when culturing adult neural stem cells. These experiments have opened new avenues for SAPNS-related research with focus on functional molecular release strategies for neural regeneration.

In addition to self-assembling peptide amphiphile nanofibers, several other types of self-assembling nanomaterials have been investigated for neural regeneration. For example, Merzlyak et al. genetically engineered M13 bacteriophages which naturally form nanofiber-like viral structures displaying signaling motifs on their protein coats for directed self-assembly of nanofibrous scaffolds (Merzlyak et al., 2009) (**Figure 14.3**). After self-assembling into nanofibrous matrices, hippocampal neural progenitor cells were seeded to verify the biological capacity of the novel phage nanobiomaterial. Cell studies exhibited the effectiveness of M13 bacteriophage nanofibrous matrices for enhanced cell viability, proliferation, and differentiation as well as directed 3D cell growth. Another type of self-assembling nanobiomaterial is rosette nanotube (RNT), which has shown substantial promise as a new type of nanobiomaterial for tissue regenerative applications (Sun et al., 2012; Zhang et al., 2010, 2009a, 2008, 2009b; Fine et al., 2009). RNTs are biologically inspired supramolecular nanomaterials formed via the self-assembly of low molecular weight DNA base pair motifs (Guanine^Cytosine, G^C, **Figure 14.4**). Nanotubular RNTs composed of a hydrophobic core and hydrophilic outer surface remain stable via electrostatic interactions, base stacking, and hydrophobic effects. Morphologically, RNTs exhibit a 3–4 nm outer diameter and several hundred nanometers in length. One critical feature of RNTs is their flexibility in designing their length, diameter, and surface chemistry. Through functionalization of the G^C motif, predefined chemical and physical properties for specific tissue regeneration can be achieved. In our recent work, we explored human bone marrow mesenchymal stem cell (MSC) adhesion, proliferation, and 4 weeks chondrogenic differentiation on twin-based RNTs with cell adherent RGDSK peptide embedded within poly-L-lactic acid scaffolds (Childs et al., 2013). Our results demonstrated that these biomimetic twin-based nanotubes can significantly enhance MSC growth and chondrogenic differentiation when compared to controls without nanotubes. Both the biomimetic nanostructure and high density of peptides with well-organized architecture contributed the greatly enhanced stem cell functions *in vitro*. Theoretically, any cell-favorable short peptides can be conjugated onto the G^C motifs to modulate surface chemistry, rendering RNTs as a biomimetic nanotemplate for a variety of tissue/organ regeneration.

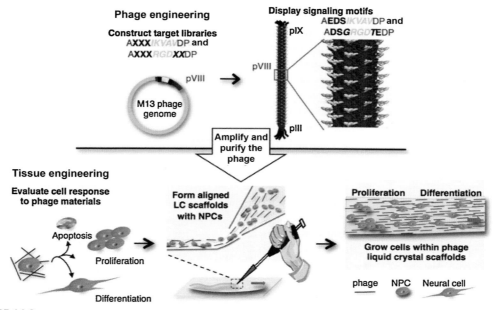

FIGURE 14.3

Schematic illustration of genetically engineered M13 phage and cell response characterization on aligned self-assembling nanofibrous matrices (pVIII, pIII and pIX represent proteins). Image is adopted from Merzlyak et al. (2009).

For instance, neurogenic peptides can be easily conjugated onto RNTs with controlled spatial distribution and density in order to improve specific protein adsorption as well as cell function during neural regeneration.

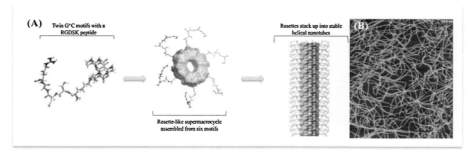

FIGURE 14.4

Schematic illustration of the self-assembly process of RNTs. (A) Six twin G^C motifs with a RGDSK peptide are self-assembled into rosette-like supermacrocycles and then stack up into stable helical nanotubes. RNTs with a 11 Å hollow core are 3–4 nm in diameter and up to several μm long. (B) Atomic force microscopy image of the RNTs with RGDSK.

14.2.2 CARBON NANOBIOMATERIALS

The unique and versatile properties of carbon nanobiomaterials, including excellent electrical, thermal, and mechanical properties, have led to greater research interest. A variety of carbon nanotubes/nanofibers (CNTs/CNFs), and graphene nanomaterials (Tran et al., 2009; Wang et al., 2011b) have attracted considerable attention for use as novel conductive scaffold materials for neural tissue regeneration.

Graphene is a carbon sheet measuring one atom in thickness composed of sp²-bonded carbon atoms condensed in a hexagonal 2D lattice (Fuhrer et al., 2010; Katsnelson, 2007). Due to this unique structure, it is the thinnest and strongest material known (Geim, 2009). As a basic building block of all graphitic forms, graphene can be wrapped into 0D fullerenes, rolled into 1D carbon nanotubes, and stacked into 3D graphite (Geim and Novoselov, 2007; Rao et al., 2009) (**Figure 14.5**). The unique physical and chemical properties of graphene and graphene-based nanomaterials, such as high planar surface area, electronic flexibility, superlative mechanical strength, and unparalleled thermal conductivity have increased their use in biological applications (Yang et al., 2013a). Graphene, graphene oxide, chemically modified grapheme, and other graphene derivatives are some of the more popular candidates for biomedical applications (Wang et al., 2011b). With regards to biocompatibility, recently Park et al. studied nerve cell growth on the surface of graphene showing comparable cell viability when compared to a polystyrene surface (Park et al., 2013a). Furthermore, Tang et al. investigated the formation and performance of neural networks on graphene films (Tang et al., 2013). It was demonstrated that graphene has the capacity to improve nerve cell performance, and promote the growth and facilitate electrical signaling of neural circuits. Such studies revealed the potential of graphene as an

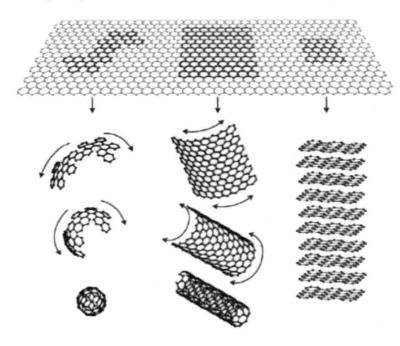

FIGURE 14.5

Graphene forms: fullerenes, carbon nanotubes, and graphite. Images are adopted from Geim and Novoselov (2007).

excellent candidate for neural tissue engineering. In addition to 2D graphene films, a study by Li et al. investigated neural stem cell growth on 3D graphene foams (Li et al., 2013). It was found that neural stem cells can actively proliferate on 3D graphene when compared to 2D graphene films. In addition, 3D graphene promoted neural stem cell differentiation to astrocytes and neurons, as well as efficiently mediated electrical stimulation for differentiated neural stem cells.

Carbon nanotubes are secondary structures composed of single- or multilayered graphene sheets that have been rolled into a tubular structure, resulting in the formation of single-walled (SWCNTs) or multiwalled (MWCNTs) carbon nanotubes (Dai, 2002). The incorporation of CNTs within nanocomposite scaffolds can create neural interfaces with increased interfacial area, conductivity, and electrochemical stability (Keefer et al., 2008). Both *in vitro* and *in vivo* experiments have investigated the cytotoxicity and biocompatibility of CNTs for potential use in neural regenerative applications. For instance, Aldinucci et al. studied the immune-modulatory action of human dendritic cells on MW-CNTs *in vitro* (Aldinucci et al., 2013). Based on their findings, differentiated and activated dendritic cells presented a lower immunogenic profile when interfaced with MWCNTs and the immune reaction modulation was related to topographical and physical features of the growth surface. It was also found that neuronal viability of postnatal mouse dorsal root ganglia was reduced when exposed in higher concentration of MWCNTs containing culture media (Gladwin et al., 2013). In that study, 250 μg/ml of MWCNT-containing media exhibited neuronal death and abnormal neurite morphology while 5 μg/ml of MWCNT-containing media presented no cytotoxicity over 14 day culture. Although there have been cytotoxicity concerns raised about CNTs, thus far the exact mechanisms of CNT's effects on cells are still not fully known. But the results presented show that nanotubes are not cytocompatible only under certain conditions (e.g. certain tube lengths, high concentration, hydrophobicity of bare nanomaterial, and dispersion of nanotubes). It was suggested that CNTs can serve as an excellent nanobiomaterial for neural regeneration through surface and structural modifications for enhanced biocompatibility and cell growth. In the following, we will briefly overview current research progress in this field.

The overall success of tissue regeneration depends heavily on initial cell-scaffold interaction, which is largely determined by the surface properties of the scaffold (Vasita et al., 2008). Modification of surface charge/chemistry as well as integration of signaling complexes on the CNT surfaces can regulate cell adhesion, proliferation, differentiation, matrix remodeling, and tissue organization (Yang et al., 2007). To date, multiple strategies including topographical alterations and chemical functionalization have been explored to modify CNT-based scaffold surface. With regards to topographical alteration, Fan et al. designed super-aligned CNT yarn scaffolds and demonstrated neurite outgrowth extended along the CNT yarns with decreased branching (Fan et al., 2012). Similarly, Béduer et al. investigated neuron growth on SiO_2 substrates patterned with double-walled CNTs (Béduer et al., 2012). It was found that neurons were able to sense the physical and chemical properties of the scaffold surface in a contact-dependent manner. Furthermore, neurons prefer to grow on substrates with patterned CNTs and directed neurite outgrowth can also be modulated by micrometric CNT patterning. It is postulated that directed neuronal outgrowth is attributed to preferential adsorption of specific culture medium proteins. In addition, the topography of CNTs impacts neural differentiation of stem cells when exposed to a chemical inducer. In a study by Park et al., MSCs cultured in neural differentiation media showed enhanced neural gene expression when grown on linear CNT networks when compared to bulk randomly-oriented CNT-based films (Park et al., 2013b). In another study, Park et al. developed a CNT network pattern for selective growth and controlled neuronal differentiation of human neural stem cells into neurons (Park et al., 2011). They illustrated that the patterned CNT network could provide synergistic cues by selective laminin adsorption and optimal nanotopography guidance. The synergistic cues successfully controlled the neuronal differentiation at the level of individual axon or neurite.

In addition to topographical modification, CNTs can be readily functionalized through various chemical methods to sites at the end, side, or induced defect sites (Peng and Wong, 2009). Treatment with strong acids (such as HNO_3 or H_2SO_4) or other oxidative agents will generate oxygenated groups (such as carboxylic acid groups) at CNT ends or defect sites which can undergo further modification by linking specific functional groups (Bekyarova et al., 2005; Banerjee et al., 2005). Chemical functionalization can alter the physical and chemical properties of CNTs including solubility, and electrical and mechanical properties (Bekyarova et al., 2005; Balasubramanian and Burghard, 2005; Hirsch, 2002; Matsumoto et al., 2010; Ni et al., 2005). Several studies have shown the potential of functionalized CNTs to improve biocompatibility and cellular responses. Hu et al. demonstrated, for the first time, the influence of functionalized CNTs with various surface charge on the outgrowth and branching pattern of neurons (Hu et al., 2004). Their studies focused on three kinds of functionalized MWCNTs fabricated by covalently conjugating –COOH, –PABS (poly-m-aminobenzene sulfonic acid), and –EN (ethylenediamine) with negative, neutral, and positive charge at physiological pH, respectively. Neurons grown on MWCNTs with positive surface charge exhibited outstanding cellular response, much more growth cones, extended neurite length, and elaborate neurite branching when compared to neurons grown upon neutral or negatively charged MWCNTs (**Figure 14.6**). Based on these findings, numerous functionalized CNTs have now been reported. A recent study by Gottipati et al. chemically functionalized SWCNT films to modulate the morphology and proliferation of astrocytes (Gottipati et al., 2013). In this work, SWCNTs were covalently linked to polyethylene glycol (PEG) by spraying into films of various thicknesses. When compared to astrocytes grown on unmodified samples, cells grown on SWCNT-PEG films were bigger and rounder in morphology, exhibited decreased immunoreactivity of glial fibrillary acidic protein, and promoted cell proliferation which is associated with the dedifferentiation of astrocytes. In a report by Jan et al., SWCNTs were functionalized with polyelectrolyte (PEI) and assembled layer-by-layer to form films (Jan and Kotov, 2007). The SWCNT-PEI induced mouse embryonic neural stem cell differentiation into neurons,

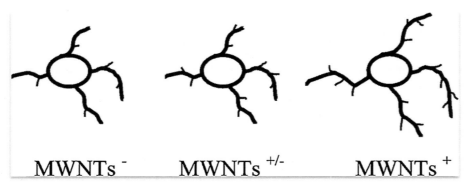

FIGURE 14.6

Schematic of neuron growth on positive, neutral, and negatively charged MWCNTs. Images are adopted from Hu et al. (2004).

astrocytes, and oligodendrocytes with clear neuritis while also exhibiting comparable biocompatibility and expression of neural markers with cells cultured on poly-L-ornithine, one of the most popular substrates for neural stem cells.

14.2.3 ELECTROSPUN POLYMERIC NANOFIBROUS NEURAL SCAFFOLD

In order to repair a complex neural network, guided neurite outgrowth is required toward the synaptic targets with spatially distributed cell populations placed into a specific pattern (Xie et al., 2009a). Aligned nanofibrous scaffolds can provide patterned cues to achieve this goal. Over the past decade, the development of innovative nanofibrous scaffolds has greatly enhanced the involvement of polymers in neural regeneration. Among various nanofibrous scaffold fabrication techniques, electrospinning has been widely investigated for neural tissue regeneration due to its versatility in implementing a variety of synthetic and natural polymeric biomaterials for the creation of highly aligned fibrous scaffolds. These scaffolds have shown promising results due to their structural similarity to fibrous proteins found in native ECM thus facilitating the outgrowth of neurites (Dahlin et al., 2011; Ma et al., 2005).

Electrospinning can be easily tailored by altering polymeric solutions and processing parameters to fabricate scaffolds with varying mechanical and biological properties, as well as fiber and gross scaffold morphology (Sill and von Recum, 2008; Theron et al., 2004; Vasita and Katti, 2006). More than 100 types of natural and synthetic polymers have been employed in electrospinning for various tissue engineering applications (Burger et al., 2006). Biodegradable polyesters, including poly-caprolactone (PCL), poly(lactic-co-glycolic acid) (PLGA), and poly(lactic acid) (PLA) are the most popular synthetic polymers currently used for neural tissue engineering applications (**Table 14.1**). These polymers can be easily tailored for specific mechanical properties and topography by altering the fiber diameter which can range from nano- to microscale for improved cell response (Keun Kwon et al., 2005; Kumbar et al., 2008). Studies have demonstrated the effects of fiber diameter on neural stem cell proliferation and differentiation. Christopherson et al. investigated hippocampus-derived adult neural stem cell response to laminin-coated electrospun polyethersulfone meshes with various fiber diameters (283 ± 45 nm, 749 ± 153 nm and 1452 ± 312 nm) (Christopherson et al., 2009). They found cells exhibited a trend of improved proliferation and spreading, and reduced aggregation with deceasing fiber diameter. In addition to fiber diameter, another topographical cue, fiber orientation, has been shown to play a critical role in neural regeneration. Oriented fibers can serve to guide neurite extension to target injured tissues or organs. Xie et al. examined the neurite outgrowth of primary dorsal root ganglia on electrospun PCL nanofibers with varied fiber orientation, structure, and surface properties (Xie et al., 2009a). They found neurites illustrated radial distribution when cultured on randomly oriented nanofibers. A parallel array of aligned nanofibers, by contrast, guided neurite extension along the direction of the nanofiber. Interestingly, when cultured at the border between aligned and random nanofibers, the same dorsal root ganglia exhibited neurite outgrowth in response to the underlying aligned and randomly oriented nanofibers, respectively (**Figures 14.7 A–B**). In order to achieve further improved cytocompatibility, synthetic polymers can be co-electrospun with natural polymers such as chitosan, fibrin, and collagen or integrated with self-assembling nanobiomaterials that mimic natural ECM chemistry. In a study by Gelain et al., a novel neural guidance channel was fabricated by combining electrospun nanofibers and self-assembling peptides (**Figures 14.7 C–E**) (Gelain et al., 2011). Self-assembling peptide RADA16-I-BMHP1 (Ac-RADARADARADARADAGGPFSSTKT-CONH$_2$) was assembled into PLGA/PCL blended electrospun microchannels followed by implantation within

Table 14.1   Several examples of electrospun polymer scaffolds for neural applications

Materials	Cells or tissues	Fiber morphology	Outcomes	References
PLGA	Rat sciatic nerve	Random nanofibrous conduits	5/11 of nerve injured animals were healed successfully without inflammatory responses after implantation	(Bini et al., 2004)
Poly(L-lactic acid) (PLLA)	Neural stem cell	Aligned fibers with nano- to microscale diameters	Aligned fibers can direct neural stem cell extension and neurites outgrowth; fiber diameters had no significant influence on cell orientation	(Yang et al., 2005)
PLA	Dorsal root ganglia	Nanofibers with aligned, random, and intermediate orientation	Neurites outgrowth was robustly guided on highly aligned fibers and the length on it was 20 and 16% longer when compared to that of random and intermediate fibers, respectively	(Corey et al., 2007)
Poly(ε-caprolactone-co-ethyl ethylene phosphate) encapsulated human glial cell-derived neurotrophic factor (GDNF)	Rat sciatic nerve	Aligned fibers were fold into a roll	Neural recovery was improved after implanting; GDNF encapsulated conduits enhanced neural regeneration	(Chew et al., 2007)
PCL	Schwann cells	Aligned and random	Cell cytoskeleton and nuclei were elongated along the fiber axes; expression of neurotrophin and neurotrophic receptors in aligned cells was down-regulated	(Chew et al., 2008)
PCL	Mouse embryonic stem cells	Aligned nanofibers	Aligned nanofibers enhanced differentiation of mouse embryonic stem cells into neural lineages and guided the neurite outgrowth.	(Xie et al., 2009b)
Silk fibroin (SF) blended poly (L-lactic acid-co-ε-caprolactone) (P(LLA-CL)	Rat sciatic nerve	Aligned nanofibers were reeled into aligned nerve guidance conduits	Better functional recovery in the SF/P(LLA-CL) nerve guidance conduit group	(Wang et al., 2011a)

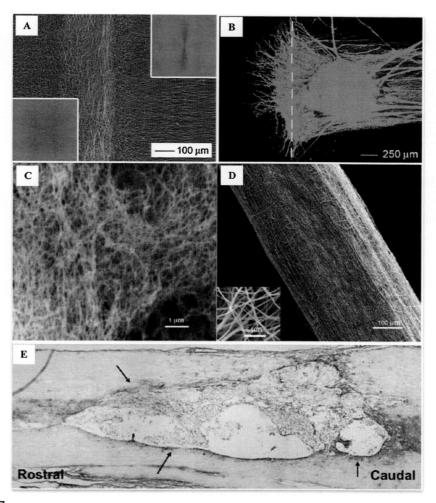

FIGURE 14.7

(A) SEM image of fiber mat with connected random and aligned fiber sand. (B) Dorsal root ganglia cell growth on the border of random and aligned nanofibers. The white dashed line represents the borderline between aligned (right side) and randomly oriented (left side) fibers. Images are adopted from Xie et al. (2009a). The neural conduit fabricated by integrating (C) self-assembled RADA16-I-BMHP1 and (D) PLGA/PCL electrospun channel, and (E) Hematoxylin-esion staining of injured spinal cord after implantation, arrows point the cyst margins. Images are adopted from Gelain et al. (2011).

a chronic SCI rat model. New tissue formation was observed and conspicuous cord reconstruction accompanied the improvement of ascending and descending motor pathways and global locomotion score after 6 months.

In addition to chemical and morphological modifications, researchers have begun to incorporate growth factors within the scaffold to further enhance neural tissue regeneration. A recent study conducted

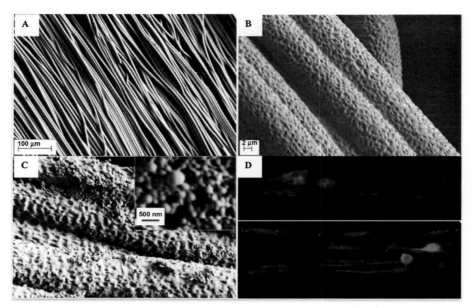

FIGURE 14.8

SEM images of (A) an aligned electrospun neural scaffold fabricated in our lab; (B) random nanopores created on the surface of the electrospun fibers via solvent evaporation; and (C) electrosprayed core-sheel nanospheres incorporated aligned scaffold. (D) Confocal images of axons extension along the direction of aligned fibers (red represents the cell skeleton and axons). A color version of this figure can be viewed online.

in our lab investigated the effects of a highly aligned fibrous neural scaffold using electrospinning in conjunction with electrosprayed growth factor encapsulated nanospheres (**Figure 14.8**). Results demonstrated that aligned scaffolds with sustained growth factor release can improve PC-12 cell growth and effectively guide neural axon extension along the orientation of fiber(s). Another intriguing feature that can improve neural generation is a scaffold's electrical conductivity. Recent studies have shown that electrical stimulation can also improve the communication between nerve and muscle cells, and further promote neural network formation and functional recovery (Ghasemi-Mobarakeh et al., 2009; Huang et al., 2012; Zhang et al., 2013). With regards to this aspect, electrically conductive polymers have offered an attractive option in neural tissue engineering (Chronakis et al., 2006). Two approaches currently employed in the fabrication of conductive electrospun nanofibrous scaffolds rely on the use of inherently electrically conductive polymeric biomaterials such as polypyrrole (PPy) and polyaniline (PANI): either directly dope conductive polymer into electrospun solution or coat nonconductive fibers with a conductive polymer (Ghasemi-Mobarakeh et al., 2009; Jin et al., 2012; Lee et al., 2009). PPy is one of the most widely studied neural biomaterials due to its ease of synthesis, excellent cytocompatibilty, and excellent conductivity (Lee et al., 2009). Both, *in vitro* and *in vivo*, studies have shown PPy-doped or coated neural substitutes are suitable for implantation (Brett Runge et al., 2010; George et al., 2005; Xu et al., 2014). It was also revealed that nanofibrous scaffolds fabricated through electrospinning PCL/gelatin doped with conductive PANI could significantly improve nerve stem cell proliferation and neurite outgrowth under electrical stimulation when compared with the absence of electrical stimulation group (Ghasemi-Mobarakeh et al., 2009).

14.3 3D BIOPRINTING FOR NEURAL TISSUE REGENERATION

3D bioprinting is achieving great popularity in tissue engineering as a means of fabricating customized 3D cellular tissue constructs. The outstanding advantage of 3D bioprinting techniques is their capacity to directly produce complex tissue scaffolds with precise spatial distribution and biomimetic architecture. These techniques can be classified into two categories: scaffold printing and cell or cell-containing material printing.

14.3.1 3D BIOPRINTING NEURAL SCAFFOLDS

Photolithography and ink-jet bioprinting are two popular 3D bioprinting techniques for the manufacture of neural scaffolds. Shepherd et al. fabricated 3D poly(2-hydroxyethyl methacrylate) (pHEMA) neural scaffolds via photolithography (Shepherd et al., 2011). In the study, a photopolymerizable hydrogel ink composed of branched pHEMA chains, HEMA monomer, comonomer, photoinitiator, and water was prepared. The ink was deposited and cross-linked under UV radiation to form a 3D interpenetrating hydrogel network for primary rat hippocampal neuron growth. Results showed scaffold architecture can be controlled precisely and the structure influenced both cell distribution and aligned extension of neurons. In addition, Melissinaki et al. explored a photocurable biodegradable PLA-based resin and fabricated scaffolds via a direct laser writing method (Melissinaki et al., 2011). They conjugated photocurable methacrylate groups to PLA resin and cross-linked using a femtosecond Ti:sapphire laser. Resultant porous scaffolds displayed a maximum resolution of 800 nm and enabled guided neuronal growth. In our lab, we have developed a novel 3D printed nanonerve scaffold through the integration of conductive graphene nanobiomaterials with 3D stereolithography (**Figure 14.9**). Our results have shown that the construct with graphene nanoplatelets has very good cytocompatibility properties. In addition, the graphene nanoplatelets can greatly improve the conductivity of the scaffold, which make the conductive scaffold promising for neural regeneration. Ink-jet printing is another convenient technique to create patterned polymeric structures for the promotion of desired cellular behavior. This technique can readily deposit cell adhesive biomaterials in a precise pattern to guide neural cell growth. A study by Sanjana et al. revealed ink-jet-printed collagen/poly-D-lysine (PDL) on a poly(ethylene) glycol surface can support rat hippocampal neurons and glial growth in defined patterns when compared to collagen/PDL absent regions (Sanjana and Fuller, 2004). With the advancement of molecular biology and development of novel cell-favorable factors, biomimetic nanomaterials could be printed on a traditional scaffolds' surface to obtain more cell-favorable features. Moreover, 3D bioprinting has also provided a means for the incorporation of electrically conductive materials within neural scaffolds. Weng et al. ink-jet-printed PPy/collagen scaffolds and incorporated electrical stimulation into the system (Weng et al., 2012). In this study, PPy and collagen were microstructured on polyarylate film by ink-jet printing for electrical stimulation of a spatially controlled system. The PPy/collagen track was illustrated to guide PC-12 adherence and growth, while electrical stimulation showed the ability to promote neurite outgrowth and orientation (**Figure 14.10**).

14.3.2 3D BIOPRINTING CELLS FOR NEURAL APPLICATIONS

In addition to the aforementioned 3D scaffold printing technologies, there is also great interest in 3D bioprinting neural cells and other native cells in precise spatial distribution for neural applications. **Figure 14.11** shows several important 3D cell bioprinting systems, including laser-based writing, ink-jet bioprinting, and extrusion-based deposition for 3D tissue and organ fabrication (Ozbolat and

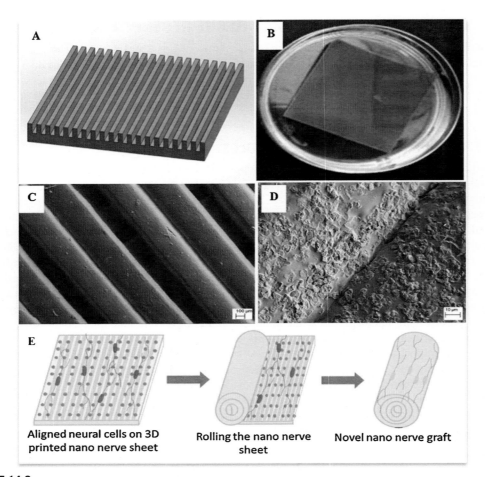

FIGURE 14.9

3D printed aligned PEG-DA neural construct sheet with highly conductive graphene nanoplatelets: (A) is a 3D CAD model of aligned neural construct sheet; (B) Photo image of 3D printed neural construct with graphene nanoplatelets; (C-D) SEM images of the 3D printed scaffold with graphene nanoplatelets at low and high magnifications; and (E) schematic illustration of the 3D nerve scaffold in implantation configuration.

Yu, 2013). These techniques provide great precision over spatial placement of cells within the internal architecture of fabricated scaffolds (Melchels et al., 2012). Multiple cell types can be printed collectively thus producing a highly biomimetic ECM microenvironment for facilitated neural tissue regeneration.

Bioinks are a key element for ink-jet bioprinting where great attention is placed on minimizing cell settling and aggregation. An ideal bioink should exhibit acceptable cell viability while meeting the physical requirements necessary for printing (Derby, 2010). In a recent study by Ferris et al., a novel bioink was created and exhibited considerable cell viability, including highly sensitive neural cells after

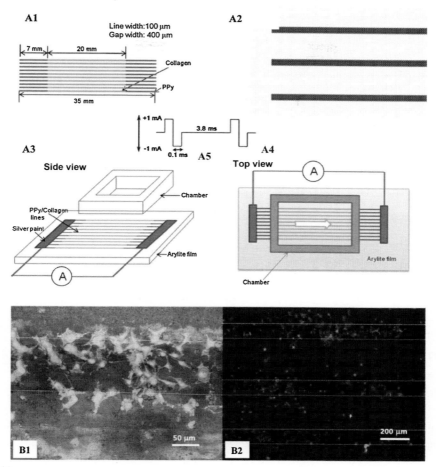

FIGURE 14.10

(A1–A2) Schematic illustration of PPy/collagen complex scaffold design and (A3–A4) PC-12 cells culture and electrical stimulation. (A5) Stimulation waveform. Fluorescence microscopy images of PC-12 cells grew on PPy/collagen scaffold with high (B1) and low (B2) magnification. Images are adopted from Weng et al. (2012).

printing (Ferris et al., 2013). The bioink was composed of endotoxin-free low-acyl gellan gum, Milli-Q water, Dulbecco's Modified Eagles Medium, Poloxamer 188 surfactant, and/or fluorosurfactant solution mixed with cells. Neural (PC-12) and skeletal muscle (C2C12) cells which contained bioinks can maintain high cell viability over extended time periods without settling and aggregation.

Time-released delivery systems are another important feature that is highly desirable with regards to tissue engineered constructs. Growth factors incorporated within tissue substitutes can support and promote cell adhesion and proliferation during the initial period of implantation when endogenic growth factors are absent (Kokai et al., 2011; Lee et al., 2003). Lee et al. developed an artificial neural tissue containing vascular endothelial growth factor (VEGF) release by printing murine neural stem cells

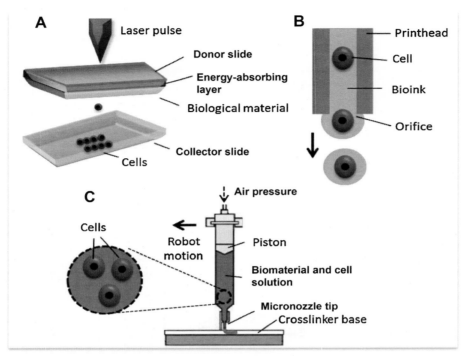

FIGURE 14.11

Typical 3D bioprinting techniques: (A) laser-based writing of cells, (B) ink-jet based systems and (C) extrusion-based deposition. Images are adopted from Ozbolat and Yu (2013).

(C17.2), collagen, and fibrin gel together (Lee et al., 2010). C17.2 cells embedded within the collagen showed comparable viability (92.89 ± 2.32%) after printing with that of manually plated cells. When C17.2 cells were printed 1 mm from border of VEGF- releasing fibrin gel, the cells tended to migrate toward the fibrin gel. This work illustrates the capability of VEGF-containing fibrin gels for sustained release. These findings help to illustrate the role of sustained growth factor delivery within a 3D bioprinted neural construct.

In addition to cellular-level investigations, some studies have investigated the capacity of bioprinted grafts in recovering sensory and motor function. In a study published by Owens et al., a cellular nerve graft containing analogous potential to an autologous graft was fabricated via bioprinting (Owens et al., 2013). Mouse bone marrow stem cells and Schwann cells were shaped into cylindrical units as bioink constituents and then printed layer-by-layer to form a neural graft (**Figure 14.12**). *In vivo* experiments demonstrated that the cellular nerve graft performed satisfactorily in recovering both motor and sensory function. Clinically, cells used for fabricating the cellular graft would be harvested from the patient thus preventing immunological rejection. With regards to defect size, bioprinting allows for the fabrication of clinically relevant grafts and the present work confirmed the superiority of a bioprinted graft when compared to an autologous graft, the current therapeutic "gold standard" for nerve injuries.

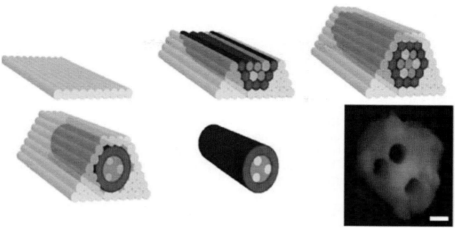

FIGURE 14.12

3D bioprinting a nerve graft. Red indicates bone marrow stem cell units, green are the cylinders composed of 90% bone marrow stem cells and 10% Schwann cells, and grey are removable agarose rods. The last picture is the cross-section view of the printed graft (Schwann cells were labeled as green color). Images are adopted from Owens et al. (2013). A color version of this figure can be viewed online.

14.4 **CONCLUSION AND FUTURE DIRECTIONS**

This chapter reviewed recent advancements with regards to the use of nanotechnology and 3D bioprinting as novel tools to fabricate bioactive constructs for improved neural regeneration. The intrinsic merit of nanotechnology to regenerative medicine is its high capacity to mimic the physical and chemical properties of natural ECM. A variety of self-assembling nanomaterials, carbon nanobiomaterials, and electrospun nanofibers currently under investigation can not only provide a suitable nanoscale environment for neural cell adhesion and growth, but also direct neural tissue repair and regeneration via topographical or chemical guidance. In addition, 3D bioprinting is an excellent approach for the fabrication of patient-specific complex neural grafts where various bioactive factors and cells can be readily incorporated and synergistically employed for improved neural regeneration.

Despite the vast improvements of nanotechnology and 3D bioprinting in neural tissue engineering, the development of ideal neural grafts is still in its infancy. Currently, grafts lack the capacity to bridge defects greater than 30 mm as well as address patients who suffer multiple traumatic injuries (Marquardt and Sakiyama-Elbert, 2013). In addition to spatial guidance of 3D neural constructs, controllable temporal release of growth factors is encouraged for future research. Moreover, current 3D bioprinting techniques also face many challenges and their ultimate success for neural applications largely relies on the development of suitable biomaterials to be used as "inks" for the fabrication of robust neural tissue. Although various traditional biomaterials have continued to be improved, they have been limited by inadequate biomimetic properties that cannot satisfy the strict requirements of 3D bioprinting for neural tissue regrowth. It is ideal to integrate 3D bioprinting and biologically inspired nanobiomaterials that mimic native cellular and ECM for a next generation of neural tissue repair and regeneration.

ACKNOWLEDGMENTS

The authors would like to thank Research Award from the Clinical and Translational Science Institute at Children's National (CTSI-CN) and support from the George Washington University Institute for Nanotechnology (GWIN).

REFERENCES

Abdul Kafi, M., EL-Said, W.A., Kim, T.-H., Choi, J.-W., 2012. Cell adhesion, spreading, and proliferation on surface functionalized with RGD nanopillar arrays. Biomaterials 33, 731–739.

Aldinucci, A., Massacesi, L., Mello, T., Scaini, D., Bianco, A., Ballerini, L., Prato, M., Ballerini, C., Turco, A., Biagioli, T., Toma, F.M., Bani, D., Guasti, D., Manuelli, C., Rizzetto, L., Cavalieri, D., 2013. Carbon nanotube scaffolds instruct human dendritic cells: modulating immune responses by contacts at the nanoscale. Nano letters 13, 6098–6105.

Balasubramanian, K., Burghard, M., 2005. Chemically functionalized carbon nanotubes. Small 1, 180–192.

Banerjee, S., Hemraj-Benny, T., Wong, S.S., 2005. Covalent surface chemistry of single-walled carbon nanotubes. Advanced Materials 17, 17–29.

Béduer, A., Seichepine, F., Flahaut, E., Loubinoux, I., Vaysse, L., Vieu, C., 2012. Elucidation of the role of carbon nanotube patterns on the development of cultured neuronal cells. Langmuir : the ACS journal of surfaces and colloids 28, 17363–17371.

Bekyarova, E., Ni, Y.C., Malarkey, E.B., Montana, V., Mcwilliams, J.L., Haddon, R.C., Parpura, V., 2005. Applications of Carbon Nanotubes in Biotechnology and Biomedicine. Journal of Biomedical Nanotechnology 1, 3–17.

Bini, T.B., Gao, S., Tan, T.C., Wang, S., Lim, A., Hai, L.B., Ramakrishna, S., 2004. Electrospun poly(L-lactide-co-glycolide) biodegradable polymer nanofibre tubes for peripheral nerve regeneration. Nanotechnology 15, 1459–1464.

Brett Runge, M., Dadsetan, M., Baltrusaitis, J., Knight, A.M., Ruesink, T., Lazcano, E.A., Lu, L., Windebank, A.J., Yaszemski, M.J., 2010. The development of electrically conductive polycaprolactone fumarate–polypyrrole composite materials for nerve regeneration. Biomaterials 31, 5916–5926.

Burger, C., Hsiao, B.S., Chu, B., 2006. Nanofibrous materials and their applications. Annual review of materials research 36, 333–368.

Cao, H.Q., Liu, T., Chew, S.Y., 2009. The application of nanofibrous scaffolds in neural tissue engineering. Advanced Drug Delivery Reviews 61, 1055–1064.

Center, N.S.C.I.S., 2013. Spinal Cord Injury Facts and Figures at a Glance. *Spinal Cord Injury Statistical Center, Birmingham.* Alabama.

Cheng, T.-Y., Chen, M.-H., Chang, W.-H., Huang, M.-Y., Wang, T.-W., 2013. Neural stem cells encapsulated in a functionalized self-assembling peptide hydrogel for brain tissue engineering. Biomaterials 34, 2005.

Chew, S.Y., Mi, R., Hoke, A., Leong, K.W., 2008. The effect of the alignment of electrospun fibrous scaffolds on Schwann cell maturation. Biomaterials 29, 653–661.

Chew, S.Y., Mi, R.F., Hoke, A., Leong, K.W., 2007. Aligned protein-polymer composite fibers enhance nerve regeneration: A potential tissue-engineering platform. Advanced Functional Materials 17, 1288–1296.

Childs, A., Hemraz, U.D., Castro, N.J., Fenniri, H., Zhang, L.G., 2013. Novel biologically-inspired rosette nanotube PLLA scaffolds for improving human mesenchymal stem cell chondrogenic differentiation. Biomed Mater 8, 065003.

Christopherson, G.T., Song, H., Mao, H.-Q., 2009. The influence of fiber diameter of electrospun substrates on neural stem cell differentiation and proliferation. Biomaterials 30, 556–564.

Chronakis, I.S., Grapenson, S., Jakob, A., 2006. Conductive polypyrrole nanofibers via electrospinning: Electrical and morphological properties. Polymer 47, 1597–1603.

Constans, A., 2004. Neural tissue engineering. Scientist 18, 40–42.

Corey, J.M., Lin, D.Y., Mycek, K.B., Chen, Q., Samuel, S., Feldman, E.L., Martin, D.C., 2007. Aligned electrospun nanofibers specify the direction of dorsal root ganglia neurite growth. Journal of Biomedical Materials Research - Part A 83, 636–645.

Cui, H., Webber, M.J., Stupp, S.I., 2010. Self-assembly of peptide amphiphiles: from molecules to nanostructures to biomaterials. Biopolymers 94, 1–18.

Cunha, C., Panseri, S., Antonini, S., 2011. Emerging nanotechnology approaches in tissue engineering for peripheral nerve regeneration. Nanomedicine: Nanotechnology, Biology, and Medicine 7, 50–59.

Dahlin, R.L., Kasper, F.K., Mikos, A.G., 2011. Polymeric nanofibers in tissue engineering. Tissue engineering. Part B, Reviews 17, 349–364.

Dai, H., 2002. Carbon nanotubes: synthesis, integration, and properties. Accounts of chemical research 35, 1035–1044.

Derby, B., 2010. Inkjet Printing of Functional and Structural Materials: Fluid Property Requirements, Feature Stability, and Resolution. Annual review of materials research, vol 40 40, 395–414.

Ellis-Behnke, R.G., Liang, Y.X., You, S.W., Tay, D.K., Zhang, S., So, K.F., Schneider, G.E., 2006. Nano neuro knitting: peptide nanofiber scaffold for brain repair and axon regeneration with functional return of vision. Proc Natl Acad Sci U S A 103, 5054–5059.

Fan, L., Feng, C., Zhao, W., Qian, L., Wang, Y., Li, Y., 2012. Directional neurite outgrowth on superaligned carbon nanotube yarn patterned substrate. Nano letters 12, 3668–3673.

Faul, M., Xu, L., Wald, M., Coronado, V., 2010. Traumatic brain injury in the United States: emergency department visits, hospitalizations, and deaths. Atlanta (GA): Centers for Disease Control and Prevention, National Center for Injury Prevention and Control.

Ferris, C.J., Gilmore, K.J., Beirne, S., Mccallum, D., Wallace, G.G., Panhuis, M.I.H., 2013. Bio-ink for on-demand printing of living cells. Biomaterials Science 1, 224–230.

Fine, E., Zhang, L., Fenniri, H., Webster, T.J., 2009. Enhanced endothelial cell functions on rosette nanotube-coated titanium vascular stents. Int J Nanomedicine 4, 91–97.

Fuhrer, M.S., Lau, C.N., Macdonald, A.H., 2010. Graphene: Materially Better Carbon. MRS Bulletin 35, 289–295.

Geim, A.K., 2009. Graphene: status and prospects. Science (New York, N.Y.) 324, 1530–1534.

Geim, A.K., Novoselov, K.S., 2007. The rise of graphene. Nature materials 6, 183–191.

Gelain, F., Baldissera, F., Vescovi, A., Panseri, S., Antonini, S., Cunha, C., Donega, M., Lowery, J., Taraballi, F., Cerri, G., Montagna, M., 2011. Transplantation of nanostructured composite scaffolds results in the regeneration of chronically injured spinal cords. ACS nano 5, 227–236.

Gelain, F., Unsworth, L.D., Zhang, S.G., 2010. Slow and sustained release of active cytokines from self-assembling peptide scaffolds. Journal of controlled release 145, 231–239.

George, P.M., Sur, M., Lyckman, A.W., Lavan, D.A., Hegde, A., Leung, Y., Avasare, R., Testa, C., Alexander, P.M., Langer, R., 2005. Fabrication and biocompatibility of polypyrrole implants suitable for neural prosthetics. Biomaterials 26, 3511–3519.

Ghasemi-Mobarakeh, L., Prabhakaran, M.P., Morshed, M., Nasr-Esfahani, M.H., Ramakrishna, S., 2009. Electrical stimulation of nerve cells using conductive nanofibrous scaffolds for nerve tissue engineering. Tissue engineering. Part A 15, 3605–3619.

Gladwin, K.M., Whitby, R.L.D., Mikhalovsky, S.V., Tomlins, P., Adu, J., 2013. In Vitro Biocompatibility of Multiwalled Carbon Nanotubes with Sensory Neurons. Advanced Healthcare Materials 2, 728–735.

Gottipati, M.K., Samuelson, J.J., Kalinina, I., Bekyarova, E., Haddon, R.C., Parpura, V., 2013. Chemically functionalized single-walled carbon nanotube films modulate the morpho-functional and proliferative characteristics of astrocytes. Nano letters 13, 4387.

Guo, J., Su, H., Zeng, Y., Liang, Y.-X., Wong, W.M., Ellis-Behnke, R.G., So, K.-F., Wu, W., 2007. Reknitting the injured spinal cord by self-assembling peptide nanofiber scaffold. Nanomedicine : nanotechnology, biology, and medicine 3, 311–321.

Hauser, C.A.E., Zhang, S., 2010. Designer self-assembling peptide nanofiber biological materials. Chemical Society reviews 39, 2780–3279.

Hirsch, A., 2002. Functionalization of single-walled carbon nanotubes. Angewandte Chemie (International ed. in English) 41, 1853–1859.

Hosseinkhani, H., Hong, P.-D., Yu, D.-S., 2013. Self-assembled proteins and peptides for regenerative medicine. Chemical reviews 113, 4837–4861.

Hu, H., Ni, Y., Montana, V., Haddon, R.C., Parpura, V., 2004. Chemically Functionalized Carbon Nanotubes as Substrates for Neuronal Growth. Nano Letters 4, 507–511.

Huang, J., Lu, L., Zhang, J., Hu, X., Zhang, Y., Liang, W., Wu, S., Luo, Z., 2012. Electrical stimulation to conductive scaffold promotes axonal regeneration and remyelination in a rat model of large nerve defect. PloS one 7, e39526.

Iwasaki, M., Wilcox, J.T., Nishimura, Y., Zweckberger, K., Suzuki, H., Wang, J., Liu, Y., Karadimas, S.K., Fehlings, M.G., 2014. Synergistic effects of self-assembling peptide and neural stem/progenitor cells to promote tissue repair and forelimb functional recovery in cervical spinal cord injury. Biomaterials 35, 2617.

Jan, E., Kotov, N.A., 2007. Successful differentiation of mouse neural stem cells on layer-by-layer assembled single-walled carbon nanotube composite. Nano letters 7, 1123–1128.

Jin, L., Feng, Z.Q., Zhu, M.L., Wang, T., Leach, M.K., Jiang, Q., 2012. A Novel Fluffy Conductive Polypyrrole Nano-Layer Coated PLLA Fibrous Scaffold for Nerve Tissue Engineering. Journal of Biomedical Nanotechnology 8, 779–785.

Katsnelson, M.I., 2007. Graphene: Carbon in two dimensions. Materials Today 10, 20–27.

Keefer, E.W., Botterman, B.R., Romero, M.I., Rossi, A.F., Gross, G.W., 2008. Carbon nanotube coating improves neuronal recordings. Nature nanotechnology 3, 434–439.

Keun Kwon, I., Kidoaki, S., Matsuda, T., 2005. Electrospun nano- to microfiber fabrics made of biodegradable copolyesters: structural characteristics, mechanical properties and cell adhesion potential. Biomaterials 26, 3929–3939.

Kim, Y.G., Lee, Y.I., Kim, J.W., Pyeon, H.J., Hyun, J.K., Hwang, J.-Y., Choi, S.-J., Lee, J.-Y., Deák, F., Kim, H.-W., 2014. Differential stimulation of neurotrophin release by the biocompatible nano-material (carbon nanotube) in primary cultured neurons. Journal of biomaterials applications 28, 790.

Ko, S.H., Su, M., Zhang, C., Ribbe, A.E., Jiang, W., Mao, C., 2010. Synergistic self-assembly of RNA and DNA molecules. Nature chemistry 2, 1050–1055.

Kokai, L.E., Bourbeau, D., Weber, D., Mcatee, J., Marra, K.G., 2011. Sustained Growth Factor Delivery Promotes Axonal Regeneration in Long Gap Peripheral Nerve Repair. Tissue engineering. Part A 17, 1263–1275.

Kumbar, S.G., James, R., Nukavarapu, S.P., Laurencin, C.T., 2008. Electrospun nanofiber scaffolds: engineering soft tissues. Biomedical Materials, 3.

Lee, A.C., Yu, V.M., Lowe, J.B., Brenner, M.J., Hunter, D.A., Mackinnon, S.E., Sakiyama-Elbert, S.E., 2003. Controlled release of nerve growth factor enhances sciatic nerve regeneration. Experimental Neurology 184, 295–303.

Lee, J.Y., Bashur, C.A., Goldstein, A.S., Schmidt, C.E., 2009. Polypyrrole-coated electrospun PLGA nanofibers for neural tissue applications. Biomaterials 30, 4325–4335.

Lee, W., Lee, Y.-B., Polio, S., Dai, G., Menon, L., Carroll, R.S., Yoo, S.-S., 2010. Bio-printing of collagen and VEGF-releasing fibrin gel scaffolds for neural stem cell culture. Experimental Neurology 223, 645–652.

Li, N., Cheng, G., Zhang, Q., Gao, S., Song, Q., Huang, R., Wang, L., Liu, L., Dai, J., Tang, M., 2013. Three-dimensional graphene foam as a biocompatible and conductive scaffold for neural stem cells. Scientific reports 3, 1604.

Liang, Y.-X., Cheung, S.W.H., Chan, K.C.W., Wu, E.X., Tay, D.K.C., Ellis-Behnke, R.G., 2011. CNS regeneration after chronic injury using a self-assembled nanomaterial and MEMRI for real-time in vivo monitoring. Nanomedicine : nanotechnology, biology, and medicine 7, 351–359.

Liu, Y., Ye, H., Satkunendrarajah, K., Yao, G.S., Bayon, Y., Fehlings, M.G., 2013. A self-assembling peptide reduces glial scarring, attenuates post-traumatic inflammation and promotes neurological recovery following spinal cord injury. Acta Biomater 9, 8075–8088.

Lock, L.L., Lacomb, M., Schwarz, K., Cheetham, A.G., Lin, Y.-A., Zhang, P., Cui, H., 2013. Self-assembly of natural and synthetic drug amphiphiles into discrete supramolecular nanostructures. Faraday discussions 166, 285–331.

Ma, Z., Kotaki, M., Inai, R., Ramakrishna, S., 2005. Potential of nanofiber matrix as tissue-engineering scaffolds. Tissue engineering 11, 101–109.

Malarkey, E.B., Fisher, K.A., Bekyarova, E., Liu, W., Haddon, R.C., Parpura, V., 2009. Conductive single-walled carbon nanotube substrates modulate neuronal growth. Nano letters 9, 264–268.

Marquardt, L.M., Sakiyama-Elbert, S.E., 2013. Engineering peripheral nerve repair. Current opinion in biotechnology 24, 887–892.

Matson, J.B., Stupp, S.I., 2011. Self-assembling peptide scaffolds for regenerative medicine. Chemical communications (Cambridge, England) 48, 26–33.

Matsumoto, K., Sato, C., Naka, Y., Whitby, R., Shimizu, N., 2010. Stimulation of neuronal neurite outgrowth using functionalized carbon nanotubes. Nanotechnology 21, 115101.

Maude, S., Ingham, E., Aggeli, A., 2013. Biomimetic self-assembling peptides as scaffolds for soft tissue engineering. Nanomedicine 8, 823–847.

Melchels, F.P.W., Domingos, M.A.N., klein, T.J., Malda, J., Bartolo, P.J., Hutmacher, D.W., 2012. Additive manufacturing of tissues and organs. Progress in Polymer Science 37, 1079–1104.

Melissinaki, V., Gill, A.A., Ortega, I., Vamvakaki, M., Ranella, A., Fotakis, C., Farsari, M. & Claeyssens, F. Direct laser writing of polylactide 3D scaffolds for neural tissue engineering applications. 2011. IEEE, 1-1.

Merzlyak, A., Indrakanti, S., Lee, S.-W., 2009. Genetically engineered nanofiber-like viruses for tissue regenerating materials. Nano letters 9, 846–852.

Ni, Y., Hu, I., Malarkey, E.B., Zhao, B., Montana, V., Haddon, R.C., Parpura, V., 2005. Chemically functionalized water soluble single-walled carbon nanotubes modulate neurite outgrowth. Journal of Nanoscience and Nanotechnology 5, 1707–1712.

Owens, C.M., Marga, F., Forgacs, G., Heesch, C.M., 2013. Biofabrication and testing of a fully cellular nerve graft. Biofabrication, 5.

Ozbolat, I.T., Yu, Y., 2013. Bioprinting toward organ fabrication: challenges and future trends. IEEE transactions on bio-medical engineering 60, 691–699.

Panseri, S., Cunha, C., Lowery, J., Del Carro, U., Taralli, F., Amadio, S., Vescovi, A., Gelain, F., 2008. Electrospun micro- and nanofiber tubes for functional nervous regeneration in sciatic nerve transections. BMC biotechnology 8, 39–139.

Park, H.-B., Nam, H.-G., Oh, H.-G., Kim, J.-H., Kim, C.-M., Song, K.-S., Jhee, K.-H., 2013a. Effect of graphene on growth of neuroblastoma cells. Journal of Microbiology and Biotechnology 23, 274–277.

Park, S.Y., Choi, D.S., Jin, H.J., Park, J., Byun, K.-E., Lee, K.-B., Hong, S., 2011. Polarization-controlled differentiation of human neural stem cells using synergistic cues from the patterns of carbon nanotube monolayer coating. ACS nano 5, 4704–4711.

Park, S.Y., Kang, B.-S., Hong, S., 2013b. Improved neural differentiation of human mesenchymal stem cells interfaced with carbon nanotube scaffolds. Nanomedicine 8, 715–723.

Peng, X.H., Wong, S.S., 2009. Functional Covalent Chemistry of Carbon Nanotube Surfaces. Advanced materials 21, 625–642.

Rao, C.N.R., Sood, A.K., Subrahmanyam, K.S., Govindaraj, A., 2009. Graphene: the new two-dimensional nanomaterial. Angewandte Chemie (International ed. in English) 48, 7752–7777.

Saito, N., Kato, H., Taruta, S., Endo, M., Usui, Y., Aoki, K., Narita, N., Shimizu, M., Hara, K., Ogiwara, N., Nakamura, K., Ishigaki, N., 2009. Carbon nanotubes: biomaterial applications. Chemical Society reviews 38, 1897–1903.

Saracino, G.A.A., Cigognini, D., Silva, D., Caprini, A., Gelain, F., 2013. Nanomaterials design and tests for neural tissue engineering. Chemical Society reviews 42, 225–262.

Schmidt, C.E., Leach, J.B., 2003. Neural tissue engineering: strategies for repair and regeneration. Annual review of biomedical engineering 5, 293–347.

Shepherd, J.N.H., Parker, S.T., Shepherd, R.F., Gillette, M.U., Lewis, J.A., Nuzzo, R.G., 2011. 3D Microperiodic Hydrogel Scaffolds for Robust Neuronal Cultures. Advanced functional materials 21, 47–54.

Shoichet, M., Schmidt, H.C., 2001. Neural tissue engineering. Biomaterials 22, 1015–1193.

Shoichet, M.S., Tate, C.C., Douglas Baumann, M., Laplaca, M.C., 2008. Strategies for Regeneration and Repair in the Injured Central Nervous System. In: Reichert, W.M. (Ed.), Indwelling Neural Implants: Strategies for Contending with the In Vivo Environment. CRC Press, Boca Raton.

Sill, T.J., Von Recum, H.A., 2008. Electrospinning: applications in drug delivery and tissue engineering. Biomaterials 29, 1989–2006.

Smith, K.H., Tejeda-Montes, E., Poch, M., Mata, A., 2011. Integrating top-down and self-assembly in the fabrication of peptide and protein-based biomedical materials. Chemical Society reviews 40, 4563.

Stephanopoulos, N., Ortony, J.H., Stupp, S.I., 2013. Self-assembly for the synthesis of functional biomaterials. Acta materialia 61, 912–930.

Subramanian, A., Krishnan, U.M., Sethuraman, S., 2009. Development of biomaterial scaffold for nerve tissue engineering: Biomaterial mediated neural regeneration. Journal of biomedical science 16, 108–1108.

Sun, L., Zhang, L., Hemraz, U.D., Fenniri, H., Webster, T.J., 2012. Bioactive rosette nanotube-hydroxyapatite nanocomposites improve osteoblast functions. Tissue engineering. Part A 18, 1741–1750.

Tang, M., Song, Q., Li, N., Jiang, Z., Huang, R., Cheng, G., 2013. Enhancement of electrical signaling in neural networks on graphene films. Biomaterials 34, 6402–6411.

Theron, S.A., Zussman, E., Yarin, A.L., 2004. Experimental investigation of the governing parameters in the electrospinning of polymer solutions. Polymer 45, 2017–2030.

Tran, P.A., Zhang, L., Webster, T.J., 2009. Carbon nanofibers and carbon nanotubes in regenerative medicine. Advanced Drug Delivery Reviews 61, 1097–1114.

Tysseling, V.M., Sahni, V., Niece, K.L., Birch, D., Czeisler, C., Fehlings, M.G., Stupp, S.I., Kessler, J.A., 2008. Self-Assembling Nanofibers Inhibit Glial Scar Formation and Promote Axon Elongation after Spinal Cord Injury. Journal of Neuroscience 28, 3814–3823.

Tysseling, V.M., Sahni, V., Pashuck, E.T., Birch, D., Hebert, A., Czeisler, C., Stupp, S.I., Kessler, J.A., 2010. Self-assembling peptide amphiphile promotes plasticity of serotonergic fibers following spinal cord injury. Journal of Neuroscience Research 88, 3161–3170.

Vasita, R., Katti, D.S., 2006. Nanofibers and their applications in tissue engineering. International journal of nanomedicine 1, 15–30.

Vasita, R., Shanmugam, K., Katti, D.S., 2008. Improved biomaterials for tissue engineering applications: Surface modification of polymers. Current topics in medicinal chemistry 8, 341–353.

Wang, C.Y., Zhang, K.H., Fan, C.Y., Mo, X.M., Ruan, H.J., Li, F.F., 2011a. Aligned natural-synthetic polyblend nanofibers for peripheral nerve regeneration. ACTA Biomaterialia 7, 634–643.

Wang, Y., Wang, J., Li, Z., Li, J., Lin, Y., 2011b. Graphene and graphene oxide: biofunctionalization and applications in biotechnology. Trends in Biotechnology 29, 205–212.

Wei, G.-J., Yao, M., Wang, Y.-S., Zhou, C.-W., Wan, D.-Y., Lei, P.-Z., Wen, J., Lei, H.-W., Dong, D.-M., 2013. Promotion of peripheral nerve regeneration of a peptide compound hydrogel scaffold. International journal of nanomedicine 8, 3217–3225.

Weng, B., Liu, X., Shepherd, R., Wallace, G.G., 2012. Inkjet printed polypyrrole/collagen scaffold: A combination of spatial control and electrical stimulation of PC12 cells. Synthetic Metals 162, 1375–1380.

Xie, J., Macewan, M.R., Li, X., Sakiyama-Elbert, S.E., Xia, Y., 2009a. Neurite Outgrowth on Nanofiber Scaffolds with Different Orders, Structures, and Surface Properties. ACS Nano 3, 1151–1159.

Xie, J., Macewan, M.R., Schwartz, A.G., Xia, Y., 2010. Electrospun nanofibers for neural tissue engineering. Nanoscale 2, 35–44.

Xie, J., Willerth, S.M., Li, X., Macewan, M.R., Rader, A., Sakiyama-Elbert, S.E., Xia, Y., 2009b. The differentiation of embryonic stem cells seeded on electrospun nanofibers into neural lineages. Biomaterials 30, 354–362.

Xu, C., Kopecˇek, J., 2007. Self-Assembling Hydrogels. Polymer Bulletin 58, 53–63.

Xu, H., Holzwarth, J.M., Yan, Y., Xu, P., Zheng, H., Yin, Y., Li, S., Ma, P.X., 2014. Conductive PPY/PDLLA conduit for peripheral nerve regeneration. Biomaterials 35, 225.

Yang, F., Murugan, R., Wang, S., Ramakrishna, S., 2005. Electrospinning of nano/micro scale poly(l-lactic acid) aligned fibers and their potential in neural tissue engineering. Biomaterials 26, 2603–2610.

Yang, W., Thordarson, P., Gooding, J.J., Ringer, S.P., Braet, F., 2007. Carbon nanotubes for biological and biomedical applications. Nanotechnology 18, 412001–1412001, (12).

Yang, Y., Asiri, A.M., Tang, Z., Du, D., Lin, Y., 2013a. Graphene based materials for biomedical applications. Materials Today 16, 365.

Yang, Y., Zhao, Z., Zhang, F., Nangreave, J., Liu, Y., Yan, H., 2013b. Self-assembly of DNA rings from scaffold-free DNA tiles. Nano letters 13, 1862–1866.

Yang, Y.L., Khoe, U., Wang, X.M., Horii, A., Yokoi, H., Zhang, S.G., 2009. Designer self-assembling peptide nanomaterials. Nano Today 4, 193–210.

Zhan, X., Zhang, W., Guo, J., Wu, W., Gao, M., Jiang, Y., Zhang, W., Wong, W.M., Yuan, Q., Su, H., Kang, X., Dai, X., 2013. Nanofiber scaffolds facilitate functional regeneration of peripheral nerve injury. Nanomedicine : nanotechnology, biology, and medicine 9, 305–315.

Zhang, L., Hemraz, U.D., Fenniri, H., Webster, T.J., 2010. Tuning cell adhesion on titanium with osteogenic rosette nanotubes. J Biomed Mater Res A 95, 550–563.

Zhang, L., Rakotondradany, F., Myles, A.J., Fenniri, H., Webster, T.J., 2009a. Arginine-glycine-aspartic acid modified rosette nanotube-hydrogel composites for bone tissue engineering. Biomaterials 30, 1309–1320.

Zhang, L., Ramsaywack, S., Fenniri, H., Webster, T.J., 2008. Enhanced osteoblast adhesion on self-assembled nanostructured hydrogel scaffolds. Tissue engineering. Part A 14, 1353–1364.

Zhang, L., Rodriguez, J., Raez, J., Myles, A.J., Fenniri, H., Webster, T.J., 2009b. Biologically inspired rosette nanotubes and nanocrystalline hydroxyapatite hydrogel nanocomposites as improved bone substitutes. Nanotechnology 20, 175101.

Zhang, L.J., Webster, T.J., 2009. Nanotechnology and nanomaterials: Promises for improved tissue regeneration. Nano Today 4, 66–80.

Zhang, X., Prasad, S., Niyogi, S., Morgan, A., Ozkan, C.S., Ozkan, M., 2005. Guided neurite growth on patterned carbon nanotubes. Sensors & Actuators: B. Chemical 106, 843–850.

Zhang, X., Xin, N., Tong, L., Tong, X.-J., 2013. Electrical stimulation enhances peripheral nerve regeneration after crush injury in rats. Molecular Medicine Reports 7, 1523–1527.

ORGAN PRINTING

Robert C. Chang and Filippos Tourlomousis

Department of Mechanical Engineering, Stevens Institute of Technology, Hoboken, NJ, USA

15.1 INTRODUCTION

Our enhanced understanding of the fundamental biological sciences, primarily the interplay between biological cells and integrative biological, chemical, and structural cues within the *in vivo* milieu or natural three-dimensional (3D) microenvironment, demands a defined role for engineering design and manufacturing to meet the challenges presented by increasingly complex biological problems. Organ printing in tissue engineering is the application of additive, computer-aided manufacturing process technologies toward the layered, patterned deposition of complex 3D cell-bearing biological structures with biomolecular and biopolymer material integration. Organ printing therefore encapsulates an ever-expanding range of enabling bioadditive manufacturing processes and strategies that show great promise in regenerative medicine applications where novel fabrication, in conjunction with rapidly evolving stem cell technologies, is envisioned to address the challenge of limited donor grafts for functional organ repair and replacement.

The term organ printing has previously been more narrowly defined as "a biomedical variant of rapid prototyping technology or computer-aided robotic layer-by-layer additive biofabrication of 3D human tissues and organs using self-assembling tissue spheroids as building blocks (Mironov et al., 2003; Mironov et al., 2008; Mironov et al., 2009; Mironov et al., 2011)." Such a definition is rooted in developmental biology principles in which the self-assembly process refers to "one in which humans are not actively involved, in which atoms, molecules, aggregates of molecules, and components arrange themselves into ordered functioning entities without human intervention (Whitesides and Grzybowski, 2002)." Accordingly, the raw material or tissue spheroid "bioink" for bioprinting is initially preprocessed by achieving high-density cell aggregates via traditional hanging drop methods, high-throughput digital microfluidics, and other scalable fabrication methods of tissue spheroids with self-organizing properties (Rezendea et al., 2013). More broadly defined in the context of tissue engineering, the autonomous organization of biological entities along the continuum of biological scales can be described conceptually as man kick-starting nature's mechanism of constructing an innately coordinated system whereby "people may design the process, and they may launch it, but once underway, it proceeds according to its own internal plan, either toward an energetically stable form or toward some system whose form and function are encoded in its individual parts (Whitesides, 1995)." Therefore, given the significant insight we've gained in systematically engineering cell-instructive microenvironments, we will adopt this latter definition of self-assembly as an organizing principle for enabling bioadditive manufacturing technologies driven by either (1) scaffold-free approaches, (2) scaffold-directed approaches, or (3) a hybrid of these prevailing approaches (Burdick and Vunjak-Novakovic, 2009). This inclusive definition of organ printing captures layer-by-layer fabrication technologies capable of satisfying the criteria of directly incorporating or seeding engineered constructs with high-density, concentrated cell suspensions toward *de novo* complex tissue form and function in organs.

15.1.1 ORGAN PRINTING TECHNIQUES

Current layer-by-layer technologies developed and implemented for the purpose of digitally printing human organs can be divided into three broad categories: (1) microextrusion-based printing, (2) ink-jet-based printing, and (3) laser-based printing. Other printing techniques have also emerged and their attributes will be briefly described in this chapter. In each of these categories, the term printing refers specifically to the direct patterning or deposition of cells either with or without a carrying matrix biopolymer. As part of the broadening bioadditive manufacturing armamentarium, these rapid prototyping (RP) or solid freeform fabrication (SFF) techniques are capable of varying degrees of resolution, reproducibly patterned architectures within a bulk construct, as well as scalable manufacturing of clinically relevant defect sizes.

15.1.1.1 Microextrustion-based Printing

A promising freeform fabrication process, which has perhaps gained the widest adoption for organ printing applications, is microextrusion-based printing, sometimes referred to in the literature as bioplotting or biodispensing. Microextrusion-based printing represents a progression from fused deposition modeling (FDM, our traditional understanding of 3D printing) that expands the catalog of candidate processed materials into fibers or filaments beyond synthetic polymer melts to include hydrogels and incorporate cells into the printing step. Furthermore, while FDM toward fabricating polycaprolactone (PCL) thermoplastic and composite material 3D scaffolds has found wide application in hard tissue applications, microextrusion-based printing is capable of building layered structures at cell-friendly temperatures amenable for processing biopolymer materials laden with cells and biological factors toward solid organ printing. Target mammalian cell types for microextrusion-based printing include liver, bone, heart, vascular lining, and cartilage (Chang et al., 2008a; Khalil and Sun, 2009; Cohen et al., 2009; Gaetani et al., 2012; Shim et al., 2011; Smith et al., 2004; Wang et al., 2006; Yan et al., 2005). A wide number of both natural and synthetic polymer hydrogel materials have been used in microextrusion-based printing including alginate, agarose, Matrigel®, fibrinogen, gelatin, collagen, chitosan, and polyethylene glycol (PEG) (Chang et al., 2008a; Khalil and Sun, 2009; Cohen et al., 2009; Gaetani et al., 2012; Shim et al., 2011; Smith et al., 2004; Wang et al., 2006; Yan et al., 2005; Fedorovich et al., 2008; Fedorovich et al., 2011a; Ahn and Koh, 2010; Smith et al., 2007; Maher et al., 2009). In microextrusion-based printing methods, cells and hydrogel precursors are dispensed from a syringe reservoir or micronozzle containing cell suspension via a syringe pump piston or pneumatic microvalve, respectively. A significant advantage of this method is the facile incorporation of cells and biological signaling molecules that can be spatially patterned within a 3D layered construct. Furthermore, foundational work has been done in experimentally quantifying and predictively modeling the postprint viability and functional retention of various cell types following process-induced mechanical loading. For example, the authors have examined the effect of dispensing pressure and nozzle tip diameter on the postprint viability of HepG2 liver cells encapsulated within alginate, demonstrating a quantifiable loss of cell viability due to process-induced mechanical damage with some cells showing recovery over a short-duration study period (Chang et al., 2008a). Others have shown with cross-validating cell damage models of printed Schwann cells and 3T3 fibroblasts that differential postprint viability is observed in tapered versus cylindrical needle tip geometries (Li et al., 2011). While microextrusion-based postprinted aggregate cell functional retention has been demonstrated in various cell types for narrow process windows, more systematic study to quantify and isolate these process-induced effects in the context of diverse material parameters (e.g. viscosity) would contribute to fundamental understanding of these mechanical processes toward widening the permissible process windows with predictable biological postprint outcomes. This understanding of process-induced mechanical forces in micro-extrusion-based printing will become even more critical as (1) the requirement for high-density cell aggregates and tissue spheroids becomes paramount in organ printing applications, and (2) stem cells sensitive to mechanical cues become a mainstay in microextrusion-based organ printing for regenerative medicine. The latter challenge of predicting the effect of process-induced mechanical cues on stem cell differentiation also represents an opportunity to reliably guide cell fates for the first time with bioadditive manufacturing systems. Compared with the other organ techniques discussed in this chapter, microextrusion-based printing offers the lowest resolution on the order of 100–200 μm.

15.1.1.2 Ink-jet-based Printing

Drop-on-demand or ink-jet-based printing has been adapted from commercial desktop printers to precisely deposit and pattern biological cells and materials with picoliter-sized volumes (Sekula et al., 2008). Relative to the microextrusion-based printing methods, ink-jet-based printers are able to achieve feature sizes on the order of 20–100 μm (Melchels et al., 2012). This approach entails re-purposing desktop printer ink cartridges by replacing the ink in the material reservoir with bioink (i.e. aqueous cell solutions) to print microscale cell droplets (Xu et al., 2005; Roth et al., 2004). Since ink-jet printing was initially aimed at fabricating two-dimensional systems (i.e. a single layer of droplets for surface patterning applications), the feature sizes have been determined to be primarily dictated by the size of the printed drop in flight and its contact angle on the surface upon impact (Stringer and Derby, 2010). The resultant resolution is that the smallest drops attainable through an ink-jet-based process are ~1 pl volumes with an equivalent radius of less than 10 μm. Therefore, on prescribed hydrophilic surfaces, minimum features sizes on the order of single-cell dimensions are achievable with this technique. In order to create a layered configuration of cells with encapsulating biopolymers, others have designed an elevator chamber model for ink-jet-based printer in which robotic control of a metal elevator rod is regulated using a step motor for precise positioning (Boland et al., 2006). In this study, precise volumes of chemical cross-linking solution are ejected from the ink-jet head or nozzle onto the motorized elevator submerged within a solution of uncrosslinked hydrogel precursor intermixed with cells homogeneously distributed. While this approach successfully circumvents the undesired effect of cells being subjected to process-induced mechanical loading (e.g. shear stresses and impact forces) through an ink-jet cartridge orifice, the ability to spatially pattern, using multiple ink-jet cartridges as well as multicellular and multimaterial constructs with functionally gradient distributions, is limited by the initial cell and prepolymer material configuration residing in the elevator chamber.

Alternatively, others have implemented electrostatically driven ink-jet-based printer systems to deposit cell suspensions with single-cell seeding control (Nakamura et al., 2010). Others have simultaneously ejected hydrogels and cells from an ink-jet nozzle into chemical cross-linking solutions (Nishiyama et al., 2009). While these high-resolution techniques enable cells to be spatially patterned with heterogeneous distribution by modulating the ejection frequency and multiple material heads, the question of cell viability from the jetting of high-density cell suspensions through a small cartridge orifice remains unanswered. Furthermore, on one hand, ink-jet printing cell suspensions in media at high ejection frequencies and smaller drop feature sizes are hampered by evaporative water losses and a lack of a material support apparatus. On the other hand, ink-jet printing cells with polymer materials place restrictions on the types of polymers that can be processed to low-viscosity, low-strength hydrogel materials with a viscosity ceiling of approximately 30 mPa*s (Melchels et al., 2012). A corollary of these material handling restrictions is the challenge with inkjet-based printing to achieve buildup in 3D for organ printing.

15.1.1.3 Laser-based Printing
Laser-based printing, sometimes referred to in the literature as laser-assisted bioprinting or laser forward transfer, is a nozzle-free printing technique that enables high-resolution patterning of cells (Guillemot et al., 2010; Gruene et al., 2011; Koch et al., 2010; Mezel et al., 2010). In laser-based printing system configurations, a collector plate with tunable wetting properties is first coated with encapsulating polymer droplets loaded with cells. Either UV or IR laser is then focused with a microscope objective lens onto an absorbing thin layer film typically composed of Au, Ag, or Ti that shields the cell-polymer droplets from direct laser interactions. Upon reaching a minimum threshold energy of the nanosecond pulsed laser, microscale droplet deposition is achieved. Key process parameters in laser-based printing that dictate droplet feature sizes include the thickness of the absorbing layer, distance between the absorbing layer and collector plate, wavelength or energy of the laser pulses, and the optical properties of the polymer material. The significant advantages of laser-based printing include the small feature sizes attainable and the ability to pattern cells without subjecting them to the mechanical forces inherent in nozzle-based, ink-jet-based, and microextrusion-based printing methods. A nozzle-free configuration also circumvents occlusion issues that beset ink-jet-based and microextrusion-based techniques. Akin to ink-jet-based printing, however, scale-up of laser-based printed constructs is challenging due to physical constraints in the Z-dimension that precludes facile layer-by-layer buildup in 3D organ printing. Laser-based printing is also restricted to the processing of low-viscosity materials and requires cells to be immobilized within a gel.

15.1.2 CHALLENGES IN ORGAN PRINTING
Based on a delineation of the aforementioned organ printing techniques and their relative virtues and limitations, it is apparent that significant manufacturing and biological challenges exist and must be overcome for printed organs to be successfully and clinically translated to benefit patients. As might be expected, many of the manufacturing challenges (e.g. resolution enhancement, increasing scalability, and widening manufacturing process windows) are in direct conflict with the biological challenges (e.g. postprint cell viability and functionality). While the incorporation of a vascular network into organs is a significant barrier in tissue engineering, this challenge has been extensively posed and addressed elsewhere, and will not be discussed here.

The authors believe that a key challenge to unlocking the full potential of organ printing is iden-tifying a manufacturing process or perhaps an integration of processes that satisfies two competing requirements: (1) scalability with large-volume patterning for commercial viability with immediate clinical translation; and (2) high-fidelity complex arrangement of cells within extracellular matrices (ECMs) designed to simulate the *in vivo* microstructural niche. On one hand, the former macroscale patterning requirement has been addressed by a category of 3D patterning bioadditive manufacturing processes. Specifically, a number of RP techniques, including microextrusion-based printing, have found significant application in 3D patterning due to their capability to produce scaffold architec-tures that replicate clinical defect sites and to produce patterned macroscale structures within the bulk construct (Chang et al., 2011; Fedorovich et al., 2011b; Duan and Wang, 2011; Butscher et al., 2011; Kasko and Wong, 2010; Melchels et al., 2010; Jakab et al., 2010). However, this class of RP processes currently lacks the resolution and processing capability needed to meet the design requirement of scaf-fold feature sizes and geometries characteristic of *in vivo* microstructural cell niches marked by the complex arrangement of cells within ECMs and embedded chemical gradients. These local niches play a pivotal role in the regulation of cellular activity. Therefore, the replication of these structures *in vitro* would allow a more detailed understanding of these microenvironments. On the other hand, the latter resolution requirement has been addressed by processes capable of creating microarchitectures within the required length scale (1–100 μm). In addition to ink-jet-based printing and laser-based printing, this category of techniques that has the resolution required to replicate microstructures also includes microinjection and holographic optical tweezers (Siniscalco et al., 2010; Bible et al., 2009a; Bible et al., 2009b; Masuzaki et al., 2010; Leach et al., 2004; Watson et al., 2004; Akselrod et al., 2006; Townes-Anderson et al., 1998). However, this class of technologies has significant limitations in terms of the maximum areal coverage that can be patterned and the process time needed to complete large-scale 3D fabrication of functional organs. To satisfy these apparently competing design requirements, scalable bioadditive manufacturing processes capable of presenting single-cell level 3D microstructural cues is required. The challenge of creating functional organs with a scalable manufacturing process first requires fundamental understanding of the key polymer rheological material parameters foundational to modeling the bioadditive manufacturing process variables. Although the authors have studied the effects of key process variables in bioadditive manufacturing and engineering 3D cell-based constructs using a microextrusion-based printer and FDM, significant barriers such as microstructural resolution and reproducibility remain and should thus be overcome to enable these bioadditive manufacturing processes to yield functional engineered organs (Chang et al., 2008a; Chang et al., 2008b; Chang et al., 2008c; Chang et al., 2010; Snyder et al., 2011; Buyukhatipoglu et al., 2010; Shor et al., 2009).

Additionally, the authors believe that the full potential of organ printing will necessitate conver-gence with stem cell technology. For example, mesenchymal stem cells (MSCs) show great promise in tissue engineering applications because of their potential to differentiate into various tissue types, including bone, cartilage, and adipose in response to external cues (Narita et al., 2008; Kim et al., 2009; Delorme et al., 2009; Prockop, 1997). Since MSCs can be isolated from adult patients, this allows for the possibility of using MSCs in regenerative medicine therapies for patient-specific repair of bone and cartilage defects with tissue insensitive to an immune response. Specifically, engineered tissues incorporating MSCs are known to be sensitive to mechanical forces capable of facilitating MSC dif-ferentiation into various mature cells (Sikavitsas et al., 2001; Sumanasinghe et al., 2009). For example, cyclic tensile strain and oscillatory fluid flow have both been reported to increase osteogenic differen-tiation and decrease adipogenic differentiation, whereas uniaxial, unconfined compression and cyclic hydrostatic pressure increase chondrogenesis (Sumanasinghe et al., 2006; Sumanasinghe et al., 2006; Sen et al., 2008; Arnsdorf et al., 2009; Li et al., 2004; Riddle et al., 2006; Haudenschild et al., 2008; Wagner et al., 2008; Castillo and Jacobs, 2010). A fundamental topic requiring study is the effect of microstructural cues from engineered scaffolds, for instance, the dimensional scale and geometry of the 3D polymer substrates. A number of micro- and nanofabrication-based cell platforms, particularly with spatially controlled features, have been advanced for investigating fundamental questions in cell behavior and tissue development (Whitesides et al., 2001; Théry, 2010; Chen et al., 1997; Quake and Scherer, 2000; Huang et al., 2008). While these micro- and nanoscale manufacturing techniques have yielded insight into structural cues on stem cell differentiation, scalable bioadditive manufacturing processes with embedded microstructural cues for creating functional tissue to meet critical organ deficit size demands have remained unexplored. It has been shown that, given the same material composition, electrospun nanofiber (<1 μm) substrates seeded with MSCs differentiate into bone-forming cells, while solid freeform fabricated (>100 μm) 3D substrates similarly seeded with MSCs yield undifferentiated cells (Farooque et al., 2014; Kumar et al., 2011; Kumar et al., 2012; Huang et al., 2013). Therefore, the author's working hypothesis is that there exists a threshold dimensional scale on the order of the single cell of presenting microstructural cues that will induce MSC differentiation into bone-forming cells. There is a need, therefore, to model and develop enabling bioadditive manufacturing processes to fill in the critical dimensional gap between polymer electrospinning and conventional RP technologies. As a global measure for the process outcome of presenting stem cells with a 3D microstructural niche for differentiation, one possibility is advancing subcellular cell shape dynamics as a quantitative metric. Although there are published reports on single-cell imaging with structural classification of 3D cell shape dimensionality, current limitations of cell shape metrology that need to be addressed include single time point imaging and accounting for the number and distribution of subcellular structural elements (Farooque et al., 2014; Kumar et al., 2011).

15.1.3 MICRO-ORGAN PRINTING AS PHYSIOLOGICAL AND DISEASE PLATFORMS

15.1.3.1 Microextusion-based Printed Liver Micro-organ on a Chip

A near-future application of organ printing techniques is the creation of cell-based physiological and disease platforms, which we will refer to in this chapter as micro-organ on a chip. Specifically, liver-on-a-chip technology has generated significant research interest in *in vitro* drug metabolism and toxicity studies. However, to date, conventional 2D monolayer cultures of primary hepatocytes cause hepatocytes to lose their morphology and liver-specific functions, including the activity of a group of metabolizing enzymes located in the endoplasmic reticulum called cytochrome P-450 oxidases (Burkhardt et al., 2013). The authors have addressed this obstacle with 3D culture models with layered microextrusion-based printing approaches. Others have also embedded matrices made from natural or synthetic hydrogels and scaffolds constructed from classical biocompatible soft polymers, which enable hepatocytes to express their phenotype in a 3D microenvironment toward increasing their viability and maintaining their functionality over longer culture time periods (Chang et al., 2008a; Chang et al., 2008b; Miranda et al., 2010). It has also been shown that hepatocyte cell physiological behavior is favored under constant perfusion conditions (Goral and Yuen, 2012). In this regard, microfluidics has been adapted through layer-by-layer biofabrication approaches, thus enabling the relatively new "organ-on-a-chip" technology (Lee et al., 2013). This development has led to the creation of microanalytical micro-organ devices, whose 3D, dynamic nature distinguishes them from the conventional static *in vitro* models as promising hepatocyte culture testbeds toward preclinical phase testing of drug development (Chang et al., 2008b; Chang et al., 2010).

The authors have implemented a layer-by-layer microextrusion-based printing technique for the creation of cell-encapsulated alginate-based liver micro-organs, which are then integrated onto a microfluidic chamber fabricated using soft lithographic techniques (Chang et al., 2008b; Chang et al., 2010). The system has shown enhanced functionality of HepG2 liver cells within the 3D micro-organ compared to traditional monolayer culture models in terms of cell viability and urea synthesis. Furthermore, the authors have also implemented the microextrusion-based printing in order to construct a cell-encapsulated multilayered tissue construct in a sinusoidal pattern to mimic the *in vivo* liver microarchitecture. The vascularized nature of the liver microenvironment was closely recapitulated by the application of continuous shear-mediated drug perfusion flow through the printed liver micro-organ after its integration within the microfabricated chamber. Drug metabolism studies showed that the system could serve as a reliable 3D *in vitro* pharmacokinetic platform for drug screening and toxicity studies.

15.1.3.2 Computational Model Setup for Perfused Printed Liver Micro-organ

Intuitively, one can surmise that perfused bioreactors with high cell density, cell encapsulation combined with parallel patterned geometry of biomimetic liver tissue constructs that have been used as microscale drug screening platforms, appear to be a rational approach for drug screening and toxicity studies. However, iterative process design requires advanced computational models that can capture the transport phenomena not only on the macroscale micro-organ level (Tan et al., 2013; Hsu et al., 2014; Hutmacher and Singh, 2008; Truscello et al., 2011) but also on the single cellular scale. Processing parameters such as flow rate, cell density, and shear stresses on the cell surface need to be correlated with concentration drug profiles and culture time for specific printed liver microarchitectures through appropriate metrics in order to set the design standards that can both predict and promote *in vivo* behavior of encapsulated hepatocytes in dynamic printed micro-organ devices under constant perfusion. The authors have therefore developed a computational modeling strategy for such *in vitro* models using a cell kinetics convection-diffusion problem, where the drug media is convected by the perfusion flow rate from the inlet port of the microscale device, diffused through the hydrogel channel walls due to its inherent porous material nature, and metabolized within the encapsulated hepatocytes.

The steady state results obtained from the simulations done for the free and porous flow regime are presented hereafter. The results obtained by the authors from the transient studies are presented and systematically analyzed elsewhere (Tourlomousis and Chang, November 2014). Using the velocity and the shear stress field computed in steady state, the drug transport and metabolism process is coupled by solving the mass transport equations for the flow and cell domains in time.

15.1.3.3 Steady State Simulations

In the first stage of modeling, the Stokes and Brinkman Equations are solved for an inlet volumetric flow rate of $Q_{in} = 0.25$ μl/h to ensure a laminar velocity profile across the microfluidic channel as shown in Figure 15.1. The aforementioned value of the inlet volumetric flow rate is prescribed as a starting point for the present study based on the author's experimental drug flow study protocol (Chang et al., 2010). Moreover, the order of magnitude is consistent with values found in previous dynamic bioreactor studies and thus serves as a reasonable starting point. The cells are initially modeled as voids in the computational domain, experience no-slip boundary condition along the walls and thereafter

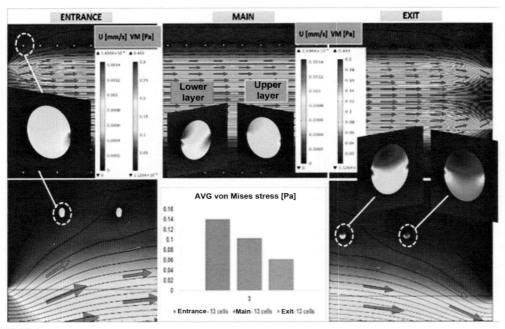

FIGURE 15.1

Velocity field and streamlines indicating a laminar flow topology. Insets: Shear stress field around cells modeled as voids.

wall shear stress (WSS) as shown in the magnified insets of Figure 15.1. The Newtonian nature of the medium allows the authors to compute the WSS as the product of the shear rate with dynamic viscosity around each cell boundary. It is evident from Figure 15.1 that the cell layer exposed directly to the free flow field experiences much higher shear stress compared to the cell layers residing closer to the hydrogel wall. Thus, this part of the study models the cell as being in direct contact with the shear-induced flow field. The average cell WSS range from 0.02 Pa to 0.01 Pa, with maximum WSS values ranging from 0.04 Pa to 0.02 Pa. Based on the geometry model and the cell average WSS computations for all the cells composing the layer of interest, three distinct regimes of the printed micro-organ can be identified in Figure 15.2: (1) the entrance, (2) the middle, and (3) the exit. Each regime contains an equivalent number of cells (13 cells), with the middle cell population experiencing the highest WSS values and the exit cell population experiencing the lowest. Thus, later in the study, one cell from each regime will be considered as representative for the assessment of the steady state results since the distribution of the shear stress magnitude experienced by the cells of each regime is nearly uniform for each regime. By studying only three distinct cells, the authors were able to further downscale the model and thereby evaluate the prototype dynamic micro-organ device.

As a second modeling step, the cells are modeled as linear elastic spheres with a Poisson ratio $\gamma = 0.74$ and elastic modulus $E = 700$ Pa (Wu et al., 2003) under the 2D assumption of plane stress conditions. Average von Mises (VM) stress computations at the cellular level, choosing one cell from each identified regime, reveals, in contrast with the aforementioned cell WSS results, that the entrance cells

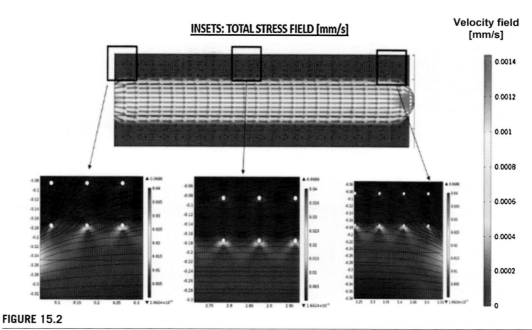

FIGURE 15.2

Identification of three distinct regimes of the printed micro-organ based on average cell WSS and maximum WSS at the single-cell level.

experience higher VM stresses than the middle cells as shown in Figure 15.3. This can be explained by the fact that the VM stress represents the sum of both the normal and shear stresses. Thus, it is evident that the entrance cells experience higher normal stresses compared to the middle cells, which are in contact and in parallel with the fully developed laminar velocity profile of the shear-induced flow field.

However, for the assessment of the dynamic micro-organ device, cell WSS metrics are used, since available literature data for optimum hepatocyte function in dynamic cell cultures is given in terms of WSS. It is reported by Tanaka et al. (Tanaka et al., 2006) that hepatocytes tend to alter their cellular morphology and metabolite function under various conditions of shear-mediated flow in a microchip. In particular, their metabolic activities in terms of albumin synthesis are decreased under shear stresses higher than 2 Pa, increased in the range of 0.14 Pa to 1 Pa and remained relatively constant from 1 to 2 Pa. Moreover, Kan et al. (Kan et al., 1998) tested the shear stress effect on the metabolic function of a coculture system of hepatocytes/parenchymal cells and reported a shear stress value of 0.47 Pa for optimum conditions.

The fact that these reported values are one to two orders of magnitude larger compared to the cell WSS obtained for the dynamic micro-organ device system compelled the authors to vary the magnitude of the inlet volumetric flow rate within the model environment to achieve cell WSS on the same order of magnitude with the results depicted in Figure 15.4. Specifying one cell from each regime of the bioprinted tissue shows that the comparatively low inlet volumetric flow rate in the postprocessing phase needs to be increased 10–20 times from the initial value of 0.25 μl/h in order for the flow field to induce cell WSS for optimal hepatocyte function.

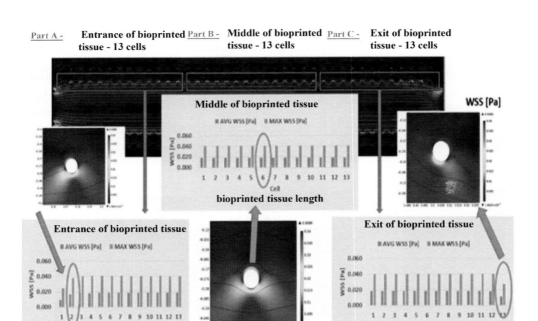

FIGURE 15.3

Velocity field and streamlines indicating a laminar flow topology. Insets: VM stress on each cell modeled as linear elastic spheres. Lower middle: Average VM stress for each regime of the printed micro-organ: (1) entrance, (2) middle, and (3) exit.

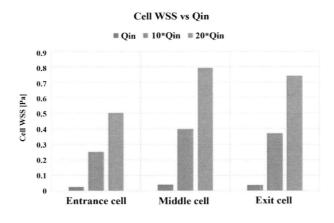

FIGURE 15.4

Effect of inlet volumetric flow rate on cell WSS.

15.1.4 FUTURE PERSPECTIVES

Based on the existing organ printing techniques and challenges presented in this chapter, the authors foresee the development of an integrated bioadditive manufacturing process to establish smaller feature sizes with prescribed geometries systematically determined by rheological characteristics, process parameters, and machine design. These capabilities will greatly enhance the presentation of precise, tunable microstructural cues in a 3D niche at the single-cell level with concomitant scalable manufacturing of functional organs. In concert with this technology development, also needed is a deeper understanding of biological cell shape metrology in the presence of tunable structural microenvironments or 3D microstructural cell niches for applications in stem cell differentiation and regenerative medicine. Each of these applications requires the creation of engineered 3D functional tissue constructs. To accomplish this, an interdisciplinary research methodology will likely implement engineering model approaches in materials processing, additive manufacturing process modeling, and metrology for benchmarking functional tissue fabrication. Such an endeavor will yield a scalable bioadditive manufacturing process capable of presenting precise, tunable microstructural cues in a 3D niche with single-cell fiber feature sizes and tunable fiber geometries by increasing the control resolution of process parameters for reproducible large-scale fabrication of 3D printed organs. Finally, in the near term, emerging organ printing and microfabrication techniques will converge to enable the direct printing of functional micro-organ-on-a- chip devices. Computational modeling of both printing processes and printed devices (e.g. micro-organ on chip) will be critical in elucidating the mechanically linked biological functional consequences that will further direct future printing process designs.

REFERENCES

Ahn, S., Koh, Y.H., Kim, G., 2010. A three-dimensional hierarchical collagen scaffold fabricated by a combined solid freeform fabrication (SFF) and electrospinning process to enhance mesenchymal stem cell (MSC) proliferation. Journal of Micromechanics and Microengineering 20, 065150.

Akselrod, G.M., Timp, W., Mirsaidov, U., Zhao, Q., Li, C., Timp, R., Timp, K., Matsudaira, P., Timp, G., 2006. Laser-guided assembly of heterotypic three-dimensional living cell microarrays. Biophysical Journal 91, 3465.

Arnsdorf, E.J., Tummala, P., Kwon, R.Y., Jacobs, C.R., 2009. Mechanically induced osteogenic differentiation: the role of RhoA, ROCKII, and cytoskeletal dynamics. Journal of Cell Science 122, 546.

Bible, E., Chau, D.Y.S., Alexander, M.R., Price, J., Shakesheff, K.M., Modo, M., 2009. Attachment of stem cells to scaffold particles for intracerebral transplantation. Nature Protocols 4, 1440.

Bible, E., Chau, D.Y.S., Alexander, M.R., Price, J., Shakesheff, K.M., Modo, M., 2009. The support of neural stem cells transplanted into stroke-induced brain cavities by PLGA particles. Biomaterials 30, 2985.

Boland, T., Xu, T., Damon, B., Cui, X., 2006. Application of inkjet printing to tissue engineering. Biotechnology Journal 1 (9), 910.

Burdick, J.A., Vunjak-Novakovic, G., 2009. Engineered microenvironments for controlled stem cell differentiation. Tissue Engineering Part A 15 (2), 205.

Burkhardt, B., Martinez-Sanchez, J.J., Bachmann, A., Ladurner, R., Nüssler, A.K., 2013. Long-term culture of primary hepatocytes: new matrices and microfluidic devices. Hepatology International 8 (1), 14.

Butscher, A., Bohner, M., Hofmann, S., Gauckler, L., R.Müller, R, 2011. Structural and material approaches to bone tissue engineering in powder-based three-dimensional printing. Acta Biomaterials 7, 907.

Buyukhatipoglu, K., Chang, R., Sun, W., Morss-Clyne, A., 2010. Bioprinted nanoparticles for tissue engineering applications. Tissue Engineering Part C 4 (16), 631.

Castillo, A.B., Jacobs, C.R., 2010. Mesenchymal stem cell mechanobiology. Current Osteoporosis Reports 8 (2), 98.

Chang, C.C., Boland, E.D., Williams, S.K., Hoying, J.B., 2011. Direct-write bioprinting three-dimensional biohybrid systems for future regenerative therapies. Journal of Biomedical Materials Research B Applied Biomaterials 98B, 160.

Chang, R., Emami, K., Wu, H., Sun, W., 2010. Biofabrication of a three-dimensional liver micro-organ as an *in vitro* drug metabolism model. Biofabrication 2 (4), 045004.

Chang, R., Nam, J., Sun, W., 2008. Computer-Aided Design, Modeling, and Freeform Fabrication of Micro-organ Flow Patterns for Pharmacokinetic Study. CAD Applications 5, 21.

Chang, R., Nam, J., Sun, W., 2008. Direct cell writing of 3D micro-organ for *in vitro* pharmacokinetic model. Tissue Engineering Part C 2 (14), 157.

Chang, R., Nam, J., Sun, W., 2008. Effects of dispensing pressure and nozzle diameter on cell survival from solid freeform fabrication-based direct cell writing. Tissue Engineering 1 (14), 41.

Chen, C.S., Mrksich, M., Huang, S., Whitesides, G.M., Ingber, D.E., 1997. Geometric control of cell life and death. Science 276 (5317), 1425.

Cohen, D.L., Malone, E., Lipson, H., Bonassar, L.J., 2009. Direct freeform fabrication of seeded hydrogels in arbitrary geometries. Tissue Engineering 12 (5), 1325.

Delorme, B., Ringe, J., Pontikoglou, C., Gaillard, J., Langornné, A., Sensebé, L., Noël, D., Jorgensen, C., Häupi, T., Charbord, P., 2009. Specific lineage-priming of bone marrow mesenchymal stem cells provides the molecular framework for their plasticity. Stem Cells 27 (5), 1142.

Duan, B., Wang, M., 2011. Selective laser sintering and its application in biomedical engineering. Materials Research Society Bulletin 36, 998.

Farooque, T.M., Camp, C.H., Tison, C.K., Kumar, G., Parekh, S.H., Simon, C.G., 2014. Measuring stem cell dimensionality in tissue scaffolds. Biomaterials 35 (9), 2558.

Fedorovich, N.E., Alblas, J., Hennink, W.E., Oner, F.C., Dhert, W.J.A., 2011b. Organ printing: the future of bone regeneration? Trends in Biotechnology 29, 601.

Fedorovich, N.E., Dewijn, J.R., Verbout, A.J., Alblas, J., Dhert, W.J., 2008. Three-dimensional fiber deposition of cell-laden, viable, patterned constructs for bone tissue printing. Tissue Engineering Part A 14 (1), 127.

Fedorovich, N.E., Kuipers, E., Gawlitta, D., Dhert, W.J., Alblas, J., 2011a. Scaffold porosity and oxygenation of printed hydrogel constructs affect functionality of embedded osteogenic progenitors. Tissue Engineering Part A 17 (19), 2473.

Gaetani, R., Doevendans, P.A., Metz, C.H., Alblas, J., Messina, E., Giacomello, A., Sluijter, J.P., 2012. Cardiac tissue engineering using tissue printing technology and human cardiac progenitor cells. Biomaterials 33 (6), 1782.

Goral, V.N., Yuen, P.K., 2012. Microfluidic platforms for hepatocyte cell culture: new technologies and applications. Annals of Biomedical Engineering 40 (6), 1244.

Gruene, M., Pflaum, M., Hess, C., Diamantouros, S., Schlie, S., Deiwick, A., Koch, L., Wilhelmi, M., Jockenhoevel, S., Haverich, A., Chichkov, B., 2011. Laser printing of three-dimensional multicellular arrays for studies of cell-cell and cell-environment interactions. Tissue Engineering Part C 17 (10), 973.

Guillemot, F., Souquet, A., Castros, S., Guillotin, B., 2010. Laser-assisted cell printing: principle, physical parameters versus cell fate and perspectives in tissue engineering. Nanomedicine 5 (3), 507.

Haudenschild, A.K., Hsieh, A.H., Kapila, S., Lotz, J.C., 2008. Pressure and distortion regulate human mesenchymal stem cell gene expression. Annals of Biomedical Engineering 37 (3), 492.

Hsu, M.N., Tan, G.D., Tania, M., Birgersson, E., Leo, H.L., 2014. Computational fluid model incorporating liver metabolic activities in perfusion bioreactor. Biotechnology and Bioengineering 111 (5), 885.

Huang, A.H., Motlekar, N.A., Stein, A., Diamond, S.L., Shore, E.M., Mauck, R.L., 2008. High-throughput screening for modulators of mesenchymal stem cell chondrogenesis. Annals of Biomedical Engineering 36 (11), 1909.

Huang, G.S., Tseng, C.S., Linju Yen, B., Hsieh, P.S., Hsu, S.H., 2013. Solid freeform-fabricated scaffolds designed to carry multicellular mesenchymal stem cell spheroids for cartilage regeneration. European Cells and Materials 26, 179.

Hutmacher, D.W., Singh, H., 2008. Computational fluid dynamics for improved bioreactor design and 3D culture. Trends in Biotechnology 26 (4), 166.

Jakab, K., Norotte, C., Marga, F., Murphy, K., Vunjak-Novakovic, G., Forgacs, G., 2010. Tissue engineering by self-assembly and bioprinting of living cells. Biofabrication 2, 022001.

Kan, P., Miyoshi, H., Yanagi, K., Ohshima, N., 1998. Effects of shear stress on metabolic function of the coculture system of hepatocyte/nonparenchymal cells for a bioartificial liver. ASAIO Journal 44 (5), M441.

Kasko, A.M., Wong, D.Y., 2010. Two-photon lithography in the future of cell-based therapeutics and regenerative medicine: a review of techniques for hydrogel patterning and controlled release. Future Medicinal Chemistry 2, 1669.

Khalil, S., Sun, W., 2009. Bioprinting endothelial cells with alginate for 3D tissue constructs. Journal of Biomechanical Engineering 131 (11), 111002.

Kim, M.R., Jeon, E.S., Kim, M.Y., Lee, J.S., Kim, J.H., 2009. Thromboxane a(2) induces differentiation of human mesenchymal stem cells to smooth muscle-like cells. Stem Cells 27 (1), 191.

Koch, L., Kuhn, S., Sorg, H., Gruene, M., Schlie, S., Gaebel, R., Polchow, B., Reimers, K., Stoelting, S., Ma, N., Vogt, P.M., Steinhoff, G., Chichkov, B., 2010. Laser printing of skin cells and human stem cells. Tissue Engineering Part C 16 (5), 847.

Kumar, G., Tison, C.K., Chatterjee, K., Pine, P.S., McDaniel, J.H., Salit, M.L., Simon, C.G., 2011. The determination of stem cell fate by 3D scaffold structures through the control of cell shape. Biomaterials 32 (35), 9188.

Kumar, G., Waters, M.S., Farooque, T.M., Young, M.F., Simon, C.G., 2012. Freeform fabricated scaffolds with roughened struts that enhance both stem cell proliferation and differentiation by controlling cell shape. Biomaterials 33 (16), 4022.

Leach, J., Sinclair, G., Jordan, P., Courtial, J., Padgett, M.J., Cooper, J., Laczik, Z., 2004. 3D manipulation of particles into crystal structures using holographic optical tweezers. Optics Express 12, 220.

Lee, J., Kim, S.H., Kim, Y.C., Choi, I., Sung, J.H., 2013. Fabrication and characterization of microfluidic liver-on-a-chip using microsomal enzymes. Enzyme and Microbial Technology 53 (3), 159.

Li, M., Tian, X., Schreyer, D.J., Chen, X., 2011. Effect of needle geometry on flow rate and cell damage in the dispensing-based biofabrication process. Biotechnology Progress 27 (6), 1777.

Li, Y.J., Batra, N.N., You, L., Meier, S.C., Coe, I.A., Yellowley, C.E., Jacobs, C.R., 2004. Oscillatory fluid flow affects human marrow stromal cell proliferation and differentiation. Journal of Orthopedic Research 22 (6), 1283.

Maher, P.S., Keatch, R.P., Donnelly, K., Mackay, R.E., Paxton, J.Z., 2009. Construction of 3D biological matrices using rapid prototyping technology. Rapid Prototyping Journal 15 (3), 204.

Masuzaki, T., Ayukawa, Y., Moriyama, Y., Jinno, Y., Atsuta, I., Ogino, Y., Koyano, K., 2010. The effect of a single remote injection of statin-impregnated poly (lactic-co-glycolic acid) microspheres on osteogenesis around titanium implants in rat tibia. Biomaterials 31, 3327.

Melchels, F.P.W., Domingos, M.A.N., Klein, T.J., Malda, J., Bartolo, P.J., Hutmacher, D.W., 2012. Additive manufacturing of tissues and organs. Progress in Polymer Science 37 (8), 1079.

Melchels, F.P.W., Feijen, J., Grijpma, D.W., 2010. A review on stereolithography and its applications in biomedical engineering. Biomaterials 31, 6121.

Mezel, C., Souquet, A., Hallo, L., Guillemot, F., 2010. Bioprinting by laser-induced forward transfer for tissue engineering applications: jet formation modeling. Biofabrication 2 (1), 014103.

Miranda, J.P., Rodrigues, A., Tostoes, R.M., Leite, S., Zimmerman, H., Carrondo, M.J., Alves, P.M., 2010. Extending hepatocyte functionality for drug-testing applications using high-viscosity alginate-encapsulated three-dimensional cultures in bioreactors. Tissue Engineering Part C 16 (6), 1223–1232.

Mironov, V., Boland, T., Trusk, T., Forgacs, G., Markwald, R.R., 2003. Organ printing: computer-aided jet-based 3D tissue engineering. Trends in Biotechnology 21 (4), 157.

Mironov, V., Kasyanov, V., Drake, C., Markwald, R.R., 2008. Organ printing: promises and challenges. Regenerative Medicine 3 (1), 93.

Mironov, V., Kasyanov, V., Markwald, R.R., 2011. Organ printing: from bioprinter to organ biofabrication line. Current Opinions in Biotechnology 22 (5), 667.

Mironov, V., Visconti, R.P., Kasyanov, V., Forgacs, G., Drake, C.J., Markwald, R.R., 2009. Organ printing: tissue spheroids as building blocks. Biomaterials 30 (12), 2164.

Nakamura, M., Iwanaga, S., Henmi, C., Arai, K., Nishiyama, Y., 2010. Biomatrices and biomaterials for future developments of bioprinting and biofabrication. Biofabrication 2 (1), 014110.

Narita, Y., Yamawaki, A., H. Kagami, H., Ueda, M., Ueda, Y., 2008. Effects of transforming growth factor-beta 1 and ascorbic acid on differentiation of human bone-marrow-derived mesenchymal stem cells into smooth muscle cell lineage. Cell and Tissue Research 333 (3), 449.

Nishiyama, Y., Nakamura, M., Henmi, C., Yamaguchi, K., Mochizuki, S., Nakagawa, H., Takiura, K., 2009. Development of a three-dimensional bioprinter: construction of cell-supporting structures using hydrogel and state-of-the-art inkjet technology. Journal of Biomechanical Engineering 131 (3), 035001.

Prockop, D.J., 1997. Marrow stromal cells as stem cells for nonhematopoietic tissues. Science 276 (5309), 71.

Quake, S.R., Scherer, A., 2000. From micro- to nanofabrication with soft materials. Science 290 (5496), 1536.

Rezendea, R.A., Pereiraa, F.D.A.S., Kasyanovb, V., Kemmoku, D.T., Maia, I., da Silvaa, J.V.L., Mironov, V., 2013. Scalable biofabrication of tissue spheroids for organ printing. Procedia CIRP 5, 276.

Riddle, R.C., Taylor, A.F., Genetos, D.C., Donahue, H.J., 2006. MAP kinase and calcium signaling mediate fluid flow-induced human mesenchymal stem cell proliferation. American Journal of Physiology and Cell Physiology 290 (3), C776.

Roth, E.A., Xu, T., Das, M., Gregory, C., Hickman, J.J., Boland, T., 2004. Inkjet printing for high-throughput cell patterning. Biomaterials 25 (17), 3707.

Sekula, S., Fuchs, J., Weg-Remers, S., Nagel, P., Schupler, S., Fragala, J., Theilacker, N., Franzreb, M., Wingren, C., Ellmark, P., Borrebaeck, C.A., Mirkin, C.A., Fuchs, H., Lenhert, S., 2008. Multiplexed lipid dip-pen nanolithography on subcellular scales for the templating of functional proteins and cell culture. Small 4 (10), 1785.

Sen, B., Xie, Z., Case, N., Ma, M., Rubin, C., Rubin, J., 2008. Mechanical strain inhibits adipogenesis in mesenchymal stem cells by stimulating a durable beta-catenin signal. Endocrinology 149 (12), 6065.

Shim, J.H., Kim, J.Y., Park, M., Park, J., Cho, D.W., 2011. Development of a hybrid scaffold with synthetic biomaterials and hydrogel using solid freeform fabrication technology. Biofabrication 3 (3), 034102.

Shor, L., Guceri, S., Chang, R., Gordon, J., Kang, Q., Hartstock, L., An, Y., Sun, W., 2009. Precision extruding deposition (PED) fabrication of poly-e-caprolectone (PCL) scaffolds for bone tissue engineering. Biofabrication 1 (1), 015003.

Sikavitsas, V.I., Temenoff, J.S., Mikos, A.G., 2001. Biomaterials and bone mechanotransduction. Biomaterials 22 (19), 2581.

Siniscalco, D., Giordano, C., Galderisi, U., L. Luongo, L., Alessio, N., Di Bernardo, G., de Novellis, V., Rossi, F., Malone, S., 2010. Intrabrain microinjection of human mesenchymal stem cells decreases allodynia in neuropathic mice. Cellular and Molecular Life Sciences 67, 655.

Smith, C.M., Christian, J.J., Warren, W.L., Williams, S.K., 2007. Characterizing environmental factors that impact the viability of tissue-engineered constructs fabricated by a direct-write bioassembly tool. Tissue Engineering 13 (2), 373.

Smith, C.M., Stone, A.L., Parkhill, R.L., Stewart, R.L., Simpkins, M.W., Kachurin, A.M., Warren, W.L., Williams, S.K., 2004. Three-dimensional bioassembly tool for generating viable tissue-engineered constructs. Tissue Engineering 10 (9), 1566.

Snyder, J., Hamid, Q., Wang, W., Chang, R., Emami, K., Wu, H., Sun, W., 2011. Bioprinting cell-laden matrigel for radioprotection study of liver by prodrug conversion in dual tissue microfluidic chip. Biofabrication 3 (3), 034112.

Stringer, J., Derby, B., 2010. Formation and stability of lines produced by inkjet printing. Langmuir 26 (12), 10365.

Sumanasinghe, R.D., Bernacki, S.H., Loboa, E.G., 2006. Osteogenic differentiation of human mesenchymal stem cells in collagen matrices: effect of uniaxial cyclic tensile strain on bone morphogenetic protein (BMP-2) mRNA expression. Tissue Engineering 12 (12), 3459.

Sumanasinghe, R.D., Pfeiler, T.W., Monteiro-Riviere, N.A., Loboa, E.G., 2009. Expression of proinflammatory cytokines by human mesenchymal stem cells in response to cyclic tensile strain. Journal of Cell Physiology 219 (1), 77.

Tan, G.D., Toh, G.W., Birgersson, E., Robens, J., van Noort, D., Leo, H.L., 2013. A thin-walled polydimethylsiloxane bioreactor for high-density hepatocyte sandwich culture. Biotechnology and Bioengineering 110 (6), 1663.

Tanaka, Y., Yamato, M., Okano, T., Kitamori, T., Sato, K., 2006. Evaluation of effects of shear stress on hepatocytes by a microchip-based system. Measurement Science and Technology 17, 3167.

Théry, M., 2010. Micropatterning as a tool to decipher cell morphogenesis and functions. Journal of Cell Science 123, 4201.

Tourlomousis, F., Chang, R.C., November 2014. Computational modeling of 3D printed tissue-on-a-chip microfluidic devices as drug screening platforms. Proceedings of ASME, IMECE, Montreal, Canada.

Townes-Anderson, E., St Jules, R.S., Sherry, D.M., Lichtenberger, J., Hassanain, M., 1998. Micromanipulation of retinal neurons by optical tweezers. Molecular Vision 4, 12.

Truscello, S., Schrooten, J., Oosterwyck Van, H., 2011. A computational tool for the upscaling of regular scaffolds during *in vitro* perfusion culture. Tissue Engineering Part C 17 (6), 619.

Wagner, D.R., Lindsey, D.P., Li, K.W., Tummala, P., Chandran, S.E., Smith, R.L., Longaker, M.T., Carter, D.R., Beaupre, G.S., 2008. Hydrostatic pressure enhances chondrogenic differentiation of human bone marrow stromal cells in osteochondrogenic medium. Annals of Biomedical Engineering 36 (5), 813.

Wang, X., Yan, Y., Pan, Y., Xiong, Z., Liu, H., Cheng, J., Liu, F., Wu, R., Zhang, R., Lu, Q., 2006. Generation of three-dimensional hepatocyte/gelatin structures with rapid prototyping system. Tissue Engineering 12 (1), 83.

Watson, D., Hagen, N., Diver, J., Marchand, P., Chachisvilis, M., 2004. Elastic light scattering from single cells: orientational dynamics in optical trap. Biophysical Journal 87, 1298.

Whitesides, G.M., 1995. Self-assembling materials. Scientific American 273 (3), 114.

Whitesides, G.M., Grzybowski, B., 2002. Self-Assembly at all scales. Science 295 (5564), 2418.

Whitesides, G.M., Ostuni, E., Takayama, S., Jiang, X., Ingber, D.E., 2001. Soft lithography in biology and biochemistry. Annual Review of Biomedical Engineering 3, 335–373.

Wu, Z.Z., Zhang, G., Long, M., Wang, H.B., Song, G.B., Cai, S.X., 2003. Comparison of the viscoelastic properties of normal hepatocytes and hepatocellular carcinoma cells under cytoskeletal perturbation. Biorheology 37 (4), 279.

Xu, T., Jin, J., Gregory, C., Hickman, J.J., Boland, T., 2005. Inkjet printing of viable mammalian cells. Biomaterials 26 (1), 93.

Yan, Y., Wang, X., Pan, Y., Liu, H., Cheng, J., Xiong, Z., Lin, F., Wu, R., Zhang, R., Lu, Q., 2005. Fabrication of viable tissue-engineered constructs with 3D cell-assembly technique. Biomaterials 26 (29), 5864.

INTELLECTUAL PROPERTY IN 3D PRINTING AND NANOTECHNOLOGY

John F. Hornick and Kai Rajan[1]

16.1 INTRODUCTION

Like many modern technologies, 3D printing and nanotechnology are improving at breakneck speed, through innovations and improvements in machines, software, methods, and materials. These innovations and improvements are made possible by inventors and researchers striving to benefit society. In addition to the satisfaction of benefitting society, the inventors and researchers, as well as the companies and institutions that fund and enable research, need assurance that the time and money spent advancing technology can be recovered by successful innovations, to fund additional research and maintain the cycle of research and innovation. Intellectual property laws provide this assurance.

Intellectual property (IP) broadly refers to "creations of the mind, such as inventions; literary and artistic works; designs; and symbols, names and images used in commerce (World Intellectual Property Organization, 2014a)." IP rights arise from laws that protect these creations of the mind from unfair exploitation, including laws for patents, copyrights, trade secrets, and trademarks. IP rights enable inventors and creators to protect their financial investments in research by securing the profits from successful and popular inventions or creations. In return, the public is enriched by the inventions and creations. By striking the right balance between protecting the interests of innovators and of the public, IP laws aim to foster an environment in which creativity and innovation can flourish.

IP laws aim to benefit society and reward innovators, but there are exceptions and limitations to protection, and there are challenges to enforcing IP rights. This chapter explores the different types of IP, how they protect 3D bioprinting and nanotechnology (particularly engineered tissue), limitations on IP, and how IP owners protect and enforce their rights.

---

[1]John Hornick is a partner and Kai Rajan is an associate with the Finnegan IP law firm, based in Washington, DC, www.finnegan.com. John is a frequent writer and lecturer on IP and 3D printing, and educates and advises clients in this space. Any opinions expressed in this article are those of the authors, not of Finnegan. This article is not legal advice.

16.2 WHY IS INTELLECTUAL PROPERTY IMPORTANT?

IP laws foster advancements in science by providing mechanisms to (1) protect research and innovation from copycats, thieves, and even accidental copying, and (2) provide recourse for IP owners. Such protections and rights incentivize innovators and investors to spend time and money to develop innovations and improvements in products and services that benefit society. Patent protection particularly enables inventors and investors to capitalize on their investments in research and development (R&D) by deterring or slowing down competitors, and forcing them to innovate in noninfringing ways (Kapmar, 2013).

The ability to protect IP is especially important in the medical industry. Anything that is used on or in the human body requires extensive testing to ensure biocompatibility, and to evaluate side effects (U.S. Food and Drug Administration, 2009, 2012). Such testing substantially raises the costs of R&D for new products such as drugs and implants. Innovators experimenting with new compounds, material configurations, products, and methods bear the burden of research costs, with the hope that some products will be successful, receive approval for sale, generate enough revenue to recoup research and development costs for both successful and unsuccessful products, and make a profit.

R&D costs in medical technologies can be prohibitively expensive for many companies, causing some companies to copy others' successful products and sell them as their own. This creates unhealthy competition that deprives the innovators and investors of profits and returns on their investments, which ultimately discourages further R&D for fear of copying. Indeed, high R&D costs coupled with lost profits from copycatting can demoralize research initiatives and stifle innovation. IP helps to prevent copycats from profiting from a free ride on others' R&D, and thus promotes healthy competition.

Our modern IP system fosters innovation, but the system and its laws are not without controversy and complications. New and emerging technologies that do not fit squarely into current IP laws cause confusion and inconsistency in IP protection. For example, in recent decades, inventors, their attorneys, the courts, and even Congress have struggled with the patentability of software, because the original patent laws and even recent amendments do not explicitly address it.[2] Similarly, despite their potential to advance science by leaps and bounds, 3D bioprinting technology, nanotechnology, and tissue engineering technology raise more questions about what aspects of these new inventions and improvements *can* be protected by IP, which *cannot* be protected, and which *should not* be protected for ethical and public policy reasons.

16.3 TYPES OF INTELLECTUAL PROPERTY

IP laws are national systems. This chapter focuses on the US IP laws. US IP laws protect innovations and creations through patents, trademarks, copyrights, and trade secrets. However, IP protection in the United States does not extend throughout the world. Rather, with the exception of the European Union, inventors must seek IP protection in each country where protection is desired, either through filing separate applications or complying with national laws. Although IP laws vary by country, the types of IP protection available throughout the world are fairly consistent.

---

[2]*Gottschalk v. Benson*, 409 U.S. 63 (1972); *Parker v. Flook*, 437 U.S. 584 (1978); *Diamond v. Diehr*, 450 U.S. 175 (1981); *Bilksi v. Kappos*, 561 U.S. __, 130 S. Ct. 3218 (2010).

16.3.1 PATENTS

Patents provide inventors of new and useful inventions the right "to exclude others from making, using, offering for sale, or selling the invention throughout the United States or importing the invention into the United States (United States Patent and Trademark Office, 2014)." This right lasts up to 20 years in the United States. There are three types of patents: utility patents; design patents; and plant patents. *Utility* patents protect new and useful machines, processes, compositions of matter, and articles of manufacture. Utility patents can also protect computer programs and software, by describing the functions performed by a computer. *Design* patents protect the decorative look (called the "ornamental design") of objects. Design patents also protect the appearance of computer icons, graphic user interfaces, and graphic animations. Finally, *plant* patents protect newly invented or discovered plant varieties.

To obtain a patent, the inventor(s) must disclose the details of their invention to the US Patent Office in a patent application, which is examined by a patent examiner to determine whether the invention will be patented (United States Patent and Trademark Office, 2013). The examination process can take years, but IP protection for granted patents reaches back to the date the patent application was filed. Once a patent is granted, the details of the invention are released to the public, to increase public knowledge and to inform the public of the inventors' exclusive rights. In return for enhancing public knowledge, the inventors are granted the exclusive right to exclude others from making, using, and selling the invention for 20 years from the filing date.

After obtaining a patent, the patent owner may choose to sell the patent, grant licenses to others who want to make, use, or sell the patented invention for limited periods of time, or enforce patent rights against infringers making, using, and/or selling the same invention without authorization. Notably, infringers can be liable to the patent owner even if they did not know the patent existed, because independent discovery is not a defense to patent infringement. Patent infringement and enforcing IP rights for 3D bioprinting and nanotechnology are addressed later in this chapter.

16.3.2 COPYRIGHTS

Copyrights protect original creative works of authorship, such as writing, music, and even software. Copyright protection for software covers the particular code written for a computer program, in contrast to software patents that protect the functions performed by a processor executing the code. A copyright provides the author of the creative work with the exclusive right to distribute and use that creative work, and prevents others from copying the creative work for up to 70 years after the author's death (United States Copyright Office, 2012a). Although this may seem like a very long time, others can often avoid copyright infringement by interpreting a work in their own creative way, such as by rewriting a work in their own words, or developing equivalent software in a different style of coding. Thus, copyright protection primarily prevents someone from simply copying the creative work of another and claiming the copied creativity as his or her own.

Copyright protection arises automatically when the work is fixed in a tangible form. Authors may also register their creative works with the U.S. Copyright Office, and can display the "©" symbol at any time. Copyright owners can sell their rights, grant licenses to others to use or reproduce the copyrighted work, and sue copycats who reproduce the copyrighted work. Obtaining a copyright license can be as simple as purchasing a computer program because authorized sales of such works often include licenses to use (and sometimes reproduce) the copyrighted work.

16.3.3 TRADEMARKS

Trademarks are words, phrases, symbols, and designs that identify the source of goods or services. Trademarks protect manufacturers, merchants, and consumers by preventing impostors from posing as reputable companies with successful products and services. Trademarks need not be registered with the U.S. Patent and Trademark Office, although registration provides additional layers of protection. Registered trademarks are distinguished by the ® symbol. Trademark protection attaches to unregistered trademarks after an individual or company that owns the trademark starts using the mark in commerce, and once the trademark  gains recognition and association with the trademark owner's goods or services. Unregistered trademarks for products can bear the ™ symbol, and unregistered service marks for services can bear the ℠ symbol (United States Patent and Trademark Office, 2013a).

To register a trademark, the owner must submit the mark to the U.S. Patent and Trademark Office, along with proof that the owner intends to use or is currently using the mark in commerce. If the mark is not already in use by another and is not too similar to other marks, it will most likely be registered. The registration process is generally much quicker than the examination procedure for patents. Those who do not wish to register their marks can use the ™ and ℠ notations at any time. However, trademark registration provides the owner of the mark with increased rights and options for stopping infringers. For example, registering a trademark places the public on notice that a particular trademark is "taken," and that copycats will be liable for trademark infringement (United States Patent and Trademark Office, 2013b).

Trademarks are mainly used for company and product branding. In tissue engineering, the products are tissue and organs. While implanted devices often have some type of marking or model identifier, printed tissue and organs may not carry logos or branding, but time will tell. At this time, trademarks are most applicable to the branding of hardware and software used in 3D bioprinting and nanotechnology.

16.3.4 TRADE SECRETS

Trade secrets (Hardin, 2012) are confidential business information that provide a competitive advantage and have some commercial value (World Intellectual Property Organization, 2014b). A trade secret can take the form of a formula, program, device, technique, manufacturing process, or any other nonpublic information that provides a competitive advantage. Trade secret protection is most

effective when successful independent discovery or reverse engineering is very difficult to accomplish without the trade secret information. Because such difficulty results in a low likelihood of successful copycats, the trade secret owner can forego the patenting process and avoid disclosing the trade secret to the public. Some of the most famous trade secrets are secret formulas and recipes, like the Coca Cola formula and the Kentucky Fried Chicken recipe (Kane, 2012). These recipes are locked away, under heavy guard, and only known to a very select few. Indeed, many have tried to decipher the recipes themselves, but no one has succeeded to the point where Coca Cola or KFC lose distinction. Therefore, they maintain a competitive advantage by keeping their recipes secret.

No formal registration or application process for trade secrets exists, and protection attaches immediately. Trade secret protection can last indefinitely, so long as the owner keeps the business information confidential (World Intellectual Property Organization, 2014c). To protect a trade secret, the owner must take appropriate steps to maintain its secrecy and avoid public disclosure or theft. The sufficiency of these steps may depend on a number of factors, such as the likelihood the secret would be stolen or the amount of interest from competitors in the trade secret. In the Coca Cola example, many beverage companies would pay millions to obtain the original formula, and many protection measures are needed to protect the formula from theft or public disclosure. Unlike patent protection, independent discovery by others is allowed. If another person successfully reverse engineers or discovers the trade secret, then trade secret protection ends, as the trade secret is deemed "public information." However, should someone attempt to steal the trade secret information, the trade secret owner can stop the thief from using the information and recover for losses from the trade secret theft, so long as the trade secret owner can prove the secret was stolen despite appropriate measures to maintain secrecy.

16.4 WHERE DOES INTELLECTUAL PROPERTY LAW ORIGINATE?

Modern IP laws originate from the United States Constitution and acts of Congress. Multiple clauses in the Constitution direct the United States government to make laws that protect the rights of individuals who contribute to the greater good through innovations and creative works. Article 1, Section 8, Clause 8 of the Constitution provides Congress with the broad power "to promote the Progress of Science and useful Arts, by securing for limited Times to Authors and Inventors the exclusive Right to their respective Writings and Discoveries." This clause forms the basis of the U.S. Patent and Copyright laws. Article 1, Section 8, Clause 3 of the Constitution provides the basis for federal trademark law, by providing Congress with the broad power "to regulate commerce with foreign nations, and among the several states, and with the Indian tribes." State laws also protect trademarks. Trade secret laws are creatures of state law, and most states have adopted the Uniform Trade Secrets Act.

16.5 WHAT ASPECTS OF 3D BIOPRINTING AND NANOTECHNOLOGY ARE PROTECTABLE?

3D bioprinting and nanotechnology are complex technologies that combine specialized computerized machinery, specialized materials, and intricate processes of additive manufacturing and nanoscale manufacturing. Many aspects of 3D bioprinting and nanotechnology for tissue engineering may be protected by IP, including improvements on technologies that are already protected, and innovative new technologies that do not fit squarely within current IP laws. To better understand what new aspects of 3D bioprinting and nanotechnology can be protected, this chapter also addresses the aspects of tissue engineering and bioengineering currently protectable under US IP laws.

Historically, implanted devices, surgical treatments and procedures, and pharmaceutical drugs have been protected by utility patents, design patents, trademarks, copyrights, and trade secrets. Implanted devices fall squarely within the "machine" category of utility patent laws, and possibly design patents for the ornamental, nonfunctional design of the implanted devices. Surgical treatments and procedures, although not protectable in some countries, fall within the "process" category of US utility patent laws. Pharmaceutical drugs have been protected by patents for "compositions of matter," where the patents are granted for the man-made chemical structures of the drugs. Other medical inventions are also afforded IP protection although they do not fall squarely within the original IP law definitions. For example, although subject to much controversy, nonhuman organisms, tissue, and even human genes have been patented, subject to limitations discussed later.

IP laws may apply to nearly all aspects of 3D bioprinting and nanotechnology, including the hardware used to print an item or handle nanoparticles, software involved in the design of tissue structures and the operation of the hardware to create the tissue, and specialized materials used to form or process the created tissue. 3D bioprinters combine cutting-edge machinery, electronic circuitry, software, materials, and processes in each creation of tissue, and the resulting tissue and organs may be protectable as well.

The most prominent forms of IP involved in 3D bioprinting and nanotechnology are patent and trade secret protection, discussed in detail later. Copyrights also protect the software executed by 3D bioprinting and nanotechnology machinery.

Section 101 of the US patent laws broadly defines a patentable invention as a "new and useful process, machine, manufacture, or composition of matter, or any new and useful improvement thereof."[3] For a long time after the US patent system was established by The Patent Act of 1790, patents were mostly granted for new and improved machines, and methods for creating or physically transforming objects. The advent of computers brought waves of patent applications for computers and for computer software. Now, software such as medical diagnostic programs, device control programs, and computer-aided design (CAD) programs, are patentable for the functions that processors perform while executing the software, and even for the physical memory or media on which the software is stored. 3D bioprinting and nanotechnology combine many known technologies with groundbreaking and unknown technologies, and open new frontiers in IP. The main categories of patentable aspects of these technologies are hardware, software, processes, and materials.

16.5.1 HARDWARE

The machines involved in creating engineered tissue with 3D bioprinting and nanotechnology fit squarely within the "machine" category of patentable subject matter. Of course, 3D printers are machines whose functionality can be protected by utility patents. Many patents can cover different parts or part combinations of a single piece of hardware. Within a single 3D bioprinter, there may be numerous systems working to move print heads, process raw materials such as the cells and binding agents,

---

[3]35 U.S.C. § 101 (added in 1952).

lay the raw materials, and perform additional processing to preserve the printed tissue. Whereas basic 3D printers heat plastic filament and lay melted plastic in patterned layers, 3D bioprinters must store, handle, and construct biological materials, a much more demanding manufacturing environment that requires the hardware needed to create live tissue, a much more difficult task than simply layering melted plastic.

Indeed, 3D bioprinters may handle many types of biological and nonbiological materials, and may require specialized reservoirs to store living cells and biological binding agents, and specialized hardware to treat printed tissue with nanomaterials or other materials to produce biocompatible tissue and organs. The nanomaterial handling hardware may be incorporated into 3D bioprinters to coat printed tissues with nanoparticles. As one would expect, the hardware involved in creating live tissue is very complex, and the innovators and investors paving the future require assurance that their investments in successful innovations will not be usurped by copycats.

Another important piece of hardware used in 3D printing is the 3D scanner. Scanners simplify and expedite the process of replicating objects, including living tissue. 3D scanners have existed for years as patentable machines (Song et al., 2000), but as tissue engineering improves with the use of 3D bioprinting and nanotechnology, scanning technology will also improve. 3D scanners allow individuals to scan a piece of living tissue, determine its cellular structure and composition, and generate instructions for a 3D bioprinter to replicate the tissue.

16.5.2 SOFTWARE

3D bioprinters and nanomaterial handling devices are highly automated computerized machines. The complex process for printing and processing living tissue involves many factors that must be continuously monitored and controlled, including temperature, acidity, wetness, oxygenation, and nutrition for the printed cells. Even relatively basic 3D printers that extrude materials to build biodegradable scaffolds for shaping cell growth must heat materials to specific temperatures, navigate one or more print heads within a three-dimensional environment, and extrude a specific amount of materials, all with high precision. These complex, repetitive steps are performed automatically by the programmed 3D bioprinter, and the software controlling the 3D bioprinter may be protectable by both patents and copyrights.

To revisit the distinctions between copyright and patent protection for software, copyright protection prevents anyone from copying source code from the author's program, and using the same source code in a different program without the author's permission. Copyright protection in this case prevents competitors from simply copying-and-pasting source code, but does not prevent competitors from developing equivalent software using different coding. Software patents can protect the functionality of the 3D printer, such as the steps performed by the 3D printer processors executing the software, and the settings used to produce a product. Patenting 3D bioprinter software allows the innovators to prevent others from simply copying the software and tweaking the code to avoid copyright infringement.

In addition to the software programs that operate 3D bioprinting equipment, software for creating digital "blueprints," and the digital blueprints themselves, may be protected by patents and copyrights. These digital blueprints can be created automatically by 3D scanners (and 3D scanner software), or manually by CAD software developed especially for tissue engineering. Again, copyrights would protect the specific source code underlying the software, whereas patents can protect the functionality of the software and a broader description of the 3D printer instructions for creating the tissue, such as dimensions, cellular arrangement, and other operational parameters.

16.5.3 **METHODS/PROCESSES**

Method patents can protect, at a high level, the entire process of creating a 3D bioprinted object, including computerized steps for controlling hardware, the physical actions performed by the hardware, and even manual steps performed before or after the printing occurs. Therefore, method patents can provide broad IP coverage of the steps involved to prepare the raw materials, construct the tissue, and perform postprocessing steps, such as treating the printed tissue with nanomaterials to enhance biocompatibility or drug delivery capabilities. Unlike patents and copyrights only covering software, method patents can cover computerized steps and also manual steps involved in the creation of printed tissue. Patents that cover the manufacturing process are usually referred to as "methods of manufacture." In addition to patents that broadly cover the entire manufacturing process, method patents can cover portions of the process, such as an improved method for storing and transporting living cells before 3D bioprinting occurs, or a particular method of laying and binding living cells, or a method of treating the printed tissue with nanomaterials.

Tissue engineering methods are ideal for trade secret protection, especially methods involving highly specialized (and also secret) hardware, or manual steps known only to the innovator. Because of the complex nature of engineered tissue and organs, reverse engineering a printed organ to determine how the cells structures were formed and to determine how the tissue was processed to increase biocompatibility may be difficult to accomplish. Accordingly, innovators who do not want to divulge any of their processes to the public in a patent can hold them as trade secrets and fully capitalize on their inventions until competitors reverse engineer their method or come up with the next best method.

16.5.4 **MATERIALS**

Tissue engineered with 3D bioprinting and nanotechnology use some entirely new materials to form the tissue structures and process the tissue for increased biocompatibility or drug delivery properties. The man-made and artificially modified materials used for 3D bioprinting and nanotechnology are patentable as "compositions of matter." Tissue printed from a 3D bioprinter is usually formed using "bio-ink," a medium of living cells suspended in liquid that protects the cells and keeps them alive, and helps the cells adhere to the tissue structure during printing. Some forms of bio-ink use natural materials or natural cells, such as stem cells. While the natural materials and cells themselves may not be patentable, artificial compounds that happen to include natural materials, as well as artificial modifications of the natural materials, can be patented. For example, some types of bio-ink use natural gelatins as the printing medium. Natural gelatin is semisolid in its natural state, but the gelatin used in bio-ink is modified to remain in liquid form during storage and printing, and only hardens during printing and post processing of the printed tissue (Maxey, 2013). The modified gelatin, which has different chemical properties, is patentable as a composition of matter. Like the modified gelatin, bio-ink that incorporates natural living cells is patentable as a composition of matter when the living cells are mixed or treated with artificial compounds to form combinations that are not found in nature. For example, the living cells in the bio-ink may be treated or processed to enhance biocompatibility, adhesion to other cells, or prolong the shelf life of the bio-ink or printed tissue.

Patents on the bio-ink, nanomaterials, and other materials used in 3D bioprinting may also tie into patents on the reservoirs and hardware that hold the bio-ink and keep the living cells alive, the mechanisms that transport the cells, and the specialized print heads that dispense bio-ink. Specialized patentable materials may also be used in these hardware components, to keep the cells alive and intact while printing tissue. Materials used to construct the 3D bioprinting hardware may be patentable, even if the bio-ink is all natural and thus unpatentable.

Some types of 3D bioprinters do not print cells to create engineered tissue and organs. Instead, the 3D bioprinter creates a biodegradable structure, called a "scaffold," from biodegradable materials, which is then covered with cultured living cells to grow into the tissue or organ structure. The living cells may not be patentable if they are simply cultured natural cells, but man-made or artificially modified scaffold materials may be patentable. Furthermore, materials and processes that improve scaffold materials to enhance biodegradability and increase adhesion to the living cells may be patentable processes and compositions of matter.

16.6 INTELLECTUAL PROPERTY PROTECTION LIMITATIONS FOR ENGINEERED TISSUE

The previous section addressed aspects of 3D bioprinting and nanotechnology that are protectable by patents, copyrights, and trade secrets. While there is broad range of IP protection available, modern laws and court decisions have carved out some exceptions to IP protection for public policy reasons. For example, the public would not benefit from granting monopolies over things that can be found in nature, or scientific principals such as $E = MC^2$. This section explores in more detail the limitations on aspects of tissue engineering with 3D bioprinting and nanotechnology that are subject to IP protection, and the exceptions to protection.

Just as the bulk of IP protection for 3D bioprinting and nanotechnology technologies is patent protection, the majority of exceptions to IP protection in these industries falls under exceptions to patentability. Section 101 of the US patent laws broadly defines what can be patented, but the U.S. Patent Office and the courts have carved out numerous exceptions and limitations (United States Patent and Trademark Office, 2014b).

The most important exception to patentability is the human body. Courts have always prohibited patents on the human body and human organisms as violations of the Constitution, because granting property rights in the human body is akin to a form of slavery (Torrance, 2013). The recently enacted America Invents Act codified the ban on human patents, stating "[n]otwithstanding any other provision of law, no patent may issue on a claim directed to or encompassing a human organism (Leahy-Smith America Invents Act, 2011)." This exception falls under the umbrella of a broader limitation on IP: that one cannot hold IP rights on things found in nature. However, nonhuman organisms, animal tissues, and even human genes have been patented for years, and are considered patentable subject matter so long as they are man-made or man-modified into something that does not occur in nature.[4]

Engineered tissue and organs created with 3D bioprinters and nanotechnology will probably remain patentable to the extent that the tissue is considered "man-made." Manufactured tissue, such as 3D bioprinted tissue, is artificially created by assembling cells, and not by natural growth. Tissue grown in Petri dishes may be deemed unpatentable because it grew by natural processes, albeit with the assistance of man. But a machine undoubtedly constructs printed tissue with the use of artificial materials

---

[4]According to *Chisum on Patents* (Chisum, Donald. Chisum on Patents, Vol. 9, Appendix 24, Updated 2014.), "To the extent that the claimed subject matter is directed to a non-human 'nonnaturally occurring manufacture or composition of matter – a product of human ingenuity' *(Diamond v. Chakrabarty)*, such claims will not be rejected under 35 U.S.C. § 101 as being directed to nonstatutory subject matter." See also Diamond v. Chakrabarty, 447 U.S. 303 (1980).

mixed with living cells. The resulting printed tissue is considered to be "man-made" for the purpose of IP protection. Ethical considerations related to IP rights on tissue and organs are discussed later.

An argument can be made that tissue grown over 3D printed scaffolding is "natural" and therefore that the grown tissue cannot be protected, but at the very least variations of the process of creating that tissue—printing the scaffolding, applying the cells, and growing the culture—is considered a patentable process.[5] Additionally, the legal restrictions on patenting human organisms do not categorically exclude human tissue, organs, and other human organism components from being protected by patents (Kapmar, 2013).

More questions about the patentability of engineered tissue arise when 3D printed tissue and organs are indistinguishable from natural organs and tissues. Should man-made tissue and organs be patentable when they are identical in form and function to natural tissue and organs? Of course, any patents on tissue and organs could only cover the man-made versions, but discerning between infringing products and naturally grown tissue could be extremely difficult when 3D printing and nanotechnology are able to produce exact replicas of natural tissue and organs. One could argue that these products of 3D printers are essentially the same as a naturally occurring organ or tissue, and should be regarded as "naturally occurring" for the purpose of patentability. Others could argue that the tissue and organs, although replicas of nature, are still "articles of manufacture" and should be patentable.

Science has yet to progress to the level where 3D bioprinted tissue is completely indistinguishable from natural tissue, but at some point this IP dilemma will emerge. One possible solution for this potential issue is marking or branding the fabricated tissue and organs to distinguish patented man-made products from naturally grown tissue and organs, such as a DNA marker (Hornick, 2014). By purposefully distinguishing man-made tissue from natural tissue, innovators may continue to protect the IP of their inventions without challenging current patent laws.

16.7 ETHICAL CONSIDERATIONS OF ENGINEERED TISSUE INTELLECTUAL PROPERTY

Despite the legality of IP protection for engineered tissue, moral implications of 3D bioprinting may lead to spirited debate over the patentability of tissue, organs, and other medical products (Gartner, 2013). Patents grant the right to exclude others from making, using, or selling the protected product or method. Some view this exclusivity as a monopoly on the protected product or method, a monopoly that is undesirable when public health and access to treatment may be adversely affected. While innovations in 3D bioprinting and nanotechnology can improve the speed and quality of tissue and organ production, should society grant the innovators the right to exclude others from creating human body parts? (Gartner, 2014).

The Food and Drug Administration (FDA), which evaluates and approves medical devices, drugs, and treatments, may impose strict guidelines for approving 3D bioprinted tissue and organs, or ban

---

[5]According to *Chisum on Patents* (§ 1.02 [7] (2009) Product of Nature-Biological Subject Matter, discussing *SC Computer Prods. v. Foxconn Int'l*, 355 F.3d 1353, 1359 (Fed. Cir. 2004)), "An invention synthesized for the first time in a laboratory is eligible for patent protection under Section 101. Processes for producing this synthetic product in the laboratory and/or for using this synthetic product may also be eligible for patent protection under Section 101. However, a natural reproduction process, whether sexual, asexual, part of a chain reaction, or a process of decay, is ineligible for patent protection under Section 101. An item reproduced by such a natural process, whether an inorganic structure or a life form, must ipso facto be ineligible for patent protection under Section 101."

them altogether when there is uncertainty regarding the consistency, quality, or safety of 3D bioprinted tissue made in hospitals and doctors' offices. Perhaps IP can help reduce this uncertainty, by granting patents to the innovators with particular tissue designs, manufacturing methods, and organ products. The patent owners' right to exclude others would mean that the FDA could focus its efforts on ensuring that the patented product or method is safe and yields consistent results, even when the tissue is printed in individual medical facilities. Indeed, the freedom to print tissue and organs in individual offices and hospitals using unregulated and untested product designs and methods may result in inconsistent healthcare with the potential for harmful results and/or heavy regulation.

As discussed earlier, limited "monopolies" created by IP are necessary in the medical and biological fields to allow inventors and investors to recoup upfront R&D expenses, and to help fund future R&D, thus furthering the progress of science. This model has driven the pharmaceutical industry for decades, granting drug makers a temporary monopoly on a particular drug before allowing generic brands to start selling the same drug (Schacht and Thomas, 2005). The patent owner, in this case a pharmaceutical company, bears the burden of accountability and ongoing testing by the FDA and health organizations, and enjoys any resulting profits from sales of the patented drug. In some cases, this allows the drug maker to set the price, and to control the drug supply, which could have the effect of limiting accessibility to affordable treatment. But without this limited monopoly, drug quality could be inconsistent and the healthcare industry could suffer from too many versions of the same drug, and accountability could be spread over many pharmaceutical companies.

Granting an innovator the right to exclude others from making or using the invention allows the public to hold the easily identified patent owner accountable if something goes wrong. The ability to hold an identifiable entity accountable for harm and negligence is very important when public health is at stake, as in the medical and healthcare industries. An IP owner is easier to identify as the responsible party when something goes wrong, whereas, in contrast, everyone points fingers at one another when companies are allowed to rip off ideas from one another and something goes wrong.

There are many arguments for and against granting IP rights for 3D printed tissue and organs. IP generally fosters innovation and can serve societal interests by yielding consistent, quality products with high accountability. Although one innovator may hold a limited monopoly over a certain tissue structure or method, competitors are free to improve upon the patent or develop their own products that may be eligible for IP protection. As tissue and organs created with 3D bioprinting and nanotechnology become more mainstream and refined, these ethical debates will surely become more frequent and prominent, and society will decide whether IP rights are suited for engineered tissue.

16.8 INTELLECTUAL PROPERTY INFRINGEMENT

IP laws are designed to prevent copycats and to advance the progress of science. However, IP laws are not 100% efficient in deterring copycats from trying to make quick profits without any contribution to society, and IP owners often need to enforce their rights and prosecute infringers.

Infringement involves the unauthorized use or sale of products and processes protected by IP. Different types of IP have different standards and requirements for proving infringement, but for any type of IP the property owners have the responsibility of enforcing their rights. The U.S. Patent and Trademark Office and the U.S. Copyright Office do not monitor and police infringing activity (United States Patent and Trademark Office, 2011; United States Copyright Office, 2012b).

16.8.1 TRADEMARK INFRINGEMENT

Trademark infringement occurs when someone uses a mark, such as a word, logo, slogan, or other identifying mark as his or her own, without permission from the owner of the mark. Infringement causes confusion among consumers, who are not able to distinguish between products and services made and sold by the owner of the trademark and products and services made and sold by the trademark infringer. The end result may be lost profits and a tarnished reputation for the trademark owner. In tissue engineering, the end product—the tissue and organs themselves—may not be subject to trademark infringement unless the tissue or organs are branded. This may or may not happen. At this time, trademark protection is more appropriate for the hardware used to create 3D printed tissue and organs.

16.8.2 TRADE SECRET MISAPPROPRIATION

Trade secret infringement (called "misappropriation") occurs when a trade secret, such as a secret method of treating tissue with nanomaterials, is stolen from the trade secret owner and used for unfair competition. To allege trade secret misappropriation, the trade secret owner must first prove that a trade secret existed by showing that (1) it was not generally known to the public, (2) it conferred a competitive advantage to the owner, and (3) reasonable steps were taken to maintain secrecy (Cornell University Law School, 2014a). However, if the accused misappropriator can show independent development of the same method without using the trade secret, such independent discovery is a complete defense, and there is no misappropriation.

Applying these principles to tissue engineering, an innovator may invent methods for creating or treating tissue to exhibit certain desirable properties, such as high biocompatibility or extraordinary drug delivery capabilities. The innovator could hold a trade secret if the method is not publicly known or apparent from simply observing the tissue. The innovator would have the ability to sue thieves who learned of the method by stealing the innovator's lab notebook, hacking into the innovator's computer where the method was documented, or leaving the innovator's employ and setting up a competing company. However, the innovator would not be able to sue another innovator who independently came up with the same method. The independent discovery defense is one of the caveats to trade secret protection. As discussed later, patent protection overcomes the independent discovery defense, albeit at the cost of disclosing the secret method to the public.

16.8.3 COPYRIGHT INFRINGEMENT

To prove copyright infringement, the copyright owner must show that its copyrighted work was reproduced, distributed, performed, publicly displayed, or made into a derivative work without the copyright owner's permission (United States Copyright Office, 2014). In 3D bioprinting and nanotechnology, copyright laws mainly protect the software used in computerized equipment, and possibly the digital blueprint files that the 3D bioprinter uses to create tissue and organs.[6] Thus, detecting copyright infringement in 3D bioprinting involves comparing the copyright owner's software to the potential infringer's software.

While detecting software copyright infringement can be as easy as comparing source code, enforcing copyrights against 3D printing software may be a daunting task. 3D printing lends itself to

---

[6]Current law probably does not favor copyright protection of the digital blueprint of functional products, such as tissue and organs, but this could change.

a decentralized manufacturing system in which individuals use their own printer to create products on-demand and on the spot. Because the products themselves may not enter a distribution chain, such as from manufacturer, to distributor, to customer, 3D printed product sales may not be regulatable like products that you would buy in stores. Instead, 3D bioprinters in hospitals or doctors' offices may be used in two scenarios: (1) the doctor designs his/her own tissue or organ using specialized software, and 3D prints the tissue/organ, or (2) the doctor downloads a digital blueprint for a certain organ or tissue, and 3D prints the tissue/organ, with or without changes.

In the first scenario, the software used to create the tissue/organ design and the software used to control the printer to create the tissue/organ are subject to copyright protection. Copyright infringement may be relatively easy to enforce in this scenario because the software would be distributed from a known source.

In the second scenario, the 3D printer software and possibly the downloaded digital blueprint are subject to copyright protection. As discussed earlier, the 3D printer software is less prone to copyright infringement when obtained from a known source. However copyrights on digital blueprints may be difficult to enforce when the files are distributed over the Internet, such as between medical professionals. If the digital blueprints are not copyrightable, copyright may not be an option for extracting profit from digital tissue and organ blueprints.

Like the problems faced with MP3 file sharing in the 1990s and 2000s, digital blueprint file sharing can include unregulated and unlicensed distribution of IP. The innovators who design successful 3D printable tissue structures and organs stand to lose profits and recoupment of R&D expenses when their digital blueprints are distributed freely and used without collecting any royalties. Like many demoralized musical artists, unchecked file sharing of digital blueprints could disincentivize the innovators in tissue engineering, especially if they are not copyrightable. But unlike MP3 file sharing issues, the end users of the 3D bioprinting digital blueprints are sophisticated hospitals, laboratories, and doctors who would stand to lose a lot from copyright infringement lawsuits. Therefore, the sophistication of the medical and healthcare industries, coupled with ever-emerging protection mechanisms in digital files (known as digital rights management, or DRM), may keep unauthorized copying at a manageable level that does not significantly stifle innovation.

16.8.4 PATENT INFRINGEMENT

Patent infringement is probably the most important type of IP enforcement in the United States and throughout the world. Proving patent infringement requires showing that an unlicensed, unauthorized entity is making, using, or selling a product or service covered by the patent. The patent owner does not need to prove that the alleged infringer knew about the patent, but proof of prior knowledge can elevate the patent infringement claim to *willful* infringement, a more serious allegation with harsher consequences. Willful infringers are true copycats who simply steal a patented idea and pass it off as their own. Under U.S. patent law, even others who independently create the patented device (and are therefore not willful infringers) are still liable for patent infringement, and independent discovery is not a defense (Cornell University Law School, 2014b). Patent protection provides the ability for the patent owner to exclude *anyone* from making, using, or selling a patented invention, as one of the incentives for convincing inventors to disclose their inventions to the public when obtaining the patent.

Patent infringement in 3D bioprinting and nanotechnology may be extremely difficult to detect. Usually, patent owners watch open markets to identify infringing products and trace the infringing product back to the patent infringer. For example, a 3D bioprinter patent owner may look for competing machines with the same parts configuration as the configuration in the patent owner's patent. Similarly, a method patent owner may check products in the open market to see whether a competitor appears to be using the patented method without authorization, and a design patent owner may enforce patent rights against infringers making or selling products that look like the patented design. 3D bioprinting, however, and more particularly 3D bioprinted tissue, presents additional obstacles to detecting patent infringement.

Patents, like any IP, are only useful when the patent owner is able to enforce IP rights and exclude others from making and selling the patented item. Enforcing IP rights requires identification of the infringing products and the infringers, usually after the infringing product enters the open market. However, 3D bioprinted tissue, like any 3D printed item, may be created on-demand and on the spot. A hospital or doctor's office could print the tissue at the time it is needed, such as immediately before surgery. The printed tissue or organ could be implanted into a patient at the same location, and thus printed tissue and organs may never enter the open market. Even once implanted, patented tissue and organs, like all patented implants, are still subject to IP laws.[7] For awhile, 3D bioprinted tissue and organs probably will be manufactured in centralized locations, but once bio-ink and tissue digital blueprints are readily available, most 3D bioprinters probably will be located in individual labs, hospitals, and doctors' offices, and tissue and organs will be printed on-demand. For patent owners, this may present an enforcement nightmare, trying to identify infringers of 3D printed tissue and organ patents. The printed tissue and organs, once implanted in patients, will be impossible to detect and analyze for possible infringement. The person sitting next to you could have implanted, infringing 3D bioprinted tissue or organs, but it would be extremely difficult, if not impossible, to detect. This obstacle to detecting infringement also applies to conventional implanted devices, such as pacemakers, but such devices may continue to be manufactured in centralized locations and distributed in the open market, increasing their visibility to the public and to patent owners, at least in the short term. Therefore, patents on engineered tissue and organs made with 3D printers and nanotechnology may be very valuable IP assets, but they may also be very difficult to enforce against infringers.

Although most 3D bioprinted tissue and organs may not be distributed in open markets, merchants will distribute raw materials, such as bio-ink and nanomaterials, 3D bioprinter and nanotechnology hardware, and control software. Patent infringement of such 3D bioprinting and nanotechnology technology may be easier to detect, and thus innovators in raw materials, hardware, and control software used in 3D bioprinting and nanotechnology may be able to track down infringers more easily, and enforce their patents more effectively. However, digital blueprints for tissue and organs, even if patentable, may be shared peer-to-peer, or tissue and organs may be made and sold from such blueprints on the black market, without any way for the patent owner to detect infringing uses of the patented products (Lipson and Kurman, 2013).

16.9 CONCLUSION

IP is an important part of the evolving technologies of 3D bioprinting and nanotechnology. As tissue engineering methods continue to improve, including hardware, software, materials, and processes, innovators will look to IP to protect their innovations from exploitation by others, and to capitalize on

---

[7]*Monsanto Co. v. McFarling*, 488 F.3d 973, 2007 U.S. App. LEXIS 12099 (Fed. Cir. 2007), ("The principles of patent law do not cease to apply when patentable inventions are incorporated within living things, either genetically or mechanically.").

their contributions to society. By enforcing their IP rights, innovators will seek to recover their financial investments to fund further research, and continue advancing the progress of science. Although there is some uncertainty as to the scope and ethical considerations of IP protection for engineered tissue, many aspects of 3D bioprinting and nanotechnology are protectable under current laws.

REFERENCES

Cornell University Law School, 2014a. Trade secret. Legal Information Institute. Available from: <http://www.law.cornell.edu/wex/trade_secret> (accessed 13.03.14).

Cornell University Law School, 2014b. Trade secret. Legal Information Institute. Available from: <http://www.law.cornell.edu/wex/patent> (accessed 25.02.14).

Gartner, Inc, 2013. Gartner reveals top predictions for IT organizations and users for 2014 and beyond. Published October 8, 2013. Available from: http://www.gartner.com/newsroom/id/2603215.

Gartner, Inc, 2014. Gartner says uses of 3D printing will ignite major debate on ethics and regulation. Published January 29, 2014. Available from: http://www.gartner.com/newsroom/id/2658315.

Hardin, D.T., 2012. Confidential top secret. Corporate compliance insights. Updated December 13, 2012. Available from: <http://www.corporatecomplianceinsights.com/want-a-cost-effective-way-to-bolster-compliance-with-export-controls-look-to-your-trade-secret-procedures/>.

Hornick, J., 2014. How to tell what's real and what's fake in a 3D printed world. 3D Printing Industry, February 5, 2014. Available from: http://3dprintingindustry.com/2014/02/05/tell-whats-real-whats-fake-3d-printed-world/?utm_source(3D(Printing(Industry(Update&utm_medium(email&utm_campaign(38c9fa9fa0-RSS_EMAIL_CAMPAIGN&utm_term(0_695d5c73dc-38c9fa9fa0-60484669.

Kane, C., 2012. 7 sought-after trade secrets. Published August 22, 2012. Available from: http://www.cnbc.com/id/48755451.

Kapmar, D., 2013. A look at the patentability of 3-D printed human organs. Published May 28, 2013. Available from: http://www.law360.com/articles/439549/a-look-at-the-patentability-of-3-d-printed-human-organs.

Leahy-Smith America Invents Act of 2011, 2011. Limitation on Issuance of Patents, §33. Public Law 112-29, 125 Stat. 284. September 16, 2011.

Lipson, H., Kurman, M., 2013. Fabricated: The New World of 3D Printing. John Wiley & Sons, Inc., pp. 1–4.

Maxey, K., 2013. Gelatin bio-ink could lead to 3D printed organs. Published November 4, 2013. Available from: http://www.engineering.com/3DPrinting/3DPrintingArticles/ArticleID/6585/Gelatin-Bio-Ink-Could-Lead-to-3D-Printed-Organs.aspx.

Schacht, W.H., Thomas, J.R., 2005. Patent law and its application to the pharmaceutical industry: an examination of the Drug Price Competition and Patent Term Restoration Act of 1984. CRS Report for Congress. Published January 10, 2005. Available from: http://www.law.umaryland.edu/marshall/crsreports/crsdocuments/rl3075601102005.pdf.

Song, L., et al., 2000. Optical 3D digitizer, system and method for digitizing an object. Inspeck Inc., assignee. Patent U.S. Patent No. 6,493,095. December 20, 2000. Print.

Torrance, A., 2013. Nothing under the sun that is made of man. Published February 7, 2013. Available from: http://www.scotusblog.com/2013/02/nothing-under-the-sun-that-is-made-of-man/.

U.S. Food and Drug Administration, 2009. U.S. Food and Drug Administration, Vaccines, blood, and biologics: tissue and tissue product questions and answers. Updated October 14. 2009, Available from: http://www.fda.gov/BiologicsBloodVaccines/TissueTissueProducts/QuestionsaboutTissues/ucm101559.htm.

U.S. Food and Drug Administration, 2012. U.S. Food and Drug Administration, Medical devices: PMA application methods. Last updated February 14, 2012. Available from: http://www.fda.gov/medicaldevices/deviceregulationandguidance/howtomarketyourdevice/premarketsubmissions/premarketapprovalpma/ucm048168.htm.

United States Copyright Office, 2012a. Copyright basics. Updated May 2012. Available from: http://www. copyright.gov.

United States Copyright Office, 2012b. Services of the copyright office. Revised March 1, 2012. Available from: http://www.copyright.gov/help/faq/faq-services.html#advice.

United States Patent and Trademark Office, 2011.General information concerning patents: infringement of patents. Updated November 2011. Available from: http://www.uspto.gov/patents/resources/general_info_concerning_ patents.jsp#heading-28.

United States Patent and Trademark Office, 2013. Process for obtaining a utility patent. Updated December 5, 2013. Available from: <http://www.uspto.gov/patents/process/index.jsp>.

United States Patent and Trademark Office, 2013a. Trademark, patent, or copyright? Updated January 18, 2013. Available from: http://www.uspto.gov/trademarks/basics/definitions.jsp.

United States Patent and Trademark Office, 2013b. Trademarks FAQs: what are the benefits of federal trademark registration? Updated April 23, 2013. Available from: http://www.uspto.gov/faq/trademarks.jsp#_ Toc275426681.

United States Patent and Trademark Office, 2014. What is a patent? Updated September 29, 2014. Available from: <http://www.uspto.gov/patents/>.

United States Patent and Trademark Office, 2014b. Patents. Updated November 7, 2014. Available from: <http:// www.uspto.gov/inventors/patents.jsp>.

United States Copyright Office, 2014. Definitions. Available from: <http://www.copyright.gov/help/faq/ faq-definitions.html> (accessed 25.02.14).

World Intellectual Property Organization, 2014a. What is intellectual property? Available from: <www.wipo.int/ about-ip/en/> (accessed 03.03.14).

World Intellectual Property Organization, 2014b. What is a trade secret? Available from: http://www.wipo.int/sme/ en/ip_business/trade_secrets/trade_secrets.htm (accessed 03.03.14).

World Intellectual Property Organization, 2014c. How are trade secrets protected? Available from: <www.wipo. int/sme/en/ip_business/trade_secrets/protection.htm> (accessed 10.03.14).

Index

Edwards Brothers Malloy
Thorofare, NJ USA
February 18, 2015